Calcium Signaling In Airway Smooth Muscle Cells

Yong-Xiao Wang
Editor

Calcium Signaling In Airway Smooth Muscle Cells

Springer

Editor
Yong-Xiao Wang
Center for Cardiovascular Sciences
Albany Medical College
Albany, NY, USA

ISBN 978-3-319-01311-4 ISBN 978-3-319-01312-1 (eBook)
DOI 10.1007/978-3-319-01312-1
Springer Cham Heidelberg New York Dordrecht London

Library of Congress Control Number: 2013951150

Printed on acid-free paper

Springer is part of Springer Science+Business Media (www.springer.com)

Preface

Smooth muscle cells (SMCs) are the major functional and structural components of the airway. In addition to maintaining air flow, these cells can proliferate, migrate, and induce cytokines, extracellular matrix proteins, growth factors, and other molecules. Airway SMCs are also key players in the development of asthma and asthma attacks by causing airflow obstruction secondary to contractile hyperresponsiveness, remodeling (e.g., hyperplasia and hypertrophy), and inflammatory responses. These physiological and pathological functions of airway SMCs are, in fact, precisely controlled and regulated by calcium signaling, which may result from extracellular calcium influx or intracellular calcium release as a result of activation or inactivation of ion channels, exchangers, and transporters on the cell plasma membrane and sarcoplasmic reticulum membrane. Scientific research in calcium signaling in airway SMCs is growing; however, to date, there has not been a comprehensive book compiling and detailing our state-of-the-art advances.

The major objective of this book is to create a valuable platform in which numerous, well-established, and emerging pioneers are able to report their recent, inspiring findings from basic, translational, and clinical research, particularly focusing on genesis, networks, microdomains, regulation, functions, and therapies of calcium signaling in airway SMCs. We also strive to include data from clinical trials exploring interventions of calcium signaling in the treatment of asthma and other related diseases. The innovative and ample contents presented will update our understanding of the role of calcium signaling and help direct future research in the field.

This book offers a broad and detailed overview for academic and industrial scientists, postdoctoral fellows, and graduate students engaged in studies of calcium signaling in airway smooth muscle and other cells, particularly in the fields of molecular biology, cell biology, biochemistry, physiology, and pharmacology. In addition, the book may also serve as a useful reference tool for clinicians, medical students, and allied health professionals.

Many of the authors of this book have played dual roles as reviewers. I would like to express my wholehearted appreciation to all of the contributors for their dedication and hard work. I also wish to thank Ms. Aleta Kalkstein, Editor of Cell Biology, and Ms. Rita Beck, Assistant Editor of Food Science, at Springer Science +Business Media for their patience and assistance.

Albany, NY, USA Yong-Xiao Wang

Contents

Contributors

Yassine Amrani Department of Infection, Immunity and Inflammation, University of Leicester, Leicester, UK

Carlos Barajas-López División de Biología Molecular, Instituto Potosino de Investigación Científica y Tecnológica, San Luis Potosí, SLP, Mexico

Charlotte K. Billington Division of Respiratory Medicine, The University of Nottingham, Nottingham, UK

Lioubov I. Brueggemann Department of Molecular Pharmacology & Therapeutics, Loyola University Chicago, Maywood, USA

Kenneth L. Byron Department of Molecular Pharmacology & Therapeutics, Loyola University Chicago, Maywood, USA

Lei Cai Department of Pathophysiology, Tongji Medical College, Huazhong University of Science and Technology, Wuhan, People's Republic of China

Pengxing Cao Department of Mathematics, University ofAuckland, Auckland, New Zealand

Philippe Delmotte Department of Physiology and Biomedical Engineering, Mayo Clinic College of Medicine, Rochester, USA

Deepak A. Deshpande Department of Medicine, Pulmonary and Critical Care Medicine, School of Medicine, University of Maryland, Baltimore, USA

Mythili Dileepan Department of Veterinary and Biomedical Sciences, College of Veterinary Medicine, University of Minnesota, St. Paul, USA

Charles W. Emala Sr. Department of Anesthesiology, Columbia University, New York, USA

Edgar Flores-Soto Departamento de Farmacología, Edificio de Investigación, sexto piso, laboratorio 3, Facultad de Medicina, Universidad Nacional Autónoma de México, México, Mexico

George Gallos Department of Anesthesiology, Columbia University, 168 New York, USA

Alonso Guedes Department of Anesthesiology, School of Veterinary Medicine, University of California, Davis, USA

Jennifer M. Haick Department of Molecular Pharmacology & Therapeutics, Loyola University Chicago, Maywood, USA

Ian P. Hall Division of Respiratory Medicine, The University of Nottingham, Nottingham, UK

Qinghua Hu Department of Pathophysiology, Tongji Medical College, Huazhong University of Science and Technology, Wuhan, People's Republic of China

Satoru Ito Department of Respiratory Medicine, Nagoya University Graduate School of Medicine, Showa-ku, Nagoya, Japan

Luke J. Janssen Firestone Institute for Respiratory Health, St. Joseph's Hospital and the Department of Medicine, McMaster University, Hamilton, Canada

Li Jia Department of Physiology and Biomedical Engineering, Mayo Clinic College of Medicine, Rochester, USA

Priyanka P. Kakad Department of Molecular Pharmacology & Therapeutics, Loyola University Chicago, Maywood, USA

Mathur S. Kannan Department of Veterinary and Biomedical Sciences, College of Veterinary Medicine, University of Minnesota, St. Paul, USA

Hiroaki Kume Department of Respiratory Medicine and Allergology, Kinki University Faculty of Medicine, Osakasayama, Japan

Stephen B. Liggett Morsani College of Medicine, University of South Florida, Tampa, USA

Qing-Hua Liu Institute for Medical Biology, College of Life Sciences, South-Central University for Nationalities, Wuhan, China

Richard J. Martin Department of Pediatrics, Division of Neonatology, Rainbow Babies Children's Hospital, Case Western Reserve University, Cleveland, USA

Hisako Matsumoto Department of Respiratory Medicine, Kyoto University, Sakyo-ku, Kyoto, Japan

Lin Mei Center for Cardiovascular Sciences (MC-8), Albany Medical College, Albany, USA

Luis M. Montaño Departamento de Farmacología, Edificio de Investigación, sexto piso, laboratorio 3, Facultad de Medicina, Universidad Nacional Autónoma de México, México, Mexico

Carl P. Nelson Division of Respiratory Medicine, The University of Nottingham, Nottingham, UK

Christina M. Pabelick Department of Anesthesiology, Department of Physiology and Biomedical Engineering, Mayo Clinic, Rochester, USA

Y.S. Prakash Department of Anesthesiology, Departments of Physiology and Biomedical Engineering, Mayo Clinic, Rochester, USA

Etienne Roux Adaptation cardiovasculaire à l'ischémie INSERM U 1034, University of Bordeaux, Pessac, France

Michael J. Sanderson Department of Microbiology and Physiological Systems, University of Massachusetts Medical School, Worcester, USA

Venkatachalem Sathish Departments of Physiology and Biomedical Engineering, Mayo Clinic, Rochester, USA

Carlo Savoia Center for Cardiovascular Sciences (MC-8), Albany Medical College, Albany, USA

Gary C. Sieck Department of Physiology and Biomedical Engineering, Mayo Clinic College of Medicine, Rochester, USA

Brij B. Singh Department of Biochemistry and Molecular Biology, School of Medicine and Health Sciences, University of North Dakota, Grand Forks, USA

James Sneyd Department of Mathematics, University of Auckland, Auckland, New Zealand

Tengyao Song Center for Cardiovascular Sciences (MC-8), Albany Medical College, Albany, USA

Yunchao Su Department of Pharmacology and Toxicology, Department of Medicine, Vascular Biology Center, Center for Biotechnology and Genomic Medicine, Medical College of Georgia, Georgia Regents University, Augusta, USA

Xiahui Tan Department of Microbiology and Physiological Systems, University of Massachusetts Medical School, Worcester, USA

Chun Ming Teoh Department of Physiology, National University of Singapore, Singapore, Singapore

Elizabeth A. Townsend Department of Anesthesiology, Columbia University, New York, USA

Thai Tran Department of Physiology, National University of Singapore, Singapore, Singapore

Timothy F. Walseth Department of Pharmacology, University of Minnesota, St. Paul, USA

Yong-Xiao Wang Center for Cardiovascular Sciences (MC-8), Albany Medical College, Albany, USA

Jun-Hua Xiao Department of Pharmacology, Tongji Medical College of Huazhong University of Science & Technology, Wuhan, P.R. China

Yun-Min Zheng Center for Cardiovascular Sciences (MC-8), Albany Medical College, Albany, USA

Ryanodine and Inositol Trisphosphate Receptors/Ca^{2+} Release Channels in Airway Smooth Muscle Cells

Lin Mei, Yun-Min Zheng, and Yong-Xiao Wang

Contents

Abstract Ryanodine receptors (RyRs) and inositol 1,4,5-trisphosphate receptors (IP_3Rs) are the most important Ca^{2+} release channels on the sarcoplasmic (or endoplasmic) reticulum (SR) in almost all types of cells. In the past several decades, the studies of RyRs and IP_3Rs have greatly facilitated our understanding of the physiological functions and pathological mechanisms for various diseases including heart failure, arrhythmias, myopathy, and seizure. Similarly, their important roles have been explored in airway smooth muscle cells (SMCs). These two receptors control intracellular Ca^{2+} release and modulate extracellular Ca^{2+} influx, thereby playing an essential role in cell contraction, relaxation, proliferation,

L. Mei • Y.-M. Zheng (✉) • Y.-X. Wang (✉)
Center for Cardiovascular Sciences (MC-8), Albany Medical College, 47 New Scotland Avenue, Albany, NY 12208, USA
e-mail: zhengy@mail.amc.edu; wangy@mail.amc.edu

Y.-X. Wang (ed.), *Calcium Signaling In Airway Smooth Muscle Cells*,
DOI 10.1007/978-3-319-01312-1_1,

migration, metabolism, and, ultimately, cell fate. The abnormality of Ca^{2+} signaling in airway SMCs may contribute to the development of multiple lung diseases, notably asthma. Concomitantly, many regulators, including Ca^{2+} itself, calmodulin, protein kinases, FK506-binding protein 12.6 (FKBP12.6), cyclic adenosine diphosphate ribose (cADPR), and redox status, are involved in the regulation of Ca^{2+} signaling and, thus, the physiological function and pathological alterations. The two SR Ca^{2+} release channels may also directly or indirectly interact with plasmalemmal and mitochondrial ion channels such as transient receptor potential cation, big-conductance Ca^{2+}-activated K^+, Ca^{2+}-activated Cl^-, and other channels, providing positive or negative feedback mechanisms to control Ca^{2+} signaling and cellular functions.

Keywords Ryanodine receptor • Inositol 1,4,5-trisphosphate receptor • Intracellular Ca^{2+} release • Extracellular Ca^{2+} influx • Big-conductance Ca^{2+}-activated K^+ channel • Ca^{2+}-activated Cl^- channel

Abbreviations

2-APB	2-aminoethoxy-diphenylborate
ACh	Acetylcholine
BK_{Ca}	Big-conductance Ca^{2+}-activated K^+ channel
cADPR	Cyclic adenosine diphosphate ribose
CaMKII	Calmodulin-dependent protein kinase II
CICR	Ca^{2+}-induced Ca^{2+} release
Cl_{Ca}	Ca^{2+}-activated Cl^- channel
CPVT	Catecholaminergic polymorphic ventricular tachycardia
DAG	Diacylglycerol
FKBP12.6	FK506-binding protein 12.6
GPX1	Glutathione peroxidase-1
H_2O_2	Hydrogen peroxide
IL-13	Interleukin-13
IP_3Rs	Inositol 1,4,5-trisphosphate receptors
IRAG	cGMP kinase substrate
LTCCs	L-type voltage-gated Ca^{2+} channels
LTD4	Leukotriene D4
M_3R	Muscarinic M_3 receptor
mACh	Methacholine
MLC	Myosin light chain
MLCK	Myosin light chain kinase
MLCP	Myosin light chain phosphatase
mTOR	Mammalian target of rapamycin
NCX	Na^+/Ca^{2+} exchanger
NO	Nitric oxide

NSCCs	Nonselective cation channels
PI3K	Phosphatidylinositol 3 kinases
PIP2	Phosphatidylinositol 4 5-bisphosphate
PKA	Protein kinase A
PKC	Protein kinase C
PKG	cGMP-dependent protein kinase
PLC	Phospholipase C
ROS	Reactive oxygen species
RyR	Ryanodine receptor
SERCA	Sarcoplasmic reticulum Ca^{2+} ATPase
SMC	Smooth muscle cell
SOCE	Store-operated calcium entry
SR	Sarcoplasmic reticulum
STIC	Spontaneous transient inward current
STOC	Spontaneous transient outward current
TNF-α	Tumor necrosis factor-α
TRP	Transient receptor potential channel
VICR	Voltage-induced Ca^{2+} release

1 Introduction

Ca^{2+} signals are precisely generated and regulated by both extracellular Ca^{2+} influx and intracellular Ca^{2+} release, which are controlled and regulated by ion channels, exchangers, and transporters. L-type voltage-gated Ca^{2+} channels (LTCCs), nonselective cation channels (NSCCs), and Na^+/Ca^{2+} exchangers (NCX) play an important role in mediating Ca^{2+} influx. The intracellular Ca^{2+} is primarily stored in the sarcoplasmic/endoplasmic reticulum (SR). In smooth muscle cells (SMCs), a Ca^{2+} signal can express in various forms, such as Ca^{2+} flashes, puffs, ripples, sparklets, sparks, waves, oscillations, and global changes in intracellular Ca^{2+} concentration ($[Ca^{2+}]_i$). These diverse Ca^{2+} signals regulate physiological and pathological functions in nearly all types of cells.

Ca^{2+} release channels on the SR can be divided into two categories: ryanodine receptors (RyRs) and inositol 1,4,5-trisphosphate receptors (IP_3Rs). RyRs were originally identified using the plant alkaloid ryanodine, in which ryanodine induced profound paralysis of cardiac and skeletal muscle cells and was bound to the SR membrane [1, 2]. Three subtypes of RyRs (RyR1, RyR2, and RyR3) have been identified. Each of them is encoded by a distinct gene in mammalian and human cells. The three subtypes of RyRs share approximately 70 % sequence homology. RyRs are the approximately 2,200 kDa homotetramer of four ~550 kDa subunits and mediate Ca^{2+} release from the SR. These Ca^{2+} release channels can be divided into three domains: N-terminal, central, and C-terminal. The C-terminal of RyRs contains the channel region; the N-terminal domain modulates the channel gating; and the central domain comprises a large cytosolic region that serves as a scaffold to interact with regulatory proteins to form a macromolecular signaling complex [3].

IP_3Rs on the SR are also tetramers with a monomeric molecular mass of approximately 300 kDa. Each subunit is composed of the N-terminal, central, and C-terminal domains. The N-terminal domain composes an IP_3-binding site, a suppressor region that inhibits IP_3 binding, and the regulatory region [4]. To date, three IP_3R subtypes (IP_3R1, IP_3R2, and IP_3R3) have been identified. Each subtype of IP_3Rs is encoded by a separate gene in mammalian and human cells. The full-length amino acid sequences of three IP_3R subtypes share 60–80 % homology overall.

Stimulation of neurotransmitter, hormone, growth factor, and other G protein-coupled receptors activates phospholipase C (PLC), which hydrolyzes the membrane lipid phosphatidylinositol 4, 5-biphosphate (PIP2) to generate diacylglycerol (DAG) and IP_3. The latter molecule can bind to its receptors and then induce Ca^{2+} release from the SR. To date, all three subtypes of IP_3Rs (IP_3R1, IP_3R2, and IP_3R3) have been discovered in SMCs; among them, IP_3R1 appears to be the most predominant subtype [5].

Depolarization of the cell membrane can cause the opening of LTCCs and extracellular Ca^{2+} influx, which gives rise to an increase in intracellular Ca^{2+} concentration ($[Ca^{2+}]_i$). The increased $[Ca^{2+}]_i$ activates RyRs and possibly IP_3R to release Ca^{2+} from the SR, a process called Ca^{2+}-induced Ca^{2+} release (CICR). Subsequently, the elevated intracellular Ca^{2+} binds to calmodulin. This complex activates myosin light chain kinase (MLCK) to phosphorylate myosin light chain (MLC). In cooperation with actin, the phosphorylated MLC initiates cross-bridge cycling and contraction of the cell. The removal of the contractile stimulation or inhibition of related contractile molecules results in cell relaxation. Concomitantly, the relaxation process requires a decrease in $[Ca^{2+}]_i$, which is mediated by the SR Ca^{2+} ATPase (SERCA) to refill the Ca^{2+} store or by NCX to pump Ca^{2+} out of the cell. This process increases MLC phosphatase (MLCP) activity to dephosphorylate the MLCs [6].

In addition to contraction, Ca^{2+} signaling plays an important role in metabolism, transcription, migration, and proliferation in SMCs in a time-dependent manner. For example, Ca^{2+} influx triggers exocytosis within microseconds at the synaptic junction, whereas the Ca^{2+} signal drives gene transcription and cell proliferation over minutes and even hours [7]. An alteration in Ca^{2+} signaling might be an essential factor for many airway diseases, as exemplified by asthma.

2 Expression of RyRs and IP_3Rs

Among the three subtypes of RyRs, RyR1 is mainly expressed in skeletal muscle cells and serves as the major Ca^{2+} release channel required for cell contraction. In addition, a low level of RyR1 expression has been reported in SMCs, cerebellum, testis, adrenal gland, spleen, and ovary. It was recently shown that RyR1 is also expressed in B-lymphocytes [8]. RyR2 is the most abundant in the heart, lung, SMCs, and pancreatic β cells. This channel also shows a high expression level in Purkinje cells of cerebellum and cerebral cortex, but a low level in stomach, kidney, adrenal glands, ovaries, thymus, and lungs [9]. RyR3 is found in the brain, spleen, heart, and testis [10]. Coronary vasculature, lung, kidney, ileum, jejunum, spleen, stomach, aorta, uterus, ureter, urinary bladder, and esophagus express RyR3 as well.

Airway SMCs express all three subtypes of RyRs; however, RyR2 is likely to be the predominant subtype. In these cells, RyR1 is localized to the periphery near the cell membrane, whereas RyR3 is more centrally localized [11]. The heterogeneous distribution of RyR1 and RyR3 may endow their multiple functions in airway SMCs.

In mammals, IP_3R is ubiquitously expressed in almost all cell types. IP_3R1 is predominantly expressed in Purkinje cells, IP_3R2 primarily in cardiac myocytes, and IP_3R3 mainly in pancreatic β cells [12]. IP_3R1 knockout mice show neurological defect and early death [13], which is consistent with its primary expression in brain. Interestingly, IP_3R2 null mice have the abolished positive ionotropic effect in atrial myocytes of the heart, revealing the potential importance of this Ca^{2+} release channel in cardiac cells [14].

SMCs express all three IP_3R subtypes. The density of IP_3Rs is approximately 100 times less in SMCs than in brain. It is intriguing to point out that the overall stoichiometric ratio of IP_3Rs to RyRs is approximately 10–12:1 in visceral SMCs but 3–4:1 in vascular SMCs [15]. IP_3R1 is the predominant subtype expressed in vascular SMCs, including aortic, cerebral artery, mesenteric artery, and portal vein myocytes [5]. IP_3R2 and IP_3R3 expression levels have been found to be higher in proliferating, neonatal, and cultured aortic SMCs [16]. IP_3R1 is also predominantly expressed in intestinal SMCs.

Western blot analysis reveals that IP_3R1, IP_3R2, and IP_3R3 are all expressed in airway SMCs [17]. The heterogeneity of tissue-dependent expression levels of IP_3R subtypes may significantly contribute to the diversities of Ca^{2+} signaling and cellular functions in different types of SMC.

3 RyR- and IP_3R-Mediated Ca^{2+} Release

RyR-mediated Ca^{2+} release is required for excitation-contraction coupling in skeletal, cardiac, and some SMCs. RyRs on the SR can be functionally coupled to LTCCs (also known as dihydropyridine receptors, DHPRs) on the plasmalemma, by which Ca^{2+} influx through LTCCs during membrane depolarization opens RyRs to induce Ca^{2+} release from the SR, i.e., CICR. RyRs may also be physically coupled to LTCCs; as such, a conformational change in LTCCs following membrane depolarization (voltage stimulation) causes the opening of associated RyRs and Ca^{2+} release, termed voltage-induced Ca^{2+} release (VICR) process. The three subtypes of RyR exhibit differences in Ca^{2+}-dependent activation and deactivation. Usually, RyR activation occurs when the concentration of Ca^{2+} in the cytosol is within approximately 0.3–10 μM, and inhibition occurs at concentrations of approximately 1 mM Ca^{2+} or higher [18].

Following stimulation of G protein-coupled receptors (GPCRs), IP_3 is produced from PIP2 by PLC. IP_3 can bind to and activate its receptors to induce SR Ca^{2+}

release and increase $[Ca^{2+}]_i$. The increased $[Ca^{2+}]_i$ may promote the binding of Ca^{2+} to IP_3Rs, which in turn forms a positive feedback to initiate the propagation of a Ca^{2+} wave through stimulation of neighboring IP_3Rs and RyRs. However, at high $[Ca^{2+}]_i$, binding of Ca^{2+} to a low-affinity site on the IP_3Rs reduces the channel open probability. This, together with a localized decrease in the SR Ca^{2+}, leads to the termination of further Ca^{2+} release and wave propagation [19].

IP_3Rs and RyRs can both mediate local and global Ca^{2+} release in SMCs. Spontaneous transient local Ca^{2+} release events, termed Ca^{2+} sparks, are due to the concerted opening of several localized RyRs. A Ca^{2+} spark has been described in SMCs of numerous tissues including the artery, vein, stomach, trachea, and bladder [20]. Interestingly, IP_3Rs are involved in RyR-mediated Ca^{2+} sparks in airway SMCs [21, 22]. Similar findings have been made in portal vein myocytes [23]. However, one study showed that IP_3R inhibition has no effect on the frequency or amplitude of spontaneous Ca^{2+} sparks in rat pulmonary artery SMCs [24]. Therefore, IP_3Rs may participate in the generation of Ca^{2+} sparks in some, but not all, types of SMCs.

Besides global and local Ca^{2+} release, RyRs and IP_3Rs generate other forms of Ca^{2+} signals in SMCs. For example, spontaneous or agonist-induced synchronous opening of approximately 30 IP_3Rs clustered within an approximately 400 nm diameter region produces a Ca^{2+} puff. Ca^{2+} flash, a rapid increase in local $[Ca^{2+}]_i$, has been observed in gallbladder SMCs [25]. This local Ca^{2+} signal is likely to be mediated by LTCCs, IP_3Rs, and RyRs. Ca^{2+} oscillation is a repetitive, non-propagating elevation in global $[Ca^{2+}]_i$ that results from periodic, pulsatile release of SR Ca^{2+} in SMCs. Ca^{2+} ripple is an IP_3R-mediated spontaneous, pro-pagating Ca^{2+} signal with a low amplitude. The activation of IP_3Rs, RyRs, or both can induce a Ca^{2+} wave, a propagating elevation in global $[Ca^{2+}]_i$ in SMCs [5]. Presumably, the tissue-specific variability of IP_3R- and RyR-mediated Ca^{2+} signal strengths and patterns are required for multiple and different physiological functions in SMCs.

The first evidence of the involvement of RyRs in Ca^{2+} signaling in airway SMCs came from the findings that the RyR antagonist ryanodine (at a high concentration) and ruthenium red could inhibit acetylcholine (ACh)-induced Ca^{2+} oscillations in porcine tracheal SMCs [26, 27]. The inhibition of RyRs by procaine and tetracaine also blocks Ca^{2+} oscillations. Similar results have been reported in human airway SMCs [28]. Exogenous application of IP_3, similar to ACh, also induces Ca^{2+} oscillations in porcine tracheal SMCs. ACh- or IP_3-induced Ca^{2+} oscillations can be inhibited by IP_3R antibodies and antagonists (e.g., heparin and 2-aminoethoxydiphenyl borate) [21, 29]. RyR inhibitors also attenuate norepinephrine- or phenylephrine-induced, IP_3R-dependent Ca^{2+} release in renal artery SMCs [30]. Thus, it appears that the initiation of Ca^{2+} oscillations is mediated by IP_3Rs, whereas the maintenance of these Ca^{2+} signals are required for the participation of RyRs.

It has been demonstrated that RyRs and IP_3Rs can localize on the same SR (Ca^{2+} store) in mesenteric and small pulmonary artery as well as portal vein SMCs. However, other studies suggest that these two Ca^{2+} release channels are present at discrete regions of the SR in mesenteric, large pulmonary artery as well as ureter myocytes [5]. This discrepancy may indicate a complicated interaction between IP_3Rs and RyRs.

The localization of RyRs and IP_3Rs on the same SR has also been observed in airway SMCs. This unique structural feature renders a distinctive platform for the synergistic role of both Ca^{2+} release channels in the accurate control of cellular functions in airway SMCs. Indeed, Ca^{2+} released from IP_3Rs readily causes the opening of RyRs and further Ca^{2+} release from the SR in airway SMCs [21, 22]. Our further studies demonstrated that this local IP_3R-mediated, RyR-dependent local CICR process is specifically attributable to the opening of RyR2 [22].

4 Interaction of RyRs and IP_3Rs with Other Ion Channels

In SMCs, IP_3Rs and RyRs may communicate with several SR and plasma membrane localized ion channels in order to regulate $[Ca^{2+}]_i$ and cellular functions.

4.1 Transient Receptor Potential (TRP) Channels

TRP channels play a major role in controlling extracellular Ca^{2+} influx in SMCs [31]. TRP channels are classified into six different subfamilies based on their activation stimuli and homology: canonical, vanilloid, melastatin, polycystin, mucolipin, and ankyrin TRP (TRPC, TRPV, TRPM, TRPP, TRPML, and TRPA) channels. The TRPC channel subfamily, which comprises seven members that are designated TRPC1-7, is perhaps the one most studied in airway SMCs. It has been shown that TRPC3 is a major member of nonselective cation channels. TRPC3 contributes to the resting $[Ca^{2+}]_i$ and is also involved in muscarinic increases in $[Ca^{2+}]_i$ in freshly isolated airway SMCs [32, 33]. Our group has also reported that the expression and activity of TRPC3 channels are significantly increased in asthmatic airway myocytes, whereas TRPC1 channel activity is also augmented but its expression remains unchanged [34].

IP_3Rs and RyRs may directly interact with TRPC channels. Studies have shown that there is a physical coupling between the IP_3R1 N-terminal and TRPC3 channel C-terminal in cerebral artery SMCs [35]. It is unclear whether TRPC1 also has a physical interaction with IP_3Rs; however, gene silencing of TRPC1 inhibits the increased $[Ca^{2+}]_i$ following stimulation of IP_3Rs with endothelin-1 in aortic SMCs [36]. TRPC6 channels in cerebral artery SMCs contain the IP_3R binding domain but do not interact physically with IP_3R1 [37]. It is also interesting to note that in the absence of IP_3 or functional IP_3Rs, TRPC channels can be activated by PLC, PIP2, and DAG in SMCs.

4.2 Big-Conductance Ca^{2+}-Activated K^+ Channels (BK_{Ca})

BK_{Ca} channels are voltage- and Ca^{2+}-sensitive and have a large single-channel conductance (100–300 pS under symmetrical extracellular and intracellular K^+ conditions). BK_{Ca} channels are formed by α-subunits and accessory β-subunits. The β-subunits play an important role in the regulation of the physiological properties of BK_{Ca} channels. The lack of β1-subunits decreases the Ca^{2+} sensitivity of BK_{Ca}, reducing the extent of plasma membrane hyperpolarization [38]. BK_{Ca} channels are acknowledged to be activated by Ca^{2+} sparks as a result of the opening of RyRs, which generates spontaneous transient outward currents (STOCs). IP_3R1 activation increases the Ca^{2+} sensitivity of BK_{Ca} channels in cerebral artery SMCs, and both IP_3R1 and BK_{Ca} channels have been shown to be physically coupled [39]. This could oppose membrane depolarization- and IP_3-induced contraction of SMCs, acting as a negative feedback mechanism to protect SMCs from overexcitation. Recently, Lifschitz et al. reported that RyR1 and RyR2, but not RyR3, formed clusters with BK_{Ca} channels and were localized near the plasmalemma of airway SMCs. Analyzing the spatial relationship between RyR2 and BK_{Ca} channels, they estimated that an average Ca^{2+} spark caused by the opening of clusters of RyR1 or RyR2 resulted in activation of two to three clusters of BK_{Ca} channels that were randomly distributed within an approximately 600 nm radius of the RyRs. Approximately eight RyRs opened to give rise to a Ca^{2+} spark, which activated approximately 15 BK_{Ca} channels to generate a STOC at 0 mV [40]. These results further support the close relationship between BK_{Ca} channels and RyRs.

4.3 Ca^{2+}-Activated Cl^- (Cl_{Ca}) Channels

Cl_{Ca} channels possibly contribute to the control of smooth muscle tone through their activation following an elevation in $[Ca^{2+}]_i$ to produce Cl^- efflux and inward membrane currents, which cause membrane depolarization. This depolarization promotes extracellular Ca^{2+} influx via LTCCs to increase muscle contraction. The properties of Cl_{Ca} channels are not completely understood. Ca^{2+} binding to the channel may directly activate the channel; however, in some tissues, activation may be involved by calmodulin-dependent protein kinase II (CaMKII) [41]. Spontaneous transient inward currents (STICs) due to the opening of Cl_{Ca} channels were first observed in tracheal myocytes [42]. The characteristics of STICs are consistent with Ca^{2+} sparks because they are abolished by chloride channel blockers in airway SMCs [43], and it has been shown that they are tightly controlled by RyR-generated Ca^{2+} sparks [44]. Recent evidence indicates that TMEM16A and TMEM16B are Cl_{Ca} channels [45]. TMEM16A is expressed in airway epithelial cells and SMCs, and knocking it out decreases Cl^- secretion in response to Ca^{2+}-dependent agonists and SMC contraction [46]. In asthmatic patients and in a mouse model of asthma, the expression of TMEM16A is upregulated in airway epithelial cells. TMEM16A inhibitors negatively regulate both epithelial mucin secretion and airway SMC contraction [47]. Moreover, Zhang et al. reported that TMEM16A, but not TMEM16B, encods Cl_{Ca} channels in airway SMCs and contributes to agonist-induced contraction. Pharmacological blockade of TMEM16A-encoded channels prevents airway hyperresponsiveness [48].

5 Regulation of RyRs and IP_3Rs

5.1 Regulators of RyRs

RyRs can form a macromolecular complex with multiple regulators and thus cause adequate cellular responses. For example, calmodulin is an ubiquitously expressed Ca^{2+}-binding protein that inhibits cardiac RyR2 activity at any level of $[Ca^{2+}]_i$. Reduced affinity for calmodulin binding to RyR2 due to phosphorylation of RyR2 by protein kinase A (PKA) is found in catecholaminergic polymorphic ventricular tachycardia (CPVT)-associated lethal arrhythmia [49].

FK506-binding protein 12 (FKBP12) and FKBP12.6 function as RyR stabilizers in several different types of cells. Quantitatively, each RyR is bound by four FKBP12 or 12.6 proteins, with a stoichiometry of one per subunit. RyR1 and RyR3 exhibit much greater affinity for FKBP12 than FKBP12.6, while RyR2 exhibits greater affinity for FKBP12.6 [50]. Dissociation of FKBP12/12.6 from RyRs increases the channel open probability, which causes Ca^{2+} leak from the SR and altered $[Ca^{2+}]_i$ in pulmonary SMCs [51]. It is well known that Ca^{2+} leak from the SR contributes to various diseases, including heart failure [52], arrhythmia [53], and aging [54].

LTCCs (Cav1.1 in skeletal muscle cells and Cav1.2 in cardiac myocytes) and RyRs play an important role in muscle excitation-contraction coupling. In skeletal muscle cells, every RyR1 channel in the junctional membrane is physically coupled to a tetrad of four Cav1.1 channels in the T-tubule membrane; as such, membrane depolarization causes activation of Cav1.1 channels and RyR1 to mediate VICR. In cardiac myocytes, RyR2-mediated Ca^{2+} release is initiated by Ca^{2+} influx via Cav1.2, i.e., CICR. These Ca^{2+} release channels are redox sensitive, and redox modifications can result in either activation or deactivation of the Cav1.2 channel. RyRs have approximately 100 cysteines per subunit and are also susceptible to redox modification by oxidation, nitrosylation, or alkylation [9]. Redox-mediated RyR modifications have been implicated in muscular dystrophy, malignant hyperthermic crisis, and heart failure. For example, our group has reported that oxidation of RyR2 causes dissociation of FKBP12.6 from RyR2, leading to the increased activity of RyR2 in pulmonary arterial SMCs. Overexpression of glutathione peroxidase-1 (GPX1), which blocks the generation of reactive oxygen species (ROSs), prevents dissociation [51]. A similar mechanism has been reported in the development of muscle weakness in aging [54] and also in a heat-induced sudden death model in RyR1 mutant mice [55]. In addition to oxidation, nitrosylation of RyR1 can also decrease the affinity of FKBP12 to RyRs, causing an SR Ca^{2+} leak that is implicated in the muscle weakness of patients with muscular dystrophy [56].

5.2 *Regulators of IP_3R*

Multiple protein kinases, regulatory proteins, and other modulators influence IP_3R activity in SMCs. cGMP-dependent protein kinase (PKG) may inhibit the activity of IP_3Rs via phosphorylation. PKG phosphorylates IP_3Rs at serine 1755 and inhibits IP_3-induced Ca^{2+} release in aortic SMCs [57]. PKG also phosphorylates IP_3Rs via IP_3-associated cGMP kinase substrate (IRAG) in tracheal SMCs, suggesting that PKG/IRAG regulation of IP_3R phosphorylation modulates airway SMC contractility [58].

Evidence indicates that FKBP12 also modulates IP_3R activity in addition to that of the RyRs. It may occur through three effector proteins: calcineurin, FK506, or mammalian target of rapamycin (mTOR). FKBP12 enhances IP_3R activity via mTOR and inhibits IP_3R activity through calcineurin in colonic SMCs, though this dual effect is tissue specific [59].

In addition to ligands, kinases, and regulators, ROSs also modulate IP_3Rs in SMCs through two main processes: IP_3 generation and IP_3R affinity. Superoxide inhibits IP_3 hydrolysis, thereby enhancing IP_3-induced Ca^{2+} release [60]. Similarly, we have recently unveiled that hydrogen peroxide (H_2O_2) stimulates IP_3R-induced Ca^{2+} release, IP_3 generation, and IP_3R affinity in pulmonary artery SMCs, which suggests that ROSs are involved in IP_3R signaling in SMCs [61]. However, ROS regulation is multimodal and may also be tissue dependent.

6 Physiological and Pathological Functions of RyRs and IP_3Rs in Airway SMCs

6.1 *Role of RyRs*

We found that the basal activity of PLC is important for the activity of RyR-mediated, spontaneous Ca^{2+} sparks in airway SMCs. There are two distinct pathways for the role of PLC. One is mediated by PLC-dependent generation of IP_3 to induce SR Ca^{2+} release, which stimulates neighboring RyRs, leading to further Ca^{2+} release and Ca^{2+} sparks. The second pathway occurs via the PLC-mediated generation of DAG and subsequent activation of protein kinase C-ε (PKCε), which inhibits the activity of RyRs and attenuates Ca^{2+} spark generation [22]. The inhibitory effect of PKCε is attributed to its specific interaction with RyR1 [62].

Further studies in our laboratory demonstrate that membrane depolarization can induce RyR-mediated local Ca^{2+} release in airway SMCs. This membrane depolarization-mediated local Ca^{2+} release can occur independently of LTCCs; however, it is due to the direct activation of muscarinic M_3 receptors (M_3Rs) in the absence of exogenous agonists. The activation of M_3Rs causes an increase in the activity of PLC, generation of IP_3, opening of IP_3Rs, Ca^{2+} release, and then activation of RyR2, inducing further Ca^{2+} release [21].

It has been reported that interleukin-13 (IL-13) or IL-4 upregulates the expression of RyRs and IP_3Rs but decreases the expression of SERCA in human bronchial SMCs. Furthermore, the enlarged ACh-induced increase in $[Ca^{2+}]_i$ is reversed by blocking IP_3Rs [63]. It has been demonstrated that SERCA2 is reduced in human asthmatic subjects. Ca^{2+} release from the SR or inhibition of SERCA-mediated Ca^{2+} reuptake is attenuated in airway SMCs from asthmatics. Knockdown of SERCA2 mediated by siRNA in airway SMCs has been shown to cause the secretory and hyperproliferative phenotypes in asthma [64].

Failure of RyR inhibitors to affect agonist-induced increase in $[Ca^{2+}]_i$ or contraction has also been observed in airway SMCs. For example, exposure to IP_3R antagonist 2-aminoethoxydiphenyl borate (2-APB) inhibits Ca^{2+} oscillation and induces airway relaxation in a concentration-dependent manner. However, the RyR antagonist ryanodine has no significant effect [65]. Tetracaine, another RyR inhibitor, relaxes agonist-induced contraction and inhibits Ca^{2+} oscillation but has no effect on IP_3-induced Ca^{2+} release or wave propagation. Conversely, both ryanodine and tetracaine completely block slow Ca^{2+} oscillation induced by KCl [66]. However, Tazzeo et al. have reported that neither ryanodine nor chloroethyl phenol inhibits the responses to KCl, cholinergic agonists, or serotonin in airway SMCs. Airway relaxation induced by agonists including isoproterenol, salmeterol, and nitric oxide (NO) is also unaffected by ryanodine [67]. Ryanodine also fails to affect histamine-induced contraction or formoterol-induced relaxation in human airways [68]. In IL-13-pretreated airway SMCs, the RyR inhibitor ryanodine or ruthenium red partially blocks leukotriene D4 (LTD4)-induced Ca^{2+} oscillations. Ca^{2+} oscillations are almost completely inhibited by 2-APB at a concentration that dominantly blocks store-operated calcium entry (SOCE) [69]. Thus, further studies are needed to fully understand the complex regulation of Ca^{2+} signaling in airway SMCs.

6.2 *Role of IP_3Rs*

Enhanced IP_3 levels in airway SMCs lead to increased Ca^{2+} mobilization, airway hyperresponsiveness and remodeling, and, finally, asthma [70]. In spontaneously asthmatic rats, IP_3-5-phosphatase, which degrades IP3, is downregulated; therefore, the elevated $[IP_3]_i$ induces Ca^{2+} release, contributing to airway hyperresponsiveness [71]. IP_3Rs may also be involved in extracellular acidification-induced remodeling of airway SMCs by increasing connective tissue growth factor production and extracellular matrix formation [70].

A recent study reported that knockdown of PIP5K1γ, which is the major source of PIP2 in airway SMCs, can prevent the development of asthma in mice. Instead, cell-permeable PIP2 increases airway contractility. These effects are mediated by integrin α9β1 and thus may lead to new treatment strategies for asthma [72].

Cytokines are known to be important factors in asthma, in which IL-13 plays a key role. It has been shown that IL-13 enhances LTD4-induced Ca^{2+} oscillation in human airway SMCs. This effect can be completely blocked by the specific IP_3R inhibitor xestospongin-C and the RyR inhibitor ryanodine or ruthenium red, which indicates that the role of IL-13 in the involvement of Ca^{2+} oscillation may be cooperatively modulated by IP_3Rs, RyRs, and, possibly, SOCE [69]. Incubation with IL-13 or IL-4 upregulates expression of RyRs, IP_3Rs, and SERCA2 and enhances ACh-induced Ca^{2+} transients in asthmatic airway SMCs. All of these effects could be reversed by inhibition of IP_3Rs [63].

Recently, Deshpande et al. reported that bitter taste receptors are expressed in human airway SMCs. Inhalation of bitter tastants decreases airway obstruction in a mouse model of asthma, which may be mediated via IP_3R-dependent localized Ca^{2+} release and activation of BK_{Ca} channels [73].

All of the investigations of the role of IP_3Rs in the regulation of intracellular Ca^{2+} homeostasis in airway SMCs show the great potential for providing new therapeutic options for the treatment of asthma.

6.3 Cyclic Adenosine Diphosphate Ribose (cADPR)/CD38

The second messenger cADPR was originally described in sea urchin eggs and can release Ca^{2+} from the SR via RyRs [74]. A similar mechanism has been reported in tracheal SMCs [75]. The levels of cADPR in airway SMCs are substantially increased by exposure to contractile agonists such as methacholine (mACh), bradykinin, endothelin, and histamine [76]. This response is blocked by RyR, but not IP_3R, inhibitors. cADPR may not only induce Ca^{2+} release [17] but also enhance the frequency of Ca^{2+} oscillation in airway SMCs [75]. 8-bromo-cADPR, an antagonist of cADPR, inhibits mACh-induced Ca^{2+} release. Furthermore, in the presence of 8-NH2-cADPR, another competitive antagonist of cADPR, the mACh-induced Ca^{2+} release in airway SMCs is blocked. In the presence of 8-bromo-cADPR, there is no further attenuation of Ca^{2+} response upon addition of ryanodine [77]. A similar effect is either mimicked or antagonized by tacrolimus (FK506), a compound that increases the activity of RyRs by dissociating its stabilizer FKBP12.6 [51]. In FKBP12.6 null mice, the ability of cADPR to induce Ca^{2+} release is lost. These results demonstrate the role of RyR2 in airway SMCs, providing a possible molecular basis for how cADPR activates RyRs and suggesting that FKBP12.6 may have a competing binding site for cADPR. It is possible that while IP_3Rs initiate Ca^{2+} oscillations in airway SMCs and determine the basal $[Ca^{2+}]_i$, cADPR and RyRs may amplify or rectify Ca^{2+} homeostasis. However, the details of molecular mechanisms are incompletely elucidated. It should also be mentioned that cADPR can directly activate RyRs in coronary artery smooth muscle [78], and calmodulin can enhance cADPR-induced Ca^{2+} release in sea urchin eggs and pancreatic β-cells [79, 80].

CD38, an enzyme that converts nicotinamide adenine dinucleotide to cADPR, has been found on the membranes of airway SMCs [81]. In CD38$^{-/-}$ mice, the contractile responsiveness of airways to mACh and endothelin are lower than that in wildtype mice [82]. In addition, the Ca^{2+} response in CD38$^{-/-}$ airway SMCs is unaffected by 8-bromo-cADPR, indicating the exclusive role of CD38 in the cADPR-mediated Ca^{2+} signaling system [82]. Moreover, upregulation of CD38 has been associated with the stimulation of cytokines under inflammatory conditions, including asthma. For example, it has been shown that tumor necrosis factor-α (TNF-α), IL-1β, and IFN-γ all increase CD38 expression in cultured human airway SMCs [83]. Similarly, IL-13, which is a potent factor promoting airway remodeling, increases CD38 expression [84]. Further work by the same group has revealed that IL-13 fails to increase the response to mACh in CD38$^{-/-}$ mice [85]. However, although cADPR does appear to influence the initial peak of Ca^{2+} response, it is not clear how this transient response transforms into sustained contraction and remodeling of airway SMCs.

Apparently, various agonists can mobilize the release of Ca^{2+} from the SR through RyRs by increasing production of cADPR in airway SMCs. Supportably, the pro-proliferative effect of TNF-α on airway SMCs is inhibited by siRNA-mediated CD38 gene knockdown, while the enhanced SOCE induced by TNF-α is blunted by CD38 siRNAs and potentiated by CD38 overexpression [86]. These results indicate a critical role for CD38 in the TNF-α signaling pathway and associated SOCE. Further work has shown that TNF-α causes a greater induction of CD38 expression in asthmatic than in nonasthmatic human airway SMCs. This effect may stem from the reduced activation of JNK/MAPK and increased activation of ERK/p38 MAPKs [87] but is not associated with a change in NF-κB or AP-1 activation [87] or phosphatidylinositol 3 kinase (PI3K) [88].

7 Conclusion and Perspective

In this chapter, we focused on recent major progress in the studies of RyR and IP_3R Ca^{2+} signaling in airway SMCs. As illustrated in Fig. 1, RyRs exhibit spontaneous functional activity; as such, a cluster of approximately 10 RyRs is sufficient to generate Ca^{2+} sparks. These local Ca^{2+} release events may activate multiple plasmalemmal ion channels including Cl_{Ca}, TRPC, and BK_{Ca} channels. The activation of Cl_{Ca} and TRPC channels cause membrane depolarization, opening of LTCCs, and a further increase in $[Ca^{2+}]_i$. Extracellular Ca^{2+} influx through LTCCs may promote and increase the opening of RyRs, inducing Ca^{2+} release from the SR, also known as a CICR process. The increased $[Ca^{2+}]_i$ mediates contraction,

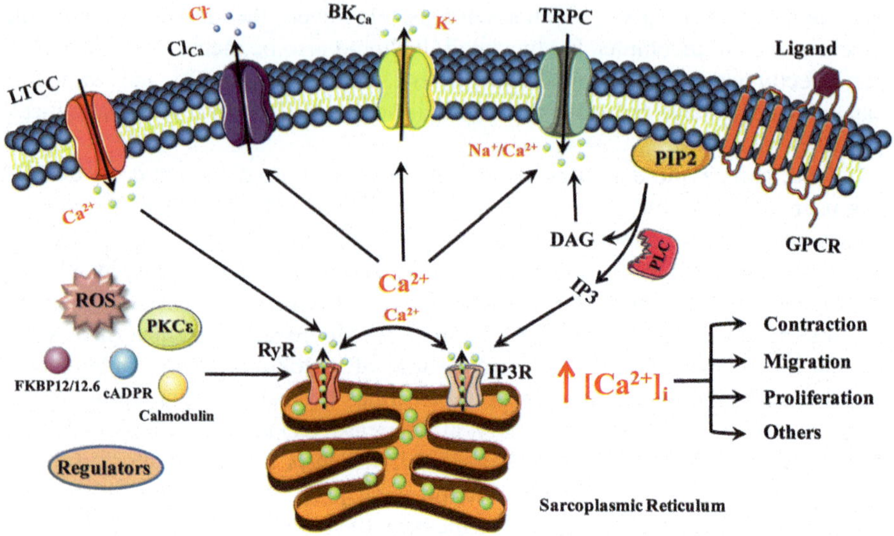

Fig. 1 *RyR and IP_3R Ca^{2+} signaling in airway SMCs.* Ryanodine receptors (RyRs) on the sarcoplasmic reticulum (SR) can generate spontaneous Ca^{2+} sparks in airway SMCs. These local Ca^{2+} release events activate multiple plasmalemmal ion channels including Ca^{2+}-activated Cl^- (Cl_{Ca}), big-conductance Ca^{2+}-activated K^+ (BK_{Ca}), and canonical transient receptor potential (TRPC) channels. The activation of Cl_{Ca} and TRPC channels causes membrane depolarization, opening of L-type voltage-gated Ca^{2+} channels (LTCCs), and extracellular Ca^{2+} influx, which promote the opening of RyRs and induce Ca^{2+} release from the SR, a process termed Ca^{2+}-induced Ca^{2+} release (CICR). The resulting increase in $[Ca^{2+}]_i$ mediates contraction, migration, proliferation, and other cellular responses. BK_{Ca} channels, if activated, lead to membrane hyperpolarization and closure of LTCCs. This provides a negative feedback mechanism to inhibit excessive Ca^{2+} signaling and associated cellular responses. RyRs can be colocalized with inositol 1,4,5-trisphosphate receptors (IP_3Rs). These two Ca^{2+} release channels may reciprocally activate each other to render a local CICR process. IP_3 is normally generated from phosphatidylinositol 4,5-bisphosphate (PIP_2) by phospholipase C (PLC), which is activated following stimulation of G protein-coupled receptors (GPCRs). In addition to IP_3, PLC also produces diacylglycerol (DAG) from PIP_2. DAG may directly activate TRPC channels. RyRs are also regulated by multiple intracellular signaling molecules such as FK506 binding protein 12.6 (FKBP12.6), cyclic adenosine diphosphate ribose (cADPR), calmodulin, protein kinase C-ε (PKC-ε), and reactive oxygen species in airway SMCs

migration, proliferation, and other cellular responses in airway SMCs. BK_{Ca} channels, if activated, lead to membrane hyperpolarization, closure of LTCCs, and a decrease in $[Ca^{2+}]_i$. This provides a negative feedback mechanism to inhibit excessive Ca^{2+} signaling and associated cellular responses.

RyRs can colocalize with IP_3Rs, by which these two colocalized Ca^{2+} release channels may reciprocally activate each other to render a local CICR process. IP_3 is normally generated from PIP_2 by PLC, which is activated following stimulation of GPCRs. In addition to IP_3, PLC also produces DAG by catalyzing PIP_2. DAG may directly activate TRPC channels, contributing to Ca^{2+} signaling in airway SMCs. RyRs are also regulated by multiple intracellular molecules such as FKBP12.6, cADPR, calmodulin, PKC-ε, and ROS in airway SMCs.

The Ca^{2+} signaling generated and regulated by RyRs and IP_3Rs is an important and sophisticated network in the initiation and maintenance of numerous physiological cellular responses in airway SMCs. On the one hand, overactivity of this Ca^{2+} signaling network has been shown to contribute to the development of asthma and, potentially, other respiratory diseases. On the other hand, the current understanding of the molecular genesis, regulatory mechanisms, and functional roles of RyRs and IP_3Rs still remain limited in airway SMCs. Indeed, significant advancements have been made in other types of cells, particularly in cardiac, skeletal, and even vascular SMCs. However, each organ or tissue has, to a greater or lesser extent, its own unique structural, functional, and molecular nature. One of the typical examples is that LTCC blockers are effective in the treatment of hypertension and other cardiovascular diseases but ineffective for asthma. Clearly, more efforts in the studies of these potentially important Ca^{2+} release channels are very necessary in the field.

It is also conceivable that the use of fluorescence resonance energy transfer, RNA interference, and other new techniques will allow us to fully elucidate the molecular mechanisms of both the RyR- and IP_3R-mediated Ca^{2+} signaling network in airway SMCs. Additionally, new transgenic animals and disease models will be valuable to future studies in this field. We believe that progress will generate novel and important findings, which might not only enhance our understanding of the functional importance of RyRs and IP_3Rs in airway SMCs but also significantly aid in the production of innovative and effective therapeutic targets for asthma, chronic obstructive pulmonary disease, and other respiratory diseases.

References

1. Kushnir A, Marks AR. The ryanodine receptor in cardiac physiology and disease. *Adv Pharmacol*. 2010;59:1–30
2. Zheng YM, Wang YX. Ryanodine receptors/Ca^{2+} release channels in pulmonary artery smooth muscle cells. *Recent Advances in Pulmonary Vascular Biology. Trivandrum: Research Signpost*. 2011:91–114
3. Zalk R, Lehnart SE, Marks AR. Modulation of the ryanodine receptor and intracellular calcium. *Annu Rev Biochem*. 2007;76:367–385
4. Michikawa T, Hamanaka H, Otsu H, Yamamoto A, Miyawaki A, Furuichi T, Tashiro Y, Mikoshiba K. Transmembrane topology and sites of N-glycosylation of inositol 1,4, 5-trisphosphate receptor. *J Biol Chem*. 1994;269:9184–9189
5. Narayanan D, Adebiyi A, Jaggar JH. Inositol trisphosphate receptors in smooth muscle cells. *Am J Physiol Heart Circ Physiol*. 2012;302:H2190–2210
6. Jude JA, Wylam ME, Walseth TF, Kannan MS. Calcium signaling in airway smooth muscle. *Proc Am Thorac Soc*. 2008;5:15–22
7. Berridge MJ, Bootman MD, Roderick HL. Calcium signalling: Dynamics, homeostasis and remodelling. *Nat Rev Mol Cell Biol*. 2003;4:517–529
8. Vukcevic M, Broman M, Islander G, Bodelsson M, Ranklev-Twetman E, Muller CR, Treves S. Functional properties of RyR1 mutations identified in swedish patients with malignant hyperthermia and central core disease. *Anesth Analg*. 2010;111:185–190

9. Lanner JT, Georgiou DK, Joshi AD, Hamilton SL. Ryanodine receptors: Structure, expression, molecular details, and function in calcium release. *Cold Spring Harb Perspect Biol.* 2010;2: a003996
10. Giannini G, Conti A, Mammarella S, Scrobogna M, Sorrentino V. The ryanodine receptor/calcium channel genes are widely and differentially expressed in murine brain and peripheral tissues. *J Cell Biol.* 1995;128:893–904
11. Du W, Stiber JA, Rosenberg PB, Meissner G, Eu JP. Ryanodine receptors in muscarinic receptor-mediated bronchoconstriction. *J Biol Chem.* 2005;280:26287–26294
12. Foskett JK, White C, Cheung KH, Mak DO. Inositol trisphosphate receptor Ca^{2+} release channels. *Physiol Rev.* 2007;87:593–658
13. Matsumoto M, Nakagawa T, Inoue T, Nagata E, Tanaka K, Takano H, Minowa O, Kuno J, Sakakibara S, Yamada M, Yoneshima H, Miyawaki A, Fukuuchi Y, Furuichi T, Okano H, Mikoshiba K, Noda T. Ataxia and epileptic seizures in mice lacking type 1 inositol 1,4,5-trisphosphate receptor. *Nature.* 1996;379:168–171
14. Li X, Zima AV, Sheikh F, Blatter LA, Chen J. Endothelin-1-induced arrhythmogenic Ca^{2+} signaling is abolished in atrial myocytes of inositol-1,4,5-trisphosphate (IP_3)-receptor type 2-deficient mice. *Circ Res.* 2005;96:1274–1281
15. Wray S, Burdyga T. Sarcoplasmic reticulum function in smooth muscle. *Physiol Rev.* 2010;90:113–178
16. Tasker PN, Taylor CW, Nixon GF. Expression and distribution of $InsP_3$ receptor subtypes in proliferating vascular smooth muscle cells. *Biochem Biophys Res Commun.* 2000;273:907–912
17. Wang YX, Zheng YM, Mei QB, Wang QS, Collier ML, Fleischer S, Xin HB, Kotlikoff MI. FKBP12.6 and cADPR regulation of Ca^{2+} release in smooth muscle cells. *Am J Physiol Cell Physiol.* 2004;286:C538–546
18. Fill M, Copello JA. Ryanodine receptor calcium release channels. *Physiol Rev.* 2002;82:893–922
19. Sanderson MJ, Delmotte P, Bai Y, Perez-Zogbhi JF. Regulation of airway smooth muscle cell contractility by Ca^{2+} signaling and sensitivity. *Proc Am Thorac Soc.* 2008;5:23–31
20. Jaggar JH, Porter VA, Lederer WJ, Nelson MT. Calcium sparks in smooth muscle. *Am J Physiol Cell Physiol.* 2000;278:C235–256
21. Liu QH, Zheng YM, Korde AS, Yadav VR, Rathore R, Wess J, Wang YX. Membrane depolarization causes a direct activation of G protein-coupled receptors leading to local Ca^{2+} release in smooth muscle. *Proc Natl Acad Sci U S A.* 2009;106:11418–11423
22. Liu QH, Zheng YM, Wang YX. Two distinct signaling pathways for regulation of spontaneous local Ca^{2+} release by phospholipase C in airway smooth muscle cells. *Pflugers Arch.* 2007;453:531–541
23. Gordienko DV, Bolton TB. Crosstalk between ryanodine receptors and IP_3 receptors as a factor shaping spontaneous Ca^{2+}-release events in rabbit portal vein myocytes. *J Physiol.* 2002;542:743–762
24. Remillard CV, Zhang WM, Shimoda LA, Sham JS. Physiological properties and functions of Ca^{2+} sparks in rat intrapulmonary arterial smooth muscle cells. *Am J Physiol Lung Cell Mol Physiol.* 2002;283:L433–444
25. Balemba OB, Salter MJ, Heppner TJ, Bonev AD, Nelson MT, Mawe GM. Spontaneous electrical rhythmicity and the role of the sarcoplasmic reticulum in the excitability of guinea pig gallbladder smooth muscle cells. *Am J Physiol Gastrointest Liver Physiol.* 2006;290: G655–664
26. Kannan MS, Prakash YS, Brenner T, Mickelson JR, Sieck GC. Role of ryanodine receptor channels in Ca^{2+} oscillations of porcine tracheal smooth muscle. *Am J Physiol.* 1997;272: L659–664
27. Prakash YS, Kannan MS, Sieck GC. Regulation of intracellular calcium oscillations in porcine tracheal smooth muscle cells. *Am J Physiol.* 1997;272:C966–975
28. Dai JM, Kuo KH, Leo JM, Pare PD, van Breemen C, Lee CH. Acetylcholine-induced asynchronous calcium waves in intact human bronchial muscle bundle. *Am J Respir Cell Mol Biol.* 2007;36:600–608

29. Tazzeo T, Zhang Y, Keshavjee S, Janssen LJ. Ryanodine receptors decant internal Ca^{2+} store in human and bovine airway smooth muscle. *Eur Respir J*. 2008;32:275–284
30. Liu X, Farley JM. Acetylcholine-induced chloride current oscillations in swine tracheal smooth muscle cells. *J Pharmacol Exp Ther*. 1996;276:178–186
31. Clapham DE, Runnels LW, Strubing C. The TRP ion channel family. *Nat Rev Neurosci*. 2001;2:387–396
32. Wang YX, Zheng YM. Molecular expression and functional role of canonical transient receptor potential channels in airway smooth muscle cells. *Adv Exp Med Biol*. 2011;704:731–747
33. White TA, Xue A, Chini EN, Thompson M, Sieck GC, Wylam ME. Role of transient receptor potential C3 in TNF-alpha-enhanced calcium influx in human airway myocytes. *Am J Respir Cell Mol Biol*. 2006;35:243–251
34. Xiao JH, Zheng YM, Liao B, Wang YX. Functional role of canonical transient receptor potential 1 and canonical transient receptor potential 3 in normal and asthmatic airway smooth muscle cells. *Am J Respir Cell Mol Biol*. 2010;43:17–25
35. Adebiyi A, Narayanan D, Jaggar JH. Caveolin-1 assembles type 1 inositol 1,4,5-trisphosphate receptors and canonical transient receptor potential 3 channels into a functional signaling complex in arterial smooth muscle cells. *J Biol Chem*. 2011;286:4341–4348
36. Tai K, Hamaide MC, Debaix H, Gailly P, Wibo M, Morel N. Agonist-evoked calcium entry in vascular smooth muscle cells requires IP_3 receptor-mediated activation of TRPC1. *Eur J Pharmacol*. 2008;583:135–147
37. Adebiyi A, Zhao G, Narayanan D, Thomas-Gatewood CM, Bannister JP, Jaggar JH. Isoform-selective physical coupling of TRPC3 channels to IP_3 receptors in smooth muscle cells regulates arterial contractility. *Circ Res*. 2010;106:1603–1612
38. Brenner R, Perez GJ, Bonev AD, Eckman DM, Kosek JC, Wiler SW, Patterson AJ, Nelson MT, Aldrich RW. Vasoregulation by the beta1 subunit of the calcium-activated potassium channel. *Nature*. 2000;407:870–876
39. Zhao G, Neeb ZP, Leo MD, Pachuau J, Adebiyi A, Ouyang K, Chen J, Jaggar JH. Type 1 IP_3 receptors activate BK_{ca} channels via local molecular coupling in arterial smooth muscle cells. *J Gen Physiol*. 2010;136:283–291
40. Lifshitz LM, Carmichael JD, Lai FA, Sorrentino V, Bellve K, Fogarty KE, ZhuGe R. Spatial organization of RyRs and BK channels underlying the activation of STOCs by Ca^{2+} sparks in airway myocytes. *J Gen Physiol*. 2011;138:195–209
41. Wang YX, Kotlikoff MI. Inactivation of calcium-activated chloride channels in smooth muscle by calcium/calmodulin-dependent protein kinase. *Proc Natl Acad Sci U S A*. 1997;94:14918–14923
42. Janssen LJ, Sims SM. Acetylcholine activates non-selective cation and chloride conductances in canine and guinea-pig tracheal myocytes. *J Physiol*. 1992;453:197–218
43. ZhuGe R, Sims SM, Tuft RA, Fogarty KE, Walsh JV, Jr. Ca^{2+} sparks activate K^+ and Cl^- channels, resulting in spontaneous transient currents in guinea-pig tracheal myocytes. *J Physiol*. 1998;513 (Pt 3):711–718
44. Bao R, Lifshitz LM, Tuft RA, Bellve K, Fogarty KE, ZhuGe R. A close association of RyRs with highly dense clusters of Ca^{2+}-activated Cl^- channels underlies the activation of STICs by Ca^{2+} sparks in mouse airway smooth muscle. *J Gen Physiol*. 2008;132:145–160
45. Caputo A, Caci E, Ferrera L, Pedemonte N, Barsanti C, Sondo E, Pfeffer U, Ravazzolo R, Zegarra-Moran O, Galietta LJ. TMEM16a, a membrane protein associated with calcium-dependent chloride channel activity. *Science*. 2008;322:590–594
46. Huang F, Rock JR, Harfe BD, Cheng T, Huang X, Jan YN, Jan LY. Studies on expression and function of the TMEM16a calcium-activated chloride channel. *Proc Natl Acad Sci U S A*. 2009;106:21413–21418
47. Huang F, Zhang H, Wu M, Yang H, Kudo M, Peters CJ, Woodruff PG, Solberg OD, Donne ML, Huang X, Sheppard D, Fahy JV, Wolters PJ, Hogan BL, Finkbeiner WE, Li M, Jan YN, Jan LY, Rock JR. Calcium-activated chloride channel TMEM16a modulates mucin secretion and airway smooth muscle contraction. *Proc Natl Acad Sci U S A*. 2012;109:16354–16359

48. Zhang CH, Li Y, Zhao W, Lifshitz LM, Li H, Harfe BD, Zhu MS, Zhuge R. The transmembrane protein 16a Ca^{2+}-activated Cl^{-} channel in airway smooth muscle contributes to airway hyperresponsiveness. *Am J Respir Crit Care Med.* 2013;187:374–381
49. Xu X, Yano M, Uchinoumi H, Hino A, Suetomi T, Ono M, Tateishi H, Oda T, Okuda S, Doi M, Kobayashi S, Yamamoto T, Ikeda Y, Ikemoto N, Matsuzaki M. Defective calmodulin binding to the cardiac ryanodine receptor plays a key role in CPVT-associated channel dysfunction. *Biochem Biophys Res Commun.* 2010;394:660–666
50. Timerman AP, Onoue H, Xin HB, Barg S, Copello J, Wiederrecht G, Fleischer S. Selective binding of FKBP12.6 by the cardiac ryanodine receptor. *J Biol Chem.* 1996;271:20385–20391
51. Liao B, Zheng YM, Yadav VR, Korde AS, Wang YX. Hypoxia induces intracellular Ca^{2+} release by causing reactive oxygen species-mediated dissociation of FK506-binding protein 12.6 from ryanodine receptor 2 in pulmonary artery myocytes. *Antioxid Redox Signal.* 2011;14:37–47
52. Shan J, Betzenhauser MJ, Kushnir A, Reiken S, Meli AC, Wronska A, Dura M, Chen BX, Marks AR. Role of chronic ryanodine receptor phosphorylation in heart failure and beta-adrenergic receptor blockade in mice. *J Clin Invest.* 2010;120:4375–4387
53. Meli AC, Refaat MM, Dura M, Reiken S, Wronska A, Wojciak J, Carroll J, Scheinman MM, Marks AR. A novel ryanodine receptor mutation linked to sudden death increases sensitivity to cytosolic calcium. *Circ Res.* 2011;109:281–290
54. Andersson DC, Betzenhauser MJ, Reiken S, Meli AC, Umanskaya A, Xie W, Shiomi T, Zalk R, Lacampagne A, Marks AR. Ryanodine receptor oxidation causes intracellular calcium leak and muscle weakness in aging. *Cell Metab.* 2011;14:196–207
55. Lanner JT, Georgiou DK, Dagnino-Acosta A, Ainbinder A, Cheng Q, Joshi AD, Chen Z, Yarotskyy V, Oakes JM, Lee CS, Monroe TO, Santillan A, Dong K, Goodyear L, Ismailov, II, Rodney GG, Dirksen RT, Hamilton SL. AICAR prevents heat-induced sudden death in RyR1 mutant mice independent of ampk activation. *Nat Med.* 2012;18:244–251
56. Bellinger AM, Reiken S, Carlson C, Mongillo M, Liu X, Rothman L, Matecki S, Lacampagne A, Marks AR. Hypernitrosylated ryanodine receptor calcium release channels are leaky in dystrophic muscle. *Nat Med.* 2009;15:325–330
57. Komalavilas P, Lincoln TM. Phosphorylation of the inositol 1,4,5-trisphosphate receptor. Cyclic GMP-dependent protein kinase mediates cAMP and cGMP dependent phosphorylation in the intact rat aorta. *J Biol Chem.* 1996;271:21933–21938
58. Schlossmann J, Ammendola A, Ashman K, Zong X, Huber A, Neubauer G, Wang GX, Allescher HD, Korth M, Wilm M, Hofmann F, Ruth P. Regulation of intracellular calcium by a signalling complex of IRAG, IP_3 receptor and cGMP kinase Ibeta. *Nature.* 2000;404:197–201
59. MacMillan D, Currie S, Bradley KN, Muir TC, McCarron JG. In smooth muscle, FK506-binding protein modulates IP_3 receptor-evoked Ca^{2+} release by mTOR and calcineurin. *J Cell Sci.* 2005;118:5443–5451
60. Suzuki YJ, Ford GD. Superoxide stimulates IP_3-induced Ca^{2+} release from vascular smooth muscle sarcoplasmic reticulum. *Am J Physiol.* 1992;262:H114–116
61. Yadav VR, Song T, Joseph L, Mei L, Zheng YM, Wang YX. Important role of PLC-gamma1 in hypoxic increase in intracellular calcium in pulmonary arterial smooth muscle cells. *Am J Physiol Lung Cell Mol Physiol.* 2013;304:L143–151
62. Liu QH, Zheng YM, Korde AS, Li XQ, Ma J, Takeshima H, Wang YX. Protein kinase C-epsilon regulates local calcium signaling in airway smooth muscle cells. *Am J Respir Cell Mol Biol.* 2009;40:663–671
63. Kellner J, Tantzscher J, Oelmez H, Edelmann M, Fischer R, Huber RM, Bergner A. Mechanisms altering airway smooth muscle cell Ca^{2+} homeostasis in two asthma models. *Respiration.* 2008;76:205–215
64. Mahn K, Hirst SJ, Ying S, Holt MR, Lavender P, Ojo OO, Siew L, Simcock DE, McVicker CG, Kanabar V, Snetkov VA, O'Connor BJ, Karner C, Cousins DJ, Macedo P, Chung KF, Corrigan CJ, Ward JP, Lee TH. Diminished sarco/endoplasmic reticulum Ca^{2+} ATPase

(SERCA) expression contributes to airway remodelling in bronchial asthma. *Proc Natl Acad Sci U S A*. 2009;106:10775–10780
65. Bai Y, Sanderson MJ. Airway smooth muscle relaxation results from a reduction in the frequency of Ca^{2+} oscillations induced by a cAMP-mediated inhibition of the IP_3 receptor. *Respir Res*. 2006;7:34
66. Sneyd J, Tsaneva-Atanasova K, Reznikov V, Bai Y, Sanderson MJ, Yule DI. A method for determining the dependence of calcium oscillations on inositol trisphosphate oscillations. *Proc Natl Acad Sci U S A*. 2006;103:1675–1680
67. White C, McGeown JG. Carbachol triggers RyR-dependent Ca^{2+} release via activation of IP_3 receptors in isolated rat gastric myocytes. *J Physiol*. 2002;542:725–733
68. Boittin FX, Macrez N, Halet G, Mironneau J. Norepinephrine-induced Ca^{2+} waves depend on $InsP_3$ and ryanodine receptor activation in vascular myocytes. *Am J Physiol*. 1999;277: C139–151
69. Matsumoto H, Hirata Y, Otsuka K, Iwata T, Inazumi A, Niimi A, Ito I, Ogawa E, Muro S, Sakai H, Chin K, Oku Y, Mishima M. Interleukin-13 enhanced Ca^{2+} oscillations in airway smooth muscle cells. *Cytokine*. 2012;57:19–24
70. Matsuzaki S, Ishizuka T, Yamada H, Kamide Y, Hisada T, Ichimonji I, Aoki H, Yatomi M, Komachi M, Tsurumaki H, Ono A, Koga Y, Dobashi K, Mogi C, Sato K, Tomura H, Mori M, Okajima F. Extracellular acidification induces connective tissue growth factor production through proton-sensing receptor OGR1 in human airway smooth muscle cells. *Biochem Biophys Res Commun*. 2011;413:499–503
71. Tao FC, Tolloczko B, Mitchell CA, Powell WS, Martin JG. Inositol (1,4,5)trisphosphate metabolism and enhanced calcium mobilization in airway smooth muscle of hyperresponsive rats. *Am J Respir Cell Mol Biol*. 2000;23:514–520
72. Chen C, Kudo M, Rutaganira F, Takano H, Lee C, Atakilit A, Robinett KS, Uede T, Wolters PJ, Shokat KM, Huang X, Sheppard D. Integrin alpha9beta1 in airway smooth muscle suppresses exaggerated airway narrowing. *J Clin Invest*. 2012;122:2916–2927
73. Deshpande DA, Wang WC, McIlmoyle EL, Robinett KS, Schillinger RM, An SS, Sham JS, Liggett SB. Bitter taste receptors on airway smooth muscle bronchodilate by localized calcium signaling and reverse obstruction. *Nat Med*. 2010;16:1299–1304
74. Clapper DL, Walseth TF, Dargie PJ, Lee HC. Pyridine nucleotide metabolites stimulate calcium release from sea urchin egg microsomes desensitized to inositol trisphosphate. *J Biol Chem*. 1987;262:9561–9568
75. Prakash YS, Kannan MS, Walseth TF, Sieck GC. Role of cyclic ADP-ribose in the regulation of $[Ca^{2+}]_i$ in porcine tracheal smooth muscle. *Am J Physiol*. 1998;274:C1653–1660
76. Kip SN, Smelter M, Iyanoye A, Chini EN, Prakash YS, Pabelick CM, Sieck GC. Agonist-induced cyclic ADP ribose production in airway smooth muscle. *Arch Biochem Biophys*. 2006;452:102–107
77. White TA, Kannan MS, Walseth TF. Intracellular calcium signaling through the cADPR pathway is agonist specific in porcine airway smooth muscle. *FASEB J*. 2003;17:482–484
78. Li PL, Tang WX, Valdivia HH, Zou AP, Campbell WB. cADP-ribose activates reconstituted ryanodine receptors from coronary arterial smooth muscle. *Am J Physiol Heart Circ Physiol*. 2001;280:H208–215
79. Lee HC, Aarhus R, Graeff R, Gurnack ME, Walseth TF. Cyclic ADP ribose activation of the ryanodine receptor is mediated by calmodulin. *Nature*. 1994;370:307–309
80. Lee HC, Aarhus R, Graeff RM. Sensitization of calcium-induced calcium release by cyclic adp-ribose and calmodulin. *J Biol Chem*. 1995;270:9060–9066
81. Franco L, Bruzzone S, Song P, Guida L, Zocchi E, Walseth TF, Crimi E, Usai C, De Flora A, Brusasco V. Extracellular cyclic ADP-ribose potentiates ACh-induced contraction in bovine tracheal smooth muscle. *Am J Physiol Lung Cell Mol Physiol*. 2001;280:L98–L106
82. Deshpande DA, White TA, Guedes AG, Milla C, Walseth TF, Lund FE, Kannan MS. Altered airway responsiveness in CD38-deficient mice. *Am J Respir Cell Mol Biol*. 2005;32:149–156
83. Deshpande DA, Walseth TF, Panettieri RA, Kannan MS. CD38/cyclic ADP-ribose-mediated Ca^{2+} signaling contributes to airway smooth muscle hyper-responsiveness. *FASEB J*. 2003;17:452–454

84. Deshpande DA, Dogan S, Walseth TF, Miller SM, Amrani Y, Panettieri RA, Kannan MS. Modulation of calcium signaling by interleukin-13 in human airway smooth muscle: Role of CD38/cyclic adenosine diphosphate ribose pathway. *Am J Respir Cell Mol Biol.* 2004;31:36–42
85. Guedes AG, Paulin J, Rivero-Nava L, Kita H, Lund FE, Kannan MS. CD38-deficient mice have reduced airway hyperresponsiveness following IL-13 challenge. *Am J Physiol Lung Cell Mol Physiol.* 2006;291:L1286–1293
86. Sieck GC, White TA, Thompson MA, Pabelick CM, Wylam ME, Prakash YS. Regulation of store-operated Ca^{2+} entry by CD38 in human airway smooth muscle. *Am J Physiol Lung Cell Mol Physiol.* 2008;294:L378–385
87. Jude JA, Solway J, Panettieri RA, Jr., Walseth TF, Kannan MS. Differential induction of CD38 expression by TNF-α in asthmatic airway smooth muscle cells. *Am J Physiol Lung Cell Mol Physiol.* 2010;299:L879–890
88. Jude JA, Tirumurugaan KG, Kang BN, Panettieri RA, Walseth TF, Kannan MS. Regulation of CD38 expression in human airway smooth muscle cells: Role of class I phosphatidylinositol 3 kinases. *Am J Respir Cell Mol Biol.* 2012;47:427–435

Kv7 (KCNQ) Potassium Channels and L-type Calcium Channels in the Regulation of Airway Diameter

Kenneth L. Byron, Lioubov I. Brueggemann, Priyanka P. Kakad, and Jennifer M. Haick

Contents

Abstract Potassium channels in airway smooth muscle cells (ASMCs) help to maintain negative membrane voltages and thereby oppose the opening of voltage-sensitive calcium channels (VSCCs). There is considerable evidence that an influx of calcium ions (Ca^{2+}) through L-type VSCCs contributes to induce contraction of ASMCs and bronchoconstriction of small airways, both physiologically and in diseases such as asthma in which exaggerated bronchoconstrictor responses are evident. Suppression of potassium channel activity is one mechanism to promote Ca^{2+} influx via VSCCs, though little is known regarding the involvement of potassium channels in bronchoconstrictor signal transduction. Pharmacological enhancement of potassium channel activity is expected to reduce Ca^{2+} influx via VSCCs and relax airways, though no K^+ channel activators have yet been successfully developed as therapies for asthma. This chapter summarizes recent evidence that Kv7 (KCNQ family) voltage-activated K^+ channels are expressed in ASMCs and that they serve as signal transduction intermediates in the actions of bronchoconstrictor agonists. The extent to which these channels contribute to regulation of airway diameter, the contributions of L-type VSCCs, and the potential to develop novel bronchodilator therapies based on these biochemical pathways are also discussed.

K.L. Byron (✉) • L.I. Brueggemann • P.P. Kakad • J.M. Haick
Department of Molecular Pharmacology & Therapeutics, Loyola University Chicago, 2160 S. First Avenue, Maywood, IL 60153, USA
e-mail: kbyron@lumc.edu

Y.-X. Wang (ed.), *Calcium Signaling In Airway Smooth Muscle Cells*,
DOI 10.1007/978-3-319-01312-1_2,

Keywords KCNQ channel • Voltage-activated potassium channel • Airway smooth muscle • Voltage-sensitive calcium channel • Bronchodilator

Potassium channels have a well-recognized role in cellular physiology. The outward flux of potassium ions (K^+ conductance) through these channels in the plasma membrane establishes a voltage drop, polarizing the plasma membrane to a negative voltage. In mammalian cells, the voltage would equal a potassium equilibrium potential of approximately −80 mV in the absence of other ionic conductances [1]. However, other ionic conductances (particularly Na^+, Ca^{2+}, and Cl^-) are present, and these tend to depolarize the membrane, rendering the membrane voltage more positive to −80 mV. In resting cells the magnitude of these other conductances is typically much less than that of the K^+ conductance, and so the membrane voltage remains very negative (typically between −65 and −45 mV) [2–6].

In airway smooth muscle cells (ASMCs), the negative membrane voltage established by the activity of potassium channels opposes the opening of L-type Ca^{2+} channels, which generally open in a steeply voltage-dependent manner at voltages positive to −40 mV [7]. Ca^{2+} influx through L-type voltage-sensitive Ca^{2+} channels (VSCCs) can activate smooth muscle contraction and in turn produce bronchoconstriction. Thus, the activity of potassium channels opposes airway smooth muscle contraction, providing an important braking mechanism that prevents excessive airway constriction.

In some pathological conditions such as asthma, airways can become hyperconstricted [8–10]. The reduction in airway diameter results in increased airway resistance, impaired breathing, and, in severe cases, death. The role of potassium channels in the etiology of these diseases is not well characterized, nor has there been much success in developing therapeutic approaches that increase the opening of potassium channels to relieve airway hyperconstriction in asthma [11, 12]. However, the very recent discovery of KCNQ (Kv7) potassium channels in airway smooth muscle [13] may provide new clues to the contributions of potassium channels and of L-type VSCCs in airway physiology and disease. KCNQ channel openers may also represent important new therapeutic agents for the relief of airway hyperconstriction. This chapter will summarize our current knowledge and provide perspective for future developments in this exciting new area of airway physiology.

1 L-type VSCCs in Airway Smooth Muscle Cells

To appreciate the importance of potassium channels in the regulation of airway smooth muscle contraction, we must first examine the role of L-type VSCCs. Although the expression of L-type VSCCs in ASMCs is well established, there is considerable disagreement as to the relative contribution of these channels to ASMC contraction and bronchoconstriction.

Several pieces of evidence support an important contribution of L-type VSCCs in ASMC contraction and bronchoconstriction. Robust Ca^{2+} currents with electrophysiological and pharmacological characteristics of L-type VSCCs activated experimentally by membrane depolarization to voltages positive to −40 mV [14–17]. Bronchoconstrictor agonists have been reported to induce membrane depolarization to voltages positive to −40 mV [1, 5, 6], providing a potential stimulus for activation of L-type VSCCs. Treatment of ASMCs with an L-type VSCC agonist (BAY K-8644) [18, 19], or with elevated external $[K^+]$ [20] to induce membrane depolarization, is sufficient to induce an elevation of $[Ca^{2+}]_{cyt}$ in ASMCs. High $[K^+]$ treatment has also been shown to induce constriction of bronchioles in precision-cut human lung slices, comparable to that induced by bronchoconstrictor agonists [21]. In many studies, ASMC $[Ca^{2+}]_{cyt}$ or tracheal/bronchial smooth muscle contractile responses induced by $G_{q/11}$-coupled receptor agonists were inhibited by specific L-type Ca^{2+} channel blockers (CCBs) [5, 22–33]. And finally, although clinical trials of CCBs produced inconsistent results, CCBs were effective at relieving airway hyperconstriction in a subset of patients in most clinical trials [25, 34].

The story is not quite as clear-cut as it may appear, however, as a number of published findings have led many investigators to conclude that L-type VSCCs play only a minor role in ASMC Ca^{2+} signaling and contraction. For example, it has been observed that Ca^{2+} responses measured in ASMCs have characteristics that indicate that the release of intracellular Ca^{2+} stores is the primary source of Ca^{2+} [35]. Moreover, near-maximal concentrations of agonists are required to depolarize the membrane to voltages that elicit submaximal L-type Ca^{2+} currents [the latter is based on current–voltage (I-V) relationships measured in voltage clamp experiments] [36]. It has also been noted that L-type VSCCs partially inactivate with sustained membrane depolarization (hundreds of milliseconds) [14, 17, 37]. The remaining "window currents" appear to be too small to deliver enough Ca^{2+} for the observed agonist-induced airway constriction under physiological conditions, during which membrane depolarization often persists for extended periods (minutes to hours). Agonists have also been reported to activate signaling pathways that are known to feed back and inhibit L-type Ca^{2+} currents in ASMCs [38].

Although these latter observations present what seems to be a compelling case against an important role for L-type VSCCs in airway constriction, there are some mitigating considerations. The relative contributions of Ca^{2+} influx and release of intracellular Ca^{2+} stores to agonist-induced ASMC contraction clearly varies with agonist concentration. As discussed earlier, there are numerous studies that have demonstrated a reduction of agonist-induced $[Ca^{2+}]_{cyt}$ and contraction responses in the presence of CCBs. Notably, the inhibition by CCBs was almost universally more prominent when submaximal agonist concentrations were tested. For example, data from Janssen [7] show that the CCB nifedipine completely blocked the contractions induced by a low concentration of carbachol (10 nM) but had no detectable effect at a maximal [carbachol] of 10 μM. CCBs also more effectively attenuated histamine- and ACh-induced constriction in dog and guinea pig tracheal smooth muscle at low doses of agonist [39, 40]. Notably, this is consistent with findings in vascular smooth muscle that suggest that agonist-induced Ca^{2+} signaling is highly concentration dependent, with low (physiological) concentrations of agonist acting predominantly via activation of L-type VSCCs and higher (pharmacological) concentrations acting predominantly via inositol 1,4,5-trisphosphate-mediated Ca^{2+} release [41, 42].

It is worthwhile considering how submaximal concentrations of agonists can appreciably increase Ca^{2+} influx through VSCCs when they induce only modest membrane depolarization. Although agonist-induced depolarization does not achieve membrane voltages near the peak of the I-V curve (measured under voltage clamp conditions), even modest membrane depolarization and activation of a small fraction of available L-type channels will, under physiological (non-voltage-clamp) conditions, lead to further membrane depolarization (due to Ca^{2+} influx) and activation of more channels. Similar mechanisms come into play in arterial circulation, where it is widely accepted that vasoconstrictor agonists activate L-type VSCCs in vascular smooth muscle cells (VSMCs) to increase Ca^{2+} influx and stimulate smooth muscle contraction. The extent of agonist-induced membrane depolarization in arterial myocytes [43, 44] is similar to that induced by bronchoconstrictor agonists in ASMCs [1, 5, 6].

Because agonist-induced membrane depolarization persists for periods in excess of hundreds of milliseconds, inactivation of L-type VSCCs invariably occurs. The remaining current after L-type VSCC inactivation (window current) is responsible for the sustained Ca^{2+} influx that determines the physiological responses of many types of excitable cells, including neurons and cardiac and vascular myocytes. Although the window current represents a relatively small influx of Ca^{2+}, this influx might nonetheless be adequate to stimulate contraction [45], either because of its subcellular localization [46] or because the response is amplified by recruitment of other Ca^{2+} sources (e.g., Ca^{2+}-induced Ca^{2+} release [16]). Alternatively (or perhaps in addition), sensitization of the contractile apparatus may occur in parallel such that actin-myosin cross-bridge formation proceeds at lower $[Ca^{2+}]_{cyt}$ [1, 45]. Feedback inhibition of L-type VSCCs may take place following agonist exposure, though this is also more likely to occur at maximal agonist concentrations compared with submaximal concentrations. Feedback pathways exist in most if not all signaling pathways; these mechanisms are generally understood to be in place to prevent overactivation of a pathway (e.g., by maximal agonist concentrations).

Taken as a whole, the available scientific literature suggests an important role of L-type VSCCs in agonist-induced ASMC contraction and bronchoconstriction, predominantly at submaximal agonist concentrations. Our own recent findings clearly support this conclusion. Measuring airway constriction in precision-cut lung slices (PCLSs) from rat, we found that a submaximal concentration of methacholine (230 nM) induces significant airway constriction that is fully reversed by addition of the CCB verapamil, whereas parallel measurements using PCLSs from the same animals revealed that airway constriction stimulated by a maximal concentration of methacholine (10 μM) was not appreciably reduced by verapamil (Fig. 1).

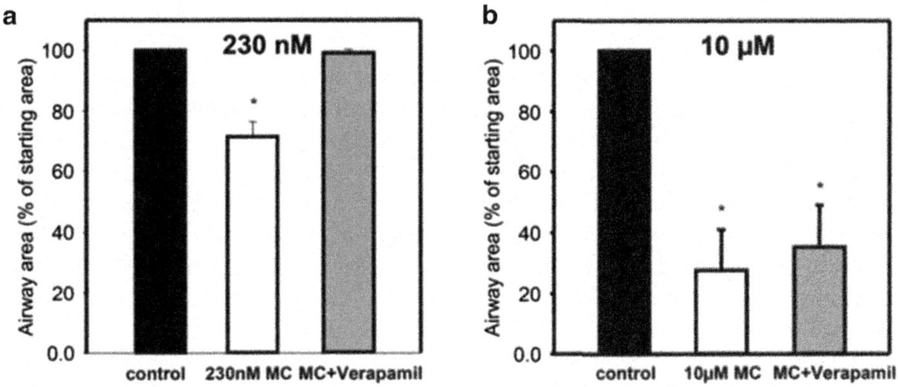

Fig. 1 *Blocking of L-type voltage-sensitive Ca^{2+} channels reverses constriction of airways induced by low, but not high, concentrations of methacholine*. Verapamil (10 μM), an L-type Ca^{2+} channel blocker, fully reversed the constriction induced by 230 nM MC (**a**) ($P < 0.001$ one-way ANOVA, $n = 6$) but was unable to reverse maximal constriction induced by 10 μM MC (**b**)

2 CCBs in Asthma Therapy

Considering the evidence that VSCCs contribute to bronchoconstrictor signal transduction presented earlier, it would seem likely that CCBs would be an effective therapy to reduce excessive bronchoconstriction in conditions such as asthma. CCBs were in fact tested in patients with asthma, but they have now largely been dismissed as ineffective therapies for this disease. A careful examination of the results of clinical trials is, however, warranted. In the early to mid-1980s a large number of clinical trials was conducted to evaluate the efficacy of CCBs in the relief of asthma symptoms. Most trials were designed to test the responses to bronchial provocation (inhalation of bronchoconstrictor agonists or cold air, or exercise-induced bronchoconstriction), comparing placebo-treated with CCB-treated patients (usually nifedipine or verapamil were administered systemically). The results were inconsistent, although almost all the clinical trials revealed at least a subset of patients who benefited from treatment with either verapamil or nifedipine [25, 34]. Greater benefits might have been obtained were it not for adverse side effects associated with systemic administration of CCBs, and limitations of formulation that prevent delivery of effective doses of verapamil or nifedipine by inhalation [25].

An important clue to the inconsistency of CCB therapy in clinical trials may be found in the observation that the protective effects of CCBs seemed to relate to the intensity of bronchial provocation. For example, in a cohort of asthmatic patients subjected to different stimulus intensities (varying rates of isocapnic hyperventilation of cold air), Solway and Fanta observed that nifedipine treatment (20 mg p.o.) significantly reduced the fall in FEV_1 (forced expiratory volume in 1 s) following low or moderate rates of respiratory heat loss but was ineffective at the highest rates tested [24]. We speculate that the majority of clinical trials utilized relatively intense bronchial provocation stimuli that tended to obscure the beneficial effects of CCBs, and this has led to the mistaken conclusion that inhibition of Ca^{2+} influx via VSCCs is not a useful therapeutic strategy.

In support of such speculation, a more consistent benefit was apparent when asthmatic patients were treated with CCBs without exogenous bronchial provocation. For example, treatment with nifedipine (10 mg t.i.d.) for 4 days significantly reduced the overnight fall in peak expiratory flow rate compared with placebo-treated control subjects [47]. Similarly, in asthmatic subjects with chronic stable asthma and baseline bronchoconstriction, nifedipine treatment significantly increased basal FEV_1 values [48], suggesting that nifedipine has a significant bronchodilating effect in subjects with chronic stable asthma who have abnormal lung function.

The pattern of results from clinical trials seems to parallel those from in vitro and preclinical studies in suggesting that L-type VSCCs play a prominent role in submaximal bronchial provocation, though this pattern may be overridden by other mechanisms when greater bronchial challenges are experienced. It is very important therapeutically to relieve airway hyperresponsiveness that occurs with submaximal bronchial provocation, which may reflect the more common experience among asthmatic patients. However, CCBs exhibit adverse side effects when administered systemically, and effective inhalational formulations have not been developed, rendering these drugs less than ideal to achieve the desired therapeutic endpoint. An alternative pharmacological approach, which would have an equivalent effect of inhibiting the activation of VSCCs, would be to administer drugs that prevent the voltage change that drives activation of VSCCs in airway myocytes.

3 Kv7 Potassium Channels in Airway Smooth Muscle Cells

Activation of VSCCs, and hence the influx of Ca^{2+}, is tightly controlled by membrane voltage. It is well established that changes in K^+ channel activity have a dramatic effect on ASMC membrane voltage: inhibition of K^+ channels results in membrane depolarization, and activation of K^+ channels causes membrane hyperpolarization [49]. Hence, regulation of K^+ channel activity is an important mechanism for the adjustment of membrane voltage to control the activity of VSCCs.

The Kv7 family of voltage-gated K^+ channels is encoded by five genes (KCNQ1-5). In excitable cells like neurons, Kv7 currents are crucial for regulating membrane excitability [50, 51]. Although Kv7 channels were long thought to be expressed predominantly in the brain, our laboratory and others recently identified a pivotal role for these channels in VSMCs. We found that subnanomolar concentrations of the vasoconstrictor hormone arginine vasopressin (AVP) suppressed Kv7 currents in arterial myocytes and provided evidence that this effect mediates the physiological vasoconstrictor actions of AVP [43, 52, 53].

Pharmacological alteration of the activity of Kv7 channels has dramatic effects on VSMC excitability, contraction, and vascular tone [43, 53–56]. We found that the Kv7 channel modulator flupirtine, a drug that is used clinically to treat acute and chronic pain [57], enhances Kv7 currents in arterial myocytes; these effects are associated with corresponding changes in vascular tone in isolated pressurized rat mesenteric arteries and changes in systemic blood pressure and mesenteric vascular resistance in vivo [43]. By screening drugs used for other clinical applications, we have also identified the cyclooxygenase-2 (COX-2) inhibitor celecoxib (Celebrex) and its COX-2-independent analog 2,5-dimethylcelecoxib as potent and effective activators of Kv7 channels and inhibitors of L-type VSCCs in vascular smooth muscle cells [58–60].

We recently extended these findings from the vasculature to the airways, demonstrating that Kv7 channels are also expressed and functional in ASMCs and that their activity can also be modulated both by bronchoconstrictor agonists and by a variety of clinically used drugs [13].

Using real-time RT-PCR, all five mammalian KCNQ subtypes (KCNQ1-5) were detected in guinea pig and human ASMCs isolated by enzymatic digestion. In the ASMCs isolated from guinea pig bronchioles, whole-cell perforated patch clamp electrophysiology revealed Kv7 currents that were larger than their VSMC counterparts but otherwise very similar, having all the expected characteristics of Kv7 currents: (1) slow kinetics of activation with no apparent inactivation during a 5 s voltage step; (2) voltage-dependent activation with a very negative threshold (approximately −60 mV), and a $V_{0.5} \sim -31$ mV; (3) reversible enhancement by the selective Kv7 channel activators flupirtine and retigabine; and (4) complete irreversible suppression by the selective Kv7 channel blockers linopirdine and XE991 [13]. We also found that the ASMC Kv7 currents were significantly enhanced by both celecoxib and 2,5-dimethylcelecoxib [13], which is similar to what we had observed for VSMC Kv7 currents [59].

Kv7 currents were also recorded from a small number of isolated human ASMCs. Current densities were relatively small in comparison with Kv7 currents measured in guinea pig ASMCs. Nevertheless, currents in human ASMCs were significantly enhanced by 10 μM retigabine (approximately threefold increase over control at −20 mV) and significantly inhibited by 10 μM XE991 (approximately 70 % inhibition from control at −20 mV) [13].

We also found that two bronchoconstrictors, methacholine (100 nM) and histamine (30 μM), both significantly suppressed Kv7 currents in guinea pig ASMCs. This effect would be predicted to depolarize the ASMC membrane and thereby stimulate Ca^{2+} influx via VSCCs. Importantly, stimulation of Na^{+} and Ca^{2+} influx via nonselective cation channels (e.g., TRPC channels) [3, 61], or activation of chloride channels [1, 62] may also contribute to membrane depolarization in response to bronchoconstrictor agonists; these effects would be greatly enhanced if K^{+} conductance were simultaneously reduced [53]. Thus, suppression of Kv7 currents in ASMCs may be a common mechanism by which bronchoconstrictor agonists stimulate membrane depolarization to activate VSCCs.

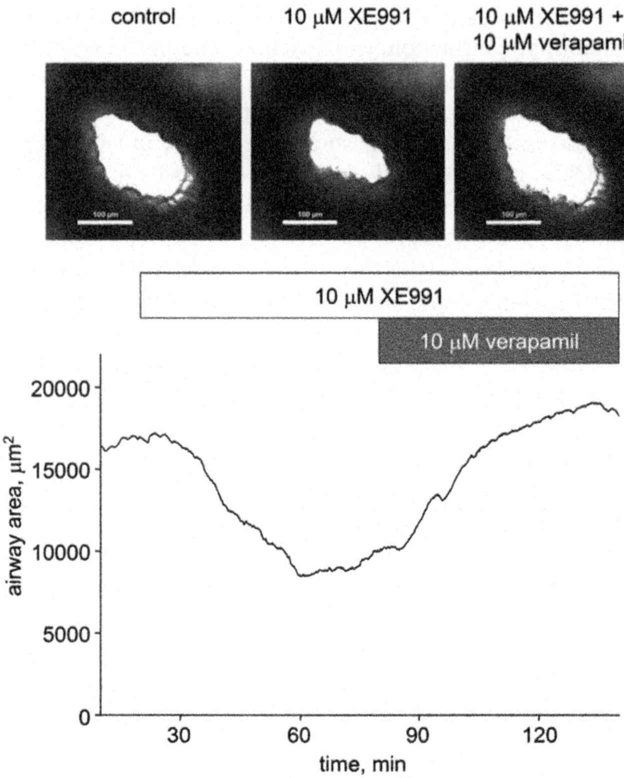

Fig. 2 *Constriction of a human bronchiole by Kv7 channel blocker XE991 is reversed by the L-type VSCC blocker verapamil. Top panels* show images of a cross section of a bronchiole in a precision-cut human lung slice before (control), during treatment with 10 μM XE991, and following relaxation with the addition of 10 μM verapamil. The graph shows the time course of bronchoconstriction and relaxation measured in the same lung slice

More importantly from a therapeutic perspective, the Kv7 channel activator flupirtine was effective in restoring ASMC Kv7 current amplitudes to near control levels in the presence of either methacholine or histamine [13]. Based on these findings we speculate that, in analogy with vascular smooth muscle, contractile agonists may target Kv7 channels as signal transduction intermediates to induce ASMC contraction and bronchoconstriction, and these effects may be attenuated by treatment with clinically used Kv7 channel activators.

Using PCLSs to investigate the physiological relevance of functional Kv7 channels in human airways, we found that the irreversible Kv7 channel blocker XE991 reproducibly induced robust airway constriction in human PCLS (averaging 60–80 %, depending on the size of the airways). This constriction was not reversed upon washout of XE991 but was fully reversed by the addition of the L-type VSCC blocker verapamil (10 μM) (Fig. 2 and [13]), suggesting that suppression of Kv7 currents is sufficient to depolarize the membrane and activate L-type VSCCs, stimulating Ca^{2+} influx, ASMC contraction, and bronchoconstriction. Application of 50 nM histamine also induced a significant constriction of small airways in human PCLSs (as estimated by a reduction of the lumenal area), but the histamine-induced constriction was significantly attenuated (by approximately 30 % on average) in the presence of 100 μM flupirtine [13]. These findings provide at least preliminary evidence that clinically used Kv7 channel activators may be useful therapeutic agents to relieve airway hyperconstriction.

4 Prospects for Future Developments

These exciting findings prompt a number of important questions that remain to be answered:

Which Kv7 channel subtypes form the functional channels in ASMCs? Kv7 channels form as tetramers of individual KCNQ gene products. They may form as homotetramers or as heterotetramers containing more than one Kv7 subtype (e.g., Kv7.2/Kv7.3 [63, 64] or Kv7.4/Kv7.5 [63, 65]). Since all five KCNQ gene products were detected in guinea pig and human ASMCs, there are many potential combinations of subunits that could contribute to the Kv7 currents measured in these cells.

Does the expression or function of Kv7 channels change during the development of airway diseases such as asthma? If ASMC Kv7 channel activity were reduced in asthma, bronchial provocation by pharmacological or mechanical stimulation would more effectively depolarize the cells to activate L-type VSCCs and enhance airway constriction. This might account for the observed airway hyperresponsiveness that is a hallmark of asthma.

What are the signal transduction mechanisms by which bronchoconstrictor agonists regulate Kv7 channel function in ASMCs? Mechanisms that could account for suppression of Kv7 currents by $G_{q/11}$-coupled receptor agonists include depletion of phosphatidylinositol 4,5-bisphosphate (PIP_2), elevation of cytosolic Ca^{2+} concentration, or activation of protein kinase C (PKC) [66]. Our previous research implicated PKC activation in the regulation of vascular smooth muscle Kv7 channels and vasopressin-induced Ca^{2+} signaling/arterial constriction [43, 52, 53, 67]. Different agonists may regulate the channels via different mechanisms, as has been observed for neuronal Kv7 channels [66, 68].

Do physiological bronchodilators (e.g., epinephrine acting on β_2-adrenergic receptors or vasoactive intestinal peptide acting on VPAC receptors) enhance the activity of Kv7 channels to relax ASMCs? These receptors couple to activation of adenylyl cyclase and activation of protein kinase A (PKA). PKA mediates the enhancement of Kv7 currents in some cell systems [69, 70], raising the possibility that this may be another physiological mechanism for regulation of Kv7 channels in ASMCs.

Which pharmacological Kv7 channel agonists are most effective in relaxation of airways, and do they exhibit any selectivity toward ASMC Kv7 channels compared with Kv7 channels expressed in other tissues (e.g., other smooth muscles, skeletal muscle, neurons, epithelia)? Identifying drugs that selectively activate Kv7 channels in ASMCs might facilitate development of bronchodilators with relatively few off-target actions.

Can aerosolized formulations of Kv7 channel activators be used to relax hyperconstricted airways and minimize off-target effects? Will these drugs be effective alone or in combination with other classes of drugs (e.g., β_2-adrenergic receptor agonists, corticosteroids) in the treatment of asthma or other airway diseases?

5 Summary

In summary, the recent discovery that Kv7 potassium channels are expressed and functional in ASMCs, and that altering their activity can influence airway diameter, suggests that these channels may play an important role in airway physiology. The availability of safe and effective clinically used Kv7 channel activators holds great promise for targeting Kv7 channels to relieve airway hyperconstriction in diseases such as asthma.

References

1. Hirota, S. et al. (2007) Ionic mechanisms and Ca^{2+} handling in airway smooth muscle, *Eur Respir J* **30** (1), 114–133.
2. Abdullah, N.A. et al. (1994) Contraction and depolarization induced by fetal bovine serum in airway smooth muscle, *Am J Physiol Lung Cell Mol Physiol* **266** (5), L528–L535.
3. Xiao, J.-H. et al. (2010) Functional Role of Canonical Transient Receptor Potential 1 and Canonical Transient Receptor Potential 3 in Normal and Asthmatic Airway Smooth Muscle Cells, *Am. J. Respir. Cell Mol. Biol.* **43** (1), 17–25.
4. Lee, H.K. et al. (1989) Electromechanical effects of leukotriene D4 on ferret tracheal muscle and its muscarinic responsiveness, *Lung* **167** (3), 173–185.
5. Bourreau, J.P. et al. (1991) Acetylcholine Ca^{2+} stores refilling directly involves a dihydropyridine-sensitive channel in dog trachea, *Am J Physiol Cell Physiol* **261** (3), C497–C505.
6. Farley, J.M. et al. (1977) Role of depolarization in acetylcholine-induced contractions of dog trachealis muscle, *J Pharmacol Exp Ther* **201** (1), 199–205.
7. Janssen, L.J. (2002) Ionic mechanisms and Ca^{2+} regulation in airway smooth muscle contraction: do the data contradict dogma?, *Am J Physiol Lung Cell Mol Physiol* **282** (6), L1161–L1178.
8. An, S.S. et al. (2007) Airway smooth muscle dynamics: a common pathway of airway obstruction in asthma, *Eur Respir J* **29** (5), 834–860.
9. Hirota, J.A. et al. (2009) Airway smooth muscle in asthma: Phenotype plasticity and function, *Pulm Pharmacol Ther.* **22** (5), 370–378.
10. Meurs, H. et al. (2008) Airway hyperresponsiveness in asthma: lessons from in vitro model systems and animal models, *Eur Respir J* **32** (2), 487–502.

11. Liu, X.S. et al. (2005) Potassium channels in airway smooth muscle and airway hyperreactivity in asthma, *Chin Med J (Engl)* **118** (7), 574–580.
12. Malerba, M. et al. (2010) The potential therapeutic role of potassium channel modulators in asthma and chronic obstructive pulmonary disease, *Journal of Biological Regulators & Homeostatic Agents* **24** (2), 123–130.
13. Brueggemann, L.I. et al. (2012) Kv7 potassium channels in airway smooth muscle cells: signal transduction intermediates and pharmacological targets for bronchodilator therapy, *Am J Physiol Lung Cell Mol Physiol* **302** (1), L120–L132.
14. Kotlikoff, M.I. (1988) Calcium currents in isolated canine airway smooth muscle cells, *Am J Physiol Cell Physiol* **254** (6), C793–C801.
15. Worley, J.F. et al. (1990) Dihydropyridine-sensitive single calcium channels in airway smooth muscle cells, *Am J Physiol Lung Cell Mol Physiol* **259** (6), L468–L480.
16. Du, W. et al. (2006) Excitation-Contraction Coupling in Airway Smooth Muscle, *J Biol Chem* **281** (40), 30143–30151.
17. Janssen, L.J. (1997) T-type and L-type Ca^{2+} currents in canine bronchial smooth muscle: characterization and physiological roles, *Am J Physiol Cell Physiol* **272** (6), C1757–C1765.
18. Wang, Y.X. et al. (1997) Modulation of maxi-K^+ channels by voltage-dependent Ca^{2+} channels and methacholine in single airway myocytes, *Am J Physiol Cell Physiol* **272** (4), C1151–C1159.
19. Liu, X. et al. (1996) Acetylcholine-induced chloride current oscillations in swine tracheal smooth muscle cells, *J Pharmacol Exp Ther* **276** (1), 178–186.
20. Kumasaka, D. et al. (1996) $MgSO_4$ relaxes porcine airway smooth muscle by reducing Ca^{2+} entry, *Am J Physiol Lung Cell Mol Physiol* **270** (3), L469–L474.
21. Ressmeyer, A.-R. et al. (2010) Human Airway Contraction and Formoterol-Induced Relaxation Is Determined by Ca^{2+} Oscillations and Ca^{2+} Sensitivity, *Am. J. Respir. Cell Mol. Biol.* **43** (2), 179–191.
22. Hay, D.W. et al. (1999) Endothelin receptors and calcium translocation pathways in human airways, *Naunyn Schmiedebergs Arch Pharmacol.* **359** (5), 404–410.
23. Kohrogi, H. et al. (1985) Nifedipine inhibits human bronchial smooth muscle contractions induced by leukotrienes C4 and D4, prostaglandin F2 alpha, and potassium, *Am Rev Respir Dis* **132** (2), 299–304.
24. Solway, J. et al. (1985) Differential inhibition of bronchoconstriction by the calcium channel blockers, verapamil and nifedipine, *Am Rev Respir Dis.* **132** (3), 666–670.
25. Fish, J.E. (1984) Calcium channel antagonists in the treatment of asthma, *J Asthma* **21** (6), 407–418.
26. Lei, Y. et al. (2011) Enhanced airway smooth muscle cell thromboxane receptor signaling via activation of JNK MAPK and extracellular calcium influx, *Eur J Pharmacol.* **650** (2–3), 629–638.
27. Dai, J.M. et al. (2007) Acetylcholine-induced asynchronous calcium waves in intact human bronchial muscle bundle, *Am J Respir Cell Mol Biol.* **36** (5), 600–608.
28. Moura, C.T. et al. (2005) Increased responsiveness to 5-hydroxytryptamine after antigenic challenge is inhibited by nifedipine and niflumic acid in rat trachea in vitro, *Clin Exp Pharmacol Physiol.* **32** (12), 1119–1123.
29. Hirota, K. et al. (2003) Effects of three different L-type Ca^{2+} entry blockers on airway constriction induced by muscarinic receptor stimulation, *Br J Anaesth.* **90** (5), 671–675.
30. Matsuda, F. et al. (2000) Comparative effect of amrinone, aminophylline and diltiazem on rat airway smooth muscle, *Acta Anaesthesiol Scand.* **44** (6), 763–766.
31. Thirstrup, S. et al. (1997) In vitro studies on the interactions of $beta_2$-adrenoceptor agonists, methylxanthines, Ca^{2+}-channel blockers, K^+-channel openers and other airway smooth muscle relaxants in isolated guinea-pig trachea, *Eur J Pharmacol.* **326** (2–3), 191–200.
32. Ahmed, T. et al. (1992) Modification of histamine- and methacholine-induced bronchoconstriction by calcium antagonist gallopamil in asthmatics, *Respiration* **59** (6), 332–338.

33. Vannier, C. et al. (1995) Inhibition of dihydropyridine-sensitive calcium entry in hypoxic relaxation of airway smooth muscle, *Am J Physiol Lung Cell Mol Physiol* **268** (2), L201–L206.
34. Fanta, C.H. (1985) Calcium-channel blockers in prophylaxis and treatment of asthma, *Am J Cardiol* **55** (3), 202B–209B.
35. Perez-Zoghbi, J.F. et al. (2009) Ion channel regulation of intracellular calcium and airway smooth muscle function, *Pulm Pharmacol Ther.* **22** (5), 388–397.
36. Janssen, L.J. (2009) Asthma therapy: how far have we come, why did we fail and where should we go next?, *Eur Respir J* **33** (1), 11–20.
37. Li, Q.J. et al. (2002) Membrane currents in canine bronchial artery and their regulation by excitatory agonists, *Am J Physiol Lung Cell Mol Physiol* **282** (6), L1358-L1365.
38. Yamakage, M. et al. (1995) Cholinergic regulation of voltage-dependent Ca^{2+} channels in porcine tracheal smooth muscle cells, *Am J Physiol Lung Cell Mol Physiol* **269** (6), L776–L782.
39. Farley, J.M. et al. (1978) The sources of calcium for acetylcholine-induced contractions of dog tracheal smooth muscle, *J Pharmacol Exp Ther* **207** (2), 340–346.
40. Barnes, P.J. (1985) Clinical studies with calcium antagonists in asthma, *Br. J. Clin. Pharmacol.* **20** (Suppl 2), 289S–298S.
41. Henderson, K.K. et al. (2007) Vasopressin-induced vasoconstriction: two concentration-dependent signaling pathways, *J Appl Physiol.* **102** (4), 1402–1409.
42. Byron, K.L. (1996) Vasopressin stimulates Ca2+ spiking activity in A7r5 vascular smooth muscle cells via activation of phospholipase A2, *Circ Res* **78** (5), 813–820.
43. Mackie, A.R. et al. (2008) Vascular KCNQ potassium channels as novel targets for the control of mesenteric artery constriction by vasopressin, based on studies in single cells, pressurized arteries, and in vivo measurements of mesenteric vascular resistance, *J Pharmacol Exp Ther* **325** (2), 475–483.
44. Mani, B.K. et al. (in press) Vascular KCNQ (Kv7) potassium channels as common signaling intermediates and therapeutic targets in cerebral vasospasm., *J. Cardiovasc. Pharmacol.*
45. Liu, C. et al. (2005) Regulation of Rho/ROCK signaling in airway smooth muscle by membrane potential and $[Ca^{2+}]_i$, *Am J Physiol Lung Cell Mol Physiol* **289** (4), L574–L582.
46. Parekh, A.B.. (2008) Ca2+ microdomains near plasma membrane Ca2+ channels: impact on cell function, *The Journal of Physiology* **586** (13), 3043–3054.
47. Patakas, D. et al. (1983) Nifedipine in bronchial asthma, *J Allergy Clin Immunol.* **72** (3), 269–273.
48. Schwartzstein, R.S. et al. (1986) Orally administered nifedipine in chronic stable asthma. Comparison with an orally administered sympathomimetic, *Am Rev Respir Dis.* **134** (2), 262–265.
49. Hall, I.P. (2000) Second messengers, ion channels and pharmacology of airway smooth muscle, *Eur Respir J* **15** (6), 1120–1127.
50. Maljevic, S. et al. (2008) Nervous system Kv7 disorders: breakdown of a subthreshold brake, *J Physiol* **586** (7), 1791–1801.
51. Peroz, D. et al. (2008) Kv7.1 (KCNQ1) properties and channelopathies, *J Physiol* **586** (7), 1785–1789.
52. Henderson, K.K. et al. (2007) Vasopressin-induced vasoconstriction: two concentration-dependent signaling pathways, *J Appl Physiol* **102** (4), 1402–1409.
53. Brueggemann, L.I. et al. (2007) Vasopressin stimulates action potential firing by protein kinase C-dependent inhibition of KCNQ5 in A7r5 rat aortic smooth muscle cells, *Am J Physiol Heart Circ Physiol* **292** (3), H1352–H1363.
54. Yeung, S.Y. et al. (2007) Molecular expression and pharmacological identification of a role for K(v)7 channels in murine vascular reactivity, *Br. J. Pharmacol.* **151** (6), 758–770.
55. Joshi, S. et al. (2006) Pulmonary vasoconstrictor action of KCNQ potassium channel blockers, *Respir Res* **7**, 31.
56. Joshi, S. et al. (2009) KCNQ Modulators Reveal a Key Role for KCNQ Potassium Channels in Regulating the Tone of Rat Pulmonary Artery Smooth Muscle, *J Pharmacol Exp Ther* **329** (1), 368–376.

57. Devulder, J. (2010) Flupirtine in pain management: pharmacological properties and clinical use, *CNS Drugs* **24** (10), 867–881.
58. Brueggemann, L.I. et al. (2010) Novel Actions of Nonsteroidal Anti-Inflammatory Drugs on Vascular Ion Channels: Accounting for Cardiovascular Side Effects and Identifying New Therapeutic Applications, *Molecular and Cellular Pharmacology* **2** (1), 15–19.
59. Brueggemann, L.I. et al. (2009) Differential effects of selective cyclooxygenase-2 inhibitors on vascular smooth muscle ion channels may account for differences in cardiovascular risk profiles, *Mol Pharmacol* **76** (5), 1053–1061.
60. Mani, B.K. et al. (2011) Activation of vascular KCNQ (Kv7) potassium channels reverses spasmogen-induced constrictor responses in rat basilar artery, *British Journal of Pharmacology* **164**, 237–249.
61. Wang, Y.X. et al. (2011) Molecular expression and functional role of canonical transient receptor potential channels in airway smooth muscle cells, *Adv Exp Med Biol* **704**, 731–747.
62. Daniel, E.E. et al. (1993) Chloride and depolarization by acetylcholine in canine airway smooth muscle, *Can J Physiol Pharmacol* **71** (3–4), 284–292.
63. Bal, M. et al. (2008) Homomeric and Heteromeric Assembly of KCNQ (Kv7) K^+ Channels Assayed by Total Internal Reflection Fluorescence/Fluorescence Resonance Energy Transfer and Patch Clamp Analysis, *J Biol Chem* **283** (45), 30668–30676.
64. Schwake, M. et al. (2003) A carboxy-terminal domain determines the subunit specificity of KCNQ K^+ channel assembly, *EMBO Rep.* **4** (1), 76–81.
65. Brueggemann, L.I. et al. (2011) Diclofenac distinguishes among homomeric and heteromeric potassium channels composed of KCNQ4 and KCNQ5 subunits, *Mol Pharmacol* **79** (1), 10–23.
66. Delmas, P. et al. (2005) Pathways modulating neural KCNQ/M (Kv7) potassium channels, *Nat Rev Neurosci* **6** (11), 850–862.
67. Fan, J. et al. (2000) Ca^{2+} signalling in rat vascular smooth muscle cells: a role for protein kinase C at physiological vasoconstrictor concentrations of vasopressin, *J Physiol* **524 Pt 3**, 821–831.
68. Hernandez, C.C. et al. (2008) Regulation of neural KCNQ channels: signalling pathways, structural motifs and functional implications, *J Physiol* **586** (7), 1811–1821.
69. Schroeder, B.C. et al. (1998) Moderate loss of function of cyclic-AMP-modulated KCNQ2/KCNQ3 K^+ channels causes epilepsy, *Nature* **396** (6712), 687–690.
70. Chambard, J.M. et al. (2005) Regulation of the voltage-gated potassium channel KCNQ4 in the auditory pathway, *Pflugers Arch* **450** (1), 34–44.

Transient Receptor Potential and Orai Channels in Airway Smooth Muscle Cells

Jun-Hua Xiao, Yong-Xiao Wang, and Yun-Min Zheng

Contents

J.-H. Xiao
Department of Pharmacology, Tongji Medical College of Huazhong University of Science and Technology, 13 Hangkong Road, Wuhan 430030, P.R. China
e-mail: xiaoj@mail.hust.edu.cn

Y.-X. Wang • Y.-M. Zheng (✉)
Center for Cardiovascular Sciences (MC-8), Albany Medical College, 47 New Scotland Avenue, Albany, NY 12208, USA
e-mail: zhengy@mail.amc.edu

Y.-X. Wang (ed.), *Calcium Signaling In Airway Smooth Muscle Cells*,
DOI 10.1007/978-3-319-01312-1_3,

Abstract Airway smooth muscle cells (ASMCs) are a crucial component of the airway passage. In ASMCs, an increase in intracellular calcium concentration ($[Ca^{2+}]_i$) acts as a key determinant of force generation, cell proliferation, and other cellular responses. $[Ca^{2+}]_i$ is generated and controlled by numerous ion channels. Recent studies demonstrate that multiple members of transient receptor potential (TRP) channels, including TRPC1-6 and TRPV1, 2, and 4, are expressed and important for the regulation of $[Ca^{2+}]_i$ in ASMCs. In particular, TRPC channels play an important role in the control of the resting $[Ca^{2+}]_i$ and extracellular Ca^{2+} influx. Three Orai molecules, known as the pore-forming subunits of specific store-operated Ca^{2+} channels (SOCCs), are expressed in ASMCs. Two stromal-interacting molecule proteins (Stim1 and Stim2), which serve as the endoplasmic reticulum Ca^{2+} sensors and thus are the major molecular components of SOCCs, are expressed in ASMCs. Identification of TRP, Orai, and Stim molecules involved in controlling and regulating $[Ca^{2+}]_i$, contractility, and proliferation in ASMCs offers the exciting prospect of new and novel therapies for the treatment of airway diseases such as asthma, chronic obstructive pulmonary disease, and others.

Keywords Transient receptor potential channels • Orai molecules • Stromal-interacting molecule proteins • Store-operated Ca^{2+} channels • Airway smooth muscle cells

1 Introduction

Airway smooth muscle cells (ASMCs) are crucial components of the airway passage, important regulators of lung function, and a determinant factor of pathological changes like asthma, chronic obstructive pulmonary disease, and pulmonary hypertension, creating important health problems [1]. A hallmark feature of asthma is variable airflow obstruction caused by contraction of airway smooth muscle (ASM). The development of more persistent airway obstruction in chronic severe asthma is attributed to the contribution of ASM hyperplasia and hypertrophy [2, 3]. In ASMCs, increases in the cytosolic free calcium concentration ($[Ca^{2+}]_i$) act as a key determinant of force generation and cell proliferation [3–5]. Ca^{2+} plays a central role in the pathophysiology of asthma.

Ion channels initiate and regulate contraction and smooth muscle tone. In the plasma membrane of ASMCs, there are at least three classes of Ca^{2+} channels: voltage-dependent Ca^{2+} channels (VDCCs), receptor-operated Ca^{2+} channels (ROCCs), and store-operated Ca^{2+} channels (SOCCs). Membrane potential plays a critical role in regulating $[Ca^{2+}]_i$ by governing Ca^{2+} influx via VDCCs. Agonist- or mitogen-induced Ca^{2+} influx is mainly caused by receptor-mediated activation of ROCCs and store depletion-mediated opening of SOCCs [6].

Ca^{2+} depletion in the sarcoplasmic reticulum (SR) triggers capacitative Ca^{2+} entry (CCE), a mechanism involved in maintaining sustained Ca^{2+} influx and refilling Ca^{2+} in the SR. The transient receptor potential (TRP) channel genes have been demonstrated to be essential for agonist-activated CCE. Expression of TRP channel genes in mammalian cells results in the formation of Ca^{2+}-permeable channels that are activated by Ca^{2+} store depletion. These findings suggest that the TRP-encoded proteins are the putative SOCCs responsible for CCE [6].

TRP superfamily members are molecular correlates of nonselective cation channels in smooth muscle cells (SMCs). However, the study of these channels in vitro in heterologous overexpression systems has given rise to a substantial body of controversial data concerning channel properties and regulation. The lack of specific channel blockers has complicated the in situ identification and characterization of TRP channels in tissues and underlines the need for gene-deficient mouse models for conclusive in vivo studies. Downregulation of TRP proteins in primary SMCs has proven to be a valuable tool in identifying native TRP channel activities [7, 8].

Mammalian cells express two stromal-interacting molecule (Stim) proteins (Stim 1 and Stim 2) and three Orai proteins (Orai1, Orai2, and Orai3) [9, 10]. Large-scale genomic screens revealed the two fundamental players of SOCCs: Stim proteins that sense Ca^{2+} levels in the endoplasmic reticulum (ER) [11, 12] and Orai proteins that constitute the basic pore-forming subunits of the specific SOCCs [13–16].

In this chapter, we highlight the important findings of studies on the expression, function, and regulation of TRP and Orai channels with the addition of Stim proteins in ASMCs.

2 Expression of TRPs, Orais, and Stims

2.1 General Structural Properties of Mammalian TRP Channels

After the identification of an expressed sequence tag (EST) with high similarity to *Drosophila trp,* the first mammalian TRP channel, classical or canonical TRP-1 (TRPC1) channel, was cloned in 1995 [17]. Subsequently, six other members of the TRPC family (TRPC2-7) were identified and other TRP families like TRPM (for melastatin), TRPV (for vanilloid receptor), TRPP (for PKD proteins), TRPML (for mucolipidins), and TRPA (for ankyrin rich proteins) were described [1].

Glycosylation scanning of heterologously expressed TRPC3 revealed the topology of TRP proteins in the plasma membrane [18]. TRP channels contain six transmembrane domains (S1–6) flanked by cytoplasmic N- and C-termini. Single-ion channel subunits are assembled as homo- and heterotetramers. A hydrophobic loop between the S5 and S6 segments of each subunit is thought

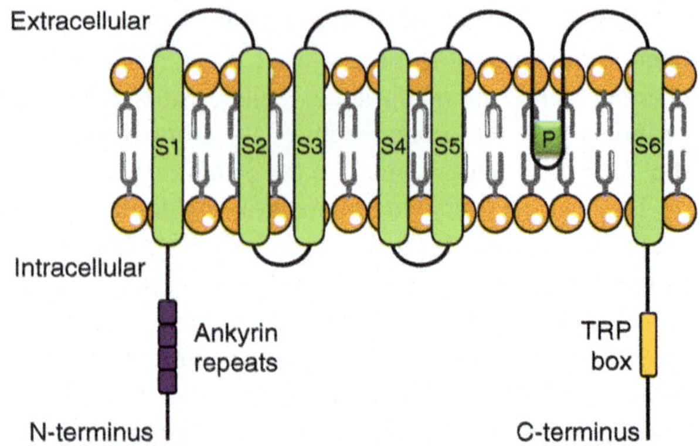

Fig. 1 Plasma membrane topology of TRPC channels; *S1–6* transmembrane domains 1–6, and *P* putative channel pore (Modified from reference [1] with permission)

to contribute to a common putative channel pore (Fig. 1). The composition of functional TRP channel complexes is governed by heteromultimerization and complex formation via scaffold proteins in vivo. Many differences identified by comparing the functional properties of TRP channels obtained in heterologous expression systems with their proposed physiological functions in different tissues may be explained by these circumstances [1, 19].

2.2 *TRPC Channels*

The seven family members can be subdivided into four subfamilies on the basis of their amino acid similarity. While TRPC1 and TRPC2 are almost unique, TRPC4 and TRPC5 share approximately 64 % amino acid identity. TRPC3, 6, and 7 form a structural and functional subfamily displaying 65–78 % identity at the amino acid level, and they share a common sensitivity toward activation by diacylglycerol (DAG) [1, 20, 21].

All TRPC family members harbor an invariant sequence in the C-terminal tail, called the TRP box (amino acid sequence: EWKFAR), as well as 3–4 NH_2-terminal ankyrin repeats. In general, TRPC channels are nonselective Ca^{2+}-permeable cation channels (NSCCs), although the selectivity ratio $P_{Ca}{}^{2+}/P_{Na}{}^{+}$ varies significantly from nonselective (TRPC1) to slightly selective (range of 1–3: TRPC3, TRPC4, TRPC5, TRPC7) to more selective (range of 5: TRPC6) (Table 1) [1, 22–24]. TRPC channels have attracted particular attention because they were the first mammalian TRP channels proposed as molecular candidates for ROCCs and SOCCs [25, 26].

As summarized in Table 2, TRPC1 mRNA and proteins are expressed in freshly isolated porcine ASM tissues, primary isolated guinea-pig ASMCs and ASM tissues, primary isolated rat ASMCs, freshly and primary isolated mouse ASMCs and ASM tissues, passaged human ASMCs using a standard *reverse transcription-polymerase* chain reaction and western blot analysis or immunofluorescence staining [6, 27–30].

Table 1 Biophysical and pharmacological properties of expressed TRPC channels in cell lines (Modified from references [1, 3, 24])

TRP	Single-channel conductance (pS)	Ion selectivity (PCa/PNa)	Constitutive activity	Blockers	Activators	Binding partners
C1	16, 24	1.0, nonselective	–	La^{3+}, Gd^{3+}, 2-APB, SKF-96365	PLC, OAG, stretch, store depletion	TRPC3, TRPC4, TRPC5, IP_3R, CaM, Homer, Stim1
C2	42	2.7, Nonselective	N/A	La^{3+}, 2-APB, SKF-96365	PLC, OAG, store depletion	IP_3R, CaM
C3	42, 66	1.5, 1.6	+	FFA, La^{3+}, Gd^{3+}, 2-APB, SKF-96365	PLC, OAG, Src kinase, IP_3, $[Ca^{2+}]_i$, store depletion	TRPC6, TRPC7, IP_3R, CaM, RyR, NCX1
C4	30, 42	1.1, 7.7	–	2-APB, SKF-96365	PLC, GTPγS, receptor operation, La^{3+}, store depletion	TRPC1,TRPC5, IP_3R, CaM
C5	38, 47, 66	1.8, 9.0	+	FFA, Gd^{3+}, 2-APB, SKF-96365	PLC, OAG, store depletion, Src kinase, IP_3, $[Ca^{2+}]_i$, PI3K, MLCK	TRPC1, TRPC4, CaM, MLCK
C6	28, 35, 37	5.0	–	FFA, La^{3+}, Gd^{3+}, 2-APB, SKF-96365	PLC, OAG, store depletion, Src, 20-HETE	TRPC3, TRPC7,CaM, FKBP12, Fyn kinase
C7	25, 50, 75	2	+	La^{3+}, Gd^{3+}, 2-APB, SKF-96365	PLC, OAG, store depletion, 20-HETE	TRPC3, TRPC6, CaM, FKBP12

2-APB 2-aminoethoxydiphenyl borate, *CaM* calmodulin, *FKBP12* FK506 binding protein 12, *FFA* flufenamic acid, *20-HETE* 20-hydroxyeicosatetraenoic acid, *IP_3R* inositol 1,4,5-trisphosphate receptors, *MLCK* myosin light chain kinase, *NCX1* Na^+/Ca^{2+} exchanger-1, *OAG* 1-oleoyl-2-acetyl-*sn*-glycerol, *PI3K* phosphoinositide-3-kinase, *PLC* phospholipase C, *RyR* ryanodine receptor, + mRNA or protein expression, – no expression, *N/A* no data available

Table 2 TRP, Orai, and Stim expression in human, guinea pig, rat, mouse, and porcine airway smooth muscle (Modified from references [3] and [24])

	TRP family member									Orai family member			Stim family member	
Species	C1	C2	C3	C4	C5	C6	C7	V2	V4	1	2	3	1	2
Human	+	–	+	+	–	+	N/A	+	+	+	+	+	+	+
Guinea pig	+	+	+	+	+	+	N/A	N/A	N/A	N/A	N/A	N/A	N/A	N/A
Rat	+	N/A	N/A	N/A	N/A	N/A	N/A	N/A	N/A	+	N/A	N/A	+	N/A
Mouse	+	N/A	+	N/A	N/A	N/A	N/A	N/A	N/A	N/A	N/A	N/A	N/A	N/A
Porcine	+	N/A	+	+	+	N/A	N/A	N/A	N/A	N/A	N/A	N/A	N/A	N/A

+ mRNA or protein expression, – no expression, *N/A* no data available

TRPC2 is unique among TRPCs because its complete gene has been lost from the Old World monkey and human genomes, in which its remnants constitute a pseudogene. TRPC2 mRNA is expressed in primary isolated guinea-pig ASMCs and not present in passaged human ASMCs [28, 31].

TRPC3 mRNA and proteins are expressed in freshly isolated porcine ASM tissues, primary isolated guinea-pig ASMCs and ASM tissues, freshly and primary isolated mouse ASMCs and ASM tissues, and passaged human ASMCs [27–33]. TRPC3 belongs to the TRPC3/6/7 subfamily of DAG-activated ROCCs [20]. Although the channel can be stimulated by DAG, TRPC3 has a high constitutive activity, in contrast to TRPC6 [1].

TRPC4 mRNA and proteins are detected in freshly isolated porcine ASM tissues, primary isolated guinea-pig ASMCs and ASM tissues, and passaged human ASMCs [27–29, 31]. In ASM, TRPC5 and TRPC4 mRNA expression appears to differ between species. For instance, TRPC5 mRNA was expressed in primary isolated guinea-pig ASMCs but not in passaged human ASMCs in an earlier study [29]. Inconsistently, TRPC5 mRNA and proteins expressions were confirmed in passaged human ASMCs in a subsequent study [31]. These data suggest that the composition of TRPC proteins in ASMCs differs between species. TRPC1 can form heteromeric complexes with either TRPC4 or TRPC5. These observations suggest different TRP isoform expression, which will determine the homomeric and heteromeric TRPC1/TRPC4/TRPC5 channels that are possible in native smooth muscle in each species and the biophysical, regulatory, and pharmacological properties of the functional channels [3, 19, 34].

TRPC6 mRNA and proteins are expressed in primary guinea-pig ASMCs and ASM tissues and passaged human ASMCs [28, 29, 31–33]. TRPC6 is a DAG-sensitive, tightly receptor-regulated, and store-independent cation channel, as opposed to TRPC3 and TRPC7, which display considerable basal activity [1, 20, 21, 35]. Some studies have shown that the expression pattern of TRPC3 and TRPC6 overlaps in most SMC tissues, and a heterotetrameric TRPC3/6 channel complex may be the native molecular correlate of the receptor-operated, nonselective cation influx into SMCs. Some findings imply that TRPC3 and 6 are functionally nonredundant and that TRPC6 plays a unique role in the control of airway and vascular smooth muscle contractility [1, 36, 37]. TRPC7 expression and physiological functions in ASM cells and tissues are vague.

2.3 TRPV1, TRPV2, and TRPV4 Channels

The TRPV subfamily, containing six members (TRPV1–6), are cation-selective channels, which are Ca^{2+}-permeable and expressed widely in both excitable and nonexcitable cells, including the heart, lung, kidney, nerves, and endothelial cells [3, 38, 39]. In contrast to the TRPC family, which has garnered a plethora of attention in the smooth muscle field, the TRPV channels are positively promiscuous in the stimuli to which they respond. Whereas TRPCs are predominantly activated as a result of Gq-coupled receptor occupation or store depletion, TRPVs can be activated by diverse stimuli, including protons, irritants, lipids, mechanical stimuli, noxious heat, and changes in cell volume [40].

TRPV1 was the first mammalian member of this family to be identified and the one that has been studied most intensively [41]. The TRPV1 channel is mainly expressed in sensory nerves. In ASMCs, TRPV1 expression is yet to be reported by direct evidence. But TRPV1 immunoreactive axons are present in the tracheal smooth muscle and the epithelium of the guinea pig and human airway [42].

TRPV2 and TRPV4 mRNA and proteins are expressed in primary cultured human ASMCs. The role of TRPV2, also expressed by human airways, remains undefined. Functional TRPV4 is expressed in human ASMCs and may act as an osmolarity sensor in the airway [43].

2.4 Other TRP Channels

To date, no data have been published on the expression or function of TRPP, TRPM, TRPML, and TRPA channels in ASMCs.

2.5 Orai Channels

Orai1 mRNA and proteins are expressed in primary cultured human and primary isolated rat ASMCs [44, 45]. Orai2 and Orai3 mRNA are expressed in primary cultured human ASMCs [45].

2.6 *Stim Proteins*

Stim1 has been identified as a single transmembrane protein that resides in both the plasma membrane and the membrane of the SR [46]. Stim1 mRNA and protein are expressed in primary cultured human and primary isolated rat ASMCs [44, 47, 48]. Stim2 mRNA and protein are expressed in primary cultured human ASMCs [47, 48].

3 Physiological Functions of TRPs, Orais, and Stims

3.1 *Controlling the Resting Membrane Potential and* $[Ca^{2+}]_i$ *in ASMCs*

The resting membrane potential in ASMCs is between −40 and −50 mV, significantly less negative than a K^+ equilibrium potential of approximately −85 mV [24]. In freshly and primary isolated mouse ASMCs, specific TRPC3 antibodies and gene silencing both result in a pronounced hyperpolarization of the resting membrane potential by approximately 14 mV. TRPC3 siRNA-mediated gene silencing significantly lowers the resting $[Ca^{2+}]_i$ in primary isolated ASMCs. Thus, TRPC3-encoded NSCCs play an important role in the control of the resting membrane potential and $[Ca^{2+}]_i$ in ASMCs [24, 30].

3.2 *Regulation of* $[Ca^{2+}]_i$ *in ASMCs*

In ASM, Ca^{2+} influx through ROCCs or SOCCs is believed to mediate long-term signals, sustained contraction, and the refilling of internal Ca^{2+} stores [6, 45]. The mechanisms for activation of ROCCs and SOCCs in ASMCs remain to be elucidated. In physiological processes of ASMCs, TRPC-encoded NSCCs play an important role in the control of extracellular Ca^{2+} influx.

Recent literature has identified the Stim and Orai proteins as key signaling players in the activation of specific SOCCs (calcium release-activated channels). Stim1 (and in some instances Stim2), a single transmembrane protein that resides in both the plasma membrane and the membrane of SR, serves as the ER Ca^{2+} sensor. Orais (Orai1, Orai2, and Orai3) function as pore-forming subunits of the specific SOCCs in ASMCs [44, 45, 47].

4 Functional Roles of TRPs, Orais, and Stims in Respiratory Diseases

4.1 Asthma

Asthma is a common disease associated with airway inflammation, bronchial hyperresponsiveness, and bronchospasm. Airway obstruction due to bronchial constriction and airway hyperreactivity is a major cause of acute respiratory incapacity in patients with asthma. ASMCs are known to switch from a contractile to a proliferative and so-called synthetic phenotype in culture upon chronic stimulation by serum and growth factors [1, 49]. Although the involvement of TRP channels in these processes is unclear, recent reports indicate that TRPC1-6, TRPV1, 2, and 4, Orai, and Stim molecules are expressed in animal and human ASMCs [6, 28–30, 43–48, 50]. TRPC1 mRNA expression was much higher in proliferating bronchial SMC than in growth-arrested cells, suggesting that TRPC1 may play a role in ASMC proliferation and airway remodeling [6]. TRPC6-deficient (TRPC6$^{-/-}$) mice displayed an increased responsiveness to methacholine and showed a 50 % reduction in respiration rate at a significantly lower methacholine concentration than wild-type littermates. A significantly stronger contraction in response to muscarinic stimulation with methacholine was detected in tracheal rings of TRPC6$^{-/-}$ mice than in tracheal rings from control mice. Upregulation of TRPC3 in the lungs of TRPC6$^{-/-}$ mice was detected in the observed phenotype [1, 51].

TRPC-encoded NSCCs in ASMCs have been shown to make a significant contribution to the development of asthma. In passaged human ASMCs, TRPC3 mRNA and protein expression was significantly increased following treatment with TNFα, an important asthma mediator [31]. These authors have further found that TRPC3 gene silencing inhibited TNFα-induced Ca^{2+} influx and an associated increase in $[Ca^{2+}]_i$ as well as a TNFα-mediated augmentation of acetylcholine-evoked increase in $[Ca^{2+}]_i$ [31]. In ovalbumin- sensitized/challenged asthmatic mice, TRPC1 protein expression level is not changed in freshly isolated asthmatic ASM tissue. On the other hand, the TRPC3 protein expression level is increased threefold. The asthmatic membrane depolarization is prevented by TRPC1 antibodies. In spite of having no clear explanation for these interesting but puzzling results, we would speculate that TRPC1-encoded NSCCs, which are dormant in normal cells, can be activated by an unknown mechanism in asthmatic cells. We have also revealed that the asthmatic increase in the activity of constitutively active NSCCs and depolarization in the membrane potential are both blocked by TRPC3 antibodies. Thus, TRPC3-encoded native constitutively active NSCCs are upregulated in both the molecular expression and functional activity, contributing to membrane depolarization and hyperresponsiveness in asthmatic airway SMCs, whereas TRPC1-encoded NSCCs are only functionally upregulated to mediate the increase cellular responses in asthmatic airway SMCs [24, 30].

A possible role for TRPV4 channels in asthma has also been proposed [1, 43, 52]. Because asthma is accompanied by a denudation of the epithelial lining of bronchi and bronchioles, bronchial SMCs and nerve endings can become exposed to bronchial fluid that is hypotonic. TRPV4 responses to systemic osmotic and somatosensory mechanical stimuli are impaired in $TRPV4^{-/-}$ mice [1, 53–55].

Stim1, having no channel-like properties, is believed to be a Ca^{2+} sensor, monitoring the Ca^{2+} content of intracellular stores [56]. Mice lacking Stim1 die perinatally, whereas mice lacking Stim2 may live for 4 weeks or so [57]. Knockout of Orai1 in mice produces a severe phenotype [58, 59], similar to that seen with knockout of Stim1, and mouse knockout models of Orai2 and Orai3 have not yet been reported [10]. In ASMCs, Orai1 and Stim1 are shown to play important roles in cell proliferation and migration [44, 48].

4.2 Chronic Obstructive Pulmonary Disease

Polymorphisms in the TRPV4 gene are associated with chronic obstructive pulmonary disease characterized by airway epithelial damage, bronchoconstriction, parenchymal destruction, and mucus hypersecretion [43].

4.3 Other Diseases

The number of TRPV1 immunoreactive axons in the trachea increases under allergic inflammatory conditions. TRPV1 expression is upregulated in ASM, but not in ASMCs, from patients with chronic cough. An increase in TRPV1 activity may play a role in the airway hypersensitivity seen in chronic cough [42].

5 Conclusion

As ASMCs express a multitude of TRP channels, Orai channels, and Stim molecules, the complexity in assessing which are the important players in generating tone, and which, if any, are redundant innocent bystanders in the process, is a real and present challenge. Heteromultimer formation, as demonstrated for members of the TRPC family, may be a double-edged sword, complicating ascribing physiological functions while at the same time offering an enhanced opportunity for tissue-specific combinations that could be used therapeutically. The relative ionic permeability of the TRP channels themselves adds to the difficulty of further understanding their physiological roles. For channels with high

Ca^{2+}-selectivity ratios, this is easier to reconcile, the most obvious conclusion being that they play a direct role in extracellular Ca^{2+} entry. However, TRP channels are much less discriminating in their permeability profile (i.e., nonselective cation channels) such as certain members of the TRPM family. Activation of these channels will, in addition to allowing Ca^{2+} influx, elicit a substantial depolarization, which in ASMCs may serve to activate VDCCs and simultaneously limit Ca^{2+} entry by attenuating the electrical driving force for Ca^{2+} entry [3].

Much remains to be done to delineate the functional roles of TRPs, Orais, and Stims in ASMCs. However, identification of their involvement in regulating contractility and proliferation in ASMCs offers the exciting prospect of new and novel therapies for the treatment of asthma, chronic obstructive pulmonary disease, and other respiratory diseases [3].

References

1. Dietrich A, Chubanov V, Kalwa H, Rost BR, Gudermann T (2006) Cation channels of the transient receptor potential superfamily: their role in physiological and pathophysiological processes of smooth muscle cells. Pharmacol Ther 112: 744–760
2. Elias JA, Zhu Z, Chupp G, Homer RJ (1999) Airway remodeling in asthma. J Clin Invest 104: 1001–1006
3. Gosling M, Poll C, Li S (2005) TRP channels in airway smooth muscle as therapeutic targets. Naunyn Schmiedeberg's Arch Pharmacol 371: 277–284
4. Rodger I (1989) Drugs affecting calcium ions in airway smooth muscle. In: Barnes PJ (ed) New Drugs for asthma. IBC, London, pp 455–456
5. Panettieri RA Jr (1998) Cellular and molecular mechanisms regulating airway smooth muscle proliferation and cell adhesion molecue expression. Am J Respir Crit Care Med 158: S133–S140
6. Sweeney M, McDaniel SS, Platoshyn O, Zhang S, Yu Y, Lapp BR, et al (2002) Role of capacitative Ca^{2+} entry in bronchial contraction and remodeling. J Appl Physiol 92:1594–1602
7. Nilius B, Owsianik G, Voets T, Peters JA (2007) Transient receptor potential cation channels in disease. Physiol Rev 87:165–217
8. Abramowitz J, Birnbaumer L (2009) Physiology and pathophysiology of canonical transient receptor potential channels. FASEB J 23:297–328
9. Williams RT, Manji SS, Parker NJ, Hancock MS, Van SL, Eid JP, et al (2001) Identification and characterization of the STIM (stromal interaction molecule) gene family: coding for a novel class of transmembrane proteins. Biochem J 357:673–685
10. Smyth JT, Putney JW (2012) Regulation of store-operated calcium entry during cell division. Biochem Soc Trans 40: 119–123
11. Roos J, DiGregorio PJ, Yeromin AV, Ohlsen K, Lioudyno M, Zhang S, et al (2005) STIM1, an essential and conserved component of store-operated Ca^{2+} channel function. J Cell Biol 169: 435–445
12. Liou J, Kim ML, Heo WD, Jones JT, Myers JW, Ferrell Jr JE, Meyer T (2005) STIM is a Ca^{2+} sensor essential for Ca^{2+} -store-depletion-triggered Ca^{2+} influx. Curr. Biol. 15, 1235–1241
13. Feske S, Gwack Y, Prakriya M, Srikanth S, Puppel SH, Tanasa B, et al. (2006) A mutation in Orai1 causes immune deficiency by abrogating CRAC channel function. Nature 441:179–185

14. Vig M, Peinelt C, Beck A, Koomoa DL, Rabah D, Koblan-Huberson M, et al (2006) CRACM1 is a plasma membrane protein essential for store-operated Ca^{2+} entry. Science 312: 1220–1223
15. Zhang SL, Yeromin AV, Zhang XH, Yu Y, Safrina O, Penna A, et al (2006) Genome-wide RNAi screen of Ca^{2+} influx identifies genes that regulate Ca^{2+} release-activated Ca^{2+} channel activity. Proc Natl Acad Sci U S A 103: 9357–9362
16. Abdullaev IF, Bisaillon JM, Potier M, Gonzalez JC, Motiani RK, Trebak M (2008) Stim1 and Orai1 mediate CRAC currents and store-operated calcium entry important for endothelial cell proliferation. Circ Res 103: 1289–1299
17. Zhu X, Chu PB, Peyton M, Birnbaumer L (1995) Molecular cloning of a widely expressed human homologue for the Drosophila trp gene. FEBS Lett 373: 193–198
18. Vannier B, Peyton M, Boulay G, Brown D, Qin N, Jiang MZ, et al (1999) Mouse trp2, the homologue of the human trpc2 pseudogene, encodes mTrp2, a store depletion-activated capacitative Ca^{2+} entry channel. Proc Natl Acad Sci U S A 96: 2060–2064
19. Hofmann T, Schaefer M, Schultz G, Gudermann T (2002) Subunit composition of mammalian transient receptor potential channels in living cells. Proc Natl Acad Sci U S A 99: 7461–7466
20. Hofmann T, Obukhov AG, Schaefer M, Harteneck C, Gudermann T, Schultz G (1999) Direct activation of human TRPC6 and TRPC3 channels by diacylglycerol. Nature 397: 259–263
21. Okada T, Inoue R, Yamazaki K, Maeda A, Kurosaki T, Yamakuni T, et al (1999) Molecular and functional characterization of a novel mouse transient receptor potential protein homologue TRP7. Ca(2+)-permeable cation channel that is constitutively activated and enhanced by stimulation of G protein-coupled receptor. J Biol Chem 274: 27359–27370
22. Dietrich A, Mederos y Schnitzler M, Kalwa H, Storch U, Gudermann T (2005) Functional characterization and physiological relevance of the TRPC3/6/7 subfamily of cation channels. Naunyn Schmiedeberg's Arch Pharmacol 371:257–265
23. Plant TD, Schaefer M (2005) Receptor-operated cation channels formed by TRPC4 and TRPC5. Naunyn Schmiedeberg's Arch Pharmacol 371: 266–276
24. Wang, YX, Zheng YM (2011) Molecular expression and functional role of canonical transient receptor potential channels in airway smooth muscle cells. Adv Exp Med Biol 704: 731–747
25. Wes PD, Chevesich J, Jeromin A, Rosenberg C, Stetten G, Montell C (1995) TRPC1, a human homolog of a Drosophila storeoperated channel. Proc Natl Acad Sci USA 92: 9652–9656
26. Birnbaumer L, Zhu X, Jiang M, Boulay G, Peyton M, Vannier B, et al (1996) On the molecular basis and regulation of cellular capacitative calcium entry: roles for Trp proteins. Proc Natl Acad Sci USA 93: 15195–15202
27. Ay B, Prakash YS, Pabelick CM, Sieck GC (2004) Store-operated Ca^{2+} entry in porcine airway smooth muscle. Am J Physiol Lung Cell Mol Physiol 286: L909–L917
28. Ong HL, Brereton HM, Harland ML, Barritt GJ (2003) Evidence for the expression of transient receptor potential proteins in guinea pig airway smooth muscle cells. Respirology 8: 23–32
29. Corteling RL, Li S, Giddings J, Westwick J, Poll C, Hall IP (2004) Expression of transient receptor potential C6 and related transient receptor potential family members in human airway smooth muscle and lung tissue. Am J Respir Cell Mol Biol 30:145–154
30. Xiao JH, Zheng YM, Liao B, Wang YX (2010) Functional role of canonical transient receptor potential 1 and canonical transient receptor potential 3 in normal and asthmatic airway smooth muscle cells. Am J Respir Cell Mol Biol 43: 17–25
31. White TA, Xue A, Chini EN, Thompson M, Sieck GC, Wylam ME (2006) Role of transient receptor potential C3 in. TNF-alpha-enhanced calcium influx in human airway myocytes. Am J Respir Cell Mol Biol 35:243–251
32. Ong HL, Chen J, Chataway T, Brereton H, Zhang L, Downs T, et al (2002) Specific detection of the endogenous transient receptor potential (TRP)-1 protein in liver and airway smooth muscle cells using immunoprecipitation and Western-blot analysis. Biochem J 364:641–648
33. Godin N, Rousseau E (2007) TRPC6 silencing in primary airway smooth muscle cells inhibits protein expression without affecting OAG-induced calcium entry. Mol Cell Biochem 296:193–201

34. Goel M, Sinkins WG, Schilling WP (2002) Selective association of TRPC channel subunits in rat brain synaptosomes. J Biol Chem 277:48303–48310
35. Vazquez G, Wedel BJ, Aziz O, Trebak M, Putney JW Jr (2004) The mammalian TRPC cation channels. Biochim Biophys Acta 1742: 21–36
36. Beech DJ, Muraki K, Flemming R (2004) Non-selective cationic channels of smooth muscle and the mammalian homologues of Drosophila TRP. J Physiol 559: 685–706
37. Gudermann T, Mederos y Schnitzler M, Dietrich A (2004). Receptor operated cation entry-more than esoteric terminology? Sci STKE, pe35
38. Vennekens R, Voets T, Bindels RJ, Droogmans G, Nilius B (2002) Current understanding of mammalian TRP homologues. Cell Calcium 31:253–264
39. O'Neil RG, Brown RC (2003) The vanilloid receptor family of calcium-permeable channels: molecular integrators of microenvironmental stimuli. News Physiol Sci 18:226–231
40. Gunthorpe MJ, Benham CD, Randall A, Davis JB (2002) The diversity in the vanilloid (TRPV) receptor family of ion channels. Trends Pharmacol Sci 23:183–191
41. Caterina MJ, Schumacher MA, Tominaga M, Rosen TA, Levine JD, Julius D (1997) The capsaicin receptor: a heat-activated ion channel in the pain pathway. Nature 389: 816–824
42. Watanabe N, Horie S, Michael GJ, Keir S, Spina D, Page CP, Priestley JV (2006) Immunohistochemical co-localization of transient receptor potential vanilloid (TRPV)1 and sensory neuropeptides in the guinea-pig respiratory system. Neuroscience 141: 1533–1543
43. Jia Y, Wang X, Varty L, Rizzo CA, Yang R, Correll CC, et al (2004) Functional TRPV4 channels are expressed in human airway smooth muscle cells. Am J Physiol Lung Cell Mol Physiol 287: L272–L278
44. Zou, JJ, Gao YD, Geng S, Yang J (2011) Role of STIM1/Orai1-mediated store- operated Ca^{2+} entry in airway smooth muscle cell proliferation. J Appl Physiol 110: 1256–1263
45. Peel SE, Liu B, Hall IP (2008) ORAI and store-operated calcium influx in human airway smooth muscle cells. Am J Respir Cell Mol Biol 38: 744–749
46. Manji SS, Parker NJ, Williams RT, van Stekelenburg L, Pearson RB, Dziadek M, Smith PJ (2000) STIM1: a novel phosphoprotein located at the cell surface. Biochim Biophys Acta 1481: 147–155
47. Peel SE, Liu B, Hall IP (2006) A key role for STIM1 in store operated calcium channel activation in airway smooth muscle. Respir Res 7: 119
48. Suganuma N, Ito S, Aso H, Kondo M, Sato M, Sokabe M, Hasegawa Y (2012) STIM1 regulates platelet-derived growth factor-induced migration and Ca^{2+} influx in human airway smooth muscle cells. PLoS One 7:e45056
49. Gosens R, Meurs H, Bromhaar MM, McKay S, Nelemans SA, Zaagsma J (2002) Functional characterization of serum- and growth factor-induced phenotypic changes in intact bovine tracheal smooth muscle. Br J Pharmacol 137: 459–466
50. Cloutier M, Campbell S, Basora N, Proteau S, Payet MD, Rousseau E (2003) 20-HETE inotropic effects involve the activation of a nonselective cationic current in airway smooth muscle. Am J Physiol Lung Cell Mol Physiol 285: L560–L568
51. Dietrich A, Gollasch M, Chubanov V, Mederos y Schnitzler M, Dubrovska G, Herz U, et al (2003) Studies on TRPC6 deficient mice reveal its nonredundant role in the regulation of smooth muscle tone. Naunyn Schmiedeberg's Arch Pharmacol 367: 238
52. Liedtke W, Simon SA (2004) A possible role for TRPV4 receptors in asthma. Am J Physiol Lung Cell Mol Physiol 287: L269–L271
53. Liedtke W, Friedman JM (2003) Abnormal osmotic regulation in trpv4−/− mice. Proc Natl Acad Sci U S A 100: 13698–13703
54. Mizuno A, Matsumoto N, Imai M, Suzuki M (2003) Impaired osmotic sensation in mice lacking TRPV4. Am J Physiol Cell Physiol 285: C96–C101
55. Suzuki M, Mizuno A, Kodaira K, Imai M (2003) Impaired pressure sensation in mice lacking TRPV4. J Biol Chem 278: 22664–22668

56. Zhang SL, Yu Y, Roos J, Kozak JA, Deerinck TJ, Ellisman MH, et al (2005) STIM1 is a Ca^{2+} sensor that activates CRAC channels and migrates from the Ca^{2+} store to the plasma membrane. Nature 437: 902–905
57. Oh-Hora M, Yamashita M, Hogan PG, Sharma S, Lamperti E, Chung W, et al (2008) Dual functions for the endoplasmic reticulum calcium sensors STIM1 and STIM2 in T cell activation and tolerance. Nat Immunol 9: 432–443
58. Vig M, DeHaven WI, Bird GS, Billingsley JM, Wang H, Rao PE, et al (2008) Defective mast cell effector functions in mice lacking the CRACM1 pore subunit of store-operated calcium release-activated calcium channels. Nat Immunol 9: 89–96
59. Gwack Y, Srikanth S, Oh-hora M, Hogan PG, Lamperti ED, Yamashita M, et al (2008) Hair loss and defective T- and B-cell function in mice lacking ORAI1. Mol Cell Biol 28: 5209–5222

Large-Conductance Calcium-Activated Potassium Channels

Hiroaki Kume

Contents

H. Kume, M.D., Ph.D. (✉)
Department of Respiratory Medicine and Allergology, Kinki University Faculty of Medicine, 377-2 Ohnohigashi, Osakasayama 589-8511, Japan
e-mail: hkume@med.kindai.ac.jp

Y.-X. Wang (ed.), *Calcium Signaling In Airway Smooth Muscle Cells*,
DOI 10.1007/978-3-319-01312-1_4, © Springer International Publishing Switzerland 2014

Abstract Large-conductanceCa^{2+}-activated K^{+} (BK$_{Ca}$, MaxiK) channels are abundantly distributed on the cell membrane and ubiquitously expressed in a variety of tissues except cardiac muscle. These channels have a conductance of approximately 250 pS and are intracellular Ca^{2+}- and membrane potential-sensitive. Activation of these channels causes membrane hyperpolarization mediated by an increase in outward K^{+} currents, leading to a reduction in tension in smooth muscle such as vessels and airways. The BK$_{Ca}$ channels are a tetramer of a pore-forming α-subunit encoded by a single gene (Slo, KCNMA1) associated with modulatory β-subunits (KCNMB1-4). The α-subunit consists of a transmembrane (TM) domain-sensing membrane potential and conducting ion, and a cytoplasmic domain-sensing Ca^{2+}. The β-subunits comprise two TMs, and their expression is tissue specific. In smooth muscle cells, the BK$_{Ca}$ channels are markedly activated due to increased local Ca^{2+} concentrations via Ca^{2+} release from the sarcoplasmic reticulum (Ca^{2+} sparks). The coupling of Ca^{2+} sparks to hundreds of BK$_{Ca}$ channels causes spontaneous outward currents (STOCs), leading to relaxation via membrane hyperpolarization. The activity and expression of the BK$_{Ca}$ channels are modulated by a variety of factors such as phosphorylation (e.g., protein kinase A, protein kinase G) and other metabolites (e.g., reactive oxygen species, estrogen, nitric oxide). Therefore, altered BK$_{Ca}$ channels play a key role in vital body functions (e.g., development, pregnancy) and diseases (e.g., hypertension, diabetes). These channels may be a therapeutic target for a variety of diseases.

Keywords BK$_{Ca}$ channels • MaxiK channels • STOCs • β-adrenergic receptors • Muscarinic receptors • Patch-clamp techniques • Protein Kinase A Protein Kinase G • Ca^{2+} sparks

Abbreviations

20-HETE	20-Hydroxyeicosatetraenoic acid
4-AP	4-aminopyridine
ACh	Acetylcholine
AF-DX 116	11-[[2-[(Diethylamino)methyl]-1-piperidinyl]acetyl]-5,11-dihydro-6H-pyrido[2,3-b][1, 4]benzodiazepin-6-one
ATP	Adenosine triphosphate
BK$_{Ca}$ channels	Large-conductance calcium-activated potassium channels
cAMP	Cyclic adenosine monophosphate
CCh	Carbachol
cGMP	Cyclic guanosine monophosphate
ChTX	Charybdotoxin
EETs	Epoxyeicosatrienoic acids
GDP-β-S	Guanosine 5′-O-(2-thio-diphoshate)
G$_i$	Inhibitory G protein of adenylyl cyclase

G_s	Stimulatory G protein of adenylyl cyclase
GTP	Guanosine triphosphate
GTP-γ-S	Guanosine 5′-O-(3-thiotriphosphate)
IbTX	Iberiotoxin
IP_3	Inositol triphosphate
MCh	Methacholine
NO	Nitric oxide
nPo	Open probability of ion channels
PKA	Protein kinase A
PKC	Protein kinase C
PKG	Protein kinase G
ROS	Reactive oxygen species
RyR	Ryanodine receptors
SR	Sarcoplasmic reticulum
STOCs	Spontaneous outward currents
TEA	Tetraethylammonium
α_s	α-subunit of G_s

1 Introduction

Potassium (K^+) current across the plasma membrane is considered to have two main physiological functions, one for determining the membrane potential and the other regulating the electrical activity. Membrane hyperpolarization reduces the excitability of the membrane by shifting the membrane potential away from the threshold for excitation and inhibits, where it exists, calcium (Ca^{2+}) influx through voltage-gated Ca^{2+} channels by closing the Ca^{2+} channel. Outward K^+ currents activated upon depolarization counteract the depolarizing action of inward currents carried either by Na^+ or Ca^{2+}. Therefore, modification of K^+ channel activity is expected to have a significant effect on the excitability of various cells including smooth muscle.

Ca^{2+}-activated K^+ (K_{Ca}) channels have an important role in cell excitability by sensing and reacting to the alteration of intracellular Ca^{2+} concentrations [1–3]. These channels are divided into three types: (1) small conductance (SK_{Ca}, 4–14 pS), (2) intermittent conductance (IK_{Ca}, 32–40 pS), and (3) large conductance (BK_{Ca}, MaxiK, 200–300 pS). Although SK_{Ca} and IK_{Ca} are membrane potential independent and intracellular Ca^{2+} dependent, BK_{Ca} channels are activated or gated both by membrane depolarization and intracellular Ca^{2+}. Activation of the BK_{Ca} channels leads to pore opening and outward K^+ flux, which shifts the membrane to a more negative potential (hyperpolarization) because of a large conductance, providing a link between the metabolic and electrical state of cells [4]. BK_{Ca} channels are densely distributed in the surface of the cell membrane in smooth muscle among various tissues and the brain and, furthermore, are observed in a wide variety of other tissue types (ovary, testis, pancreas, adrenal glands), except cardiac myocytes.

The physiological role of BK_{Ca} channels is considered to play a key role in regulating tone in smooth muscle [5] and transmitter release and spike shaping in neurons [6]. Therefore, BK_{Ca} channels may be involved in the pathophysiology of various diseases such as hypertension [7–9] epilepsy [10], cancer [11], and asthma [12]. The existence of BK_{Ca} channels was reported for the first time in the *Drosophilia slowpoke* mutant [13]. The studies concerning K_{Ca} channels have advanced by leaps and bounds since single-channel recordings were established [14] and the crystal structure of mammalian voltage-gated K^+ channels were clarified [15]. This chapter will focus on the characteristics, function, and regulation of BK_{Ca} expressed on the plasmalemma of smooth muscle.

2 Structure and Function of BK_{Ca} Channels

2.1 α-Subunits

BK_{Ca} channels are composed of tetramers formed by pore-forming α-subunits along with accessory β-subunits (Fig. 1b), activated by an increase in the membrane potential and by an increase in concentrations of intracellular Ca^{2+}. The ubiquitous expression of the BK_{Ca} channels causes multiple abnormalities observed in mice with targeted deletion of the BK_{Ca} channel gene [16–19]. The α-subunit is ubiquitously expressed by mammalian tissues and encoded by a single gene (Slo, KCNMA1) [20–22]. The α-subunit transmembrane (TM) domain comprises a seven-membrane-spanning segment (S0–S6) with extracellular loops and shares homology with all voltage-gated K^+ channels with six TM domains (S1–S6) and a pore helix (Fig. 1a). The S1–S4 domains are arranged in a bundle that carries out the

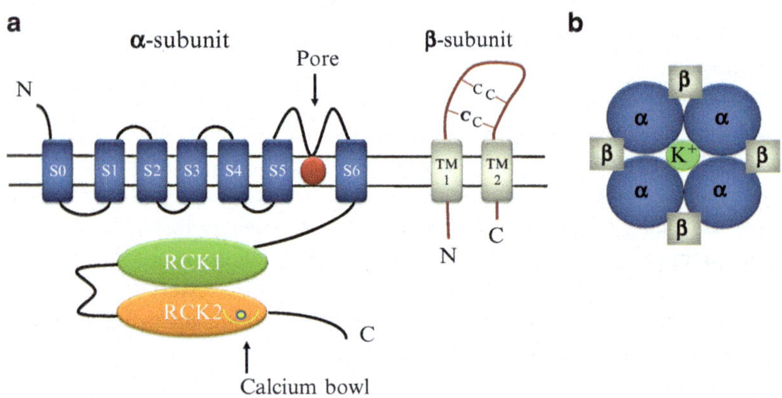

Fig. 1 Structure of BK_{Ca} channel. (**a**) Schema representing topological characteristics of α and β subunits of BK_{Ca} channel as described in text. (**b**) Model of tetrameric assembly of α subunit showing K^+ in pore with β subunits. Regulate the conductance of K^+ (RCK), N-terminal (N), C-terminal (C)

voltage-sensing function, and the S5–S6 and pore helices contribute to the pore formation and the K^+ selective filter [23]. The BK_{Ca} channels are unique among voltage-gated K^+ channels because they contain an additional S0 to the S1 in the N-terminal [24] (Fig. 1a). The S0 and N-terminal segment to the S0 are required for the β1-subunit to modulate the channel.

The C-terminal tail confers the Ca^{2+}-sensing ability of the BK_{Ca} channels, involving a pair of Ca^{2+}-sensing domains that regulate the conductance of K^+ (RCK), i.e., RCK1 and RCK2 [25] (Fig. 1a). The RCK2 domain contains a Ca^{2+} bowl domain that is contributed to the sensitivity to intracellular Ca^{2+} [25a, 25b, 25c, 25d, 25e]. However, the RCK1 domain also has a key role in the high-affinity binding site for Ca^{2+} [25f]. The crystal structure of the C-terminus of these channels was resolved recently. The octamer of RCK domains forms an intracellular Ca^{2+}-gating ring on the cytosolic side [26], coupled to channel opening by tugging on the pore through a spring like S6-RCK1 linker in response to intracellular Ca^{2+} [27]. On the other hand, the C-terminus may have the ability to regulate BK_{Ca} channel activity by modulating the voltage sensor [26, 27a, 27b].

Although the Ca^{2+} sensor of BK_{Ca} channels has a high specificity for Ca^{2+}, other factors, including divalent cations, also influence the opening of these channel. Magnesium (Mg^{2+}) enhances activation of these channels via a distinct binding site involving the voltage sensor and RCK1 domain [28, 29, 29a]. In contrast, an intracellular proton (H^+) attenuates the opening of BK_{Ca} channels [25e, 31] (see Sect. 5). Moreover, the redox state influences the gating of these channels [32] (see Sect. 5).

2.2 *β-subunits*

BK_{Ca} channels associate with modulatory β-subunits, which are expressed in a cell-specific manner and have unique regulatory action on these channels. The β-subunits bring about the diversity of BK_{Ca} channels. There are four distinct β-subunits (Fig. 1b): β1–4 encoded by KCNMB1, KCNMB2, KCNMB3, and KCNMB4. These β-subunits in BK_{Ca} channels consist of two TM domains with intracellular N- and C-termini and a long extracellular loop (Fig. 1a). The β1-subunit was the first β-subunit to be cloned and is primarily expressed in smooth muscle [27b, 33, 35, 36]. The first TM domain of a β-subunit is located near the S1 domains of an α-subunit, and the second TM domain is located near the S0 domain [36a, 36b, 36c, 36d]. The β-subunit exits by a neighboring voltage-sensing bundle of the α-subunit. The extracellular loop of the β-subunit is contributed to variable toxin resistance, and the intracellular N- and C-termini contribute to variable membrane potentials [36e, 36f, 36g, 36h]. The β1-subunit influences the gating kinetics of BK_{Ca} channels to favor a more open state mediated by activating the voltage sensor directly [37, 37a], resulting in less Ca^{2+} being required to open the BK_{Ca} channels (increasing sensitivity to intracellular Ca^{2+}).

In excitation-contraction coupling of muscle cells, local increases in Ca^{2+} concentrations occur due to focal releases of Ca^{2+} through ryanodine receptors

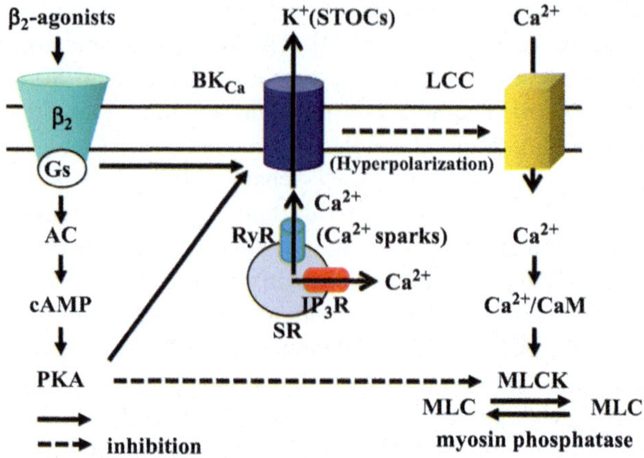

Fig. 2 Intracellular mechanisms of BK_{Ca} channel activation. BK_{Ca} channel activity is enhanced by a localized increase in intracellular Ca^{2+} (*Ca^{2+} sparks*) in the vicinity of these channels via Ca^{2+} release through ryanodine receptor (*RyR*) or inositol triphosphate receptor (*IP_3R*) on sarcoplasmic reticulum (SR). Spontaneous outward currents (*STOCs*) occur due to hundreds of opened BK_{Ca} channels via coupling of Ca^{2+} sparks to these channels and result in an inhibition of L-type Ca^{2+} channel (*LCC*) mediated via membrane hyperpolarization. BK_{Ca} channels are also activated by the dual pathway, i.e., (1) direct action of stimulatory G protein of adenylyl cyclase (G_s), (2) protein kinase A (PKA)-dependent phosphorylation, in β_2_adreneceptor agonist (β_2_agonist)-induced stimulation. *AC* adenylyl cyclase, *CaM* calmodulin, *MLCK* myosin light chain kinase, *MLC* myosin light chain

(RyR) from the sarcoplasmic reticulum (SR), termed Ca^{2+} sparks [2, 3]. Hundreds of BK_{Ca} channels are opened by the Ca^{2+} sparks from SR closed to the sarcolemma, causing spontaneous outward currents (STOCs) (Fig. 2; see Sect. 4). The coupling of ryanosine-mediated Ca^{2+} sparks to the BK_{Ca} channel-mediated STOCs is enhanced by a β1-subunit, resulting in hyperpolarization of smooth muscle cells and subsequently reducing Ca^{2+} influx and relaxation. Deletion of the β1-subunit in mice causes an increase in contraction of vascular smooth muscle and an elevation of systemic blood pressure [7, 8]. In mice in which the BK_{Ca} channel β_1-subunit has been knocked out CCh-induced contraction of trachea was enhanced compared with wild-type mice and not just the single-channel activity of the BK_{Ca} channel in an inside-out patch; in addition, STOCs in a whole-cell configuration were markedly attenuated in tracheal smooth muscle cells compared with wild-type mice [38].

3 Effects of Outward K^+ Currents on Smooth Muscle Tone

Airway smooth muscles isolated from guinea pig and human spontaneously generate a sustained contraction (muscle tone), sometimes accompanied by small regular or irregular rhythmic fluctuations of membrane potential (guinea pig [39–42], human [43–48]). The muscle tone in guinea-pig trachea can be abolished by blocking endogenous production of prostaglandins with indomethacin or aspirin

[46, 49–53]. On the other hand, in human trachea, the tone is inhibited by a lipoxygenase inhibitor (AA 861 [48]) or a leukotriene antagonist (FPL 55712 [46]). Since spontaneously generated muscle tone is easily eliminated by removal of Ca^{2+} from the external medium but is not significantly affected by L-type Ca^{2+} channel blocking agents, such as verapamil and nifedipine [54–57], it is possible that Ca^{2+} influx through the receptor-operated pathway, activated by endogenous products of arachidonic acids, is responsible for maintaining the tone. It is still not clear whether receptor-operated Ca^{2+} pathways are affected by membrane potential, but there is some evidence that the production of second messengers, such as IP_3, is decreased by membrane hyperpolarization [58]. It is therefore possible that membrane potential not only directly affects ionic channel activity but also modifies production of second messengers by a potential-sensitive enzyme system in the plasma membrane, secondarily resulting in contraction or relaxation.

In tracheal muscle, sustained contractions elicited by 5-hydroxytryptamine (0.5–1 μM) were inhibited in a graded fashion depending on the degree of membrane hyperpolarization produced by anodal currents applied using the sucrose-gap method [59]. A decrease in Ca^{2+} influx through voltage-gated Ca^{2+} channels may be responsible for this relaxation, but it is also possible, at least partially, that a receptor-operated Ca^{2+} influx pathway activated by 5-hydroxytryptamine is inhibited by membrane hyperpolarization. Further careful experiments are necessary to examine the relationship between membrane potential and Ca^{2+} influx caused by receptor activation in order to clarify a possible contribution of a second messenger system.

Typical action potentials have not been found in airway muscle under physiological conditions. This has been considered to be due to a marked increase in K^+ conductance of the plasma membrane upon depolarization [60]. Thus, when the K^+ conductance of the membrane is reduced by blocking K^+ channels, one would expect an increase in excitability. It has actually been shown, in airway muscles that are only weakly excitable, that spontaneous phasic contractions can be initiated in company with electrical activities by applying several K^+ channel blocking agents: tetraethylammonium (TEA), 4-aminopyridine (4-AP), charybdotoxin (ChTX), or iberiotoxin (IbTX) [61–63]. ChTX and IbTX, toxins in scorpion venom, are selective inhibitors of BK_{Ca} channels. Some of these have been demonstrated to be accompanied by electrical activity. These results suggest that outward K^+ currents passing through BK_{Ca} channels may serve an important regulatory function in these smooth muscles.

4 Electrophysiological Characteristics of BK_{Ca} Channels

4.1 Single-Channel Recording

K_{Ca} channels activated by intracellular Ca^{2+} and membrane depolarization are ubiquitously distributed in smooth muscle, including airway muscle. The dominant BK_{Ca} channels have a large conductance (approximately 250 pS in symmetrical

135–150 mM K^+ medium). Single-channel currents of BK_{Ca} channels have been recorded using the patch-clamp method from the trachealis of guinea pig [64], dog [65, 66], ox [67, 68], pig [69–71], rabbit [25e, 72], and equine [73]. BK_{Ca} channels are highly selective for K^+ despite a large conductance [74]. The sensitivity to Ca^{2+} apparently differs to some degree in different types of smooth muscle, but this could be due, at least partially, to differences in experimental conditions. Ca^{2+} sensitivity may be increased by intracellular Mg^{2+}, as is the case in vascular muscle [75], by intracellular alkalinization, as reported for rabbit tracheal muscle [25e], and it is likely to be modified by second messengers. Single-channel currents of the BK_{Ca} channels in guinea-pig tracheal muscle, studied in outside-out patches (140 mM K^+, 0.1 μM Ca^{2+} inside and 6 mM K^+, 1.2 mM Ca^{2+} outside), were reversibly blocked by external application of 100 nM ChTX, not as a result of reduced current amplitude, but by reducing the open probability (nP_o), a fraction of the time during which the channel is open [64].

The BK_{Ca} channels in myocytes isolated from canine trachealis have been studied (with inside-out patches in symmetrical 135 mM K^+ medium, I μM Ca^{2+} on the cytosolic side). External application of TEA (10 mM) or application of Cs^+ to the cytosolic side of the membrane reduced the single-channel conductance [66]. The conductance of the BK_{Ca} channel studied in the outside-out configuration (140 mM K^+ inside and 5.9 mM K^+ outside, 0 to + 30 mV) was reduced from 154 to 19.9 pS by TEA (2 mM), but it was not affected by 4-AP (1 mM) applied externally [76]. When examined on the BK_{Ca} channels prepared from canine and bovine tracheal muscles and incorporated into planar lipid bilayers, ChTX (10–50 nM) applied at the external surface decreased nP_o without affecting single-channel conductance, and this effect was irreversible [77]. Thus, the blocking kinetics are similar to, but the reversibility is different from, the results obtained from single cells of guinea-pig tracheal muscle [64].

In bovine trachealis, TEA (1 mM) applied externally strongly reduced the amplitude of single BK_{Ca} channel current, whereas ChTX (100 nM) markedly decreased the P_o of the channel without affecting current amplitude [68]. The effect of ChTX was reversible. In contrast, the BK_{Ca} channels were not affected by 4-AP (1 mM) applied internally and (2 mM) externally. In swine tracheal muscle, the BK_{Ca} channels predominate, but channels having a conductance of 90 pS at + 80 mV in symmetrical 140 mM K^+ medium (IK_{Ca}) were also found [69]. Near the resting potential, the BK_{Ca} channels are considered to be mostly in a closed state, whereas the IK_{Ca} channels remain active, judging from the sensitivity to Ca^{2+} and the independence of membrane potential. When studied in inside-out patches in symmetrical 140 mM K^+ medium, the BK_{Ca} channels were inhibited by extracellular TEA with an IC_{50} of approximately 0.6 mM [71]. On the other hand, intracellular TEA (20 mM) partially inhibited outward but not inward currents.

In freshly isolated human bronchial smooth muscle cells, single currents of the BK_{Ca} channels were also recorded in a cell-attached patch, inside-out patch, and outside-out patch [77a, 77b]. These channels have a conductance approximately 210 pS in symmetrical 140 mM K^+ medium. The activity of these channels was enhanced by the membrane potential and cytosolic concentrations of Ca^{2+} and attenuated by external applications of TEA, ChTX, and IbTX.

The effects of intracellular pH (pH_i) on BK_{Ca} channels have been studied in rabbit tracheal muscle using inside-out patches [25e]. BK_{Ca} channel activity was markedly inhibited by intracellular acidification, by reducing the sensitivity to Ca^{2+}, and by shortening the open state of the channel. On the other hand, intracellular alkalinization had an opposite effect (increasing Ca^{2+} sensitivity and lengthening the open state of the channel). A decrease in smooth muscle tone with intracellular alkalinization might be caused at least partially by activation of the BK_{Ca} channels [78].

The effects of some agents known to inhibit calmodulin have been examined on BK_{Ca} channels in canine trachealis [79]. Haloperidol, trifluoperazine, and chlorpromazine blocked the open channel with dissociation constants of 1.0, 1.4, and 2.0 μM, respectively. Since this order of the drugs differs from their potency as calmodulin antagonists (trifluoperazine > chlorpromazine > haloperidol [80]), calmodulin is not considered to be involved in channel activation with Ca^{2+}. A similar conclusion was reached regarding the BK_{Ca} channel in rat myometrium using W-7, trifluoperazine, W-5, and calmodulin [81].

4.2 Whole-Cell Clamp Recording

In single cells from guinea-pig trachea, three different outward currents have been recoded using the whole-cell clamp method: STOCs, transient outward current (I_T), and steady outward currents (I_S) [82]. STOCs were originally demonstrated in smooth muscle cells of rabbit jejunum and ear artery and have been thought to represent the opening of BK_{Ca} channels in response to Ca^{2+} released from intracellular stores (Fig. 2) [83]. All outward currents in guinea-pig tracheal muscle were blocked by replacing intrapipette K^+ (140 mM) with Cs^+ and also by external application of TEA (10 mM). I_T was more susceptible to TEA than STOCs [82]. I_T and STOCs, but not I_S, were suppressed by reducing intrapipette Ca^{2+} concentration, suggesting that the BK_{Ca} channel is the pathway of these currents. I_T and STOCs were also inhibited by 4-AP (10 mM) and nifedipine (100 nM), a blocker of L-type Ca^{2+} channels. None of the currents was affected by apamin, a bee venom toxin.

Three different types of outward current have also been recorded in canine tracheal muscle [76]. I_T and STOCs, but not I_S, were sensitive to Ca^{2+} and easily blocked by TEA (IC_{50}: 0.3 mM). I_S was less susceptible to TEA (IC_{50}: 3 mM) but selectively inhibited by 4-AP (IC_{50}: 3 mM). This current is likely to flow through a delayed rectifier K^+ channel. The susceptibility of I_S to 4-AP appears to differ in guinea pig and dog, the latter being more sensitive to 4-AP than the former. In the presence of TEA (1 mM) and 4-AP, an action potential with overshoot could be elicited under current clamp conditions. Slowly inactivated outward currents, similar to I_T in guinea-pig and dog tracheal smooth muscle, have also been recorded in cells isolated from human bronchus [84].

In bovine trachealis, outward currents recorded in a whole-cell clamp configuration were enhanced by increasing intrapipette Ca^{2+} concentrations and by increasing depolarization beyond −20 to 0 mV [68]. ChTX (50 nM–1 μM) partially inhibited outward currents elicited at a large depolarization (beyond + 10 mV). The results suggest that BK_{Ca} channels might not be active at the resting potential and that it is doubtful whether these channels become significantly activated even during agonist-induced contraction in bovine trachea. As the authors suggest [68], however, since the experiments were carried out at 22 °C, it is possible that in the bovine BK_{Ca} channels play an inhibitory role, even under physiological conditions.

In swine tracheal muscle, large and small populations of STOCs could be observed, probably reflecting the activity of BK_{Ca} and SK_{Ca} channels, respectively, found in inside-out patches [69]. The small STOCs were present at a much wider range of membrane potential than the large STOCs, which were recruited as the membrane was depolarized. Therefore, near the resting potential STOCs consist mainly of the small type. STOCs (probably both types) could be recorded in solution in which Ca^{2+} was substituted by Mn^{2+} (2.5 mM) for 1 h, suggesting that Ca^{2+} influx is not necessary for STOCs. This differs from the result in guinea-pig tracheal muscle, where the STOCs were blocked by nifedipine [82]. STOCs in swine trachealis (which appear to be mainly of the large type) were concentration dependently inhibited with TEA with an IC_{50} of around 1 mM, a value similar to that obtained (0.6 mM) with inside-out patches [71].

5 Chemical Modulation of BK_{Ca} Channels

5.1 Effects Mediated Through Ca^{2+} Release

Intracellular second messengers may affect the BK_{Ca} channels, either indirectly by increasing intracellular Ca^{2+} concentrations or by directly affecting channel activity. In single cells isolated from porcine tracheal muscle, the BK_{Ca} channels were studied in a cell-attached configuration after permeabilization of the plasma membrane with β-escin, a major constituent of saponins [85]. The channels' properties were found to be very similar to those studied in inside-out patches. The BK_{Ca} channels were activated by acetylcholine (ACh, 30 μM), substance P (0.1 μM), or IP_3 (2.4–20 μM), as well as by caffeine (5 mM), suggesting that the activity was due to Ca^{2+} released from intracellular stores. These activations with the agonists and IP_3 were markedly and reversibly reduced by heparin (50–100 μg/ml), which is known to inhibit IP_3 binding to its receptors in the SR. Activation of the BK_{Ca} channels by ACh (0.01 μM) was also observed in swine tracheal muscle cells in a cell-attached patch-clamp configuration [71]. Furthermore, in cultured human bronchial smooth bradykinin (0.01–1 μM), an inflammatory mediator caused bronchoconstriction and activated the BK_{Ca} channels in a concentration-dependent manner, and the augmented currents were inhibited by heparin (10 mg/ml) [85a]. Ca^{2+} release from the SR via stimulation of IP_3 receptors causes an increase in activation of BK_{Ca} channels in smooth muscle, including airways and vessels. Two pathways participate in Ca^{2+} release from the SR, i.e., (1) the RyR and (2) inositol triphosphate (IP_3) receptor. In smooth muscle cells, the IP_3 receptor is more abundant than ryanodine receptor and reacts to IP_3 generated by the activation of G protein-coupled receptors and phospholipase C.

5.2 *Stimulatory Effects Mediated by β-Adrenoceptor/G Protein/Cyclic Adenosine Monophosphate Processes*

In rabbit tracheal smooth muscle, the involvement of cyclic adenosine monophosphate (cAMP)-dependent processes in BK_{Ca} channel regulation has been examined [72]. In the presence of cAMP and adenosine triphosphate (ATP) (0.3 mM), internal application of protein kinase A (PKA, 10 units/ml) to the cytosolic side of inside-out membrane patches reversibly increased the nP_o of the BK_{Ca} channels without changing the amplitude of single-channel currents, and the recovery from this activation was significantly delayed by okadaic acid, an inhibitor of protein phosphatases (Figs. 2, 3, and 4) [72, 86, 87]. A similar effect was observed with the catalytic subunit of PKA (10 units/ml), indicating that phosphorylation of a BK_{Ca} channel protein leads to an enhancement of the open state of the channel. External application of isoprenaline (0.2 μM) and okadaic acid (10 μM) also increased activation of the BK_{Ca} channels in a cell-attached patch-clamp configuration [72]. In Xenopus oocytes, similar results were observed in β-adrenergic action [88]. Moreover, external application of forskolin (10 μM) activated the BK_{Ca} channels in tracheal smooth muscle cells [89]. These results are in accordance with those obtained in cultured smooth muscle cells of rat aorta using isoprenaline (10 μM), forskolin (10 μM), and dibutyryl cyclic AMP (100 μM) in cell-attached patches and using PKA (0.5 μM) and cyclic AMP (1 μM) in inside-out patches [90]. Similar results were also obtained in the circular muscle cells of canine colon using the catalytic subunit of PKA and phosphatase inhibitors, calyculin A and okadaic acid [91].

Activation of BK_{Ca} channels by isoprenaline has been shown to be mediated by the α-subunit (α_s) of the stimulatory G protein of adenylyl cyclase (G_s), independently of cyclic AMP-dependent protein phosphorylation (Figs. 2 and 4) [92, 93]. In porcine, canine, and ferret tracheal muscle cells, isoprenaline increased activation of BK_{Ca} channels in outside-out patches when guanosine triphosphate (GTP) (100 μM) was present at the cytosolic side of the patches. A similar increase in activity was observed even when phosphorylation was inhibited by the nonmetabolizable ATP analog adenosine 5′-[β, γ-imido] diphosphate (ATP [β; γ NH], AMP-PNP (1 mM)). In inside-out patches with a patch pipette containing

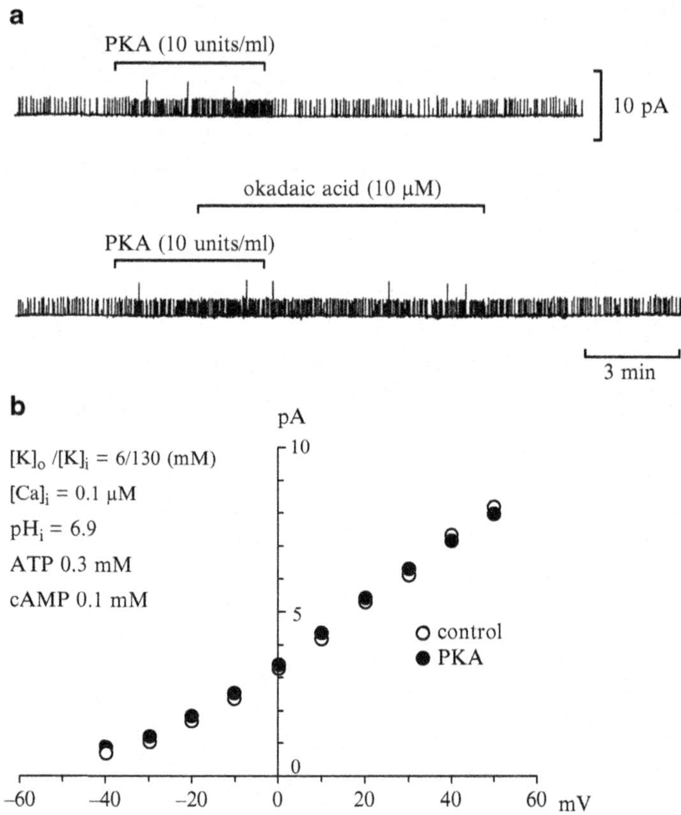

Fig. 3 (**a**) A typical record of single BK_{Ca} channels enhanced by protein kinase A (PKA) in an inside-out patch used rabbit tracheal smooth muscle cells at a holding potential of 0 mV. Internal application of PKA (10 units/mL) reversibly increased the BK_{Ca} channel activity with no change in the amplitude (*upper trace*). When PKA was washed out in the presence of okadaic acid (5 μM), the recovery from the activated state was markedly prolonged (*lower trace*). (**b**) Current–voltage relationship for BK_{Ca} channels in presence (*closed circles*) and absence (*open circles*) of PKA. The maximal slope conductance was approximately 120 pS at 10–20 mV

isoprenaline (1 μM), the nonhydrolyzable GTP analog guanosine 5′-O-(3-thiotriphosphate) (GTP-γ-S, 100 μM) similarly potentiated BK_{Ca} channel activity. An α_s preincubated with GTP-γ-S (recombinant α_s proteins, 100–1,000 pM) increased the channel activity in a concentration-dependent manner when applied to the cytosolic side of inside-out patches. In contrast, an α_s preincubated with guanosine 5′-O-(2-thio-diphoshate) (GDP-β-S) had no effect. These results show that BK_{Ca} channels are stimulated directly by α_s and that cyclic AMP-dependent phosphorylation is not required. A direct action of G_s protein on BK_{Ca} channels has also been demonstrated in channels from rat or pig myometrium, incorporated into planar lipid bilayers, using GTP-γ-S and AMP-PNP [94]. At least two mechanisms are involved in the activation of BK_{Ca} channels following β-adrenoceptor

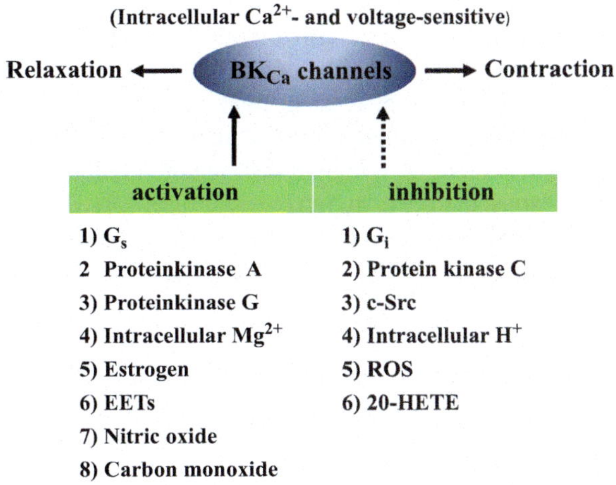

Fig. 4 A variety of factors for regulating BK_{Ca} channel functions. Stimulatory G protein of adenylyl cyclase (*G_s*), inhibitory G protein of adenylyl cyclase (*G_i*), epoxyeicosatrienoic acids (*EETs*), 20-hydroxyeicosatetraenoic acid (*20-HETE*), reactive oxygen species (*ROS*)

stimulation; one is mediated through cAMP-dependent channel phosphorylation and the other through direct regulation by G_s protein (cAMP independent) [93].

Whatever the mechanism, activation of BK_{Ca} channels is likely to underlie the membrane hyperpolarization produced by isoprenaline observed with intracellular microelectrodes in guinea-pig [41, 42], dog [95], and human tracheal muscles [47]. This idea is supported by the observations in guinea-pig trachealis that relaxation by noradrenaline (1 μM) of carbachol (CCh)-induced contraction was nearly blocked by ChTX (50 nM) [77] and the concentration-relaxation curves to isoprenaline and salbutamol were selectively shifted to the right by ChTX [64, 96, 97]. After G_s activity was irreversibly augmented by incubation with cholera toxin in guinea-pig trachea, methacholine (MCh)-induced contraction was significantly attenuated, and this effect by G_s was reversed in the presence of ChTX [98]. The G_s/BK_{Ca} stimulatory linkage could be involved in the β-adrenergic relaxation in airway smooth muscle. When isoprenaline was repeatedly applied in cell-attached patches to tracheal smooth muscle cells for 5 min every 15 min, the BK_{Ca} channel activity was gradually reduced with no change in unitary amplitude [99]. When isoprenaline was repeatedly applied to fura-2 loaded tissues of tracheal smooth muscle, the relaxant action of isoprenaline was also attenuated after repeated application of this agonist by elevating intracellular Ca^{2+}, and this reduced responsiveness to isoprenaline was markedly potentiated in the presence of IbTX, whereas it was markedly prevented in the presence of verapamil, an L-type Ca^{2+} channel blocker [99, 100]. Desensitization of β-adrenoceptors could be involved in the BK_{Ca}/L-type Ca^{2+} channel linkage.

Although relaxation produced by β-adrenoceptor activation is generally accompanied by membrane hyperpolarization, for which activation of the BK_{Ca} channel is thought to be responsible as described previously, the hyperpolarization may not be entirely responsible for the relaxation. Lower concentrations of isoprenaline are known to produce relaxation without clear hyperpolarization [41], and salmetrol, a β_2-adrenoceptor agonist, has recently been shown to have relaxant effects without hyperpolarization [101]. These results suggest that a direct inhibition of the contractile machinery may play a more important role than membrane hyperpolarization, at least in β-adrenoceptor-mediated relaxation. In the simultaneous record of tension and intracellular Ca^{2+}, isoprenaline caused relaxation of tracheal smooth muscle by reducing not only the concentration of intracellular Ca^{2+} but also sensitivity to intracellular Ca^{2+} [102]. Further studies are needed to clarify this.

5.3 Inhibitory Effects Mediated by Muscarinic Receptor/G Protein Processes

Stimulation of muscarinic cholinoceptors inhibits BK_{Ca} channel activity [70, 103]. When MCh (50 μM) was applied to outside-out patches of porcine or canine tracheal muscle cells, the nP_o of the BK_{Ca} channel was markedly decreased without changes in the amplitude of single-channel currents. The decrease in nP_o is due to a reduction in channel open times, probably reflecting a decrease in the Ca^{2+} sensitivity of the channel. The muscarinic inhibition of BK_{Ca} channels, similar to that found in airway smooth muscle, has been reported for the circular muscle of canine colon [104]. The lack of the effect of ChTX- on CCh-induced contractions in guinea-pig trachealis [77] is in agreement with the finding that muscarinic cholinoceptor stimulation inhibits BK_{Ca} channels. The inhibition of channels through muscarinic cholinoceptor activation in guinea-pig [82] and swine tracheal muscle cells [71] may be partly responsible for a prolonged suppression by ACh of STOCs following a transient increase. This has been observed in longitudinal muscle cells of the rabbit jejunum [83]. As discussed by Saunders and Farley [71], this inhibition is difficult to explain by the depletion of intracellular Ca^{2+} stores because it occurs even during elevated Ca^{2+} concentration (dog trachealis [105], porcine trachealis [106]). In porcine and canine trachealis, the inhibition of BK_{Ca} channels produced by muscarinic cholinoceptor stimulation was potentiated by cytosolic application of GTP (100 μM) and strong, irreversible potentiation was obtained with GTP-γ-S (100 μM) [70]. On the other hand, when GDP-β-S (1 mM) was applied to the cytosolic side, the muscarinic cholinoceptor inhibition was not observed. Incubation (4–6 h) of airway smooth muscle cells with pertussis toxin (0.1–1.0 μg/ml), which blocks signal transduction through ADP ribosylation of G_i, the inhibitory G protein of adenylyl cyclase, abolished the channel inhibition by MCh without reducing channel activity in the control state (Fig. 4). After incubation of tracheal smooth muscle with pertussis toxin 1.0 μg/ml for 6 h, MCh-induced contraction was significantly attenuated, and this effect by pertussis toxin was reversed in the presence of ChTX [98]. The G_i/BK_{Ca} inhibitory linkage may be involved in muscarinic-induced contraction in airway smooth muscle. Similar modification of the muscarinic cholinoceptor inhibition with GTP and pertussis toxin has been reported for canine colonic muscle [107]. From a functional point of view, it would be favorable to reduce the K^+ conductance of the plasma membrane to produce excitation by agonists such as ACh. G_i protein coupled to the M_2 subtype of muscarinic cholinoceptors, leading to an inhibition in cAMP. These M_2 receptors exist on the surface in airway smooth muscle cells. A selective M_2 receptor antagonist (AF-DX 116, a benzodiazepine derivative) potentiated relaxation induced by isoprenaline and forskolin in MCh-precontracted tracheal muscle [98, 108, 109]. AF-DX116 had no effect on isoprenaline-induced relaxation when the preparation was precontracted with histamine. The functional antagonism between isoprenaline (or forskolin) and M_2 receptor stimulation may not only be simply mediated by inhibition of adenylyl cyclase through M_2 receptors [108, 109] but also be exerted by the direct inhibition of BK_{Ca} channels by pertussis toxin-sensitive G_i protein through activation of muscarinic cholinoceptors since there is evidence that activation of BK_{Ca} channels is involved in the relaxation induced by forskolin and isoprenaline, as already described. Furthermore, M_2 receptors inhibited activity of BK_{Ca} channels via the dual pathway of a direct membrane-delimited interaction of $G\beta\gamma$ and activation of phospholipase C/protein kinase C (PKC) [110]. In knockout mice of the BK_{Ca} channel β1-subunit, contraction and membrane depolarization induced by CCh were enhanced in tracheal smooth muscle compared with wild-type mice, and these augmented effects of CCh were inhibited in the presence of AF-DX116 [98, 110a]. These results indicate that the BK_{Ca} channel β1-subunit plays a functional role in opposing M_2 muscarinic receptor signaling.

5.4 Stimulatory Effects Mediated by Cyclic Guanosine Monophosphate Processes

The nitric oxide (NO)/cGMP pathway plays an important role in the relaxation of smooth muscle, including vessels and airways. BK_{Ca} channels were markedly enhanced by cGMP-mediated processes, suggesting that an augmentation of BK_{Ca} channel activity leads to cGMP-induced relaxation of smooth muscle [111, 112]. The BK_{Ca} channel in α-subunit-null mice caused an increase in contraction in tibial arteries compared with wild-type mice [18]. This phenomenon was due to impaired response to cGMP-dependent vasorelaxation, indicating that the BK_{Ca} channel is an important effector for cGMP-mediated action. NO also augmented BK_{Ca} channel activity in various vascular smooth muscles (Fig. 4) [113, 114]. Protein kinase G (PKG) was involved in this activation of BK_{Ca} channels via the NO/cGMP pathway (Fig. 4) [115–117]. Activation of BK_{Ca} channels via dopamine receptors was caused through PKG and has been shown to mediate relaxation in coronary and renal arteries [118]. PKG may be cross-activated by cyclic AMP to stimulate BK_{Ca} channels [119]. Moreover, the dual pathway of modulation of the BK_{Ca} channel by NO has been demonstrated, i.e., (1) PKG dependent [113, 114,

120] and (2) direct activation of NO with a channel protein [121, 122]. Since the stimulatory effect of NO on BK_{Ca} channels is abolished by knockdown of the β-subunit with antisense, the β-subunit acts as a mediator of NO [123].

5.5 Regulation Mediated by Other Factors

As described earlier, BK_{Ca} channels are regulated by not only membrane potential and cations (Ca^{2+}, H^+, Mg^{2+}) but also cAMP- and cGMP-dependent processes. However, these channels may be influenced by other factors shown below and be related to their physiological roles. The effects of various factors on BK_{Ca} channel activity is summarized in Fig. 4.

5.5.1 Protein Kinase C (PKC)

BK_{Ca} channels are activated via phosphorylation of their channels by PKA and PKG as described earlier (see Sect. 5). However, the effects of PKC on these channels are still controversial. PKC enhanced activity of the BK_{Ca} channels in rat pulmonary arterial smooth muscle [124]. In contrast, PKC reversed cAMP-induced activation of these channels (Fig. 4) [125, 126]. Phosphorylation by PKC acts on the BK_{Ca} channels as follows: (1) direct inhibition and (2) a switch to influence the effects of PKA and PKG [127, 128]. In addition to these, c-Src, proto-oncogene tyrosin-protein kinase, suppressed activity of the BK_{Ca} channels in coronary and aortic myocytes (Fig. 4) [129], whereas c-Src induced phosphorylation augmented these channels in HEK 293 cells [130].

5.5.2 Estrogen

Steroid hormones alter cardiovascular function at least in part by regulating BK_{Ca} channel activity. In electrophysiological methods, estrogen and xenoestrogen caused an increase in the BK_{Ca} channel activity in coronary [131–133], uterine [134], and cerebral [135] arterial smooth muscle cells (Fig. 4), resulting in dilatation of these vessels [114, 136]. This stimulatory action of estrogen in BK_{Ca} channels was mediated by interaction with the β1-subunit [137], and NO generated by estrogen enhanced activity of these channels through cGMP/PKG signaling processes [131, 134]. As another mechanism independent of NO synthesis, G protein coupled with estrogen receptors activated BK_{Ca} channels, leading to vasorelaxation [133].

5.5.3 Redox and Reactive Oxygen Species

Reactive oxygen species (ROSs), synthesized in endothelial and smooth muscle cells, exert physiological and pathophysiological effects on smooth muscle by altering the intracellular reduction or oxygen (redox) status [138]. Excessive generation of ROSs is observed in various conditions and diseases such as aging, diabetes, hypoxia, and hypertension. Redox state was shown to influence the gating of BK_{Ca} channels [32]. However, the effects of redox are complex. Preferential oxidation of methionine increased activity of BK_{Ca} channels, whereas oxidation of cysteines reduced this channel activity (Fig. 4) [139–142]. The activity of BK_{Ca} channels was enhanced by hydrogen peroxide (H_2O_2) in pulmonary arterial smooth muscle, resulting in vasodilation mediated by membrane hyperpolarization [143, 144]. H_2O_2 may directly bind to BK_{Ca} channels to regulate them. In addition, H_2O_2-induced activation of BK_{Ca} channels was involved in phospholipase A_2-arachidonic acid pathway and metabolites of lipoxygenase [145]. On the other hand, H_2O_2 caused contraction of tracheal smooth muscle in a concentration-dependent fashion by elevating concentrations of intracellular Ca^{2+} [146]. Moreover, peroxynitrite (OONO-), an oxidant generated by the reaction of NO and superoxide, caused contraction of cerebral artery mediated by inhibiting BK_{Ca} channel activity [147, 148].

5.5.4 Arachidonic Acid

Arachidonic acid and its metabolites, such as 20-hydroxyeicosatetraenoic acid (20-HETE) and epoxyeicosatrienoic acids (EETs), play an important role in the regulation of vascular tone. Arachidonic acid and EETs caused dilation of vessels mediated by increasing the BK_{Ca} channel activity in smooth muscle cells (Fig. 4) [149–151]. The vasodilation induced by these two agents was attenuated via suppression of this channel activity [152, 153]. In airway smooth muscle, 20-HETE also caused relaxation with membrane hyperpolarization via activation of the BK_{Ca} channels [154]. On the other hand, 20-HETE is a vasoconstrictor. The BK_{Ca} channel activity was inhibited by 20-HETE, and this phenomenon was mediated by PKC (Fig. 4) [155, 156]. The vasoconstriction induced by 20-HETE was also attenuated by blocking this channel activity [155]. Acute hypoxia reduced generation of 20-HETE, and the subsequent inhibitory action of 20-HETE on the BK_{Ca} channels was removed in cerebral arterial smooth muscle cells [157].

5.5.5 Cholesterol

In human brain cells, cholesterol has been reported to increase after chronic exposure to ethanol. Ethanol enhanced BK_{Ca} channel activity, and cholesterol abolished this ethanol-induced augmentation of this channel activity [157a].

Caveolae are cholesterol-rich plasmmalemmal microdomains characterized by the presence of the scaffolding protein caveolin-1 (Cav-1). In cultured endothelial cells the BK_{Ca} channels were inhibited by Cav-1, which has a high affinity for cholesterol [157b]. Cholesterol repletion also suppressed this channel activity [157c]. On the other hand, bile acids have been known to cause relaxation of smooth muscle. In mesenteric arterial cells, bile acids activated the BK_{Ca} channels [157d], indicating that this phenomenon explains their action on smooth muscle.

5.5.6 NO, CO

NO, which is generated primarily by nitric oxide synthase (NOS) in the endothelium and causes relaxation of vascular smooth muscle cells via hyperpolarization of the cell membrane [158, 159]. NO-induced vasodilation was attenuated by blockade of the BK_{Ca} channel activity (Fig. 4) [113, 114]. As described earlier, NO activates the BK_{Ca} channel mediated by cGMP/PKG processes (see Sect. 5).

Carbon monoxide (CO) is synthesized in both endothelial and smooth muscle cells in vessels by metabolism of heme [160]. CO also acts on vessels as a vasodilator, similar to NO. The BK_{Ca} channels were activated by CO independent of PKG (Fig. 4) [161]. Sensitivity to intracellular Ca^{2+} was augmented via an enhancement of coupling Ca^{2+} sparks to the BK_{Ca} channels [162]. This CO-induced activation of the BK_{Ca} channels was involved in the α-subunit [163–165], not in the β-subunit [123].

5.5.7 pH

As described earlier, concentrations of intracellular H^+ influence the function of smooth muscle, including airways and vessels (see Sect. 4). However, the effects of pH on smooth muscle tone are controversial. Although it is generally considered that acidification leads to the relaxation of smooth muscle, whereas alkalization leads to contraction [166], a previous report indicated the opposite results [78]. BK_{Ca} channels were activated by increasing the concentration of intracellular H^+ in coronary and basilar arterial smooth muscle cells [167, 168], suggesting that acidosis-induced relaxation is due to the activation of these channels by intracellular H^+. In contrast, acidification inhibited the activity of the BK_{Ca} channels in internal mammary and tail arterial smooth muscles [169, 170], similar to tracheal smooth muscle cells [25e]. Little is known about the physiological relevance of intracellular proton-induced regulation of BK_{Ca} channels on smooth muscle tone.

6 Effects of BK_{Ca} Channels on Physiological and Pathophysiological Conditions

6.1 Pregnancy

The myogenic tone of the ovine uterine artery is reduced during pregnancy, and this decreased tone was abolished by blocking BK_{Ca} channels [171], indicating that enhanced activity of BK_{Ca} channels plays an important functional role in the adaptive alteration of uterine circulation. BK_{Ca} channel activity was enhanced by increasing the expression of the β1-subunit during pregnancy [171, 172]. Moreover, endothelial NO synthases, cGMP, and PKG were activated in the uterine artery during pregnancy [172, 173]. Thus, NO/cGMP/PKG processes are involved in the augmented activity of BK_{Ca} channels induced by pregnancy [172]. Uterine blood flow is augmented during the follicular phase of the ovarian cycle [174]. This phenomenon is also due to an augmentation in BK_{Ca} channel activity mediated by estrogen via cGMP/PKG processes [134].

6.2 Development and Aging

In smooth muscle cells, including vessels, structure and function are complexly affected by development and aging. Vasorelaxation was attenuated by the advance of age [175], and activation of BK_{Ca} channels markedly decreased with aging [176]. In coronary arterial smooth muscle cells, the current density of the BK_{Ca} channels was significantly attenuated in old rats compared with young rats [177, 178]. The expression of BK_{Ca} channels was also markedly reduced in aged human coronary arterial myocytes [179]. Moreover, iberiotoxin-vasoconstriction in the coronary artery was blunted in old rats compared with young rats [177, 178], suggesting that aging causes a significant decrease in the contribution of the BK_{Ca} channels to regulation of vascular tone. These results indicate that loss of the BK_{Ca} function may be related to an increased risk of coronary spasm and myocardial ischemia in the elderly. Resistance in pulmonary vessels is reduced at birth, and vasodilation causes blood to perfuse to the lung. BK_{Ca} channels may be involved in this phenomenon. In fetal ovine and rabbit pulmonary arterial smooth muscle cells, BK_{Ca} channel activity was markedly enhanced by elevating the oxygen partial pressure [180, 181].

In freshly isolated human fatal airway smooth muscle cells, three types of BK_{Ca} channel were recorded in an inside-out patch in a symmetrical 140 mM K^+ medium, i.e., (1) 200-, (2) 150-, (3) 100-pS channels [181a]. The BK_{Ca} channels of 200 pS had characteristics identical to those found in freshly isolated human adult bronchial smooth muscle cells. In contrast, two other BK_{Ca} channels did not exist in these freshly dispersed adult bronchial myocytes but were observed in adult bronchial myocytes proliferating in culture [77b]. These results suggest that these two BK_{Ca} channels of smaller conductance are associated with the proliferation and growth of human airway smooth muscle cells.

6.3 Hypoxia

It is generally considered that acute hypoxia causes constriction in pulmonary vessels and, in contrast, dilation in systemic vessels [182]. Under conditions of acute hypoxia, the BK_{Ca} channels were inhibited in pulmonary arterial myocytes [183, 184]; in contrast, these channels were potentiated in cerebral and pial arterial smooth muscle cells [157, 185]. In chronic hypoxia, impaired BK_{Ca} channel function was also observed in rat and human pulmonary arteries [186, 187]. Voltage-gated K^+ channels are the predominant K^+ currents in regulating resting membrane in adult pulmonary arterial smooth muscle [188]. However, prolonged exposure to hypoxia causes contraction of these myocytes by reducing the BK_{Ca} channel function, leading to pulmonary hypertension. Chronic hypoxia resulted in a reduction in expression of not only the α-subunit [189] but also the β1-subunit [190] of the BK_{Ca} channels, indicating that the dysfunction of these channels is mediated by both voltage- and Ca^{2+}-sensing processes under the condition of chronic hypoxia [186]. However, the physiological relevance of hypoxia-induced regulation of these channels remains to be elucidated. Since the generation of ROSs and 20-HETE is altered during exposure to hypoxia, the BK_{Ca} channel activity may be influenced complexly.

6.4 Shock

In both endotoxic and hemorrhagic shock, the membrane potential of vascular smooth muscle cells was hyperpolarized when mediated by an increase in the BK_{Ca} channel activity [191, 192]. Although reduced responsiveness to norepinephrine occurs in vessels under the condition of shock, BK_{Ca} channel blockers restored vascular contractility in response to epinephrine, suggesting that BK_{Ca} channel activity is involved in the development of hypotension. Endotoxin lipopolysaccharide enhanced the activity of BK_{Ca} channels [193]; furthermore, this channel activity was augmented in hemorrhage shock animals [192].

6.5 Hypertension

Since activation of BK_{Ca} channels leads to hyperpolarization in cell membranes and relaxation in smooth muscles, dysfunction of these channels in vessels may contribute to the development of hypertension. However, the role of BK_{Ca} channels in causing hypertension is unclear. In gene knockout methods, it has been reported that targeted gene deletion of both the α-and β1-subunit of these channels caused increases in vascular tone and systemic blood pressure [7, 8, 19]. In spontaneously hypertensive rats (SHR), expression of the β1-subunit was decreased in cerebral arterial myocytes [194], suggesting impairment of the coupling of Ca^{2+} sparks to the BK_{Ca} channels. However, hypertension did not develop in β1 null mice [195], indicating that deletion of the β1-subunit may be insufficient to cause hypertension. Involvement of BK_{Ca} channel β1-subunits in hypertension is still controversial. On the other hand, decreased expression of the α-subunit was observed in vascular smooth muscle cells from rats made hypertensive with N^{ω}-nitro-L-arginine [196]. Furthermore, in hypertensive animal models, other reports demonstrated that the BK_{Ca} channel function was conversely upregulated [197, 198] and that the expression of α- or β1-subunits was also augmented [199, 200]. It was thought that the upregulated BK_{Ca} channel function might be caused as a compensatory mechanism for excessive vasoconstriction in hypertension [199].

6.6 Diabetes

Myogenic tone was increased in cerebral artery from type 1 and type 2 diabetic animals [201, 202], suggesting that vascular dysfunction develops in this disease. As the mechanism of this phenomenon, impairment of BK_{Ca} channel function is proposed. In Zucker diabetic fatty rats, iberiotoxin did not significantly enhance agonist-induced contraction in mesenteric artery, and NS 1619, an opener of BK_{Ca} channels, also did not markedly hyperpolarize the cell membrane compared with nondiabetic animals [203]. In streptozotocin-induce diabetic rats, similar results were observed [201, 204, 205]. Moreover, loss of the BK_{Ca} channel function was demonstrated in both type 1 [201, 204] and type 2 [206, 207] diabetic vascular smooth muscle cells using electrophysiological methods. These results indicate that the ability of these channels to resist vasoconstriction is diminished in this disease. The expression of the β1-, but not α-, subunit of the BK_{Ca} channels was attenuated in type 1 [208, 209] and type 2 [204, 207] diabetes, and a diminished β1-subunit caused a reduction in Ca^{2+} and voltage sensitivity in these channels [201, 204, 207, 208], leading to impairment of the coupling of Ca^{2+} sparks to these channels [201]. Since diabetes enhanced the generation of ROSs in vascular smooth muscle cells [210], BK_{Ca} channels could be influenced by ROSs in this disease. ROSs suppressed the BK_{Ca} channel activity with a response similar to the deletion of the β1-subunit [211]. Hyperglycemia also has an effect on the development of BK_{Ca} channel dysfunction in diabetes.

6.7 *Bronchial Asthma*

In airway smooth muscle cells, BK_{Ca} channels were markedly activated by β-adrenoceptor action [72, 92, 93, 96–98], suggesting that these channels contribute to suppressing bronchoconstriction in this disease. Moreover, the BK_{Ca} channels were inhibited by muscarinic receptor stimulation [70, 92, 98, 103, 110], suggesting that these channels contribute to airway hyperresponsiveness implicated in the pathophysiology of this disease. On the other hand, interleukin (IL)-4 and IL-13 play an important role in the pathophysiology of asthma as inflammatory mediators. In human bronchial smooth muscle cells, IL-4 caused a marked increase in BK_{Ca} channel activity, whereas IL-13 caused modest activation of these channels [212]. These results indicate that BK_{Ca} channels may be involved in a variety of pathophysiologies in bronchial asthma. However, the role of these channels in this awaits elucidation in future studies.

7 Conclusions

In this chapter, the electrophysiological and structural characteristics of BK_{Ca} channels were described in detail. These channels are abundantly distributed on the cell surface in ubiquitous tissues and regulate membrane potential via intracellular Ca^{2+}- and voltage-sensitive outward currents (large conductance). Hence, BK_{Ca} channels might play a unique role in Ca^{2+} signaling related to mechanisms underlying various vital bodily functions and diseases. Recently, an activator of BK_{Ca} channels was clinically used as a medicine for glaucoma because this agent inhibits an increase in the concentrations of intracellular Ca^{2+} in nerve cells. These channels also cause membrane hyperpolarization and relaxation via a reduction in intracellular Ca^{2+} in smooth muscle cells. However, little is currently known about the clinical relevance of BK_{Ca} channel activity in hypertension, bronchial asthma, and other diseases. Furthermore, experiments are needed to precisely clarify the involvement of BK_{Ca} channels in the physiological and pathophysiological conditions. Novel studies such as on the polymorphisms of the BK_{Ca} channel β1-subunit may foster a gradual advance toward a solution to these problems.

References

1. Berkefeld H, Fakler B, Schulte U (2010) Ca^{2+}-activated K^+ channels: from protein complexes to function. Physiol Rev 90:1437–59.
2. Wu RS, Marx SO (2010) The BK potassium channel in the vascular smooth muscle and kidney: α- and β-subunits. Kidney Int 78: 963–74.
3. Hu XQ, Zhang L (2012) Function and regulation of large conductance Ca^{2+}-activated K^+ channel in vascular smooth muscle cells. Drug Discov Toda 17: 974–87.

4. Tomita T, Kume H (1994) Electrophysiology of potassium channels in airways smooth muscle. In Raeburn D, Giembycz MA (ed) Airway smooth muscle: Development and regulation of contractility, Birkhauser Verlag, Basel.
5. Nelson MT, Cheng H, Rubart M, Santana LF, Bonev AD, Knot HJ, Lederer WJ (1995) Relaxation of arterial smooth muscle by calcium sparks. Science 270: 633–7.
6. Kaczorowski GJ, Knaus HG, Leonard RJ, McManus OB, Garcia ML (1996) High-conductance calcium-activated potassium channels; structure, pharmacology, and function. J Bioenerg Biomembr 28: 255–67.
7. Brenner R, Peréz GJ, Bonev AD, Eckman DM, Kosek JC, Wiler SW, Patterson AJ, Nelson MT, Aldrich RW (2000) Vasoregulation by the β1 subunit of the calcium-activated potassium channel. Nature 407: 870–6.
8. Plüger S, Faulhaber J, Fürstenau M, Löhn M, Waldschütz R, Gollasch M, Haller H, Luft FC, Ehmke H, Pongs O (2000) Mice with disrupted BK channel β1 subunit gene feature abnormal Ca^{2+} spark/STOC coupling and elevated blood pressure. Circ Res 87: E53–60.
9. Fernández-Fernández JM, Tomás M, Vázquez E, Orio P, Latorre R, Sentí M, Marrugat J, Valverde MA (2004) Gain-of-function mutation in the KCNMB1 potassium channel subunit is associated with low prevalence of diastolic hypertension. J Clin Invest 113: 1032–9.
10. Du W, Bautista JF, Yang H, Diez-Sampedro A, You SA, Wang L, Kotagal P, Lüders HO, Shi J, Cui J, Richerson GB, Wang QK. (2005) Calcium-sensitive potassium channelopathy in human epilepsy and paroxysmal movement disorder. Nat Genet 37: 733–8.
11. Bloch M, Ousingsawat J, Simon R, Schraml P, Gasser TC, Mihatsch MJ, Kunzelmann K, Bubendorf L (2007) KCNMA1 gene amplification promotes tumor cell proliferation in human prostate cancer. Oncogene 26: 2525–34.
12. Seibold MA, Wang B, Eng C, Kumar G, Beckman KB, Sen S, Choudhry S, Meade K, Lenoir M, Watson HG, Thyne S, Williams LK, Kumar R, Weiss KB, Grammer LC, Avila PC, Schleimer RP, Burchard EG, Brenner R (2008) An african-specific functional polymorphism in KCNMB1 shows sex-specific association with asthma severity. Hum Mol Genet 17: 2681–90.
13. Elkins T, Ganetzky B, Wu CF (1986) A Drosophila mutation that eliminates a calcium-dependent potassium current. Proc Natl Acad Sci USA 83: 8415–9.
14. Neher E, Sakmann B (1976) Single-channel currents recorded from membrane of denervated frog muscle fibres. Nature 260:799–802.
15. Long SB, Campbell EB, Mackinnon R (2005) Crystal structure of a mammalian voltage-dependent Shaker family K^+ channel. Science 309: 897–903.
16. Sausbier M, Hu H, Arntz C, Feil S, Kamm S, Adelsberger H, Sausbier U, Sailer CA, Feil R, Hofmann F, Korth M, Shipston MJ, Knaus HG, Wolfer DP, Pedroarena CM, Storm JF, Ruth P (2004) Cerebellar ataxia and Purkinje cell dysfunction caused by Ca^{2+}-activated K^+ channel deficiency. Proc Natl Acad Sci USA 101: 9474–8.
17. Rüttiger L, Sausbier M, Zimmermann U, Winter H, Braig C, Engel J, Knirsch M, Arntz C, Langer P, Hirt B, Müller M, Köpschall I, Pfister M, Münkner S, Rohbock K, Pfaff I, Rüsch A, Ruth P, Knipper M (2004) Deletion of the Ca^{2+}-activated potassium (BK) α-subunit but not the BKbeta1-subunit leads to progressive hearing loss. Proc Natl Acad Sci USA 101: 12922–7.
18. Meredith AL, Thorneloe KS, Werner ME, Nelson MT, Aldrich RW (2004) Overactive bladder and incontinence in the absence of the BK large conductance Ca^{2+}-activated K^+ channel. J Biol Chem 279: 36746–52.
19. Sausbier M, Arntz C, Bucurenciu I, Zhao H, Zhou XB, Sausbier U, Feil S, Kamm S, Essin K, Sailer CA, Abdullah U, Krippeit-Drews P, Feil R, Hofmann F, Knaus HG, Kenyon C, Shipston MJ, Storm JF, Neuhuber W, Korth M, Schubert R, Gollasch M, Ruth P (2005) Elevated blood pressure linked to primary hyperaldosteronism and impaired vasodilation in BK channel-deficient mice. Circulation 112: 60–8.
20. Atkinson NS, Robertson GA, Ganetzky B (1991) A component of calcium-activated potassium channels encoded by the *Drosophila* slo locus. Science 253: 551–5.

21. Adelman JP, Shen KZ, Kavanaugh MP, Warren RA, Wu YN, Lagrutta A, Bond CT, North RA (1992) Calcium-activated potassium channels expressed from cloned complementary DNAs. Neuron 9: 209–16.
22. Butler A, Tsunoda S, McCobb DP, Wei A, Salkoff L (1993) mSlo, a complex mouse gene encoding "maxi" calcium-activated potassium channels. Science 261: 221–4.
23. Wallner M, Meera P, Toro L (1996) Determinant for beta-subunit regulation in high-conductance voltage-activated and Ca^{2+}-sensitive K^+ channels: an additional transmembrane region at the N terminus. Proc Natl Acad Sci USA 93: 14922–7.
24. Ma Z, Lou XJ, Horrigan FT (2006) Role of charged residues in the S1-S4 voltage sensor of BK channels. J Gen Physiol 127: 309–28.
25. Jiang Y, Pico A, Cadene M, Chait BT, MacKinnon R (2001) Structure of the RCK domain from the E. coli K^+ channel and demonstration of its presence in the human BK channel. Neuron 29: 593–601.
25a. Jiang Y, Lee A, Chen J, Cadene M, Chait BT, MacKinnon R (2002) Crystal structure and mechanism of a calcium-gated potassium channel. Nature 417: 515–22.
25b. Schreiber M, Salkoff L (1997) A novel calcium-sensing domain in the BK channel. Biophys J 73: 1355–63.
25c. Kim HJ, Lim HH, Rho SH, Eom SH, Park CS (2006) Hydrophobic interface between two regulators of K^+ conductance domains critical for calcium-dependent activation of large conductance Ca^{2+}-activated K^+ channels. J Biol Chem 281: 38573–81.
25d. Yusifov T, Savalli N, Gandhi CS, Ottolia M, Olcese R (2008) The RCK2 domain of the human BK_{Ca} channel is a calcium sensor. Proc Natl Acad Sci U S A 105: 376–81.
25e. Zhou Y, Zeng XH, Lingle CJ (2012) Barium ions selectively activate BK channels via the Ca^{2+}-bowl site. Proc Natl Acad Sci U S A 109: 11413–8.
25f. Sweet TB, Cox DH (2008) Measurements of the BK_{Ca} channel's high-affinity Ca_{2+} binding constants: effects of membrane voltage. J Gen Physiol 132:491–505
26. Yuan P, Leonetti MD, Pico AR, Hsiung Y, MacKinnon R (2010) Structure of the human BK channel Ca^{2+}-activation apparatus at 3.0 A resolution. Science 329: 182–6.
27. Niu X, Qian X, Magleby KL (2004) Linker-gating ring complex as passive spring and Ca^{2+}-dependent machine for a voltage- and Ca^{2+}-activated potassium channel. Neuron 42: 745–56.
27a. Yang H, Shi J, Zhang G, Yang J, Delaloye K, Cui J (2008) Activation of Slo1 BK channels by Mg^{2+} coordinated between the voltage sensor and RCK1 domains. Nat Struct Mol Biol 15: 1152–9.
27b. Savalli N, Pantazis A, Yusifov T, Sigg D, Olcese R (1012) The contribution of RCK domains to human BK channel allosteric activation. J Biol Chem 287: 21741–50.
28. Xia XM, Zeng X, Lingle CJ (2002) Multiple regulatory sites in large-conductance calcium-activated potassium channels. Nature 418: 880–4.
29. Shi J, Krishnamoorthy G, Yang Y, Hu L, Chaturvedi N, Harilal D, Qin J, Cui J (2002) Mechanism of magnesium activation of calcium-activated potassium channels. Nature 418: 876–80.
29a. Zhang X, Solaro CR, Lingle CJ (2001) Allosteric regulation of BK channel gating by Ca^{2+} and Mg^{2+} through a nonselective, low affinity divalent cation site. J Gen Physiol 118: 607–36.
30. Kume H, Takagi K, Satake T, Tokuno H, Tomita T (1990) Effects of intracellular pH on calcium-activated potassium channels in rabbit tracheal smooth muscle. J Physiol 424: 445–57.
31. Park JK, Kim YC, Sim JH, Choi MY, Choi W, Hwang KK, Cho MC, Kim KW, Lim SW, Lee SJ (2007) Regulation of membrane excitability by intracellular pH (pHi) changers through Ca^{2+}-activated K^+ current (BK channel) in single smooth muscle cells from rabbit basilar artery. Pflügers Arch 454: 307–19.
32. Wang ZW, Nara M, Wang YX, Kotlikoff MI (1997) Redox regulation of large conductance Ca^{2+}-activated K^+ channels in smooth muscle cells. J Gen Physiol 110: 35–44.

33. Jiang Z, Wallner M, Meera P, Toro L (1999) Human and rodent MaxiK channel β-subunit genes: cloning and characterization. Genomics 55: 57–67.
34. Knaus HG, Folander K, Garcia-Calvo M, Garcia ML, Kaczorowski GJ, Smith M, Swanson R (1994) Primary sequence and immunological characterization of β-subunit of high conductance Ca^{2+}-activated K^+ channel from smooth muscle. J Biol Chem 269: 17274–8.
35. Meera P, Wallner M, Jiang Z, Toro L (1996) A calcium switch for the functional coupling between alpha (hslo) and β subunits (KV,Ca β) of maxi K channels. FEBS Lett 382: 84–8.
36. Vogalis F, Vincent T, Qureshi I, Schmalz F, Ward MW, Sanders KM, Horowitz B (1996) Cloning and expression of the large-conductance Ca^{2+}-activated K^+ channel from colonic smooth muscle. Am J Physiol 271: G629–39.
36a. Liu G, Zakharov SI, Yang L, Wu RS, Deng SX, Landry DW, Karlin A, Marx SO (2008) Locations of the beta1 transmembrane helices in the BK potassium channel. Proc Natl Acad Sci U S A 105: 10727–32.
36b. Wu RS, Chudasama N, Zakharov SI, Doshi D, Motoike H, Liu G, Yao Y, Niu X, Deng SX, Landry DW, Karlin A, Marx SO (2009) Location of the beta 4 transmembrane helices in the BK potassium channel. J Neurosci 29: 8321–8.
36c. Liu G, Niu X, Wu RS, Chudasama N, Yao Y, Jin X, Weinberg R, Zakharov SI, Motoike H, Marx SO, Karlin A (2010) Location of modulatory beta subunits in BK potassium channels. J Gen Physiol 135: 449–59.
36d. Wu RS, Liu G, Zakharov SI, Chudasama N, Motoike H, Karlin A, Marx SO (2013) Positions of β2 and β3 subunits in the large-conductance calcium- and voltage-activated BK potassium channel. J Gen Physiol 141: 105–17.
36e. Knaus HG, Eberhart A, Kaczorowski GJ, Garcia ML (1994) Covalent attachment of charybdotoxin to the beta-subunit of the high conductance Ca^{2+}-activated K+ channel. identification of the site of incorporation and implications for channel topology. J Biol Chem 269: 23336–41.
36f. Xia XM, Ding JP, Zeng XH, Duan KL, Lingle CJ (2000) Rectification and rapid activation at low Ca^{2+} of Ca^{2+}-activated, voltage-dependent BK currents: consequences of rapid inactivation by a novel beta subunit. J Neurosci 20:4890–903.
36g. Meera P, Wallner M, Toro L (2000) A neuronal beta subunit (KCNMB4) makes the large conductance, voltage- and Ca^{2+}-activated K^+ channel resistant to charybdotoxin and iberiotoxin. Proc Natl Acad Sci U S A 97: 5562–7.
36h. Orio P, Torres Y, Rojas P, Carvacho I, Garcia ML, Toro L, Valverde MA, Latorre R (2006) Structural determinants for functional coupling between the beta and alpha subunits in the Ca^{2+}-activated K^+ (BK) channel. J Gen Physiol 127:191–204.
37. Bao L, Cox DH (2005) Gating and ionic currents reveal how the BK_{Ca} channel's Ca^{2+} sensitivity is enhanced by its β1 subunit. J Gen Physiol 126: 393–412.
37a. Wang B, Brenner R (2006) An S6 mutation in BK channels reveals beta1 subunit effects on intrinsic and voltage-dependent gating. J Gen Physiol 128: 731–44.
38. Semenov I, Wang B, Herlihy JT, Brenner R (2006) BK channel β1-subunit regulation of calcium handling and constriction in tracheal smooth muscle. Am J Physiol Lung Cell Mol Physiol 291: L802–10.
39. Souhrada JF, Dickey DW (1976) Mechanical activities of trachea as measured in vitro and in vivo. Respir Physiol 26: 27–40.
40. Small RC (1982) Electrical slow waves and tone of guinia-pig trachealis muscle: effects of drugs and temperature changes. Br J Pharmacol 72: 45–54.
41. Allen SL, Beech DJ, Foster RW, Morgan GP, Small RC (1985) Electrophysiological and other aspects of relaxant action of isoprenaline in guinea-pig isolated trachealis. Br J pharmacol 86: 843–854.
42. Honda K, Satake T, Takagi, K, Tomita T (1986) Effects of relaxant on electrical and mechanicl activities in the guinea-pig tracheal muscle. Br J Pharmacol 87: 665–671.
43. Brink C, Grimand C, Guillot C, orehe J (1980) The interaction between indomethacin and contractile agents on human isolated airway muscle. Br J pharmacol 69: 383–388.

44. Davis C, kannan MS, Jones TR, Daniel EE (1982) Control of human airway smooth muscle: in vitro studies. J Appl Physiol 1982; 53: 1080–1087.
45. Hutas I, Hadhazy P, Debreczeni L, Visi ES (1981) Relaxation of human isolated bronchial smooth muscle. Lung 159: 153–161.
46. Ito M, Baba K, Takagi K, Tomita T (1985) Some properties of calcium-induced contraction in the isolated human and guinea-pig tracheal smooth muscle. Respir Physiol 59: 143–153.
47. Honda K, Tomita T (1987) Electivity in isolated human tracheal muscle. Jpn J Physiol 37: 333–336.
48. Ito Y, Suzuki H, Aizawa h, Hakoda H, Hirose T (1989) The spontaneous elestical and mechanical activity of human bronchial smooth muscle its modulation by drugs. Br J Pharmacol 98: 1249–1260.
49. Farmer JB, Farrar DG, Wilson J (1974) Antagonisim of tone and prostaglandin mediated responses in a tracheal preparation by indomethacin and SC-19220. Br J Pharmacol 52: 559–565.
50. Orehek J, Douglas JS, Bouhuy A (1975) Contractile responses of the guinea pig trachea in vitro: modification by prostaglandin synthesis-inhibiting drugs. J Pharmacol Exp ther 194: 554–564.
51. Brink C, Duncan PG, Douglas JS (1981) Histamine, endogenous prostaglandins and cyclic nucleotides in the regulation of airway muscle responses in guinea-pig. Prostaglandins 22: 729–738.
52. Sead MH, Burka JF (1983) Effects of immunological sensitization on the response and sensitivity of guinea-pig airways to bronchonstrictors. Modulation by selective inhibition by arachidonic acid metabolism. Can j Physiol Pharmacol 61: 876–887.
53. Mansour S, Daniel EE (1986) Maintenance of tone, role of arachidonate metabolites, and effects of sensitization in guinea-pig trachea. Can J Physiol pharmacol 64: 1096–1103.
54. Arch JRS, Buckle DR, Bumstead J, Clarke GD, Taylor SG (1988) Evaluation of the potassium channel activator cromakalim (BRL 34915) as a bronchodilator in the guinea-pig: comparison with nifedipine. Br J Pharmacol 95: 763–770.
55. Foster RW, Okpalugo BI, Small RC (1984) Antagonism of Ca^{2+} and other actions of verapamil in guinea-pig isolated trachealis, Br J pharmacol 81: 199–507.
56. Ahmed F, Foster RW, Small RC (1985) Some effects of nifedipine in guinea-pig isolated trachealis. Br J Pharmacol 84: 861–869.
57. Baba K, Satake T, Takagi K, Tomita T (1986) Effects of verapamil on the response of the guinea-pig tracheal muscle. Br J Pharmacol 88: 441–449.
58. Ito S, Kajikuri J, Itoh T, Kuriyama H (1991) Effects of lemakalim on changes in Ca^{2+} concentration and mechanical activity induced by noradrenaline in the rabbit mesenteric artery. Br J Pharmacol 104: 227–233.
59. Coburn RF, Yamaguchi T (1977) Membrane potential-dependent and-independent tension in the canine tracheal muscle. J Pharmacol Exp Ther 201: 276–284.
60. Kirkpatrick CT (1981) Tracheobronchial smooth muscle. In; Smooth Muscle; An Assessment of Current Knowledge. E Bulbrung, AF Brading, Aww Jones, T Tomita eds. Edward Arnold, London 385–395.
61. Smith C, Phillips M, Miller C (1986) Purification of charybdotoxin, a specific inhibitor of the high-conductance Ca^{2+} -activated K+ channel. J Biol Chem 231: 14607–14613.
62. Gimenez-Gallego G, Navia MA, Reuben JP, Katz GM, Kaczorowski GJ, Garcia ML (1988) Purification, sequence, and model structure of charybdotoxin, a potent selective inhibitor of calcium-activated potassium channels. Proc Natl Acard Sci USA 85: 3329–3333.
63. Isaac L, McArdle S, Miller NM, Foster RW, Small RC (1996) Effects of some K^+-channel inhibitors on the electrical behaviour of guinea-pig isolated trachealis and on its responses to spasmogenic drugs. Br J Pharmacol 117: 1653–62.

64. Murray MA, Berry JL, Cook SJ, Foster RW, Green KA, Small RC. Guinea-pig isolated trachealis; the effects of charybdotoxin on mechanical activity, membrane potential changes and the activity of plasmalemmal K^+-channels. Br J Pharmacol 1991; 103: 1814–1818.
65. Stockbridge L, French AS, Man SFP (1991) Subconductance states in activated potassium channels from canine airway smooth muscle. Biochim Biophys Acta 1064: 212–218.
66. McCann JD, Welsh MJ (1986) Calcium-activated potassium channels in canine airway smooth muscle. J Physiol 372: 113–127.
67. Berry JL, Elliott KRF, Foster RW, Green KA, Murray MA, Small RC (1991) Mechanical, biochemical and electrophysiological studies of RP 49356 and cromakalim in guinea-pig and bovine trachealis muscle. Pulm Phamacol 4: 91–98.
68. Green KA, Foster RW, Small RC (1991) A patch-clamp study of K^+ channel activity in bovine isolated tracheal smooth muscle cells. Br J Pharmacol 102: 871–878.
69. Saunders H-MH, Farley JM (1991) Spontaneous transient outward currents and Ca^{++} -activated K^+ channels in swine tracheal smooth muscle cells. J Pharmacol Exp Ther 257: 1114–1120.
70. Kume H, Kotlikooff MI (1991) Muscarinic inhibition off single K_{Ca} channels in smooth muscle cells by a pertussis-sensitive G protein. Am J Physiol 261: C1204–C1207.
71. Saunders HH, Farley JM (1992) Pharmacological properties of potassium of potassium currents in swine tracheal smooth muscle. J Pharmacol Exp Ther 260: 1038–1044.
72. Kume H, Takai A, Tokuno H, Tomita T (1989) Regulation of Ca^{2+}-dependent K^+-channel activity in tracheal myocytes by phosphorylation. Nature 341: 152–154.
73. Wang ZW, Kotlikoff MI (1996) Activation of K_{Ca} channels in airway smooth muscle cells by endogenous protein kinase A. Am J Physiol 271: L100–5.
74. Lattorre R, Oberhauser A, Labarca P, Alvarez O (1989) Varieties of calcium-activated potassium channels. Annu Rev Physiol 51: 385–399.
75. Trieschmann U, Isenberg G (1989) Ca^{2+}-activated K^+ channels contribute to the resting potential of vascular myocytes. Ca^{2+}-sensitivity is increased by intracellular Mg^{2+}-ions. Pflugers Arch 414: S183–S184.
76. Muraki K, ImaizumiY, Kojima T, Kawai T, Watanabe M (1990) Effects of tetraethylammonium and 4-aminopyridine on outward and excitability in canine tracheal smooth muscle cells. Br J Pharmacol 100: 507–515.
77. Savaria D, Lanoue C, Cadieux A, Rosseau E (1992) Large conducting potassium channel reconstituted from airway smooth muscle. Am J Physiol 262: L327–L336.
77a. Snetkov VA, Hirst SJ, Twort CH, Ward JP (1995) Potassium currents in human freshly isolated bronchial smooth muscle cells. Br J Pharmacol 115: 1117–25.
77b. Snetkov VA, Hirst SJ, Ward JP (1996) Ion channels in freshly isolated and cultured human bronchial smooth muscle cells. Exp Physiol 81: 791–804.
78. Way S (1988) Smooth muscle intracellular pH: measurement, regulation and function. Am J Physiol 254: C213–C225.
79. McCann JD, Welsh MJ (1987) Neuroleptics antagonize a calcium-activated potassium channel in airway smooth muscle. J Gen Physipl 89: 339–352.
80. Levin RM, Weiss B (1979) Selective binding of antipsychotics and other psychoactive agents to the calcium-dependent activator of cyclic nucleotide phosphodiesterase. J Pharmacol Exp Ther 208: 454–459.
81. Kihira M, madtuzawa K, Tokuno H, Tomita T (1990) Effect of calmodulin antagonists on calcium-activated potassium channels in pregnant rat myometrium. Br J Pharmacol 100: 353–359.
82. Hisada T, Kurachi Y, Sugimoto T (1990) Properties of membrance currents in isolated smooth muscle from guinea-pig trachea. Pflugers Arch 416: 151–161.
83. Benhan CD, Bolton TB (1986) Spontaneous transient outward currents in single visceral and vascular smooth muscle cells of the rabbit. J Physiol 381: 385–406.
84. Marthan R, Martin C, Amedee T, Mironneau J (1989) Calcium channel currents in isolated smooth muscle cells from human bronchus. J Appl Physiol 66: 1706–1714.

85. Muraki K, Imaizumi Y, watanwbe M (1992) Ca-dependent K channels in smotth muscle cells permeabilized by β−escin recorded using the cell-attached patch-clamp technique. Pflugers Arch 420: 461–469.
85a. Liu B, Freyer AM, Hall IP (2007) Bradykinin activates calcium-dependent potassium channels in cultured human airway smooth muscle cells. Am J Physiol Lung Cell Mol Physiol 292: L898–907.
86. Ingebritsen TS, Cohen P (1983) The protein phosphatases involved in celluar regulation. 1. Classification and substrate specificity. Eur J Biochm 132: 255–261.
87. Bialojan C, Takagi A (1988) Inhibitory effect of a marine-sponge toxin, okadaic acid, on protein phosphatases. Specificity and kinetics. Biochem J 256: 283–290.
88. Nara M, Dhulipala PD, Wang YX, Kotlikoff MI (1998) Reconstitution of β-adrenergic modulation of large conductance, calcium-activated potassium (maxi-K) channels in Xenopus oocytes. Identification of the cAMP-dependent protein kinase phosphorylation site. J Biol Chem 273: 14920–4.
89. Hiramatsu T, Kume H, Kotlikoff MI, Takagi K (1994) Role of calcium-activated potassium channels in the relaxation of tracheal smooth muscles by forskolin. Clin Exp Pharmacol Physiol 21: 367–75.
90. Sadoshima J, Akaike N, Kanaide H, Nakamura M (1988) Cyclic AMP modulates Ca^{2+}-activated K^{+} channel in cultured smooth muscle cells of rat aortas. Am J Physiol 255: H754–H759.
91. Carl A, Kenyon JL, Uemura D, Fuestani N, Sanders KM (1991) Regulation of Ca^{2+}-activated K^{+} channels by protein kinase-A and phosphatase inhibitors. Am J Physiol 261: C387–C392.
92. Kume H, Graziano MP, Kotlikoff MI (1992) Stimulatory and inhibitory regulation of calcium-activated potassium channels by guanine nucleotide proteins. Proc Natl Acad Sci USA 89: 11051–11055
93. Kume H, Hall IP, Washabau RJ, Takagi K, Kotlikoff MI (1994) β-Adrenergic agonists regulate K_{Ca} channels in airway smooth muscle by cAMP-dependent and -independent mechanisms. J Clin Invest 93: 371–9.
94. Toro L, Stefani E, Erulkar S (1990) Hormonal regulation of potassium current in single myometrium cells. Proc Natl Acad Sci USA 87: 2892–2895.
95. Ito Y, Tajima K (1982) Dual effects of catecholamines on pre- and post-junctional membranes in the dog trachea. Br J Pharmacol 75: 433–440.
96. Jones TR, Charette L, Garcia ML, Kaczorowski GJ (1990) Selective inhibition of relaxation of guinea-pig trachea by charybdotoxin, a potent Ca^{2+}-activated K^{+} channel inhibitor. J Pharmacol Exp Ther 255: 697–706.
97. Miura M, Belvisi MG, Stretton CD, Yacoub MH, Barnes PJ (1992) Role of potassium channels in bronchodilator responses in human airways. Am Rev Respir Dis 146: 132–6.
98. Kume H, Mikawa K, Takagi K, Kotlikoff MI (1995) Role of G proteins and K_{Ca} channels in the muscarinic and β-adrenergic regulation of airway smooth muscle. Am J Physiol 268: L221–9.
99. Kume H, Ishikawa T, Oguma T, Ito S, Shimokata K, Kotlikoff MI (2003) Involvement of Ca^{2+} mobilization in tachyphylaxis to β-adrenergic receptors in trachealis. Am J Respir Cell Mol Biol 29: 359–66.
100. Kume H, Takagi K (1999) Inhibition of β-adrenergic desensitization by K_{Ca} channels in human trachealis. Am J Respir Crit Care Med 159: 452–60.
101. Cook SJ, Small RC, Berry JL, Chiu P, Downing SJ, Foster RW (1993) β-Adrenoceptor subtypes and the opening of plasmalemmal K^{+} -channels muscle: electrophysiology and mechanical studies in guinea-pig tissue. Br J Pharmacol 109: 1140–114.
102. Oguma T, Kume H, Ito S, Takeda N, Honjo H, Kodama I, Shimokata K, Kamiya K (2006) Involvement of reduced sensitivity to Ca^{2+} in β-adrenergic action on airway smooth muscle. Clin Exp Allergy 36:183–91.

103. Wang YX, Fleischmann BK, Kotlikoff MI (1997) Modulation of maxi-K^+ channels by voltage-dependent Ca^{2+} channels and methacholine in single airway myocytes. Am J Physiol 272: C1151–9.
104. Cole WC, Carl A, Sanders KM (1989) Muscarinic suppression of Ca^{2+}-dependent K^+ current in colonic smooth muscle. Am J Physiol 257: C481–C487.
105. Ozaki H, Kwon SC, Tajimi M, Karaki H (1990) Changes in cytosolic Ca^{2+} and contraction induced by various and relaxant in canine tracheal smooth muscle. Pflugers Arch 416: 351–359.
106. Shieh CC, Petrini MF, Dwyer TM, Farley JM (1992) Cromakalim effcts on acetylcholine-induced changers in cytosolic calcium and tension in swine trachealis. J Pharmacol Exp Ther 260: 261–268.
107. Cole WC, Sanders KM (1989) G protein mediate suppression of Ca^{2+} -activated K current by acetylcholine in smooth muscle cells. Am J Physiol 257: C596–C600.
108. Torphy TJ, Rinard GA, Reitow MG, Mayer SE (1983) Functional antagonism in canine tracheal smooth muscle; inhibition by methacholine of the mechanical and biochemical response to isoproterenol. J Pharmacol Exp Ther 227: 694–699.
109. Torphy TJ, Zheng C, Peterson SM, Fiscus RR, Ranrd GA, Mayer SE (1985) Inhibitory effect of methacholine on drug-induced relaxation, cyclic AMP accumulation, and cyclic AMP-dependent protein kinase activation in canine tracheal smooth muscle. J Pharmacol Exp Ther 233: 409–417.
110. Zhou XB, Wulfsen I, Lutz S, Utku E, Sausbier U, Ruth P, Wieland T, Korth M (2008) M_2 muscarinic receptors induce airway smooth muscle activation via a dual, Gβγ-mediated inhibition of large conductance Ca^{2+}-activated K^+ channel activity. J Biol Chem 283: 21036–44.
110a. Semenov I, Wang B, Herlihy JT, Brenner R (2011) BK channel β1 subunits regulate airway contraction secondary to M2 muscarinic acetylcholine receptor mediated depolarization. J Physiol 589: 1803–17.
111. Mikawa K, Kume H, Takagi K (1997) Effects of BK_{Ca} channels on the reduction of cytosolic Ca^{2+} in cGMP-induced relaxation of guinea-pig trachea. Clin Exp Pharmacol Physiol 24: 175–81.
112. Nara M, Dhulipala PD, Ji GJ, Kamasani UR, Wang YX, Matalon S, Kotlikoff MI (2000) Guanylyl cyclase stimulatory coupling to K_{Ca} channels. Am J Physiol Cell Physiol 279: C1938–45.
113. Archer SL, Huang JM, Hampl V, Nelson DP, Shultz PJ, Weir EK (1994) Nitric oxide and cGMP cause vasorelaxation by activation of a charybdotoxin-sensitive K channel by cGMP-dependent protein kinase. Proc Natl Acad Sci USA 91: 7583–7.
114. Wellman GC, Bonev AD, Nelson MT, Brayden JE (1996) Gender differences in coronary artery diameter involve estrogen, nitric oxide, and Ca^{2+}-dependent K^+ channels. Circ Res 79: 1024–30.
115. White RE, Lee AB, Shcherbatko AD, Lincoln TM, Schonbrunn A, Armstrong DL (1993) Potassium channel stimulation by natriuretic peptides through cGMP-dependent dephosphorylation. Nature 361: 263–6.
116. Stockand JD, Sansom SC (1996) Mechanism of activation by cGMP-dependent protein kinase of large Ca^{2+}-activated K^+ channels in mesangial cells. Am J Physiol 271: C1669–77.
117. White RE, Darkow DJ, Lang JL (1995) Estrogen relaxes coronary arteries by opening BK_{Ca} channels through a cGMP-dependent mechanism. Circ Res 77: 936–42.
118. Han G, Kryman JP, McMillin PJ, White RE, Carrier GO (1999) A novel transduction mechanism mediating dopamine-induced vascular relaxation: opening of BK_{Ca} channels by cyclic AMP-induced stimulation of the cyclic GMP-dependent protein kinase. J Cardiovasc Pharmacol 34: 619–27.
119. White RE, Kryman JP, El-Mowafy AM, Han G, Carrier GO (2000) cAMP-dependent vasodilators cross-activate the cGMP-dependent protein kinase to stimulate BK_{Ca} channel activity in coronary artery smooth muscle cells. Circ Res 86: 897–905.

120. Peng W, Hoidal JR, Farrukh IS (1996) Regulation of Ca^{2+}-activated K^+ channels in pulmonary vascular smooth muscle cells: role of nitric oxide. J Appl Physiol 81: 1264–72.
121. Bolotina VM, Najibi S, Palacino JJ, Pagano PJ, Cohen RA (1994) Nitric oxide directly activates calcium-dependent potassium channels in vascular smooth muscle. Nature 368: 850–3.
122. Mistry DK, Garland CJ (1998) Characteristics of single, large-conductance calcium-dependent potassium channels (BK_{Ca}) from smooth muscle cells isolated from the rabbit mesenteric artery. J Membr Biol 164: 125–38.
123. Wu L, Cao K, Lu Y, Wang R (2002) Different mechanisms underlying the stimulation of K_{Ca} channels by nitric oxide and carbon monoxide. J Clin Invest 110: 691–700.
124. Barman SA, Zhu S, White RE (2004) PKC activates BK_{Ca} channels in rat pulmonary arterial smooth muscle via cGMP-dependent protein kinase. Am J Physiol Lung Cell Mol Physiol 286: L1275–81.
125. Minami K, Fukuzawa K, Nakaya Y (1993) Protein kinase C inhibits the Ca^{2+}-activated K^+ channel of cultured porcine coronary artery smooth muscle cells. Biochem Biophys Res Commun 190: 263–9.
126. Barman SA, Zhu S, White RE (2004) Protein kinase C inhibits BK_{Ca} channel activity in pulmonary arterial smooth muscle. Am J Physiol Lung Cell Mol Physiol 286: L149–55.
127. Zhou XB, Arntz C, Kamm S, Motejlek K, Sausbier U, Wang GX, Ruth P, Korth M (2001) A molecular switch for specific stimulation of the BK_{Ca} channel by cGMP and cAMP kinase. J Biol Chem 276: 43239–45.
128. Zhou XB, Wulfsen I, Utku E, Sausbier U, Sausbier M, Wieland T, Ruth P, Korth M (2010) Dual role of protein kinase C on BK channel regulation. Proc Natl Acad Sci USA 107: 8005–10.
129. Alioua A, Mahajan A, Nishimaru K, Zarei MM, Stefani E, Toro L (2002) Coupling of c-Src to large conductance voltage- and Ca^{2+}-activated K^+ channels as a new mechanism of agonist-induced vasoconstriction. Proc Natl Acad Sci U A 99: 14560–5.
130. Ling S, Woronuk G, Sy L, Lev S, Braun AP (2000) Enhanced activity of a large conductance, calcium-sensitive K^+ channel in the presence of Src tyrosine kinase. J Biol Chem 275: 30683–9.
131. White RE, Darkow DJ, Lang JL (1995) Estrogen relaxes coronary arteries by opening BK_{Ca} channels through a cGMP-dependent mechanism. Circ Res 77: 936–42.
132. Han G, Yu X, Lu L, Li S, Ma H, Zhu S, Cui X, White RE (2006) Estrogen receptor alpha mediates acute potassium channel stimulation in human coronary artery smooth muscle cells. J Pharmacol Exp Ther 316: 1025–30.
133. Yu X, Ma H, Barman SA, Liu AT, Sellers M, Stallone JN, Prossnitz ER, White RE, Han G (2011) Activation of G protein-coupled estrogen receptor induces endothelium-independent relaxation of coronary artery smooth muscle. Am J Physiol Endocrinol Metab 301: E882–8.
134. Rosenfeld CR, White RE, Roy T, Cox BE (2000) Calcium-activated potassium channels and nitric oxide coregulate estrogen-induced vasodilation. Am J Physiol Heart Circ Physiol 279: H319–28.
135. Zhang HT, Wang Y, Deng XL, Dong MQ, Zhao LM, Wang YW (2010) Daidzein relaxes rat cerebral basilar artery via activation of large-conductance Ca^{2+}-activated K^+ channels in vascular smooth muscle cells. Eur J Pharmacol 630:100–6.
136. Tep-areenan P, Kendall DA, Randall MD (2003) Mechanisms of vasorelaxation to 17-β-oestradiol in rat arteries. Eur J Pharmacol 476: 139–49.
137. Valverde MA, Rojas P, Amigo J, Cosmelli D, Orio P, Bahamonde MI, Mann GE, Vergara C, Latorre R (1999) Acute activation of Maxi-K channels (hSlo) by estradiol binding to the β subunit. Science 285:1929–31.
138. Wolin MS (2009) Reactive oxygen species and the control of vascular function. Am J Physiol Heart Circ Physiol 296: H539–49.

139. Tang XD, Daggett H, Hanner M, Garcia ML, McManus OB, Brot N, Weissbach H, Heinemann SH, Hoshi T (2001) Oxidative regulation of large conductance calcium-activated potassium channels. J Gen Physiol 117: 253–74.
140. Soto MA, González C, Lissi E, Vergara C, Latorre R (2002) Ca^{2+}-activated K^+ channel inhibition by reactive oxygen species. Am J Physiol Cell Physiol 282: C461–71.
141. Zeng XH, Xia XM, Lingle CJ (2003) Redox-sensitive extracellular gates formed by auxiliary β subunits of calcium-activated potassium channels. Nat Struct Biol 10: 448–54.
142. Santarelli LC, Wassef R, Heinemann SH, Hoshi T (2006) Three methionine residues located within the regulator of conductance for K^+ (RCK) domains confer oxidative sensitivity to large-conductance Ca^{2+}-activated K^+ channels. J Physiol 571: 329–48.
143. Hayabuchi Y, Nakaya Y, Matsuoka S, Kuroda Y (1998) Hydrogen peroxide-induced vascular relaxation in porcine coronary arteries is mediated by Ca^{2+}-activated K^+ channels. Heart Vessels 13: 9–17.
144. Miura H, Bosnjak JJ, Ning G, Saito T, Miura M, Gutterman DD (2003) Role for hydrogen peroxide in flow-induced dilation of human coronary arterioles. Circ Res 92:e31–40.
145. Barlow RS, El-Mowafy AM, White RE (2000) H_2O_2 opens BK_{Ca} channels via the PLA_2-arachidonic acid signaling cascade in coronary artery smooth muscle. Am J Physiol Heart Circ Physiol 279: H475–83.
146. Kojima K, Kume H, Ito S, Oguma T, Shiraki A, Kondo M, Ito Y, Shimokata K (2007) Direct effects of hydrogen peroxide on airway smooth muscle tone: roles of Ca^{2+} influx and Rho-kinase. Eur J Pharmacol 556: 151–6.
147. Brzezinska AK, Gebremedhin D, Chilian WM, Kalyanaraman B, Elliott SJ (2000) Peroxynitrite reversibly inhibits Ca^{2+}-activated K+ channels in rat cerebral artery smooth muscle cells. Am J Physiol Heart Circ Physiol 278: H1883–90.
148. Liu Y, Terata K, Chai Q, Li H, Kleinman LH, Gutterman DD (2002) Peroxynitrite inhibits Ca^{2+}-activated K^+ channel activity in smooth muscle of human coronary arterioles. Circ Res 91: 1070–6.
149. Kirber MT, Ordway RW, Clapp LH, Walsh JV Jr, Singer JJ (1992) Both membrane stretch and fatty acids directly activate large conductance Ca^{2+}-activated K^+ channels in vascular smooth muscle cells. FEBS Lett 297: 24–8.
150. Roman RJ (2002) P-450 metabolites of arachidonic acid in the control of cardiovascular function. Physiol Rev 82: 131–85.
151. Clarke AL, Petrou S, Walsh JV Jr, Singer JJ (2002) Modulation of BK_{Ca} channel activity by fatty acids: structural requirements and mechanism of action. Am J Physiol Cell Physiol 283: C1441–53.
152. Gebremedhin D, Ma YH, Falck JR, Roman RJ, VanRollins M, Harder DR (1992) Mechanism of action of cerebral epoxyeicosatrienoic acids on cerebral arterial smooth muscle. Am J Physiol 263: H519–25.
153. Miura H, Gutterman DD (1998) Human coronary arteriolar dilation to arachidonic acid depends on cytochrome P-450 monooxygenase and Ca^{2+}-activated K^+ channels. Circ Res 83: 501–7.
154. Morin C, Sirois M, Echave V, Gomes MM, Rousseau E (2007) Functional effects of 20-HETE on human bronchi: hyperpolarization and relaxation due to BK_{Ca} channel activation. Am J Physiol Lung Cell Mol Physiol 293: L1037–44.
155. Zou AP, Fleming JT, Falck JR, Jacobs ER, Gebremedhin D, Harder DR, Roman RJ (1996) 20-HETE is an endogenous inhibitor of the large-conductance Ca^{2+}-activated K^+ channel in renal arterioles. Am J Physiol 270: R228–37.
156. Lange A, Gebremedhin D, Narayanan J, Harder D (1997) 20-Hydroxyeicosatetraenoic acid-induced vasoconstriction and inhibition of potassium current in cerebral vascular smooth muscle is dependent on activation of protein kinase C. J Biol Chem 272: 27345–52.
157. Gebremedhin D, Yamaura K, Harder DR (2008) Role of 20-HETE in the hypoxia-induced activation of Ca^{2+}-activated K^+ channel currents in rat cerebral arterial muscle cells. J Physiol Heart Circ Physiol 294: H107–20.

157a. Crowley JJ, Treistman SN, Dopico AM (2003) Cholesterol antagonizes ethanol potentiation of human brain BK_{Ca} channels reconstituted into phospholipid bilayers. Mol Pharmacol 64: 365–72.
157b. Wang XL, Ye D, Peterson TE, Cao S, Shah VH, Katusic ZS, Sieck GC, Lee HC (2005) Caveolae targeting and regulation of large conductance Ca^{2+}-activated K^+ channels in vascular endothelial cells. J Biol Chem 280: 11656–64.
157c. Riddle MA, Hughes JM, Walker BR (2011) Role of caveolin-1 in endothelial BK_{Ca} channel regulation of vasoreactivity. Am J Physiol Cell Physiol 301: C1404–14.
157d. Dopico AM, Walsh JV Jr, Singer JJ (2002) Natural bile acids and synthetic analogues modulate large conductance Ca^{2+}-activated K^+ (BK_{Ca}) channel activity in smooth muscle cells. J Gen Physiol 119: 251–73.
158. Mitchell JA, Ali F, Bailey L, Moreno L, Harrington LS (2008) Role of nitric oxide and prostacyclin as vasoactive hormones released by the endothelium. Exp Physiol 9: 141–7.
159. Tare M, Parkington HC, Coleman HA, Neild TO, Dusting GJ (1990) Hyperpolarization and relaxation of arterial smooth muscle caused by nitric oxide derived from the endothelium. Nature 346: 69–71.
160. Zhang F, Kaide JI, Rodriguez-Mulero F, Abraham NG, Nasjletti A (2001) Vasoregulatory function of the heme-heme oxygenase-carbon monoxide system. Am J Hypertens 14: 62S–67S.
161. Wang R, Wu L, Wang Z (1997) The direct effect of carbon monoxide on K_{Ca} channels in vascular smooth muscle cells. Pflugers Arch 434: 285–91.
162. Jaggar JH, Leffler CW, Cheranov SY, Tcheranova D, E S, Cheng X (2002) Carbon monoxide dilates cerebral arterioles by enhancing the coupling of Ca^{2+} sparks to Ca^{2+}-activated K^+ channels. Circ Res 91: 610–7.
163. Xi Q, Tcheranova D, Parfenova H, Horowitz B, Leffler CW, Jaggar JH (2004) Carbon monoxide activates K_{Ca} channels in newborn arteriole smooth muscle cells by increasing apparent Ca^{2+} sensitivity of α-subunits. Am J Physiol Heart Circ Physiol 286: H610–8.
164. Hou S, Xu R, Heinemann SH, Hoshi T (2008) The RCK1 high-affinity Ca^{2+} sensor confers carbon monoxide sensitivity to Slo1 BK channels. Proc Natl Acad Sci USA. 105: 4039-43.
165. Telezhkin V, Brazier SP, Mears R, Müller CT, Riccardi D, Kemp PJ (2011) Cysteine residue 911 in C-terminal tail of human BK_{Ca} channel α subunit is crucial for its activation by carbon monoxide. Pflugers Arch 461: 665–75.
166. Wray S, Smith RD (2004) Mechanisms of action of pH-induced effects on vascular smooth muscle. Mol Cell Biochem 263: 163–72.
167. Hayabuchi Y, Nakaya Y, Matsuoka S, Kuroda Y (1998) Effect of acidosis on Ca^{2+}-activated K^+ channels in cultured porcine coronary artery smooth muscle cells. Pflugers Arch 436: 509–14.
168. Park JK, Kim YC, Sim JH, Choi MY, Choi W, Hwang KK, Cho MC, Kim KW, Lim SW, Lee SJ (2007) Regulation of membrane excitability by intracellular pH (pHi) changers through Ca^{2+}-activated K^+ current (BK channel) in single smooth muscle cells from rabbit basilar artery. Pflugers Arch 454: 307–19.
169. Schubert R, Krien U, Gagov H (2001) Protons inhibit the BK_{Ca} channel of rat small artery smooth muscle cells. J Vasc Res 38: 30–8.
170. Brelidze TI, Magleby KL (2004) Protons block BK channels by competitive inhibition with K^+ and contribute to the limits of unitary currents at high voltages. J Gen Physiol 123: 305–19.
171. Hu XQ, Xiao D, Zhu R, Huang X, Yang S, Wilson S, Zhang L (2011) Pregnancy upregulates large-conductance Ca^{2+}-activated K^+ channel activity and attenuates myogenic tone in uterine arteries. Hypertension 58: 1132–9.
172. Rosenfeld CR, Liu XT, DeSpain K (2009) Pregnancy modifies the large conductance Ca^{2+}-activated K^+ channel and cGMP-dependent signaling pathway in uterine vascular smooth muscle. Am J Physiol Heart Circ Physiol 296: H1878–87.

173. Magness RR, Shaw CE, Phernetton TM, Zheng J, Bird IM (1 Magness RR, Shaw CE, Phernetton TM, Zheng J, Bird IM (1997) Endothelial vasodilator production by uterine and systemic arteries. II. Pregnancy effects on NO synthase expression. Am J Physiol 272: H1730–40.
174. Khan LH, Rosenfeld CR, Liu XT, Magness RR (2010) Regulation of the cGMP-cPKG pathway and large-conductance Ca^{2+}-activated K^+ channels in uterine arteries during the ovine ovarian cycle Am J Physiol Endocrinol Metab 298: E222–8.
175. van Drongelen J, Ploemen IH, Pertijs J, Gooi JH, Sweep FC, Lotgering FK, Spaanderman ME, Smits P (2011) Am J Physiol Heart Circ Physiol 300: H1609–15.
176. Pang L, Rusch NJ (2009) High-conductance, Ca^{2+}-activated K^+ channels: altered expression profiles in aging and cardiovascular disease. Mol Interv 9: 230–3.
177. Nishimaru K, Eghbali M, Lu R, Marijic J, Stefani E, Toro L (2004) Functional and molecular evidence of MaxiK channel β1 subunit decrease with coronary artery ageing in the rat. J Physiol 559: 849–62.
178. Albarwani S, Al-Siyabi S, Baomar H, Hassan MO (2010) Exercise training attenuates ageing-induced BK_{Ca} channel downregulation in rat coronary arteries. Exp Physiol 95: 746–55.
179. Marijic J, Li Q, Song M, Nishimaru K, Stefani E, Toro L (2001) Decreased expression of voltage- and Ca^{2+}-activated K^+ channels in coronary smooth muscle during aging. Circ Res 88: 210–6.
180. Cornfield DN, Reeve HL, Tolarova S, Weir EK, Archer S (1996) Oxygen causes fetal pulmonary vasodilation through activation of a calcium-dependent potassium channel. Proc Natl Acad Sci USA 93: 8089–94.
181. Hong Z, Weir EK, Varghese A, Olschewski A (2005) Effect of normoxia and hypoxia on K^+ current and resting membrane potential of fetal rabbit pulmonary artery smooth muscle. Physiol Res 54: 175–84.
181a. Snetkov VA, Pandya H, Hirst SJ, Ward JP (1998) Potassium channels in human fetal airway smooth muscle cells. Pediatr Res 43: 548–54.
182. Shimoda LA, Polak J (2011) Hypoxia. 4. Hypoxia and ion channel function. Am J Physiol Cell Physiol 300: C951–67.
183. Post JM, Hume JR, Archer SL, Weir EK (1992) Direct role for potassium channel inhibition in hypoxic pulmonary vasoconstriction. Am J Physiol 262: C882–90.
184. Vandier C, Delpech M, Bonnet P (1998) Spontaneous transient outward currents and delayed rectifier K^+ current: effects of hypoxia. Am J Physiol 275: L145–54.
185. Gebremedhin D, Bonnet P, Greene AS, England SK, Rusch NJ, Lombard JH, Harder DR (1994) Hypoxia increases the activity of Ca^{2+}-sensitive K^+ channels in cat cerebral arterial muscle cell membranes. Pflugers Arch 428: 621–30.
186. Peng W, Hoidal JR, Karwande SV, Farrukh IS (1997) Effect of chronic hypoxia on K^+ channels: regulation in human pulmonary vascular smooth muscle cells. Am J Physiol 272: C1271–8.
187. Dubuis E, Potier M, Wang R, Vandier C (2005) Continuous inhalation of carbon monoxide attenuates hypoxic pulmonary hypertension development presumably through activation of BK_{Ca} channels. Cardiovasc Res 65: 751–61.
188. Reeve HL, Weir EK, Archer SL, Cornfield DN (1998) A maturational shift in pulmonary K^+ channels, from Ca^{2+} sensitive to voltage dependent. Am J Physiol 275: L1019–25.
189. Bonnet S, Dumas-de-La-Roque E, Bégueret H, Marthan R, Fayon M, Dos Santos P, Savineau JP, Baulieu EE (2003) Dehydroepiandrosterone (DHEA) prevents and reverses chronic hypoxic pulmonary hypertension. Proc Natl Acad Sci USA 100: 9488–93.
190. Navarro-Antolín J, Levitsky KL, Calderón E, Ordóñez A, López-Barneo J (2005) Decreased expression of maxi-K^+ channel β1-subunit and altered vasoregulation in hypoxia. Circulation 112: 1309–15.

191. Chen SJ, Wu CC, Yang SN, Lin CI, Yen MH (2000) Abnormal activation of K^+ channels in aortic smooth muscle of rats with endotoxic shock: electrophysiological and functional evidence. Br J Pharmacol 131: 213–22.
192. Zhao G, Zhao Y, Pan B, Liu J, Huang X, Zhang X, Cao C, Hou N, Wu C, Zhao KS, Cheng H (2007) Hypersensitivity of BK_{Ca} to Ca^{2+} sparks underlies hyporeactivity of arterial smooth muscle in shock. Circ Res 101:493–502.
193. Hoang LM, Mathers DA (1998) Internally applied endotoxin and the activation of BK channels in cerebral artery smooth muscle via a nitric oxide-like pathway. Br J Pharmacol 123: 5–12.
194. Amberg GC, Santana LF (2003) Downregulation of the BK channel β1 subunit in genetic hypertension. Circ Res 93: 965–71.
195. Xu H, Garver H, Galligan JJ, Fink GD (2011) Large-conductance Ca^{2+}-activated K^+ channel β1-subunit knockout mice are not hypertensive. Am J Physiol Heart Circ Physiol 300: H476–85.
196. Bratz IN, Dick GM, Partridge LD, Kanagy NL (2005) Reduced molecular expression of K^+ channel proteins in vascular smooth muscle from rats made hypertensive with N^{ω}-nitro-L-arginine. Am J Physiol Heart Circ Physiol 289: H1277–83.
197. Rusch NJ, De Lucena RG, Wooldridge TA, England SK, Cowley AW Jr (1992) A Ca^{2+}-dependent K^+ current is enhanced in arterial membranes of hypertensive rats. Hypertension 19: 301–7.
198. Kiyoshi H, Yamazaki D, Ohya S, Kitsukawa M, Muraki K, Saito SY, Ohizumi Y, Imaizumi Y (2006) Molecular and electrophysiological characteristics of K^+ conductance sensitive to acidic pH in aortic smooth muscle cells of WKY and SHR. Am J Physiol Heart Circ Physiol 291: H2723–34.
199. Liu Y, Pleyte K, Knaus HG, Rusch NJ (1997) Increased expression of Ca^{2+}-sensitive K^+ channels in aorta of hypertensive rats. Hypertension 30: 1403–9.
200. Xu H, Bian X, Watts SW, Hlavacova A (2005) Activation of vascular BK channel by tempol in DOCA-salt hypertensive rats. Hypertension 46: 1154–62.
201. Dong L, Zheng YM, Van Riper D, Rathore R, Liu QH, Singer HA, Wang YX (2008) Functional and molecular evidence for impairment of calcium-activated potassium channels in type-1 diabetic cerebral artery smooth muscle cells. J Cereb Blood Flow Metab 28: 377–86.
202. Jarajapu YP, Guberski DL, Grant MB, Knot HJ (2008) Myogenic tone and reactivity of cerebral arteries in type II diabetic BBZDR/Wor rat. Eur J Pharmacol 579: 298–307.
203. Burnham MP, Johnson IT, Weston AH (2006) Reduced Ca^{2+}-dependent activation of large-conductance Ca^{2+}-activated K^+ channels from arteries of Type 2 diabetic Zucker diabetic fatty rats. Am J Physiol Heart Circ Physiol 290: H1520–7.
204. Wang Y, Zhang HT, Su XL, Deng XL, Yuan BX, Zhang W, Wang XF, Yang YB (2010) Experimental diabetes mellitus down-regulates large-conductance Ca^{2+}-activated K^+ channels in cerebral artery smooth muscle and alters functional conductance. Curr Neurovasc 7: 75–84.
205. Mori A, Suzuki S, Sakamoto K, Nakahara T, Ishii K (2011) Vasodilation of retinal arterioles induced by activation of BK_{Ca} channels is attenuated in diabetic rats. Eur J Pharmacol 669: 94–9.
206. Dimitropoulou C, Han G, Miller AW, Molero M, Fuchs LC, White RE, Carrier GO (2002) Potassium (BK_{Ca}) currents are reduced in microvascular smooth muscle cells from insulin-resistant rats. Am J Physiol Heart Circ Physiol 282: H908–17.
207. Lu T, Ye D, He T, Wang XL, Wang HL, Lee HC (2008) Impaired Ca^{2+}-dependent activation of large-conductance Ca^{2+}-activated K^+ channels in the coronary artery smooth muscle cells of Zucker Diabetic Fatty rats. Biophys J. Dec;95(11):5165–77.
208. McGahon MK, Dash DP, Arora A, Wall N, Dawicki J, Simpson DA, Scholfield CN, McGeown JG, Curtis TM (2007) Diabetes downregulates large-conductance Ca^{2+}-activated potassium β 1 channel subunit in retinal arteriolar smooth muscle. Circ Res 100: 703–11.

209. Zhang DM, He T, Katusic ZS, Lee HC, Lu T (2010) Muscle-specific f-box only proteins facilitate BK channel β1 subunit downregulation in vascular smooth muscle cells of diabetes mellitus. Circ Res 107:1454–9.
210. Dong L, Xie MJ, Zhang P, Ji LL, Liu WC, Dong MQ, Gao F (2009) Rotenone partially reverses decreased BK_{Ca} currents in cerebral artery smooth muscle cells from streptozotocin-induced diabetic mice. Clin Exp Pharmacol Physiol 36: E57–64.
211. Tang XD, Garcia ML, Heinemann SH, Hoshi T (2004) Reactive oxygen species impair Slo1 BK channel function by altering cysteine-mediated calcium sensing. Nat Struct Mol Biol 11: 171–8.
212. Martin G, O'Connell RJ, Pietrzykowski AZ, Treistman SN, Ethier MF, Madison JM (2008) Interleukin-4 activates large-conductance, calcium-activated potassium (BK_{Ca}) channels in human airway smooth muscle cells. Exp Physiol 93: 908–18.

Calcium-Activated Chloride Channels

George Gallos and Charles W. Emala Sr.

Contents

Abstract Emerging functional evidence demonstrates the importance of membrane potential in the regulation of many intracellular signaling processes. The efflux of chloride through the plasma membrane has been identified as a major contributor to plasma membrane depolarization in airway smooth muscle. Early studies demonstrated that calcium arising from intracellular sources and released by ryanodine or IP3 receptor activation on the sarcoplasmic-reticulum-induced plasma membrane chloride currents. Moreover, external calcium entry through voltage-dependent calcium channels was shown to augment calcium-activated chloride currents. One of the earliest studies in this area suggested a role for chloride influencing uptake and release of calcium from the sarcoplasmic reticulum in addition to chloride flux's effect on plasma membrane electrical potential. Recently the elusive proteins responsible for calcium-activated chloride currents in many

---

G. Gallos • C.W. Emala Sr. (✉)
Department of Anesthesiology, Columbia University, 622 West 168th Street, PH 505, New York, NY 10032, USA
e-mail: cwe5@cume.columbia.edu

Y.-X. Wang (ed.), *Calcium Signaling In Airway Smooth Muscle Cells*,
DOI 10.1007/978-3-319-01312-1_5,

cells (TMEM16/anoctamin family) were cloned, which has renewed interest in the field of calcium-activated chloride channels (CaCCs). Anoctamin 1 has been identified on the apical side of airway epithelium, is critical to fluid secretion, and has been associated with increased mucin secretion in asthmatics. Anoctamin 1 is critical to the development of the trachea as global knockout mice exhibit severe tracheomalacia. Anoctamin 1 has been immunochemically localized to airway smooth muscle and human bronchi were shown to contract less effectively in the presence of benzbromarone, an antagonist of these channels. Studies of the TMEM16/anoctamin family of CaCCs are revolutionizing the understanding of calcium-activated chloride currents in many cell types, and emerging evidence suggests that this channel also contributes to the regulation of airway smooth muscle tone.

Keywords TMEM16/Anoctamin • Electrophysiology • Niflumic acid • Benzbromarone • Membrane potential • Sarcoplasmic reticulum

1 Introduction

Emerging functional evidence is demonstrating the importance of membrane potential in the regulation of many intracellular signaling processes. While the contribution of K^+ and Ca^{2+} cation channels to membrane potential and airway smooth muscle function has been extensively studied, the contribution of anion (i.e., Cl^-) channels has received less scrutiny. In particular, although the efflux of chloride through the plasma membrane has been identified as the major contributor to plasma membrane depolarization, the fundamental importance of chloride flux to airway smooth muscle membrane potential and contractile tone has been questioned [1, 2]. However, recent advances in molecular cloning identifying the elusive protein responsible for calcium-activated chloride currents has renewed interest in the field of calcium-activated chloride channels (CaCCs). As such, there is new emerging functional evidence that suggests these channels may contribute to contractile tone in airway smooth muscle.

2 Background: Importance of Membrane Potential and Chloride Flux in Airway Smooth Muscle Function

The importance of plasma membrane chloride ion channels in airway smooth muscle must be considered in the context of the contribution of membrane electrical potential and ion flux to the control of contractile function. Skepticism concerning the relative importance of membrane potential to contractile force originated in part from early work directed at voltage-dependent L-type calcium channels, one of the earliest discovered voltage-dependent channels in airway smooth muscle [3]. Unlike vascular smooth muscle, dihydropyridine therapies directed against this channel ultimately proved ineffective at treating asthma [4, 5] despite early promising effects in exercise-induced asthma [6–10] and ex vivo studies in human airway smooth muscle [11]. This dismissal of the therapeutic potential of the voltage-sensitive calcium channel occurred despite the accepted principle that external calcium entry is required to maintain contraction (likely via both voltage- gated and store-operated calcium entry). This external calcium is required to refill internal stores and perhaps amplify the intracellular calcium concentrations achieved by calcium release from IP3 receptor and ryanodine-receptor-mediated SR stores. Moreover, extensive ex vivo functional (organ bath) [12] and electrophysiologic [13] evidence demonstrates a role for voltage-sensitive calcium channels in both function and membrane potential gating in airway smooth muscle. Perhaps the failure of therapy directed toward this voltage-gated calcium channel in asthma was due to the relative contribution of intracellular calcium stores versus extracellular calcium in the initiation of airway smooth muscle contraction [14].

Decades of research has demonstrated that airway smooth muscle is different from other smooth muscle beds (e.g., vascular) in terms of the relative importance of membrane potential and coupling to voltage-sensitive ion channels. For example, cholinergic contraction occurs despite blockade of voltage-sensitive calcium channels [15] or chloride channels [16] and antagonists of voltage-dependent calcium channels were ineffective in clinical trials in asthma [4, 5]. However, this does not mean that membrane potential is not important in airway smooth muscle, particularly during the maintenance phase of a contraction. In an elegant review, Hirota et al. suggest that depolarization to −40 mV is optimal for a persistent low-level, voltage-dependent calcium influx and for calcium entry via the reverse mode of the $Na+/Ca^2+$ exchanger [17]. Moreover, the true effect of membrane potential changes may be upon receptor and second messenger systems not directly linked to ion channels.

At least six findings continue to support a mechanistic importance for membrane potential in the control of airway smooth muscle tone: (a) a component of β-*agonist* relaxation of airway smooth muscle involves K^+ efflux through calcium-activated potassium channels (K_{Ca}) inducing relative plasma membrane hyperpolarization [18]; (b) refilling of sarcoplasmic reticulum (SR) with Ca^{2+} following a classic contractile agonist exposure involves opening of voltage-sensitive plasma membrane Ca^{2+} channels [stored-operated calcium entry (SOCE)] likely including nonselective cation channels of the transient receptor potential (TRP) family of which both TRP-C [19] and TRP-V [20] members were recently identified on airway smooth muscle; (c) the recent discovery that bitter taste receptors relax airway smooth muscle despite an increase in intracellular Ca^{2+} through a hypothesized opening of K_{Ca} (a membrane hyperpolarizing event) [21]; (d) changes in membrane potential have been shown to activate both M3 muscarinic receptors (Gq-coupled) in airway smooth muscle cells [22] and M2 muscarinic receptors [23, 24] (Gi-coupled) independent of receptor occupancy by ligand; (e) T-type calcium channels [25] that are activated within membrane potential ranges achieved during agonist-induced airway smooth muscle contraction and depolarization have been characterized in airway smooth muscle [15]; (f) our laboratory recently discovered the expression of both $GABA_A$ [26] and glycine [27] chloride channels on airway smooth muscle that modulate maintenance of contractile tone.

Moreover, evidence that inositol phosphates can in turn regulate membrane potential [28, 29] and that IP release of SR calcium can activate plasma membrane chloride channels suggests that second messenger regulation of intracellular SR calcium release and plasma membrane potential do not operate in isolation. Furthermore, membrane potential regulation of Rho kinase activity has been described. Rho kinase-mediated deactivation of myosin light-chain phosphatase increases contractile sensitivity to cytoplasmic calcium influx regardless of whether the calcium arises from extracellular or intracellular sources. Depolarization induced by electric field stimulation [30] or pharmacologically (i.e., potassium chloride; [31]) has demonstrated increases in Rho kinase activity and increases in muscle force generation. Thus, depolarization of the cytoplasmic membrane not only has a traditional role in plasma membrane ion channel activation but also regulates key G-protein-coupled intracellular calcium pathways and phosphorylation of contractile proteins sensitive to calcium, resulting in enhanced muscle force generation.

3 Families of Chloride Channels and Functional Heterogeneity

Perhaps the most widely studied plasma membrane chloride channel is the ligand-gated $GABA_A$ channel, which mediates the predominant inhibitory ion current in mature neurons. However, proteins that move chloride (including channels, transporters, exchangers, and even secreted proteins), represent a wide array of proteins that play critical roles in a myriad of physiologic processes. At least eight families of chloride channels have been identified as membrane or intracellular chloride channels/binding proteins; they include the ligand-gated chloride channels (e.g., $GABA_A$ and glycine), cystic fibrosis transmembrane conductance regulator (CFTR), CLC [32], bestrophins [33], calcium-activated chloride channel regulator (CLCA) [34], chloride intracellular channel (CLIC), Tweety, and the most recently characterized TMEM16/anoctamin family. However, only four of these chloride channel families contain members regulated by calcium (CLC, bestrophins, anoctamins, and CLCA) [35] and members of only one family (anoctamin) contain members (ANO1 and ANO2) that display the electrophysiologic properties of calcium and voltage regulation long attributed to CaCCs. While a great deal of research has focused on these protein families, there still exists considerable discrepancy between features observed experimentally and the molecular identity of a given family. This apparent discrepancy exists because many assigned members of these families have poorly defined functions as chloride channels. CLIC are intracellular chloride channels, bestrophin-1 may be an SR chloride channel coupling to STIM and Orai-1 proteins modulating SOCE [36], and some of the CLCA members are soluble and secreted proteins [37]. This seeming paradox exists in part because advances in electrophysiology preceded and have outpaced molecular cloning efforts. Thus, while sophisticated electrophysiology helped characterize broad differences in gating and kinetic properties, pharmacological responsiveness, and variable modes of regulation of what is presumed to be distinct chloride channel families, the actual molecular identity for many of these chloride channel families remains unclear.

4 Electrophysiologic and Functional Identity of Calcium-Activated Chloride Channels

One particular class of chloride channels that has proved particularly elusive and has been the source of considerable controversy involves the CaCC family. Despite its clear involvement in several important physiological functions (including olfaction, smooth muscle contraction, photoreception, and fluid secretion), many attempts to accurately identify the protein family that possessed all of the observed electrophysiologic features associated with the classic description of a Ca^{2+}-activated Cl^- channel family have failed. The original description of a CaCC dates back to the early 1980s. In 1982, Miledi et al. described an outward chloride current in *Xenopus laevis* oocytes [38]. They demonstrated that under voltage clamp conditions, when a voltage ramp was applied, an outward current appeared at approximately −20 mV. They also demonstrated that while reducing external calcium did diminish this outward current, increasing intracellular calcium greatly enhanced the current. In the intervening year after this initial publication, both [39] and [40] extended these electrophysiological findings to show antagonism of the current under calcium chelation or competitive ionic antagonism. This work established certain electrophysiological criteria for CaCCs, namely, that activation of the observed chloride current occurred under induced membrane potential changes and that intracellular Ca^{2+} concentrations played a critical role in enhancing the current. Subsequent studies also established that CaCCs display a sensitivity to Cl^- channel blockers, show small channel conductances, and exhibit a voltage-dependent Ca^{2+} sensitivity [41] and a preferential halide flux that follows the sequence $I^- > Br^- > Cl^- > F^-$ [42]. Time and again molecular candidates were put forward that failed to fully satisfy these electrophysiologic criteria [43, 44].

Perhaps the most unique feature from this list of attributes concerns characteristics related to CaCC calcium sensitivity and kinetics. With regard to calcium sensitivity, many studies have illustrated that CaCC currents can be evoked when intracellular Ca^{2+} ranges from 0.2 to 5 uM [39, 45]. However, the kinetics displayed vary depending on the concentration of intracellular Ca^{2+}, with a notably slow activation time required to achieve steady state at lower concentrations (approximately 2 s when intracellular Ca^{2+} concentrations are less than 1 uM). This intracellular Ca^{2+} threshold also exhibits coupling to voltage, where CaCC current is voltage-dependent when the intracellular Ca^{2+} concentration is less than 1 uM. Furthermore, this current/voltage (I/V) relationship exhibits outward rectification at intracellular Ca^{2+} of less than 0.5 uM, which greatly diminishes when intracellular Ca^{2+} concentrations exceed 1 uM [40].

In addition, over the years, several other studies followed that have expanded our knowledge concerning the modulation and regulation of CaCC currents. There is convincing evidence that CaCC enhancement of current is feasible from both direct and indirect effects of calcium. These earlier studies employed three different approaches: lysed cell patch electrophysiology, enzymatic blockade of calcium-/calmodulin-dependent protein kinase type II (CaMKII), or photo-induced release of caged calcium. Operating on the premise that "excised cell" or "inside-out" membrane preparations are devoid of ATP, these studies examined whether CaCC activity displayed any attenuation over time, as would be expected if calcium-mediated activation was dependent upon a modulating enzymatic process. Given the observed maintenance of calcium-induced chloride current in these specialized electrophysiologic membrane preparations, it appears that at least in cardiac muscle [46], mesangial kidney [47], pulmonary endothelium [48], salivary glands [49], and liver cells [50], there is evidence for direct gating. Confirmatory findings in support of direct gating by calcium were also observed in rat parotid cells where CaCC currents induced by ionomycin were maintained under conditions of CaMKII or calmodulin blockade [51] as well as in acinar cells, where local apical release of caged Ca^{2+} led to immediate CaCC current [52].

Despite strong support for the idea that calcium has a direct effect on CaCC activation, there is also compelling evidence that in some tissues CaCC activity is also regulated in part by other second messengers. The majority of this work has focused on the role of phosphorylation induced by CaMKII. In studies performed by Wang et al. using equine airway myocytes, these investigators observed discordance in decay rates of CaCC current induced by caffeine-mediated calcium levels in the presence or absence of dialyzed ATP. Using nonhydrolyzable ATP analogs and direct and indirect inhibitors of CaMKII, these investigators provided evidence that CaMKII phosphorylation of CaCC results in the inactivation of these channels [53]. Furthermore, by employing a separate group of studies under sustained elevated calcium release mediated by ionomycin (in the presence or absence of phosphorylation inhibition), they uncovered dual mechanisms affecting CaCC closed-state probability: (1) a closed state is induced by reductions in intracellular Ca^{2+} (in the absence of CaMKII activity) and (2) CaMKII phosphorylation of CaCC leads to a closed state (in the absence of a decline in intracellular Ca^{2+}). Interestingly, in other tissues CaMKII activity has been shown to be a positive activator of CaCC currents. Tissues in which CaMKII has been shown to enhance CaCC current include epithelium (from the airway [54], pancreas[55], and biliary ducts [56]), human immune cells (T-cells [57], macrophages [58]), and colonic tumor cells [59, 60]. These studies employed either CaMKII inhibitors (showing reductions in current) or, alternatively, showed that dialyzing purified preparations of CaMKII led to an enhancement of CaCC current. Interestingly, variability of enzymatic regulation as a function of tissue type was also illustrated in a study by Arreola et al., which showed differential effects of CaMKII blockade on CaCC currents obtained in T84 colonic tumor cells as compared with those found in parotid acinar cells [51]. While CaCC current was suppressed by either direct CaMKII inhibitors (KN-62) or by using a calmodulin-binding domain peptide (290–309) in T84 cells, these approaches failed to suppress CaCC current in parotid cells. The variability among these studies suggests that either different protein candidates (e.g., anoctamin 1 versus anoctamin 2) or splice variants in a given protein (e.g., splice forms of anoctamin 1 [61]) likely underlie this disparity in experimental observations.

5 Role of Chloride Flux in Airway Smooth Muscle

While calcium-activated chloride currents have been measured in airway smooth muscle cells since the early 1990s, incriminating CaCCs in the control of airway smooth muscle tone has historically been hampered by limitations in the specificity of pharmacologic reagents. In an excellent review by Large et al., the inherent lack of specificity of many reagents touted as CaCC blockers (e.g., NPPB, niflumic acid, etc.) is illustrated [62]. The lack of specific CaCC antagonists has limited the enthusiasm for research targeting CaCCs and other Cl^- channels and has been a major reason why anion-directed research has being neglected for so long. Yet, the functional importance of these currents for smooth muscle function and the recent molecular identity of the proteins responsible for these currents warrant renewed inquiry into the fundamental roles that CaCCs play in airway smooth muscle physiology.

Early studies demonstrated that calcium arising from intracellular sources and released by ryanodine or IP3 receptor activation on the SR induced plasma membrane chloride currents. Moreover, external calcium entry through voltage-dependent calcium channels was shown to augment calcium-activated chloride currents [13]. However, even one of the earliest studies in this area suggested a role for chloride influencing uptake and release of calcium from the SR in addition to chloride flux's effect on plasma membrane electrical potential [16]. Although acetylcholine has been the most commonly studied Gq-coupled ligand inducing chloride flux in airway smooth muscle [16, 63–66], histamine [67] and substance P [68] also induce chloride currents. Calcium-induced chloride currents have been demonstrated in canine [16], porcine [66], equine [64, 69], guinea pig [16, 70], murine [71], and human [63] airway smooth muscle. A key cellular signaling event responsible for the termination of these chloride currents in airway smooth muscle is phosphorylation mediated by Ca^{2+}/calmodulin-dependent kinase II [53].

The resting membrane potential of airway smooth muscle is approximately −60 mV, and since the chloride equilibrium potential (E_{Cl}) is in the range of −40 to −20 mV [72], the opening of a chloride channel at resting membrane potential

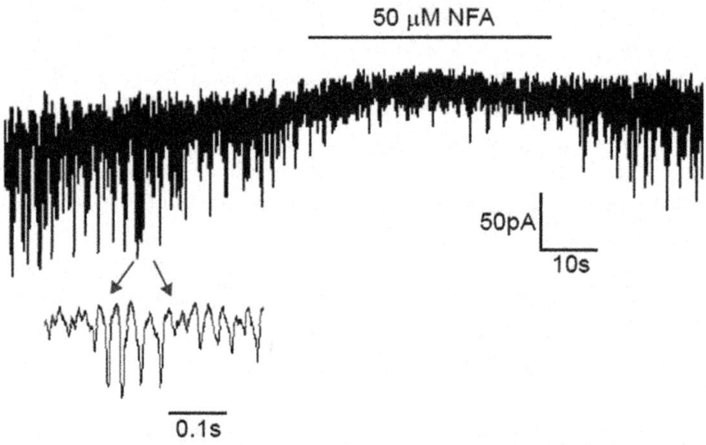

Fig. 1 Representative tracing of spontaneous transient inward current attributed to calcium-activated chloride channels in human airway smooth muscle cell under current clamp at ¯60 mV. This spontaneous current is reversibly inhibited by the chloride channel blocker niflumic acid (50 uM)

would lead to chloride efflux and contribute to plasma membrane depolarization [63]. If the cell has depolarized to membrane potentials more positive than E_{Cl}, then channel opening would allow for chloride influx and a relative membrane hyperpolarization.

Chloride channels have been shown to play a central role in airway smooth muscle electrophysiologic events including spontaneous transient inward currents (Fig. 1), [16] depolarization-induced oscillations in membrane potential [73], and in some cases associated with oscillations of intracellular calcium [66] mediated by inositol phosphate receptors located on the SR [74]. The depolarization phase of these slow waves is mediated by voltage-dependent calcium channels while the repolarization phase is mediated by CaCCs [75]. These electrical slow waves in airway smooth muscle are remarkably similar to electrical slow waves in gut smooth muscle, which are thought to underlie phasic contractions. Specialized cells, known as interstitial cells of Cajal, are interposed between nerve terminals and gut smooth muscle cells and are the origin of this spontaneous electrical activity [76]. The anoctamin 1 CaCCs are critical to the generation of this slow wave activity from cells of Cajal [77]. In airway smooth muscle electrophysiological recordings at resting membrane potential (−60 mV) have demonstrated transient spontaneous outward currents [thought to be mediated by calcium-activated potassium (K_{Ca}) channels] alternating with transient spontaneous inward currents (STICs) thought to be activated by calcium-activated chloride currents [78], both of which are increased by intracellular calcium. In mouse tracheal myocytes, STICs were shown to be spatiotemporally regulated by calcium sparks that activate chloride channels located in close proximity to ryanodine receptors on the SR [71].

While a large amount of research has focused on the electrophysiologic effects of calcium regulation of chloride flux at the plasma membrane, considerably less is known regarding the contribution of these currents to excitation-contraction coupling. Acetylcholine contractions were shown to occur despite pretreatment with the chloride channel blocker niflumic acid in isolated guinea pig tracheal myocytes. However, the proportion of cells that could repetitively contract in response to acetylcholine was influenced by the intracellular concentration of chloride [16]. In a subsequent study these same authors demonstrated that STICs (attributed to chloride channels) were associated with rhythmic contractile behavior in a limited number of isolated guinea pig or canine tracheal smooth muscle cells studied at 25 °C [78]. Intact porcine or bovine tracheal smooth muscle pretreated with two different chloride channel blockers, niflumic acid or NPPB, contracted normally to subsequent carbachol-induced contractions, but repetitive acetylcholine-induced contractions were inhibited by NPPB but not niflumic acid [12]. Moreover, while both nifedipine and niflumic acid suppressed caffeine-induced contractile responses, implicating a role for Ca^{2+} influx in these responses, NPPB was more potent than niflumic acid in suppressing caffeine-induced contractions.

Although the role of plasma membrane chloride channels in the electrophysiological events of STICs and membrane electrical slow waves has been studied, considerably less attention has focused on a possible role of chloride channels in intracellular organelles. The movement of calcium into and out of the SR is electrogenic requiring a commensurate neutralization of the potential difference. Chloride is a candidate anion that could move to neutralize the potential charge generation occurring with calcium movement as demonstrated by the depletion of intracellular chloride resulting in impairment in repetitive acetylcholine contractions in intact airway smooth muscle [12]. Indeed calcium movement into the SR of permeabilized gastric smooth muscle was inhibited by NPPB [79]. The identification of chloride channels on the SR of airway smooth muscle has not been directly demonstrated, but if such a channel needs to balance the charge generation of calcium movement, a chloride channel stimulated by calcium (a CaCC) would make functional sense.

Members of the CLCA family of chloride channels were initially postulated to be the chloride channels in airway smooth muscle responsible for calcium-induced chloride currents. The mouse CLCA4 channel is expressed in a high percentage of smooth muscle cells from many organs including bronchi, and heterologous expression of this channel resulted in robust calcium-activated chloride currents [80]. Another member of this family (mCLCA3/hCLCA1) generated a great amount of research interest since it was upregulated in mouse models of asthma [81], and hCLCA1 is upregulated in the airway epithelium of asthmatics [82–84]. Genetic reduction of these channels by antisense [85] or pharmacologic blockade with niflumic acid [86] prevented allergen-induced mucus overproduction and airway hyperresponsiveness. However, deletion of mCLCA3 did not change airway hyperresponsiveness, and this protein has been discovered to be a secreted protein rather than a membrane protein functioning as a chloride channel [37]. Thus, despite decades of electrophysiologic and functional evidence for a role of CaCCs in airway smooth muscle, the molecular identity of this channel remained elusive.

6 Revolution in Molecular Identity of TMEM16A (Anoctamin1, DOG1) in 2008

Since the 1980s, when the first description of CaCC currents was published, a series of attempts to correlate potential chloride channel candidates with the experimentally observed electrophysiologic characteristics followed. However, since CaCCs play a fundamental role in fertilization to prevent polyspermia, the *Xenopus* oocyte proved an unacceptable expression system and therefore delayed cloning efforts. Yet even after adopting other cloning expression systems the search for the putative CaCC protein family spanned many years as several investigators sought to incriminate the CLCA, CLC-3, bestrophin, and 20 family members [87]. Unfortunately, in reconstituted cell systems these proteins did not demonstrate requisite activation under induced membrane potential changes or an outward rectifying current at submaximal Ca^{2+} concentrations. Within the airway smooth muscle literature, these studies also created considerable controversy as to the putative channel responsible for the chloride-mediated contribution to contractile force reported during an airway smooth muscle contraction [12]. This also paralleled extensive electrophysiologic studies in airway smooth muscle showing "classic" calcium-induced chloride efflux contributing to the depolarization of the plasma membrane during the initiation phase of contraction [16, 63–66]. Furthermore, a wide variety of ligands (acetylcholine, histamine, substance P) and regulators of intracellular calcium (e.g., caffeine) were similarly shown to induce calcium-dependent chloride efflux, suggesting a significant procontractile role for CaCCs in airway smooth muscle.

In 2008, three separate laboratories independently discovered that one member of the TMEM16 or anoctamin family of proteins demonstrated electrophysiologic and pharmacologic characteristics typical of classic CaCCs, a discovery that has transformed CaCC studies in many cell types [41, 88, 89]. These three labs each took unique experimental approaches to independently arrive at the same conclusion. Extrapolating the observation that most salamander axolotl oocytes permit polyspermia (and since CaCCs are thought to be critical determinants for the suppression of polyspermia in *Xenopus* oocytes) Schroeder et al. demonstrated that no CaCC current was induced following UV flash activation of caged IP3. Using size-fractionated *Xenopus* oocyte polyA RNA these investigators employed a subtractive cloning approach to determine the fraction of *Xenopus* RNA that induced CaCC currents in axolotl oocytes. A directional cDNA library then yielded a single clone that when expressed was found to induce voltage-dependent CaCC currents, exhibit outward rectifying currents at submaximal intracellular calcium levels and a linear I/V relationship at high Ca levels, to be antagonized by classic CaCC blockers (NFA, DIDS), and demonstrate a large anion preference for flux. Equally elegant, yet experimentally distinct, Caputo et al.'s approach capitalized on the finding that bronchial epithelial cells exposed to IL-4 exhibit increased calcium-dependent chloride channel activity. Using focused microarray gene analysis limited to membrane proteins regulated by IL-4 coupled to siRNA knockdown in a pancreatic cell line with robust, endogenous CaCC activity, this group of investigators was able to incriminate anoctamin (ANO) 1 by anion flux assays utilizing halide-sensitive yellow fluorescent protein and whole-cell patch techniques that recapitulated many of the classic CaCC features. Screening gene databases for a candidate with multiple transmembrane-spanning domains (consistent with an ion channel) that also possessed several known human homologs with sequence tags from tissues known to have CaCC activity, Yang et al. used a heterologous HEK expression system to confirm that ANO1 possessed classic CaCC features. Of particular interest to the airway, they also showed convincing data that GCPR agonists (including carbachol and histamine) were capable of generating CaCC currents in HEK cells and, moreover, that the ANO1 protein was present in epithelial cells of the peripheral mouse lung.

The cloning of the ANO1 CaCC in 2008 transformed the molecular understanding of calcium regulation of chloride flux in many cell types. The anoctamin family has ten members, each of which is predicted to have eight transmembrane domains, and most (all except ANO8 and ANO10) are predicted to have a large intracellular loop between transmembrane domains 5 and 6, believed to play a role in ion selectivity [89]. ANO3–7 were shown to be expressed intracellularly, and these members failed to produce chloride currents in response to calcium in transfected HEK293 cells [90]. ANO9 and 10 depressed baseline Cl^- conductances, and coexpression of ANO1 and 9 abolished ANO1 activity in transfected thyroid cells [91]. Special features of some of the anoctamin family members include a 20-amino-acid stretch of negative charges (glutamic and aspartic acid residues) within the cytosolic loop between transmembrane domains 5 and 6, while its cytosolic carboxyl-terminus is unusually abundant in prolines (56 out of 282 total residues). ANO5 is mutated in a rare disease of bone calcification, gnathodiaphyseal dysplasia (also known as GDD1) [92], and has been linked to several limb-girdle muscular dystrophies [93–95]. ANO7 has been implicated in prostate cancer [96].

Despite the relative similarity among these ten members, it is somewhat revealing that only two members (ANO1 and 2) have clearly been shown to function as plasma-membrane-associated CaCCs [87, 97]. Transient transfection of anoctamins in HEK293 cells yield conflicting results regarding the subcellular localization of ANO4, 6, and 7 [90, 98], and some anoctamin family members function as lipid scramblases [97, 99]. There may be a phylogenetic basis to this difference in functionality since anoctamin 1 and 2 exhibit a distinct divergence from the other family members and indeed share a common branch point in the family tree [41, 100]. Given the lack of homology with other anion channels altogether, perhaps ANO1 and 2 are the molecular equivalent to what is more commonly termed in comparative biology as convergence (where the same functional trait emerges from distinct and genetically unrelated lineages). Could the ANO1 and 2 family members be the analogous equivalent to the K_{Ca} channel (with regard to voltage and calcium sensing), as is observed in birds, insects, and bats (with regard to their shared capacity for flight)? Darwinian considerations aside, ANO1 and 2 do demonstrate functional differences in chloride conductance and sensitivity to calcium. The unitary conductance of ANO2 and the native olfactory CaCC [101] is approximately eightfold lower than the value of 8.3 pS shown for ANO1 [102], the kinetics of anoctamin 2 is faster, and the affinity of ANO2 for calcium is less affected by membrane potential than ANO1 [103, 104]. At positive membrane potentials, the EC_{50} for calcium activation is in the micromolar versus nanomolar ranges from ANO2 versus ANO1.

With regard to this heightened calcium sensitivity it is interesting that the molecular motifs of the originally reported sequence for ANO1 were seemingly devoid of the typical direct calcium binding sites described for voltage-gated calcium channels or calmodulin (EF hands and IQ motif, respectively) [41]. However, given the lack of sequence homology ANO1 shares with other channels, it should not be surprising that ANO1 may possess unique calcium-sensing mechanisms. Recent work by Xiao et al. [105] reveals that a novel amino acid motif in the first intracellular loop plays a critical role in ANO1's calcium-mediated activation. Since calcium sensing in two other unrelated calcium-activated channels [Best-1 and large conductance K_{Ca} (BK)] are composed of a high percentage of acidic amino acids, these researchers screened the ANO1 sequence for a region that resembled this sequence. They found a similar domain in the first intracellular loop (Fig. 2) that was composed of the following sequence: $_{466}$EEEEEAVK$_{473}$. While other studies have shown that individual amino acids can influence ion permeability [41] or voltage dependence [88], this site is unique in that this region contains components important for both calcium and voltage sensing. In particular, they found that deleting $_{470}$EAVK decreased calcium sensitivity, while deleting the adjacent segment $_{466}$EEE removed the intrinsic voltage dependence without altering calcium sensitivity. While it does remain unclear whether $_{466}$EEEEEAVK$_{473}$ is a component of the Ca^{2+}/voltage sensor or a region affecting these biophysical properties, these findings are important because they provide a mechanistic link for the observed coupling between membrane potential and calcium, which is one of the distinct hallmarks attributed to CaCCs. Another example of the complexity that governs ANO1 calcium sensing involves the functional impact that alternative splicing may exert on ANO1 calcium sensitivity. For example, Ferrera et al. demonstrate a reduction in Ca^{2+} sensitivity of ANO1 when the b splice form is expressed [61] (Fig. 2). While the possibility that alternative splicing can functionally alter the biophysical properties of the channel is exciting, the exact mechanism by which this occurs is still incompletely understood and may involve potential interactions with calmodulin-binding domains [106]. Furthermore, the effect of calcium and membrane potential on ANO1 underscores the possible importance of this protein in tissues that exhibit electromechanical coupling (e.g., airway smooth muscle), whereby it may act as an interlink between metabotropic and ionotropic signaling.

Other potential features that have remained largely unexplored that might affect the functionality of ANO1 include findings derived from sequence analysis of this protein. Yang et al. confirmed that consensus analysis revealed multiple protein kinase (A, C, and G) as well as casein kinase phosphorylation sites at intracellular protein segments [41]. In addition, there is a conserved Asn predicted to be a glycosylation site [107], all of which illustrates that the regulation of this CaCC may be far more complex than we currently appreciate. Even more complexity of functional regulation is apparent when considering the structure of ANO1.

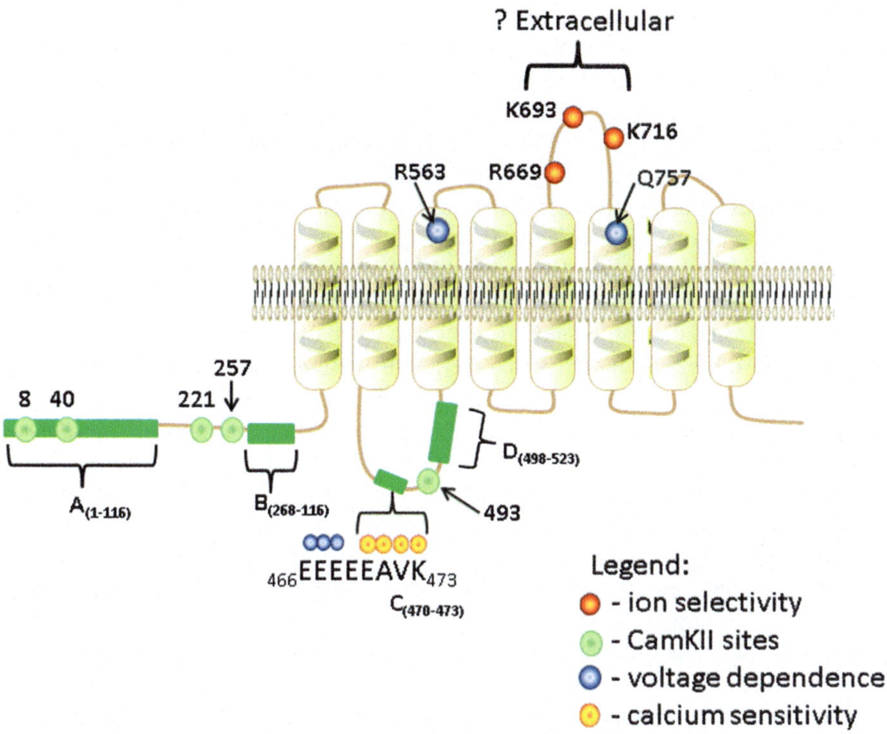

Fig. 2 Proposed topology of TMEM16A (anoctamin 1) calcium-activated chloride channel. At least four splice variants exist denoted as *A*, *B*, *C*, and *D*. Functional contributions of individual amino acids to voltage [87, 105], ion selectivity [87], and calcium sensing [105] are highlighted. In addition, putative phosphorylation sites for CaMKII are included [36]. Numbering refers to full-length (1,008 amino acids) sequence published as supplemental data in [88]. It should be noted that studies by Yu et al. suggest that the span of residues illustrated to form the third extracellular loop (denoted here as ? extracellular) between transmembrane domains 5 and 6 may instead be located in the cytoplasm with the pore domain lying between transmembrane domains 6 and 7 [126]

Two recent studies provide convincing evidence that ANO1 forms a homodimer that is maintained by noncovalent interactions [108, 109] that likely involve a dimerization domain in the N-terminal region of this protein [110].

7 Evidence for ANO1 Expression and Function in Many Cell Types

While it has been irrefutably shown that TMEM16A does possess classic features of a CaCC, this does not eliminate the possibility that other protein targets might also exist that possess these features. Despite the large number of EST clones thought to be splice forms of ANO1 that have been deposited from over 20 distinct tissues [61], protein expression or functional confirmation of ANO1 has not been convincingly established in many of the tissues where mRNA expression has been documented. However, there is good evidence linking this protein in various tissues with the respective CaCC activity previously documented in these tissues. In their landmark study defining ANO1 as a CaCC, Yang et al. showed immunohistochemical evidence for ANO1 in epithelial cells of the kidney and lung, acinar cells of the pancreas and salivary gland, and sensory neurons [41]. Furthermore, they provided functional evidence that siRNA knockdown of ANO1 reduced both native CaCC currents and pilocarpine-induced saliva production in mice. Caputo et al. also showed that endogenous CaCC activity in pancreatic CFPAC-1 cells was functionally inhibited following treatment with siRNA against ANO1 [88].

Huang et al. found robust expression of ANO1 on epithelial cells (of exocrine origin), the smooth muscle cells of the oviduct and ductus epididymis, as well as on the interstitial cells of Cajal (pacemaker cells in the gastrointestinal tract) [111]. They also demonstrated the potential importance of ANO1 in gastrointestinal peristalsis by demonstrating reduced rhythmic contractions in gastric smooth muscle from TMEM16A knockout mice [111]. Correlative findings by Espinosa et al. utilizing a monoclonal antibody directed against DOG-1 (aka ANO1) also found ANO1 to be a sensitive marker for GIST tumors [which are derived from gastrointestinal pacing cells (interstitial cells of Cajal (ICC))] [112]. Moreover, ANO1 has also been linked to proliferation and possibly metastatic progression of certain cancers including prostate [113], squamous cell tumors of the head and neck [114], human pancreatic cancer cell line CFPAC-1[115], and GIST tumors [116].

8 ANO1 in Airway Epithelium

ANO1 has been identified on the apical side of airway epithelium [111], is critical to fluid secretion, and has been associated with increased mucin secretion in asthmatics [117]. Airway epithelium recovered from ANO1 knockout mice studied at 3–5 days of age demonstrated a reduced UTP-induced short-circuit current [118] and absent calcium-stimulated chloride secretion in response to carbachol [119]. Studies using pharmacologic inhibitors of ANO1 suggested that this channel was responsible for a small component of calcium-induced chloride conductances under basal conditions [120, 121] in airway bronchial epithelial cells. However, in primary human bronchial epithelial cells, siRNA knockdown of ANO1 was sufficient to markedly inhibit calcium-activated chloride currents [122]. However, following IL-4 exposure, the expression and function of ANO1 was remarkably upregulated and modulated by activators and inhibitors of ANO1 in primary human bronchial epithelial cells [120]. Airway secretory cells appear to have even greater abundance and basal function of ANO1. Under basal conditions airway submucosal glands were responsive to activators and inhibitors of ANO1 [120]. Airway goblet cells demonstrated an increase in both ANO1 and MUC5AC following IL-4 exposure [122]. The link between IL-4 exposure and increased ANO1 expression and function is of particular interest in asthmatic airways because this cytokine has been linked to airway hyperresponsiveness [123]. Indeed, the seminal discovery of ANO1 by one group followed the IL-4-induced increase in bronchial epithelial cell chloride currents [88]. In agreement with the finding of greater expression of ANO1 in secretory cells as opposed to bronchial epithelium, asthmatics were found to have a greater expression of ANO1 in secretory cells of the airway epithelium. Moreover, IL-13-induced mucin secretion in normal human bronchial epithelial cells grown at an air/liquid interface was reduced by ANO1 blockers [117].

9 Anoctamins in Airway Smooth Muscle

ANO1 is critical to the development of the trachea because global ANO1 knockout mice die within 1 month of birth and exhibit severe tracheomalacia, including a defect in airway smooth muscle development on the posterior wall of the trachea and a failure of airway epithelial stratification [124]. ANO1 has been immunochemically localized to mouse airway smooth muscle colocalizing with α-smooth muscle actin [111]. Human bronchi were shown to contract less effectively in the presence of benzobromarone, an antagonist of ANO1 channels [117]. However, many pharmacologic ligands used as inhibitors of ANO1 have not been shown to be devoid of antagonism at other chloride or other ion channels. More recently, Zhang et al. used an ovalbumin-sensitized mouse model of chronic asthma to demonstrate selective upregulation of TMEM16A. This group also showed that treatment with niflumic acid or benzbromarone attenuated methacholine-induced airway contraction and correlated this finding with an absence of calcium spark-induced transient inward currents in isolated airway smooth muscle cells from neonatal TMEM16A knockout mice [125]. While this study suggests a fundamental link between sarcoplasmic calcium release and procontractile electrophysiologic phenomena, it still remains unclear whether these observations apply to human airway smooth muscle physiology.

10 Concluding Remarks

Plasma membrane electrical potential regulates multiple intracellular signaling events known to mediate airway smooth muscle tone. Chloride channels join their K^+ and Ca^{2+} cation channel cousins as contributors to airway smooth muscle membrane potential. While much ambiguity remains in the classification and understanding of many families of chloride channels, the seminal discovery of TMEM16A/ANO1 finally identified a long sought after molecular candidate of calcium-activated chloride currents. Studies of ANO1 are revolutionizing our understanding of calcium-activated chloride currents in many cell types, and emerging evidence suggests that this channel also contributes to the regulation of airway smooth muscle tone.

References

1. Janssen LJ. 2002; Ionic mechanisms and Ca(2+) regulation in airway smooth muscle contraction: do the data contradict dogma? *Am J Physiol Lung Cell Mol Physiol* 282: L1161–L1178.
2. Janssen LJ. 2012; Airway smooth muscle electrophysiology in a state of flux? *Am J Physiol Lung Cell Mol Physiol* 302:L730–L732.
3. Kotlikoff MI, Kume H, Tomasic M. 1992; Muscarinic regulation of membrane ion channels in airway smooth muscle cells. *Biochem Pharmacol* 43:5–10.
4. Gupta S, Meena HS, Chopra R, Bidani R. 1993; Effect of aerosolised verapamil and nifedipine in bronchial asthma. *J Assoc Physicians India* 41:425–427.
5. Talwar D, Jindal SK. 1993; Effect of calcium channel antagonists on cholinergic bronchial responsiveness in asthma. *J Assoc Physicians India* 41:279–280.
6. Cerrina J, Denjean A, Alexander G, Lockhart A, Duroux A. 1981; Inhibition of exercise-induced asthma by a calcium antagonist, nifedipine. *Am Rev Respir Dis* 123:156–160.
7. Patel KR. 1981; The effect of verapamil on histamine and methacholine-induced bronchoconstriction. *Clin Allergy* 11:441–447.
8. Patel KR. 1981; Calcium antagonists in exercise-induced asthma. *Br Med J (Clin Res Ed)* 282:932–933.
9. Barnes PJ, Wilson NM, Brown MJ. 1981; A calcium antagonist, nifedipine, modifies exercise-induced asthma. *Thorax* 36:726–730.
10. Corris PA, Nariman S, Gibson GJ. 1983; Nifedipine in the prevention of asthma induced by exercise and histamine. *Am Rev Respir Dis* 128:991–992.
11. Drazen JM, Fanta CH, Lacouture PG. 1983; Effect of nifedipine on constriction of human tracheal strips in vitro. *Br J Pharmacol* 78:687–691.
12. Hirota S, Trimble N, Pertens E, Janssen LJ. 2006; Intracellular Cl- fluxes play a novel role in Ca2+ handling in airway smooth muscle. *Am J Physiol Lung Cell Mol Physiol* 290: L1146–L1153.
13. Janssen LJ, Sims SM. 1995; Ca(2+)-dependent Cl- current in canine tracheal smooth muscle cells. *Am J Physiol* 269:C163–C169.
14. Fanta CH. 1985; Calcium-channel blockers in prophylaxis and treatment of asthma. *Am J Cardiol* 55:202B–209B.
15. Janssen LJ. 1997; T-type and L-type Ca2+ currents in canine bronchial smooth muscle: characterization and physiological roles. *Am J Physiol* 272:C1757–C1765.
16. Janssen LJ, Sims SM. 1992; Acetylcholine activates non-selective cation and chloride conductances in canine and guinea-pig tracheal myocytes. *J Physiol* 453:197–218.
17. Hirota S, Helli P, Janssen LJ. 2007; Ionic mechanisms and Ca2+ handling in airway smooth muscle. *Eur Respir J* 30:114–133.
18. Kume H, Hall IP, Washabau RJ, Takagi K, Kotlikoff MI. 1994; Beta-adrenergic agonists regulate KCa channels in airway smooth muscle by cAMP-dependent and -independent mechanisms. *J Clin Invest* 93:371–379.
19. Corteling RL, Li S, Giddings J, Westwick J, Poll C, Hall IP. 2004; Expression of transient receptor potential C6 and related transient receptor potential family members in human airway smooth muscle and lung tissue. *Am J Respir Cell Mol Biol* 30:145–154.

20. Jia Y, Wang X, Varty L, Rizzo CA, Yang R, Correll CC, Phelps PT, Egan RW, Hey JA. 2004; Functional TRPV4 channels are expressed in human airway smooth muscle cells. *Am J Physiol Lung Cell Mol Physiol* 287:L272–L278.
21. Deshpande DA, Wang WC, McIlmoyle EL, Robinett KS, Schillinger RM, An SS, Sham JS, Liggett SB. 2010; Bitter taste receptors on airway smooth muscle bronchodilate by localized calcium signaling and reverse obstruction. *Nat Med* 16:1299–1304.
22. Billups D, Billups B, Challiss RA, Nahorski SR. 2006; Modulation of Gq-protein-coupled inositol trisphosphate and Ca2+ signaling by the membrane potential. *J Neurosci* 26:9983–9995.
23. Ben-Chaim Y, Chanda B, Dascal N, Bezanilla F, Parnas I, Parnas H. 2006; Movement of 'gating charge' is coupled to ligand binding in a G-protein-coupled receptor. *Nature* 444:106–109.
24. Mahaut-Smith MP, Martinez-Pinna J, Gurung IS. 2008; A role for membrane potential in regulating GPCRs? *Trends Pharmacol Sci* 29:421–429.
25. Yamakage M, Chen X, Tsujiguchi N, Kamada Y, Namiki A. 2001; Different inhibitory effects of volatile anesthetics on T- and L-type voltage-dependent Ca^{2+} channels in porcine tracheal and bronchial smooth muscles. *Anesthesiology* 94:683–693.
26. Gallos G, Gleason NR, Zhang Y, Pak SW, Sonett JR, Yang J, Emala CW. 2008; Activation of endogenous GABAA channels on airway smooth muscle potentiates isoproterenol mediated relaxation. *Am J Physiol Lung Cell Mol Physiol* 295:L1040–L1047.
27. Yim PD, Gallos G, Xu D, Zhang Y, Emala CW. 2011; Novel expression of a functional glycine receptor chloride channel that attenuates contraction in airway smooth muscle. *FASEB J.* 25:1706–1717
28. Zhang GH, Melvin JE. 1993; Membrane potential regulates Ca2+ uptake and inositol phosphate generation in rat sublingual mucous acini. *Cell Calcium* 14:551–562.
29. Gromada J, Dissing S. 1996; Membrane potential and cytosolic free calcium levels modulate acetylcholine-induced inositol phosphate production in insulin-secreting BTC3 cells. *Biochim Biophys Acta* 1310:145–148.
30. Liu C, Zuo J, Pertens E, Helli PB, Janssen LJ. 2005; Regulation of Rho/ROCK signaling in airway smooth muscle by membrane potential and [Ca2+]i. *Am J Physiol Lung Cell Mol Physiol* 289:L574–L582.
31. Liu C, Zuo J, Janssen LJ. 2006; Regulation of airway smooth muscle RhoA/ROCK activities by cholinergic and bronchodilator stimuli. *Eur Respir J* 28:703–711.
32. Jentsch TJ, Friedrich T, Schriever A, Yamada H. 1999; The CLC chloride channel family. *Pflugers Arch* 437:783–795.
33. Tsunenari T, Sun H, Williams J, Cahill H, Smallwood P, Yau KW, Nathans J. 2003; Structure-function analysis of the bestrophin family of anion channels. *J Biol Chem* 278:41114–41125.
34. Winpenny JP, Gray MA. 2012; The anoctamin (TMEM16) gene family: calcium-activated chloride channels come of age. *Exp Physiol* 97:175–176.
35. Duran C, Thompson CH, Xiao Q, Hartzell HC. 2010; Chloride channels: often enigmatic, rarely predictable. *Annu Rev Physiol* 72:95–121.
36. Kunzelmann K, Kongsuphol P, Aldehni F, Tian Y, Ousingsawat J, Warth R, Schreiber R. 2009; Bestrophin and TMEM16-Ca(2+) activated Cl(−) channels with different functions. *Cell Calcium* 46:233–241.
37. Mundhenk L, Alfalah M, Elble RC, Pauli BU, Naim HY, Gruber AD. 2006; Both cleavage products of the mCLCA3 protein are secreted soluble proteins. *J Biol Chem* 281:30072–30080.
38. Miledi R. 1982; A calcium-dependent transient outward current in Xenopus laevis oocytes. *Proc R Soc Lond B Biol Sci.* 215:491–4977.
39. Miledi R, Parker I. 1984; Chloride current induced by injection of calcium into Xenopus oocytes. *J Physiol.* 357:173–183.

40. Barish ME. 1983; A transient calcium-dependent chloride current in the immature Xenopus oocyte. *J Physiol.* 342:309–325.
41. Yang YD, Cho H, Koo JY, Tak MH, Cho Y, Shim WS, Park SP, Lee J, Lee B, Kim BM, Raouf R, Shin YK, Oh U. 2008; TMEM16A confers receptor-activated calcium-dependent chloride conductance. *Nature* 455:1210–1215.
42. Hartzell C, Putzier I, Arreola J. 2005; Calcium-activated chloride channels. *Annu Rev Physiol* 67:719–758.
43. Qu Z, Hartzell HC. 2000; Anion permeation in Ca(2+)-activated Cl(−) channels. *J Gen Physiol* 116:825–844.
44. Pusch M. 2004; Ca(2+)-activated chloride channels go molecular. *J Gen Physiol* 123:323–325.
45. Zhou Y, Shapiro M, Dong Q, Louahed J, Weiss C, Wan S, Chen Q, Dragwa C, Savio D, Huang M, Fuller C, Tomer Y, Nicolaides NC, McLane M, Levitt RC. 2002; A calcium-activated chloride channel blocker inhibits goblet cell metaplasia and mucus overproduction. *Novartis Found Symp* 248:150–165.
46. Collier ML, Levesque PC, Kenyon JL, Hume JR. 1996; Unitary Cl− channels activated by cytoplasmic Ca2+ in canine ventricular myocytes. *Circ Res* 78:936–944.
47. Ling BN, Seal EE, Eaton DC. 1993; Regulation of mesangial cell ion channels by insulin and angiotensin II. Possible role in diabetic glomerular hyperfiltration. *J Clin Invest* 92:2141–2151.
48. Nilius B, Prenen J, Szucs G, Wei L, Tanzi F, Voets T, Droogmans G. 1997; Calcium-activated chloride channels in bovine pulmonary artery endothelial cells. *J Physiol* 498 (Pt 2):381–396.
49. Martin DK. 1993; Small conductance chloride channels in acinar cells from the rat mandibular salivary gland are directly controlled by a G-protein. *Biochem Biophys Res Commun* 192:1266–1273.
50. Koumi S, Sato R, Aramaki T. 1994; Characterization of the calcium-activated chloride channel in isolated guinea-pig hepatocytes. *J Gen Physiol* 104:357–373.
51. Arreola J, Melvin JE, Begenisich T. 1998; Differences in regulation of Ca(2+)-activated Cl-channels in colonic and parotid secretory cells. *Am J Physiol* 274:C161–C166.
52. Park MK, Lomax RB, Tepikin AV, Petersen OH. 2001; Local uncaging of caged Ca(2+) reveals distribution of Ca(2+)-activated Cl(−) channels in pancreatic acinar cells. *Proc Natl Acad Sci U S A* 98:10948–10953.
53. Wang YX, Kotlikoff MI. 1997; Inactivation of calcium-activated chloride channels in smooth muscle by calcium/calmodulin-dependent protein kinase. *Proc Natl Acad Sci U S A* 94:14918–14923.
54. Wagner JA, Cozens AL, Schulman H, Gruenert DC, Stryer L, Gardner P. 1991; Activation of chloride channels in normal and cystic fibrosis airway epithelial cells by multifunctional calcium/calmodulin-dependent protein kinase. *Nature* 349:793–796.
55. Chao AC, Kouyama K, Heist EK, Dong YJ, Gardner P. 1995; Calcium- and CaMKII-dependent chloride secretion induced by the microsomal Ca(2+)-ATPase inhibitor 2,5-di-(tert-butyl)-1,4-hydroquinone in cystic fibrosis pancreatic epithelial cells. *J Clin Invest* 96:1794–1801.
56. Schlenker T, Fitz JG. 1996; Ca(2+)-activated C1- channels in a human biliary cell line: regulation by Ca2+/calmodulin-dependent protein kinase. *Am J Physiol* 271:G304–G310.
57. Nishimoto I, Wagner JA, Schulman H, Gardner P. 1991; Regulation of Cl- channels by multifunctional CaM kinase. *Neuron* 6:547–555.
58. Holevinsky KO, Jow F, Nelson DJ. 1994; Elevation in intracellular calcium activates both chloride and proton currents in human macrophages. *J Membr Biol* 140:13–30.
59. Xie W, Kaetzel MA, Bruzik KS, Dedman JR, Shears SB, Nelson DJ. 1996; Inositol 3,4,5,6-tetrakisphosphate inhibits the calmodulin-dependent protein kinase II-activated chloride conductance in T84 colonic epithelial cells. *J Biol Chem* 271:14092–14097.

60. Xie W, Solomons KR, Freeman S, Kaetzel MA, Bruzik KS, Nelson DJ, Shears SB. 1998; Regulation of Ca2+-dependent Cl- conductance in a human colonic epithelial cell line (T84): cross-talk between Ins(3,4,5,6)P4 and protein phosphatases. *J Physiol* 510 (Pt 3): 661–673.
61. Ferrera L, Caputo A, Ubby I, Bussani E, Zegarra-Moran O, Ravazzolo R, Pagani F, Galietta LJ. 2009; Regulation of TMEM16A chloride channel properties by alternative splicing. *J Biol Chem* 284:33360–33368.
62. Large WA, Wang Q. 1996; Characteristics and physiological role of the Ca(2+)-activated Cl– conductance in smooth muscle. *Am J Physiol* 271:C435–C454.
63. Liu C, Tazzeo T, Lippton H, Janssen LJ. 2007; Role of tyrosine phosphorylation in U46619-induced vasoconstriction of pulmonary vasculature and its modulation by genistein, daidzein, and equol. *J Cardiovasc Pharmacol* 50:441–448.
64. Wang YX, Kotlikoff MI. 1997; Muscarinic signaling pathway for calcium release and calcium-activated chloride current in smooth muscle. *Am J Physiol* 273:C509–C519.
65. Kotlikoff MI, Wang YX. 1998; Calcium release and calcium-activated chloride channels in airway smooth muscle cells. *Am J Respir Crit Care Med* 158:S109–S114.
66. Liu X, Farley JM. 1996; Acetylcholine-induced chloride current oscillations in swine tracheal smooth muscle cells. *J Pharmacol Exp Ther* 276:178–186.
67. Janssen LJ, Sims SM. 1993; Histamine activates Cl- and K+ currents in guinea-pig tracheal myocytes: convergence with muscarinic signalling pathway. *J Physiol* 465:661–677.
68. Janssen LJ, Sims SM. 1994; Substance P activates Cl- and K+ conductances in guinea-pig tracheal smooth muscle cells. *Can J Physiol Pharmacol* 72:705–710.
69. Wang YX, Kotlikoff MI. 2000; Signalling pathway for histamine activation of non-selective cation channels in equine tracheal myocytes. *J Physiol* 523 Pt 1:131–138.
70. Henmi S, Imaizumi Y, Muraki K, Watanabe M. 1996; Time course of Ca(2+)-dependent K+ and Cl- currents in single smooth muscle cells of guinea-pig trachea. *Eur J Pharmacol* 306:227–236.
71. Bao R, Lifshitz LM, Tuft RA, Bellve K, Fogarty KE, ZhuGe R. 2008; A close association of RyRs with highly dense clusters of Ca2+-activated Cl- channels underlies the activation of STICs by Ca2+ sparks in mouse airway smooth muscle. *J Gen Physiol* 132:145–160.
72. Aickin CC, Vermue NA. 1983; Microelectrode measurement of intracellular chloride activity in smooth muscle cells of guinea-pig ureter. *Pflugers Arch* 397:25–28.
73. Janssen LJ, Daniel EE. 1991; Depolarizing agents induce oscillations in canine bronchial smooth muscle membrane potential: possible mechanisms. *J Pharmacol Exp Ther* 259:110–117.
74. Liu X, Farley JM. 1996; Acetylcholine-induced Ca++-dependent chloride current oscillations are mediated by inositol 1,4,5-trisphosphate in tracheal myocytes. *J Pharmacol Exp Ther* 277:796–804.
75. Janssen LJ, Hague C, Nana R. 1998; Ionic mechanisms underlying electrical slow waves in canine airway smooth muscle. *Am J Physiol* 275:L516–L523.
76. Ward SM, Burns AJ, Torihashi S, Sanders KM. 1994; Mutation of the proto-oncogene c-kit blocks development of interstitial cells and electrical rhythmicity in murine intestine. *J Physiol* 480 (Pt 1):91–97.
77. Hwang SJ, Blair PJ, Britton FC, O'Driscoll KE, Hennig G, Bayguinov YR, Rock JR, Harfe BD, Sanders KM, Ward SM. 2009; Expression of anoctamin 1/TMEM16A by interstitial cells of Cajal is fundamental for slow wave activity in gastrointestinal muscles. *J Physiol* 587:4887–4904.
78. Janssen LJ, Sims SM. 1994; Spontaneous transient inward currents and rhythmicity in canine and guinea-pig tracheal smooth muscle cells. *Pflugers Arch* 427:473–480.
79. Pollock NS, Kargacin ME, Kargacin GJ. 1998; Chloride channel blockers inhibit Ca2+ uptake by the smooth muscle sarcoplasmic reticulum. *Biophys J* 75:1759–1766.

80. Elble RC, Ji G, Nehrke K, DeBiasio J, Kingsley PD, Kotlikoff MI, Pauli BU. 2002; Molecular and functional characterization of a murine calcium-activated chloride channel expressed in smooth muscle. *J Biol Chem* 277:18586–18591.
81. Zhou Y, Dong Q, Louahed J, Dragwa C, Savio D, Huang M, Weiss C, Tomer Y, McLane MP, Nicolaides NC, Levitt RC. 2001; Characterization of a calcium-activated chloride channel as a shared target of Th2 cytokine pathways and its potential involvement in asthma. *Am J Respir Cell Mol Biol* 25:486–491.
82. Woodruff PG, Boushey HA, Dolganov GM, Barker CS, Yang YH, Donnelly S, Ellwanger A, Sidhu SS, Dao-Pick TP, Pantoja C, Erle DJ, Yamamoto KR, Fahy JV. 2007; Genome-wide profiling identifies epithelial cell genes associated with asthma and with treatment response to corticosteroids. *Proc Natl Acad Sci U S A* 104:15858–15863.
83. Toda M, Tulic MK, Levitt RC, Hamid Q. 2002; A calcium-activated chloride channel (HCLCA1) is strongly related to IL-9 expression and mucus production in bronchial epithelium of patients with asthma. *J Allergy Clin Immunol* 109:246–250.
84. Hoshino M, Morita S, Iwashita H, Sagiya Y, Nagi T, Nakanishi A, Ashida Y, Nishimura O, Fujisawa Y, Fujino M. 2002; Increased expression of the human Ca2+−activated Cl- channel 1 (CaCC1) gene in the asthmatic airway. *Am J Respir Crit Care Med* 165:1132–1136.
85. Nakanishi A, Morita S, Iwashita H, Sagiya Y, Ashida Y, Shirafuji H, Fujisawa Y, Nishimura O, Fujino M. 2001; Role of gob-5 in mucus overproduction and airway hyperresponsiveness in asthma. *Proc Natl Acad Sci U S A* 98:5175–5180.
86. Nakano T, Inoue H, Fukuyama S, Matsumoto K, Matsumura M, Tsuda M, Matsumoto T, Aizawa H, Nakanishi Y. 2006; Niflumic acid suppresses interleukin-13-induced asthma phenotypes. *Am J Respir Crit Care Med* 173:1216–1221.
87. Galietta LJ. 2009; The TMEM16 protein family: a new class of chloride channels? *Biophys J* 97:3047–3053.
88. Caputo A, Caci E, Ferrera L, Pedemonte N, Barsanti C, Sondo E, Pfeffer U, Ravazzolo R, Zegarra-Moran O, Galietta LJ. 2008; TMEM16A, a membrane protein associated with calcium-dependent chloride channel activity. *Science* 322:590–594.
89. Schroeder BC, Cheng T, Jan YN, Jan LY. 2008; Expression cloning of TMEM16A as a calcium-activated chloride channel subunit. *Cell* 134:1019–1029.
90. Duran C, Qu Z, Osunkoya AO, Cui Y, Hartzell HC. 2012; ANOs 3–7 in the anoctamin/Tmem16 Cl- channel family are intracellular proteins. *Am J Physiol Cell Physiol* 302: C482–C493.
91. Schreiber R, Uliyakina I, Kongsuphol P, Warth R, Mirza M, Martins JR, Kunzelmann K. 2010; Expression and function of epithelial anoctamins. *J Biol Chem* 285:7838–7845.
92. Tsutsumi S, Kamata N, Vokes TJ, Maruoka Y, Nakakuki K, Enomoto S, Omura K, Amagasa T, Nagayama M, Saito-Ohara F, Inazawa J, Moritani M, Yamaoka T, Inoue H, Itakura M. 2004; The novel gene encoding a putative transmembrane protein is mutated in gnathodiaphyseal dysplasia (GDD). *Am J Hum Genet* 74:1255–1261.
93. Bolduc V, Marlow G, Boycott KM, Saleki K, Inoue H, Kroon J, Itakura M, Robitaille Y, Parent L, Baas F, Mizuta K, Kamata N, Richard I, Linssen WH, Mahjneh I, de VM, Bashir R, Brais B. 2010; Recessive mutations in the putative calcium-activated chloride channel Anoctamin 5 cause proximal LGMD2L and distal MMD3 muscular dystrophies. *Am J Hum Genet* 86:213–221.
94. Hicks D, Sarkozy A, Muelas N, Koehler K, Huebner A, Hudson G, Chinnery PF, Barresi R, Eagle M, Polvikoski T, Bailey G, Miller J, Radunovic A, Hughes PJ, Roberts R, Krause S, Walter MC, Laval SH, Straub V, Lochmuller H, Bushby K. 2011; A founder mutation in Anoctamin 5 is a major cause of limb-girdle muscular dystrophy. *Brain* 134:171–182.
95. Mahjneh I, Jaiswal J, Lamminen A, Somer M, Marlow G, Kiuru-Enari S, Bashir R. 2010; A new distal myopathy with mutation in anoctamin 5. *Neuromuscul Disord* 20:791–795.
96. Kiessling A, Weigle B, Fuessel S, Ebner R, Meye A, Rieger MA, Schmitz M, Temme A, Bachmann M, Wirth MP, Rieber EP. 2005; D-TMPP: a novel androgen-regulated gene

preferentially expressed in prostate and prostate cancer that is the first characterized member of an eukaryotic gene family. *Prostate* 64:387–400.
97. Suzuki J, Fujii T, Imao T, Ishihara K, Kuba H, Nagata S. 2013; Calcium-dependent Phospholipid Scramblase Activity of TMEM16 Family Members. *J Biol Chem.* 288:13305–13316.
98. Tian Y, Schreiber R, Kunzelmann K. 2012; Anoctamins are a family of Ca2+−activated Cl-channels. *J Cell Sci* 125:4991–4998.
99. Suzuki J, Umeda M, Sims PJ, Nagata S. 2010; Calcium-dependent phospholipid scrambling by TMEM16F. *Nature* 468:834–838.
100. Scudieri P, Sondo E, Ferrera L, Galietta LJ. 2012; The anoctamin family: TMEM16A and TMEM16B as calcium-activated chloride channels. *Exp Physiol* 97:177–183.
101. Reisert J, Bauer PJ, Yau KW, Frings S. 2003; The Ca-activated Cl channel and its control in rat olfactory receptor neurons. *J Gen Physiol* 122:349–363.
102. Strotmann J, Levai O, Fleischer J, Schwarzenbacher K, Breer H. 2004; Olfactory receptor proteins in axonal processes of chemosensory neurons. *J Neurosci* 24:7754–7761.
103. Pifferi S, Dibattista M, Menini A. 2009; TMEM16B induces chloride currents activated by calcium in mammalian cells. *Pflugers Arch* 458:1023–1038.
104. Stephan AB, Shum EY, Hirsh S, Cygnar KD, Reisert J, Zhao H. 2009; ANO2 is the cilial calcium-activated chloride channel that may mediate olfactory amplification. *Proc Natl Acad Sci U S A* 106:11776–11781.
105. Xiao Q, Yu K, Perez-Cornejo P, Cui Y, Arreola J, Hartzell HC. 2011; Voltage- and calcium-dependent gating of TMEM16A/Ano1 chloride channels are physically coupled by the first intracellular loop. *Proc Natl Acad Sci U S A* 108:8891–8896.
106. Tian Y, Kongsuphol P, Hug M, Ousingsawat J, Witzgall R, Schreiber R, Kunzelmann K. 2011; Calmodulin-dependent activation of the epithelial calcium-dependent chloride channel TMEM16A. *FASEB J* 25:1058–1068.
107. Katoh M, Katoh M. 2004; Identification and characterization of TMEM16E and TMEM16F genes in silico. *Int J Oncol* 24:1345–1349.
108. Fallah G, Romer T, tro-Dassen S, Braam U, Markwardt F, Schmalzing G. 2011; TMEM16A (a)/anoctamin-1 shares a homodimeric architecture with CLC chloride channels. *Mol Cell Proteomics* 10:M110.
109. Sheridan JT, Worthington EN, Yu K, Gabriel SE, Hartzell HC, Tarran R. 2011; Characterization of the oligomeric structure of the Ca(2+)-activated Cl- channel Ano1/TMEM16A. *J Biol Chem* 286:1381–1388.
110. Tien J, Lee HY, Minor DL, Jr., Jan YN, Jan LY. 2013; Identification of a dimerization domain in the TMEM16A calcium-activated chloride channel (CaCC). *Proc Natl Acad Sci U S A* 110:6352–6357.
111. Huang F, Rock JR, Harfe BD, Cheng T, Huang X, Jan YN, Jan LY. 2009; Studies on expression and function of the TMEM16A calcium-activated chloride channel. *Proc Natl Acad Sci U S A* 106:21413–21418.
112. Espinosa I, Lee CH, Kim MK, Rouse BT, Subramanian S, Montgomery K, Varma S, Corless CL, Heinrich MC, Smith KS, Wang Z, Rubin B, Nielsen TO, Seitz RS, Ross DT, West RB, Cleary ML, van de RM. 2008; A novel monoclonal antibody against DOG1 is a sensitive and specific marker for gastrointestinal stromal tumors. *Am J Surg Pathol* 32:210–218.
113. Liu W, Lu M, Liu B, Huang Y, Wang K. 2012; Inhibition of Ca(2+)-activated Cl(−) channel ANO1/TMEM16A expression suppresses tumor growth and invasiveness in human prostate carcinoma. *Cancer Lett* 326:41–51.
114. Ayoub C, Wasylyk C, Li Y, Thomas E, Marisa L, Robe A, Roux M, Abecassis J, de RA, Wasylyk B. 2010; ANO1 amplification and expression in HNSCC with a high propensity for future distant metastasis and its functions in HNSCC cell lines. *Br J Cancer* 103:715–726.
115. Mazzone A, Eisenman ST, Strege PR, Yao Z, Ordog T, Gibbons SJ, Farrugia G. 2012; Inhibition of cell proliferation by a selective inhibitor of the Ca(2+)-activated Cl(−) channel, Ano1. *Biochem Biophys Res Commun* 427:248–253.

116. Stanich JE, Gibbons SJ, Eisenman ST, Bardsley MR, Rock JR, Harfe BD, Ordog T, Farrugia G. 2011; Ano1 as a regulator of proliferation. *Am J Physiol Gastrointest Liver Physiol* 301: G1044–G1051.
117. Huang F, Zhang H, Wu M, Yang H, Kudo M, Peters CJ, Woodruff PG, Solberg OD, Donne ML, Huang X, Sheppard D, Fahy JV, Wolters PJ, Hogan BL, Finkbeiner WE, Li M, Jan YN, Jan LY, Rock JR. 2012; Calcium-activated chloride channel TMEM16A modulates mucin secretion and airway smooth muscle contraction. *Proc Natl Acad Sci U S A*. 109:16354–16359
118. Rock JR, O'Neal WK, Gabriel SE, Randell SH, Harfe BD, Boucher RC, Grubb BR. 2009; Transmembrane protein 16A (TMEM16A) is a Ca2+–regulated Cl- secretory channel in mouse airways. *J Biol Chem* 284:14875–14880.
119. Ousingsawat J, Martins JR, Schreiber R, Rock JR, Harfe BD, Kunzelmann K. 2009; Loss of TMEM16A causes a defect in epithelial Ca2+–dependent chloride transport. *J Biol Chem* 284:28698–28703.
120. Namkung W, Yao Z, Finkbeiner WE, Verkman AS. 2011; Small-molecule activators of TMEM16A, a calcium-activated chloride channel, stimulate epithelial chloride secretion and intestinal contraction. *FASEB J* 25:4048–4062.
121. Namkung W, Phuan PW, Verkman AS. 2011; TMEM16A inhibitors reveal TMEM16A as a minor component of calcium-activated chloride channel conductance in airway and intestinal epithelial cells. *J Biol Chem* 286:2365–2374.
122. Scudieri P, Caci E, Bruno S, Ferrera L, Schiavon M, Sondo E, Tomati V, Gianotti A, Zegarra-Moran O, Pedemonte N, Rea F, Ravazzolo R, Galietta LJ. 2012; Association of TMEM16A chloride channel overexpression with airway goblet cell metaplasia. *J Physiol* 590:6141–6155.
123. Shore SA. 2004; Direct effects of Th2 cytokines on airway smooth muscle. *Curr Opin Pharmacol* 4:235–240.
124. Rock JR, Futtner CR, Harfe BD. 2008; The transmembrane protein TMEM16A is required for normal development of the murine trachea. *Dev Biol* 321:141–149.
125. Zhang CH, Li Y, Zhao W, Lifshitz LM, Li H, Harfe BD, Zhu MS, ZhuGe R. 2013; The transmembrane protein 16A Ca(2+)-activated Cl- channel in airway smooth muscle contributes to airway hyperresponsiveness. *Am J Respir Crit Care Med* 187:374–381.
126. Yu K, Duran C, Qu Z, Cui YY, Hartzell HC. 2012; Explaining calcium-dependent gating of anoctamin-1 chloride channels requires a revised topology. *Circ Res* 110:990–999.

Local Calcium Signaling in Airway Smooth Muscle Cells

Qing-Hua Liu, Carlo Savoia, Yong-Xiao Wang, and Yun-Min Zheng

Contents

Abstract Potentially fatal asthma attacks may result from airway hyperresponsiveness (AHR), which is the exaggerated contractile response of airway smooth muscle cells (ASMCs) to nonspecific stimuli. A better understanding of Ca^{2+} signaling in ASMC contraction can help develop advanced therapeutics for asthma. A common elementary form of Ca^{2+} signaling is the Ca^{2+} spark (i.e., a local transient Ca^{2+} release event). Ca^{2+} sparks occur as a result of the coordinated opening of a cluster of ryanodine receptors (RyRs) and play a fundamental role in skeletal, cardiac, and smooth muscle cells. This chapter summarizes the recent advances from our work and that of others in studies of Ca^{2+} sparks in ASMCs. Ca^{2+} sparks have been observed in equine, porcine, guinea-pig, and mouse ASMCs. Classical parasympathetic stimulation or membrane depolarization will activate native Gq protein-coupled muscarinic M3 receptors (M_3Rs) and phospholipase C (PLC), generating inositol 1,4,5-trisphosphate (IP_3) and diacylglycerol (DAG) from

Q.-H. Liu (✉)
Institute for Medical Biology, College of Life Sciences, South-Central University for Nationalities, Wuhan 430074, China
e-mail: liu258q@yahoo.com

C. Savoia • Y.-X. Wang • Y.-M. Zheng (✉)
Center for Cardiovascular Sciences (MC-8), Albany Medical College, 47 New Scotland Avenue, Albany, NY 12208, USA
e-mail: zhengy@mail.amc.edu

Y.-X. Wang (ed.), *Calcium Signaling In Airway Smooth Muscle Cells*,
DOI 10.1007/978-3-319-01312-1_6, © Springer International Publishing Switzerland 2014

phosphatidylinositol 4,5-bisphosphate (PIP_2) in ASMCs. IP_3 will activate IP_3 receptors (IP_3Rs), inducing Ca^{2+} release, which may locally induce further Ca^{2+} release from RyR2, increasing Ca^{2+} sparks and associated contraction. Meanwhile, DAG activates protein kinase C-ε (PKCε), which inhibits Ca^{2+} sparks and contraction through RyR1. Calcineurin (CaN) promotes Ca^{2+} sparks and contraction through RyR1, in contrast to the role of PKCε. In ASMCs, basal Ca^{2+} sparks directly mediate a contractile force, as seen during RyR activation. These local Ca^{2+} events are also capable of regulating membrane potential through spontaneous transient inward currents (STICs) and spontaneous transient outward currents (STOCs). At rest with the membrane potential closer to K^+ equilibrium potential (E_K), Ca^{2+} sparks will preferentially activate TMEM16A-encoded Ca^{2+}-activated Cl^- channels, depolarizing the cell membrane and promoting contraction. As the membrane potential rises closer to Cl^- equilibrium potential (E_{Cl}), Ca^{2+} sparks will begin to activate big-conductance Ca^{2+}-activated K^+ (BK) channels, leading to cell hyperpolarization and preventing contraction. A disruption in this balancing of cell excitability could play a role in asthmatic AHR. During asthma, Ca^{2+} sparks, TMEM16A expression, and STICs are increased, providing a mechanistic setting for AHR, whereas the STOC pathway cannot maintain balance and a lower level of cell excitability, resulting in excessive contraction. Therefore, Ca^{2+} sparks and the associated signaling axis in ASMCs may become new and effective targets for asthma therapeutics.

Keywords Ca^{2+} sparks • Ryanodine receptor • Inositol 1,4,5-trisphosphate receptor • Protein kinase C-ε, calcineurin • Spontaneous transient outward current • Spontaneous transient inward current

Abbreviation

ACh	Acetylcholine
AHR	Airway hyperresponsiveness
AP	Action potential
ASMC	Airway smooth muscle cell
BiMPT	Biphasic membrane potential transient
BK	Big-conductance Ca^{2+}-activated K^+ channel
Ca^{2+}	Calcium
$[Ca^{2+}]_i$	Intracellular calcium concentration
cADPR	Cyclic ADP-ribose
CaMKII	Ca^{2+}/calmodulin-dependent kinase II
CaN	Calcineurin
CICR	Calcium-induced calcium release
Cl_{Ca}	Calcium-activated chloride channel
CRU	Calcium release unit
DAG	Diacylglycerol
E_{Cl}	Chloride equilibration potential

EGTA	Ethylene glycol tetraacetic acid
E_K	Potassium equilibration potential
FKBP12/12.6	FK506 binding protein 12/12.6
IP_3	Inositol 1,4,5-trisphosphate
IP_3R	Inositol 1,4,5-trisphosphate receptor
M_3R	Muscarinic type 3 receptor
mACH	Methacholine
PIP_2	Phosphatidylinositol 4,5-bisphosphate
PKC	Protein kinase C
PLC	Phospholipase C
RyR	Ryanodine receptor
SR	Sarcoplasmic reticulum
STIC	Spontaneous transient inward current
STOC	Spontaneous transient outward current
STOIC	Spontaneous transient outward/inward currents
VDCC	Voltage-dependent calcium channel

1 Introduction

Respiratory diseases are a widespread and significant cause of death around the world. Biomedical research is needed to further our understanding of the human respiratory system and discover the molecular nature of the various pathologies we suffer from. Asthma, considered to be the most common respiratory disease among children, is currently a large active area of investigation. Potentially fatal asthma attacks result from airway hyperresponsiveness (AHR), which is the exaggerated contractile response of airway smooth muscle cells (ASMCs) to nonspecific normally nonharmful stimuli. To better prevent and treat AHR, we need to study ASMC contraction.

Up until the early 1990s, muscle contraction was thought to be dependent on large-scale, cell-wide fluctuations in intracellular calcium (Ca^{2+}). However, the biomedical world was confounded as to how Ca^{2+} could affect multiple other cellular signaling processes within a muscle cell. Then, in 1993, subcellular local transient Ca^{2+} release events, termed Ca^{2+} sparks, were first observed in single rat heart cells using a laser scanning confocal microscope [1]. Ca^{2+} sparks generate from the spontaneous opening of ryanodine receptors (RyRs) on the sarcoplasmic reticulum (SR). When triggered, they can summate, generating a propagating wave of Ca^{2+}, resulting in cardiomyocyte contraction. Over two decades of research in "Ca^{2+} sparkology" has revealed their importance in many high-threshold, Ca^{2+}-dependent processes like cardiac and skeletal muscle excitation-contraction coupling, vascular tone regulation, membrane excitability, and neuronal secretion, making them a very active area of biomedical research [2]. Since their initial discovery, numerous other types of local Ca^{2+} signals have been observed, such as quarks, embers, spikes, sparklets, puffs, blips, marks, scraps, blinks, skraps, and pulsars [3, 4]. These phenomena are generated by ion channels on the cell, SR, or mitochondrial membrane; however, they are not well studied in ASMCs and thus not the main focus of this chapter.

To better understand ASMC contraction and the pathological mechanism underlying AHR, we needed to investigate the potential molecular nature, regulation, and role of Ca^{2+} sparks in ASMCs. Although Ca^{2+} sparks have been studied in many smooth muscle tissues, differences exist from one type to another, requiring investigation at a tissue-specific level. Ca^{2+} sparks have been observed in equine [5], porcine [6, 7], guinea-pig [8], and mouse ASMCs [9], and it is the primary intent of this chapter to review the significant progress made by our laboratory and other investigators in the field.

2 Molecular Nature of Ca^{2+} Sparks

As stated previously, muscle contraction is dependent on an increase in intracellular Ca^{2+} concentration ($[Ca^{2+}]_i$). At rest $[Ca^{2+}]_i$ is approximately 0.1 μM, with extracellular Ca^{2+} approximately 2 mM, roughly four orders of magnitude higher. The main cellular Ca^{2+} store is the SR, which holds Ca^{2+} at levels roughly equivalent to the extracellular fluid. The SR possesses two Ca^{2+}-selective ion channels, the RyR and inositol 1,4,5,-trisphosphate receptor (IP_3R). Mammalian RyRs exist in three isoforms encoded by three distinct genes sharing approximately 65 % homology. All three isoforms exist as homotetramers of four 565 kDa subunits [10, 11]. Skeletal muscle cells mainly express RyR1, heart muscle cells predominantly express RyR2, and smooth muscle cells express all three subtypes (RyR1, RyR2, and RyR3). The RyR is a high-conductance Ca^{2+} release channel (approximately 100 pS for pCa^{2+}) opening for longer durations than the IP_3R, releasing about 20-fold more Ca^{2+} per individual channel opening [12]. Ca^{2+} sparks have been observed in all three types of muscle cell and generate from the spontaneous opening of RyRs as distinct, isolated, and localized events, usually covering <3 μm [13].

All three subtypes of RyR are likely to be expressed in ASMCs, as is evident by the discovery of RyR1, RyR2, and RyR3 mRNAs in porcine tracheal smooth muscle tissues using standard reverse transcription-polymerase chain reaction (RT-PCR) [14]. We have run real-time quantitative RT-PCR and obtained similar observations for RyR1, RyR2, and RyR3 mRNA levels in mouse ASMCs (unpublished data). Western blot analysis and immunofluorescence staining indicate that RyR1, RyR2, and RyR3 proteins are present as well in ASMCs [5, 15, 16]. Within a given SMC, there are multiple spark-generating sites termed Ca^{2+} release units (CRUs) discharging at random, with each locus made up of a heterogeneous mix of RyR1 and RyR2 (but perhaps not RyR3). Interestingly, it is noted that individual Ca^{2+} spark currents ranged over fivefold, suggesting the involvement of anywhere from 1 to 50 RyRs in a single CRU [17]. Along these lines, the number of RyRs present in a CRU dictates the extent of Ca^{2+} current and amplitude. The amplitude of a spark at a given CRU is generally constant; however, the spark-site-to-spark-site variance is significantly different [18]. The average number of Ca^{2+} spark sites is thought to be on the order of 300–400 per cell, making their interplay and propensity for spark formation a topic of great interest [19].

3 Regulation of Ca^{2+} Sparks

Ca^{2+} sparks occur spontaneously in ASMCs at rest as a result of RyR channel opening. RyRs are localized subcellularly and can undergo posttranscriptional and translational modifications, generating structural and functional diversity [20]. Roughly, four-fifths of the RyR amino acid sequence at the N-terminus comprises an enormous cytoplasmic domain that serves as a scaffold for accessory proteins, while the remaining fifth at the C-terminus forms the channel pore within the SR [21]. The ability of the RyR to interact with a variety of accessory proteins that are capable of augmenting the channel function allows the RyR to act as a convergence point for many intracellular signal pathways. One example of accessory proteins is the FK506 binding protein 12.6 (FKBP12.6), which associates with and functionally modulates RyR2 channel function. Treatment with FK506 to remove FKBP12.6 from RyR2 results in an increase in spontaneous Ca^{2+} sparks, and treatment with cyclic ADP-ribose (cADPR) produces a similar effect. FK506 or cADPR is proposed to bind to and remove FKBP12.6 from RyR2, effectively increasing its channel gating properties, which physiologically results in an increase in contraction to methacholine (mACH) in ASMCs [5].

Another mechanism for regulating Ca^{2+} sparks in ASMCs is the action of the IP_3R, of which there are three subtypes [1–3, 22]. In 1996, it was shown that IP_3Rs are necessary for acetylcholine (ACh)-induced contraction in porcine ASMCs [23, 24]. We have found that within mouse ASMCs, the basal activity of G protein subunit Gqα results in phospholipase C (PLC) activation, which in turn generates inositol 1,4,5-trisphosphate (IP_3) from phosphatidylinositol 4,5-bisphosphate (PIP2). IP_3 then binds with IP_3Rs, resulting in a Ca^{2+} release that subsequently activates neighboring RyRs through a local Ca^{2+}-induced Ca^{2+}-release (CICR) mechanism promoting Ca^{2+} spark formation [9]. ACh activates the classic muscarinic M_3 receptors (M_3Rs) through Gqα to induce cellwide transients (Ca^{2+} waves) in ASMCs via the IP_3R-mediated CICR. The frequency and amplitude of these waves are dependent on the SR Ca^{2+} stores and the time required to refill the stores [25]. The mechanism of IP_3R-mediated Ca^{2+} release has been further supported by experiments in which photolytic release of caged IP_3 in the absence of agonists causes the generation of Ca^{2+} waves and contraction in ASMCs [26, 27].

Moreover, mACH treatment or membrane depolarization selectively activates RyR2 via the Gqα-PLC-IP_3-IP_3R pathway [15]. Previous studies indicated that membrane depolarization activated voltage-dependent calcium channels (VDCCs), resulting in Ca^{2+} influx, which in turn triggers Ca^{2+} release from RyRs. Our results, however, demonstrate that such a mechanism is not operative in ASMCs. Our data indicate that depolarization first activates M_3R, which then triggers the Gqα-PLC-IP_3-IP_3R signaling pathway and, finally, selectively activates the RyR_2 subtype (but not RyR1 or RyR3) in ASMCs, resulting in an increase in Ca^{2+} sparks, leading to contraction.

Another pathway, PLC/diacylglycerol (DAG)/protein kinase C (PKC), is also involved in the regulation of Ca^{2+} sparks. In mouse ASMCs, DAG, generated by PLC activity, activates PKC subtype-epsilon (PKCε). As seen in Fig. 1, PKCε selectively inhibits Ca^{2+} sparks through its specific effects on RyR1, but not RyR2 or RyR3. This inhibition of Ca^{2+} sparks reduces the contractile response of ASMCs to mACH. We have further found that FKBP12.6 is not involved in this process [28]. Taken together, Ca^{2+} spark activity is independent of extracellular Ca^{2+} influx and regulated by Gq/PLC/IP_3/IP_3R-RyR2 (excitatory) and Gq/PLC/DAG/PKCε/RyR1 (inhibitory) signaling pathways. Unlike in arterial smooth muscle cells, Ca^{2+} sparks result in airway muscle contraction. Furthermore, our new unpublished data, as shown in Fig. 2, demonstrate that the protein phosphatase calcineurin (CaN) can upregulate Ca^{2+} sparks through the specific RyR1 subtype in a coordinated kinase-phosphatase mechanism opposing the action of PKCε on mouse ASMC Ca^{2+} sparks. Inhibition of CaN with the specific synthetic CaN-autoinhibitory peptide (CAIP) reduces maximal contraction of isolated tracheal rings to 10 μM mACH. These findings provide new insights into the regulatory pathways governing ASMC Ca^{2+} spark frequency.

RyRs are gated by a wide range of the endogenous ligand Ca^{2+}, which activates RyRs in the low hundreds of nanomolar range and inhibits in the ~1 mM range; thus, Ca^{2+} release from one RyR can promote Ca^{2+} release from a neighboring RyR, until an array of RyRs at one loci is triggered to open synchronously, generating a detectable spark from a functional CRU [29, 30]. It has been noted that removal of extracellular Ca^{2+} using nominally Ca^{2+}-free bath solution with 50 μM ethylene glycol tetraacetic acid (EGTA) resulted in no significant decrease in Ca^{2+} sparks. Similarly, removal of extracellular Ca^{2+} did not affect spontaneous Ca^{2+} spark frequency in porcine ASMCs [6]. The lack of a role for extracellular Ca^{2+} on Ca^{2+} spark frequency suggests the importance of internal store, SR content.

Noticeably, there is evidence that increases in SR Ca^{2+} content will result in an enhancement in Ca^{2+} spark frequency in ventricular myocytes of rabbit, rat, and mouse [31, 32]. Thus, Ca^{2+} sparks are regulated by SR load and its electrochemical gradient for Ca^{2+} movement out of the SR through RyR-based CRUs.

Furthermore, various other studies align themselves with or propose alternatives to the current findings. Within the heart, FK506 and rapamycin, which are macrolide immunosuppressants capable of disrupting interactions between RyRs and FKBP12/12.6, resulted in increased RyR channel leakiness, providing support for the similar role of FKBP12.6 seen in ASMCs [33]. Similarly to ASMC, IP_3Rs were also found to promote Ca^{2+} spark formation in rabbit portal vein smooth muscle cells. Tonic activation of PLC at the plasma membrane will generate cytosolic IP_3. IP_3 will then function as a Ca^{2+} mobilizing second messenger by binding to the IP_3R activating it, generating Ca^{2+} puffs. These small IP_3R-generated Ca^{2+} transients will trigger local RyRs coexisting in the SR to activate and open

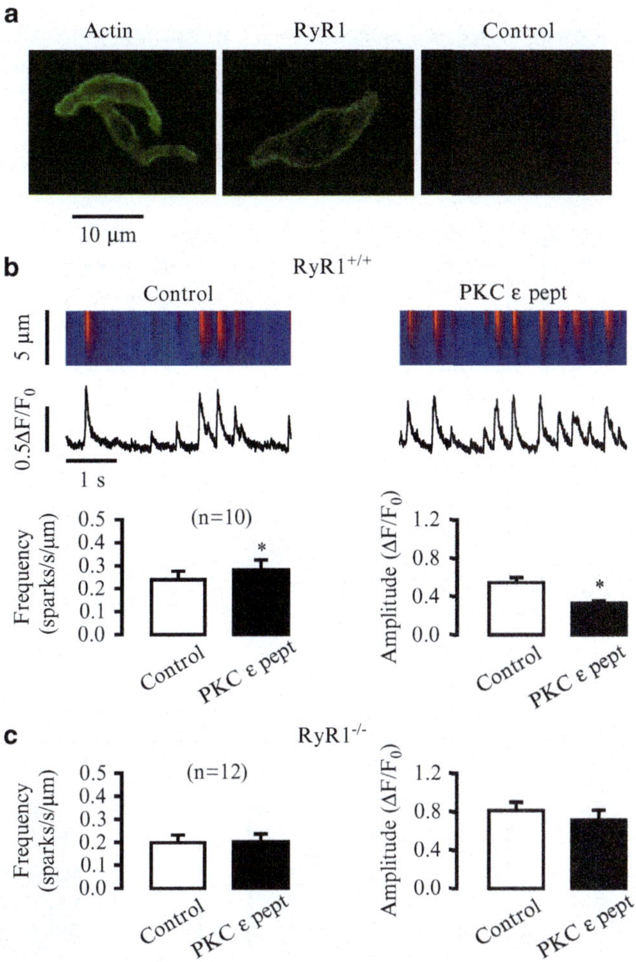

Fig. 1 *RyR1 gene deletion abolishes the effect of specific PKCε inhibition on Ca^{2+} sparks in airway SMCs in the absence of functional IP_3Rs.* (**a**) Expression of smooth muscle-specific actin and RyR1 in cells isolated from embryonic RyR1$^{+/+}$ mice. Cells were initially incubated with anti-RyR1 antibody, anti-actin (smooth muscle) antibody or without either antibody (control), and then with Alexa Fluor 488–conjugated anti-mouse antibody. Immunofluorescent staining was examined using a Zeiss laser scanning confocal microscope. (**b**) Effect of PKCε peptide inhibitor on Ca^{2+} sparks in embryonic RyR1$^{+/+}$ cells after treatment with xestospongin-C to functionally remove IP_3Rs. Original recordings of Ca^{2+} sparks were made in an RyR1$^{+/+}$ cell before and after application of 100 μM PKCε peptide inhibitor (PKCε pept) for 8 min in the presence of xestospongin-C (10 μM). Bar graphs summarize the effects of PKCε peptide inhibitor on the frequency and amplitude of Ca^{2+} sparks in RyR1$^{+/+}$ cells. $^*P < 0.05$ compared with control (before application of PKCε pept). Numbers in parentheses indicate the number of cells examined. (**c**) Effect of PKCε peptide inhibitor (100 μM) on Ca^{2+} sparks in embryonic RyR1$^{+/+}$ cells after treatment with xestospongin-C (10 μM) for 8 min (Adapted from Liu et al. [28], 663–671)

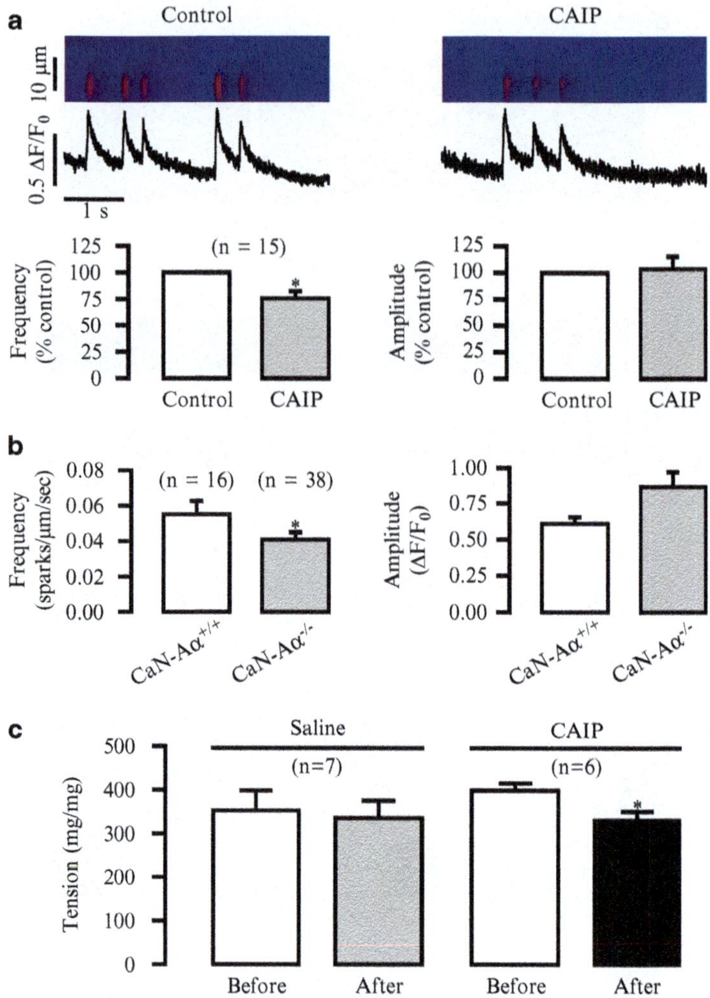

Fig. 2 *Calcineurin upregulates Ca^{2+} sparks and contraction in airway SMCs.* (**a**) Original confocal line scan images of Ca^{2+} sparks in an ASMC before and after treatment with CAIP (10 μM) for 8 min in the presence of xestospongin-C (10 μM). *Bar graph* represents the average frequency and amplitude of Ca^{2+} sparks before and after CAIP in the presence of xestospongin-C. $^{*}P < 0.05$ compared with control. Numbers in parentheses indicate the number of cells examined. (**b**) *Bar graph* represents the average frequency and amplitude Ca^{2+} sparks from CaN-A$\alpha^{+/+}$ and CaN-A$\alpha^{-/-}$ cells. (**c**) Muscle tension generation to 10 μM mACH in isolated tracheal ring before and after treatment with saline or CAIP (20 μM) for 45 min

increasing Ca^{2+} spark formation, releasing more Ca^{2+} in the cytosol through a local CICR mechanism [34]. However, reported inhibition of PLC by U73122 does not inhibit the frequency of spontaneous Ca^{2+} sparks in pulmonary arterial smooth muscle cells, indicating that tonic activation of PLC will not enhance spontaneous Ca^{2+} sparks in this tissue type [35]. Another possibility was seen in guinea-pig vas deferens, whereby basal activation of IP_3Rs inhibited Ca^{2+} spark formation [36]. Therefore, the regulatory pathways differ in different cell types and need to be elucidated individually. Another example of these tissue-specific differences is seen by the varying role of PKC in regulating Ca^{2+} sparks. Within cerebral arterial smooth muscle cells, 1-oleoyl-2-acetyl-glycerol (OAG) and phorbol-12-myristate-13-acetate (PMA), which are PKC activators, decreased the frequency of spontaneous Ca^{2+} sparks [37]. On the other hand, in pulmonary arterial smooth muscle cells, the PKC inhibitor staurosporine and the activator OAG had no effect on spontaneous Ca^{2+} sparks [35].

Another mechanism involved in the regulation of Ca^{2+} sparks not mentioned so far is the role of VDCCs. Skeletal, cardiac, and smooth muscle cells all express VDCCs, which will mediate Ca^{2+} influx when triggered by a depolarizing current. In skeletal muscle cells, VDCCs physically couple with RyRs such that the activation of VDCCs will directly affect the structural conformation of the RyR activating it, resulting in the appearance of Ca^{2+} sparks [38, 39]. In heart cells, membrane depolarization will activate VDCCs, generating extracellular Ca^{2+} influx in the form of Ca^{2+} sparklets, which will then participate in CICR-activating RyRs, resulting in the formation of Ca^{2+} sparks [40]. In smooth muscle cells, VDCCs and RyRs are loosely coupled. For example, in the case of bladder smooth muscle cells, Ca^{2+} influx via VDCCs results in a global Ca^{2+} rise, which will then activate RyRs to generate Ca^{2+} sparks [41].

Studies within ASMCs have shown a specific role for FKBP12.6, IP_3Rs, PKC, CaN, and Ca^{2+} store content in regulating Ca^{2+} sparks; however, this is only scratching the surface. The RyR is a true convergence point for numerous cellular signaling pathways, including Ca^{2+}, cAMP, and nitric oxide [42], while a low pH [43], ATP [44], high temperature [45], cADPR [46], and reactive oxygen species [47] promote sparks, whereas Mg^{2+} inhibits spark formation [48]. Further work is needed to better understand the regulatory mechanisms governing ASMC Ca^{2+} sparks and how they play into the contractile response of airways.

4 Role of Ca^{2+} Sparks

Prior to the original discovery of Ca^{2+} sparks in 1993, Benham and Bolton in 1986 recorded spontaneous transient outward currents (STOCs) in rabbit jejunum and ear artery smooth muscle cells. These STOCs are generated by big-conductance Ca^{2+}-activated K^+ (BK) channels and can be abolished by dialysis of EGTA into the cell and enhanced by extracellular application of caffeine, ACh, or noradrenalin. Thus, the occurrence of STOCs is primarily dependent on a rise in intracellular Ca^{2+} released from RyRs on the SR [49]. Such STOCs were then noted in portal vein SMCs [50–52]. STOCs were thought to be dependent on a spontaneous rise in intracellular Ca^{2+}. Indeed, Nelson et al. in 1995 confirmed that STOCs were activated by Ca^{2+} sparks and found that they resulted in the relaxation of rat cerebral arterial SMCs [53]. Thus, Ca^{2+} sparks activate BK channels, allowing K^+

efflux to generate STOCs, which results in membrane hyperpolarization; this in turn leads to VDCC inactivation, preventing Ca^{2+} influx, thereby causing relaxation. Studies within frog stomach smooth muscle cells showed that at a holding potential of 0 mV, application of caffeine at a low concentration (0.5 mM) increased spark frequency by 2.2-fold, which correlated well with a 2.4-fold increase in STOC frequency; however, not every spark always causes a STOC [17]. Relatively small sparks can generate large STOCs and relatively large sparks might fail to elicit STOCs, as seen by approximately 21 % of Ca^{2+} sparks failing to elicit STOCs in the aforementioned frog stomach smooth muscle cells. An extension of these studies within the frog stomach demonstrated that a STOC event results from an exposure to >10 μM Ca^{2+} with an upper limit of approximately 150 μM in an area less than 1 μm^2 roughly 150–300 nm^2 ~ 0.28 μm^2 [18]. This spark-STOC microdomain is perfectly capable of having local Ca^{2+} elevations sufficient to activate localized BK channels while not changing the resting intracellular Ca^{2+}.

Similar to STOCs, spontaneous transient inward currents (STICs) have also been observed in guinea-pig mesenteric and rabbit portal vein smooth muscle cells, and STICs are caused by Ca^{2+}-activated chloride (Cl_{Ca}) channels triggered by release of Ca^{2+} from the SR [54, 55]. During a STIC, Cl^- efflux depolarizes the cell membrane, making it easier to activate VDCCs, promoting Ca^{2+} release and contraction. The ability of smooth muscle cell Ca^{2+} sparks to regulate STIC and STOC currents, which play a role in regulating membrane excitability, has led us to question whether any or all of these findings are applicable to those in ASMCs.

We have found that STOCs generate in ASMCs with a high frequency at a more depolarized membrane potential and are significantly increased by the activation of VDCCs; [56] more interestingly, muscarinic stimulation inhibits the generation of STOCs, presumably causing membrane depolarization and then promoting cell contraction [56]. In support of these findings, excised inside-out patch clamp recordings have revealed that the open probability of single BK channels is significantly decreased in ASMCs following muscarinic stimulation [57, 58], and functional studies indicate that muscarinic contraction is due, at least in part, to the inhibition of BK channels [58]. This unique inhibition of BK channels by muscarinic stimulation in ASMCs is likely to be mediated by the muscarinic M2 receptor/BK-channel-dependent coupling mechanism [59].

STICs take place with STOCs in ASMCs [60, 61]. The former preferentially occur at membrane potentials more negative to −40 mV, whereas the latter events happen at membrane potentials more positive to −30 mV [62]. The activity of STICs is controlled by Ca^{2+}/calmodulin-dependent kinase II (CaMKII) [63]; as such, the activity of CaMKII is likely to be enhanced following asthmatic stimuli, which causes STICs, membrane depolarization, extracellular Ca^{2+} influx, and cell contraction, contributing to the development of asthma.

Within guinea-pig ASMCs using the perforated patch clamp in conjunction with wide-field Ca^{2+} imaging [8], at a holding potential of −96 mV, roughly E_K spontaneous Ca^{2+} sparks synchronously triggered STICs. These STICs were prevented by treatment with niflumic acid, a known blocker of Cl_{Ca} channels, suggesting that a single Ca^{2+} spark will activate STICs by initiating Cl^- efflux through ${Cl_{Ca}}^{2+}$

channels. At 0 mV, Ca^{2+} sparks were accompanied by STOCs, mediated by the BK channel. At holding potentials between E_K and E_{Cl} approximately −30 mV, a single spontaneous Ca^{2+} spark triggered a biphasic current with an outward cation efflux believed to be the result of BK channel activation followed by anion efflux resulting from Cl_{Ca} activation. A single spark site could trigger STOCs, STICs, and biphasic transients termed STOICs (because the outward preceded the inward current) depending on the membrane potential. It was also interesting to note that the peak STOC current was reached before the Ca^{2+} spark peak and decayed much faster, too, whereas the STIC current more closely paralleled the rise and fall of the Ca^{2+} spark yet occurred second in measured STOICs, suggesting differences in the spatial arrangement, coupling, and activation of BK and Cl_{Ca} channels to Ca^{2+} sparks. This work showed the possibility for a single spark at a single spark site to elicit both BK-mediated STOCs and Cl_{Ca} channel STICs, suggesting that Ca^{2+} sparks in guinea-pig ASMCs function to stabilize membrane potential. At membrane potentials closer to the resting membrane potential, a STIC will dominate and depolarize the cell membrane, but at less negative potentials closer to E_{Cl} (usually around −30 to −20 mV in smooth muscle cells) the STOC may dominate, providing negative feedback hyperpolarizing the cell membrane. The ability of Ca^{2+} sparks to induce depolarizing and hyperpolarizing currents suggests a significant role in regulating a cell's contractile state.

In mouse ASMCs, Lifshitz et al. [64] showed that Cl_{Ca} channels have an activation threshold greater than 120 nM Ca^{2+} and are exposed to 2.4 μM Ca^{2+} or greater during a spark with an EC_{50} of 3.3 μM (similar to the Cl_{Ca} currents seen in *Xenopus* oocytes and olfactory neurons), with the spark Ca^{2+} current spreading out with a 600 nm radius. The STIC amplitude does not correlate with spark amplitude, allowing both large and small STICs to be generated from variable sized sparks, and not all sparks generate STICs such that the maximum current of the STIC is not necessarily dependent on the spark size but on the number of Cl_{Ca} channels present near the CRU. This leads us to believe that Cl_{Ca} channels are distributed nonuniformly along the cell membrane. Furthermore, it was suggested that all Cl_{Ca} channels present in a mouse ASMC on the order of approximately 20,000 would localize within this 600 nm radius with a density as high as 300 channels/μm^2, albeit a little farther away from the CRU than BK channel clusters. The number of RyRs in a CRU is suggested to be in a range of approximately 10, and unlike in skeletal and cardiac muscles, these mouse ASMC CRU-based RyRs do not open in concert. In comparison to the localized coupling of Cl_{Ca} to Ca^{2+} sparks as just shown, dual immunocytochemistry staining was used to visualize the distribution of RyR isoforms and BK channels. In ASMCs, RyR1 and RyR2 form clusters at the cell periphery, with RyR3 at the cell interior. These clusters of RyRs in a CRU will activate two to three clusters of BK channels within a similar 600 nm radius, as stated in the previously cited studies on Cl_{Ca} channel coupling.

Ca^{2+} sparks have been found to be associated with biphasic membrane potential transients (BiMPTs) in mouse ASMCs [65]. At a resting membrane potential of −46.3 mV, a single Ca^{2+} spark hyperpolarized the membrane by −20 mV and successively depolarized the membrane by 10 mV, thus regulating the membrane potential by approximately 30 mV. Similar to the studies from frog stomach smooth muscle cells years earlier, there was a weak correlation between Ca^{2+} spark signal strength and the voltage changes just mentioned, suggesting a large diversity in the coupling of Ca^{2+} sparks and their target channels. Paxilline and niflumic acid were able to block the hyperpolarization and depolarization phases, respectively, indicating the role of BK and Cl_{Ca} channels in BiMPTs. No single BiMPT was observed to depolarize beyond E_{Cl} −15 mV or hyperpolarize below E_K −80 mV, thereby acting to prevent extreme changes in membrane potential. ASMCs are known not to fire action potentials (APs), and the application of short depolarizing currents failed to elicit an AP; however, in the presence of ryanodine, to effectively block Ca^{2+} sparks and BiMPTs, that same stimulus generated an AP that was subsequently blocked by nifedipine, indicating a role for VDCCs in the generation of APs. Disrupting the balance between BK and Cl_{Ca} channel activity can hinder cells to regulate and manage membrane potential excitability, leading to alterations in contractility and possible pathological consequences.

STICs are dependent on a spontaneous rise in intracellular Ca^{2+} and are Cl^- transients passing through the TMEM16A channels in ASMCs [66–68]. These Ca^{2+} spark-activated STICs will induce membrane depolarization and can activate VDCCs, resulting in Ca^{2+} influx, leading to contraction. It has been shown that cADPR treatment to remove FKBP12.6 from RyR2 results in an increase in STICs in ASMCs and increased contractile responses to mACH [5]. TMEM16A is expressed in ASMCs, and selective inhibition with benzbromarone inhibited the contractile response of tracheal rings to mACH in a range of 1–100 μM, but not to KCl, suggesting that TMEM16A is specifically activated and participates in ASMC contraction due to Ca^{2+} release from G-protein-coupled muscarinic receptor activation [69]. By tipping the balance of activation in favor of BK or TMEM16A channels, the relaxed or contracted state, respectively, might be favored, and it is therefore an essential part of asthma and other related respiratory diseases.

5 Conclusions and Perspectives

The original discovery of Ca^{2+} sparks in cardiac muscle cells has illustrated an innovative example of so-called microdomain Ca^{2+} signaling. When a cardiomyocyte is stimulated, individual Ca^{2+} sparks will fuse together, summating as a large local elevation in $[Ca^{2+}]_i$ and generating a propagating Ca^{2+} wave across the cell, inducing contraction. Ca^{2+} sparks are caused by the opening of RyR2 in cardiac cells. Distinctively, ASMCs express all three subtypes of RyRs. As illustrated in Fig. 3, at least RyR1 and RyR2 are involved in the generation of Ca^{2+} sparks in ASMCs. Interestingly, Ca^{2+} sparks due to the opening of RyR1 are downregulated by PKCε and upregulated by CaN. This novel kinase/phosphatase regulatory mechanism is important for the adequate generation of Ca^{2+} sparks and associated cellular functions. It should be pointed out that RyR1-mediated Ca^{2+} sparks are negatively regulated by the basal activity of PLC/DAG-dependent PKCε. RyR2-mediated Ca^{2+}

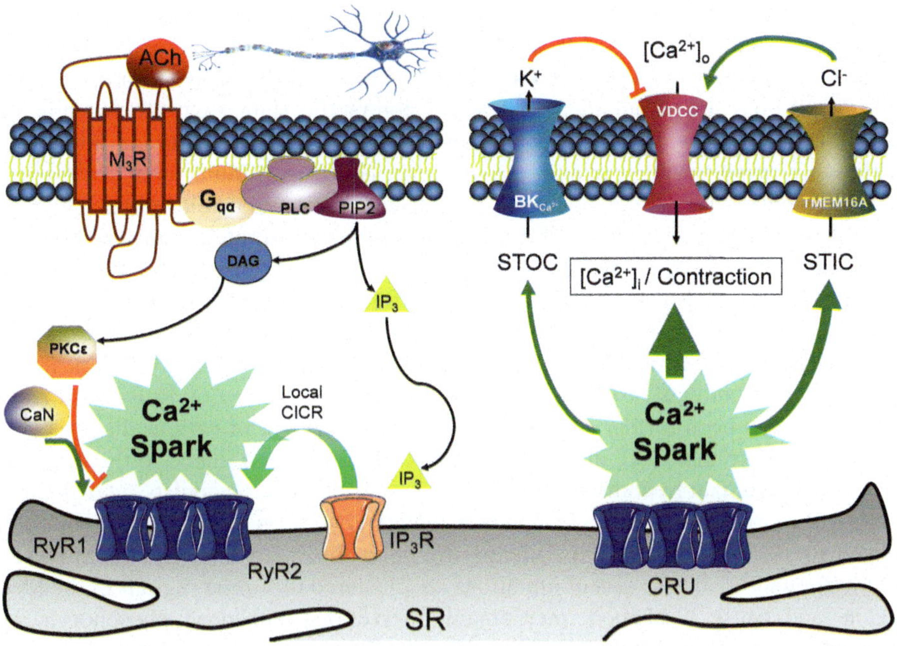

Fig. 3 *Role and regulation of local Ca^{2+} signaling in ASMCs*. The *left panel* portrays a summary of our work in the studies of the regulatory mechanisms of Ca^{2+} sparks. Activation of classic M_3Rs by the neurotransmitter ACh from parasympathetic nerves will trigger a Gqα-coupled activation of PLC, resulting in the generation of IP_3 and DAG from PIP2. IP_3 will bind to IP_3Rs inducing Ca^{2+} release. This Ca^{2+} will trigger a local CICR through RyR2 increasing Ca^{2+} sparks and then inducing contraction. At the same time, DAG will activate PKCε which will inhibit Ca^{2+} sparks through RyR1 and muscarinic contraction. Opposing PKCε, CaN promotes Ca^{2+} sparks and contraction through RyR1. In the *right panel*, Ca^{2+} sparks generate from the coordinated opening of RyRs in a functional Ca^{2+} release unit (*CRU*). This local Ca^{2+} signaling will participate in the activation of contractile machinery by effectively regulating the cells membrane potential. At rest, Ca^{2+} sparks will preferentially activate TMEM16A-encoded Ca^{2+}-activated Cl^- channels, leading to Cl^- efflux, depolarizing the membrane. As the cell membrane is depolarized, Ca^{2+} sparks activate BK channels, resulting in K^+ efflux, which hyperpolarizes the membrane. A disruption in this balance can change the cell excitability. As suggested by our own work and the recent publications by others, it is believed that in asthma, local Ca^{2+} signaling is increased. This scenario can provide a mechanistic setting for the excessive contraction to nonspecific stimuli, airway hyperresponsiveness and asthma attacks. The STOC pathway is no longer able to stabilize membrane transients, resulting in the increased responsiveness

sparks are significantly upregulated by Ca^{2+} release from neighboring IP_3Rs. This local CICR plays a significant role in cell force generation. Stimulation of muscarinic receptors by parasympathetic nerves, which provides a major neural control of ASMCs, also remarkably enhances RyR2-mediated Ca^{2+} sparks, thereby making a significant contribution in muscarinic cellular functions. Stimulation of muscarinic receptors may also activate PKCε through the PLC/DAG pathway and then inhibit Ca^{2+} sparks through RyR1 to regulate muscarinic responses.

At rest, Ca^{2+} sparks spontaneously occur and preferentially generate depolarizing STICs in ASMCs. The spontaneous Ca^{2+} release and subsequent Ca^{2+} influx associated with STICs could synergistically cause contraction; as such, ASMCs are maintained at a slight contractile status. On the other hand, Ca^{2+} sparks produce STOCs as well at more depolarized membrane potentials. These hyperpolarizing currents prevent excessive Ca^{2+} influx and contraction, thereby preserving a dynamic state of cell contraction and relaxation. It is also worth stating that stimulation of muscarinic and other G-protein-coupled receptors triggers a spatial and temporal integration of local Ca^{2+} release events through RyR channels, inducing Ca^{2+} oscillations that crisscross the entire cell, generating a contractile tone [7]. Thus, specific regulation of RyR1 and RyR2 to control Ca^{2+} sparks may serve as a novel signaling mechanism for multiple physiological and pathological cellular responses in ASMCs. The importance of RyR-based Ca^{2+} signaling has been further appreciated by its involvement in various diseases. For example, more than 100 mutations within RyR1 have been linked to the development of malignant hypothermia, making it hypersensitive to agonists such as Ca^{2+}, cADPR, caffeine, and halogenated anesthetics [70, 71]. Over 80 mutations have been found in the RyR2 gene underlying heart disorders such as stress-induced arrhythmia [72]. These sets of mutations are clustered in similar regions of the RyR gene and cause weakened interdomain interactions, enhanced phosphorylation by protein kinases, altered sensitivity to Ca^{2+}, and modified interactions with accessory proteins [33]. Furthermore, the RyR-related models of heart failure are dependent on a Ca^{2+} leak pathway, reducing systolic contractile effects [73]. However, there is no report on the occurrence of RyR1 or RyR2 mutations in ASCMs and their association with asthma and other relevant lung diseases. Potentially life-threatening asthma attacks are caused by the exaggerated contractile response to nonspecific stimuli. Our very recent studies have shown that Ca^{2+} sparks are significantly upregulated in ASMCs from an ovalbumin-induced mouse model of asthma, and the increased frequency of Ca^{2+} sparks strongly correlates with augmented airway resistance (data not published). These results reveal that the increased Ca^{2+} signaling is critical for the excessive contractile responses in ASMCs and asthma. The increased Ca^{2+} sparks also cause more STICs, contributing to the development of asthma. Our work further indicates that RyR1 gene deletion prevents the development of ovalbumin-induced asthma in mice. In line with our findings, a recent report showed that the Cl_{Ca} channel gene TMEM16A is upregulated in airway smooth muscle from ovalbumin-evoked asthmatic mice [74]. Blockers of Cl_{Ca} channels such as niflumic acid and benzbromarone are able to block airway hyperresponsiveness; deletion of TMEM16A reduces caffeine- and mACH-induced cell shortening. Without a doubt, it would be intriguing to decipher the relative contributions of RyR-dependent Ca^{2+} sparks and Cl_{Ca} channels. Ca^{2+} sparks in ASMCs are an ongoing area of biomedical research, and while great strides have been made, diligent focus and efforts must be employed to bring these findings to the bedside with the identification of novel drug targets for asthma and other respiratory diseases.

References

1. Cheng H, Lederer WJ, Cannell MB. Calcium sparks: Elementary events underlying excitation-contraction coupling in heart muscle. *Science* 1993;262:740–744.
2. Cheng H, Lederer WJ. Calcium sparks. *Physiological reviews* 2008;88:1491–1545.
3. Ledoux J, Taylor MS, Bonev AD, Hannah RM, Solodushko V, Shui B, Tallini Y, Kotlikoff MI, Nelson MT. Functional architecture of inositol 1,4,5-trisphosphate signaling in restricted spaces of myoendothelial projections. *Proceedings of the National Academy of Sciences of the United States of America* 2008;105:9627–9632.
4. Niggli E, Shirokova N. A guide to sparkology: The taxonomy of elementary cellular ca2+ signaling events. *Cell calcium* 2007;42:379–387.
5. Wang YX, Zheng YM, Mei QB, Wang QS, Collier ML, Fleischer S, Xin HB, Kotlikoff MI. Fkbp12.6 and cadpr regulation of Ca^{2+} release in smooth muscle cells. *American journal of physiology Cell physiology* 2004;286:C538–546.
6. Sieck GC, Kannan MS, Prakash YS. Heterogeneity in dynamic regulation of intracellular calcium in airway smooth muscle cells. *Canadian journal of physiology and pharmacology* 1997;75:878–888.
7. Pabelick CM, Prakash YS, Kannan MS, Sieck GC. Spatial and temporal aspects of calcium sparks in porcine tracheal smooth muscle cells. *The American journal of physiology* 1999;277: L1018–1025.
8. ZhuGe R, Sims SM, Tuft RA, Fogarty KE, Walsh JV, Jr. Ca^{2+} sparks activate K^+ and Cl^- channels, resulting in spontaneous transient currents in guinea-pig tracheal myocytes. *The Journal of physiology* 1998;513 (Pt 3):711–718.
9. Liu QH, Zheng YM, Wang YX. Two distinct signaling pathways for regulation of spontaneous local Ca^{2+} release by phospholipase c in airway smooth muscle cells. *Pflugers Archiv : European journal of physiology* 2007;453:531–541.
10. Campbell KP, Knudson CM, Imagawa T, Leung AT, Sutko JL, Kahl SD, Raab CR, Madson L. Identification and characterization of the high affinity [^{3}H]ryanodine receptor of the junctional sarcoplasmic reticulum Ca^{2+} release channel. *The Journal of biological chemistry* 1987;262:6460–6463.
11. Lai FA, Erickson HP, Rousseau E, Liu QY, Meissner G. Purification and reconstitution of the calcium release channel from skeletal muscle. *Nature* 1988;331:315–319.
12. Ehrlich BE, Kaftan E, Bezprozvannaya S, Bezprozvanny I. The pharmacology of intracellular Ca^{2+}-release channels. *Trends in pharmacological sciences* 1994;15:145–149.
13. ZhuGe R, Tuft RA, Fogarty KE, Bellve K, Fay FS, Walsh JV, Jr. The influence of sarcoplasmic reticulum Ca^{2+} concentration on Ca^{2+} sparks and spontaneous transient outward currents in single smooth muscle cells. *The Journal of general physiology* 1999;113:215–228.
14. Kannan MS, Prakash YS, Brenner T, Mickelson JR, Sieck GC. Role of ryanodine receptor channels in Ca^{2+} oscillations of porcine tracheal smooth muscle. *The American journal of physiology* 1997;272:L659–664.
15. Liu QH, Zheng YM, Korde AS, Yadav VR, Rathore R, Wess J, Wang YX. Membrane depolarization causes a direct activation of g protein-coupled receptors leading to local Ca^{2+} release in smooth muscle. *Proceedings of the National Academy of Sciences of the United States of America* 2009;106:11418–11423.
16. Du W, McMahon TJ, Zhang ZS, Stiber JA, Meissner G, Eu JP. Excitation-contraction coupling in airway smooth muscle. *The Journal of biological chemistry* 2006;281:30143–30151.
17. ZhuGe R, Fogarty KE, Tuft RA, Lifshitz LM, Sayar K, Walsh JV, Jr. Dynamics of signaling between Ca^{2+} sparks and Ca^{2+}-activated K^+ channels studied with a novel image-based method for direct intracellular measurement of ryanodine receptor Ca^{2+} current. *The Journal of general physiology* 2000;116:845–864.
18. Zhuge R, Fogarty KE, Tuft RA, Walsh JV, Jr. Spontaneous transient outward currents arise from microdomains where bk channels are exposed to a mean Ca^{2+} concentration on the order of 10 microm during a Ca^{2+} spark. *The Journal of general physiology* 2002;120:15–27.

19. Zhuge R, Fogarty KE, Baker SP, McCarron JG, Tuft RA, Lifshitz LM, Walsh JV, Jr. Ca^{2+} spark sites in smooth muscle cells are numerous and differ in number of ryanodine receptors, large-conductance K^+ channels, and coupling ratio between them. *American journal of physiology Cell physiology* 2004;287:C1577–1588.
20. Sorrentino V, Barone V, Rossi D. Intracellular Ca^{2+} release channels in evolution. *Current opinion in genetics & development* 2000;10:662–667.
21. Zalk R, Lehnart SE, Marks AR. Modulation of the ryanodine receptor and intracellular calcium. *Annual review of biochemistry* 2007;76:367–385.
22. Berridge MJ. Inositol trisphosphate and calcium oscillations. *Biochemical Society symposium* 2007:1–7.
23. Liu X, Farley JM. Acetylcholine-induced chloride current oscillations in swine tracheal smooth muscle cells. *The Journal of pharmacology and experimental therapeutics* 1996;276:178–186.
24. Liu X, Farley JM. Acetylcholine-induced Ca^{++}-dependent chloride current oscillations are mediated by inositol 1,4,5-trisphosphate in tracheal myocytes. *The Journal of pharmacology and experimental therapeutics* 1996;277:796–804.
25. Sneyd J, Tsaneva-Atanasova K, Yule DI, Thompson JL, Shuttleworth TJ. Control of calcium oscillations by membrane fluxes. *Proceedings of the National Academy of Sciences of the United States of America* 2004;101:1392–1396.
26. Sneyd J, Tsaneva-Atanasova K, Reznikov V, Bai Y, Sanderson MJ, Yule DI. A method for determining the dependence of calcium oscillations on inositol trisphosphate oscillations. *Proceedings of the National Academy of Sciences of the United States of America* 2006;103:1675–1680.
27. Bai Y, Sanderson MJ. Airway smooth muscle relaxation results from a reduction in the frequency of Ca^{2+} oscillations induced by a camp-mediated inhibition of the ip3 receptor. *Respiratory research* 2006;7:34.
28. Liu QH, Zheng YM, Korde AS, Li XQ, Ma J, Takeshima H, Wang YX. Protein kinase c-epsilon regulates local calcium signaling in airway smooth muscle cells. *American journal of respiratory cell and molecular biology* 2009;40:663–671.
29. Bezprozvanny I, Watras J, Ehrlich BE. Bell-shaped calcium-response curves of ins (1,4,5) p3- and calcium-gated channels from endoplasmic reticulum of cerebellum. *Nature* 1991; 351:751–754.
30. Fill M, Copello JA. Ryanodine receptor calcium release channels. *Physiological reviews* 2002;82:893–922.
31. Satoh H, Blatter LA, Bers DM. Effects of $[Ca^{2+}]_i$, sr Ca^{2+} load, and rest on Ca^{2+} spark frequency in ventricular myocytes. *The American journal of physiology* 1997;272:H657–668.
32. Santana LF, Kranias EG, Lederer WJ. Calcium sparks and excitation-contraction coupling in phospholamban-deficient mouse ventricular myocytes. *The Journal of physiology* 1997;503 (Pt 1):21–29.
33. Blayney LM, Lai FA. Ryanodine receptor-mediated arrhythmias and sudden cardiac death. *Pharmacology & therapeutics* 2009;123:151–177.
34. Gordienko DV, Bolton TB. Crosstalk between ryanodine receptors and IP_3 receptors as a factor shaping spontaneous Ca^{2+}-release events in rabbit portal vein myocytes. *The Journal of physiology* 2002;542:743–762.
35. Zhang WM, Yip KP, Lin MJ, Shimoda LA, Li WH, Sham JS. Et-1 activates Ca^{2+} sparks in pasmc: Local Ca^{2+} signaling between inositol trisphosphate and ryanodine receptors. *American journal of physiology Lung cellular and molecular physiology* 2003;285:L680–690.
36. White C, McGeown JG. Inositol 1,4,5-trisphosphate receptors modulate Ca^{2+} sparks and Ca^{2+} store content in vas deferens myocytes. *American journal of physiology Cell physiology* 2003;285:C195–204.
37. Bonev AD, Jaggar JH, Rubart M, Nelson MT. Activators of protein kinase c decrease Ca^{2+} spark frequency in smooth muscle cells from cerebral arteries. *The American journal of physiology* 1997;273:C2090–2095.

38. Tsugorka A, Rios E, Blatter LA. Imaging elementary events of calcium release in skeletal muscle cells. *Science* 1995;269:1723–1726.
39. Klein MG, Cheng H, Santana LF, Jiang YH, Lederer WJ, Schneider MF. Two mechanisms of quantized calcium release in skeletal muscle. *Nature* 1996;379:455–458.
40. Wang SQ, Song LS, Lakatta EG, Cheng H. Ca^{2+} signalling between single l-type Ca^{2+} channels and ryanodine receptors in heart cells. *Nature* 2001;410:592–596.
41. Collier ML, Ji G, Wang Y, Kotlikoff MI. Calcium-induced calcium release in smooth muscle: Loose coupling between the action potential and calcium release. *The Journal of general physiology* 2000;115:653–662.
42. MacKrill JJ. Protein-protein interactions in intracellular Ca^{2+}-release channel function. *The Biochemical journal* 1999;337 (Pt 3):345–361.
43. Dabertrand F, Nelson MT, Brayden JE. Acidosis dilates brain parenchymal arterioles by conversion of calcium waves to sparks to activate bk channels. *Circulation research* 2012;110:285–294.
44. Yang Z, Steele DS. Effects of cytosolic atp on Ca^{2+} sparks and sr Ca^{2+} content in permeabilized cardiac myocytes. *Circulation research* 2001;89:526–533.
45. Fu Y, Zhang GQ, Hao XM, Wu CH, Chai Z, Wang SQ. Temperature dependence and thermodynamic properties of Ca^{2+} sparks in rat cardiomyocytes. *Biophysical journal* 2005;89:2533–2541.
46. Zhang X, Tallini YN, Chen Z, Gan L, Wei B, Doran R, Miao L, Xin HB, Kotlikoff MI, Ji G. Dissociation of fkbp12.6 from ryanodine receptor type 2 is regulated by cyclic adp-ribose but not beta-adrenergic stimulation in mouse cardiomyocytes. *Cardiovascular research* 2009;84:253–262.
47. Yan Y, Liu J, Wei C, Li K, Xie W, Wang Y, Cheng H. Bidirectional regulation of Ca^{2+} sparks by mitochondria-derived reactive oxygen species in cardiac myocytes. *Cardiovascular research* 2008;77:432–441.
48. Zhou J, Launikonis BS, Rios E, Brum G. Regulation of Ca^{2+} sparks by Ca^{2+} and mg2+ in mammalian and amphibian muscle. An ryr isoform-specific role in excitation-contraction coupling? *The Journal of general physiology* 2004;124:409–428.
49. Benham CD, Bolton TB. Spontaneous transient outward currents in single visceral and vascular smooth muscle cells of the rabbit. *The Journal of physiology* 1986;381:385–406.
50. Ohya Y, Terada K, Yamaguchi K, Inoue R, Okabe K, Kitamura K, Hirata M, Kuriyama H. Effects of inositol phosphates on the membrane activity of smooth muscle cells of the rabbit portal vein. *Pflugers Archiv : European journal of physiology* 1988;412:382–389.
51. Hume JR, Leblanc N. Macroscopic K^+ currents in single smooth muscle cells of the rabbit portal vein. *The Journal of physiology* 1989;413:49–73.
52. Beech DJ, Bolton TB. Two components of potassium current activated by depolarization of single smooth muscle cells from the rabbit portal vein. *The Journal of physiology* 1989;418:293–309.
53. Nelson MT, Cheng H, Rubart M, Santana LF, Bonev AD, Knot HJ, Lederer WJ. Relaxation of arterial smooth muscle by calcium sparks. *Science* 1995;270:633–637.
54. Van Helden DF. Spontaneous and noradrenaline-induced transient depolarizations in the smooth muscle of guinea-pig mesenteric vein. *The Journal of physiology* 1991;437:511–541.
55. Wang Q, Hogg RC, Large WA. Properties of spontaneous inward currents recorded in smooth muscle cells isolated from the rabbit portal vein. *The Journal of physiology* 1992;451:525–537.
56. Wang YX, Fleischmann BK, Kotlikoff MI. Modulation of maxi-K^+ channels by voltage-dependent Ca^{2+} channels and methacholine in single airway myocytes. *The American journal of physiology* 1997;272:C1151–1159.
57. Kotlikoff MI, Kume H, Tomasic M. Muscarinic regulation of membrane ion channels in airway smooth muscle cells. *Biochemical pharmacology* 1992;43:5–10.
58. Kume H, Mikawa K, Takagi K, Kotlikoff MI. Role of g proteins and kca channels in the muscarinic and beta-adrenergic regulation of airway smooth muscle. *The American journal of physiology* 1995;268:L221–229.

59. Semenov I, Wang B, Herlihy JT, Brenner R. Bk channel beta1 subunits regulate airway contraction secondary to m2 muscarinic acetylcholine receptor mediated depolarization. *The Journal of physiology* 2011;589:1803–1817.
60. Wang YX, Kotlikoff MI. Muscarinic signaling pathway for calcium release and calcium-activated chloride current in smooth muscle. *The American journal of physiology* 1997;273: C509–519.
61. Bao R, Lifshitz LM, Tuft RA, Bellve K, Fogarty KE, ZhuGe R. A close association of ryrs with highly dense clusters of Ca^{2+}-activated Cl^{-} channels underlies the activation of stics by Ca^{2+} sparks in mouse airway smooth muscle. *The Journal of general physiology* 2008;132:145–160.
62. Kotlikoff MI, Wang YX. Calcium release and calcium-activated chloride channels in airway smooth muscle cells. *American journal of respiratory and critical care medicine* 1998;158: S109–114.
63. Wang YX, Kotlikoff MI. Inactivation of calcium-activated chloride channels in smooth muscle by calcium/calmodulin-dependent protein kinase. *Proceedings of the National Academy of Sciences of the United States of America* 1997;94:14918–14923.
64. Lifshitz LM, Carmichael JD, Lai FA, Sorrentino V, Bellve K, Fogarty KE, ZhuGe R. Spatial organization of ryrs and bk channels underlying the activation of stocs by Ca^{2+} sparks in airway myocytes. *The Journal of general physiology* 2011;138:195–209.
65. Zhuge R, Bao R, Fogarty KE, Lifshitz LM. Ca^{2+} sparks act as potent regulators of excitation-contraction coupling in airway smooth muscle. *The Journal of biological chemistry* 2010;285:2203–2210.
66. Schroeder BC, Cheng T, Jan YN, Jan LY. Expression cloning of tmem16a as a calcium-activated chloride channel subunit. *Cell* 2008;134:1019–1029.
67. Caputo A, Caci E, Ferrera L, Pedemonte N, Barsanti C, Sondo E, Pfeffer U, Ravazzolo R, Zegarra-Moran O, Galietta LJ. Tmem16a, a membrane protein associated with calcium-dependent chloride channel activity. *Science* 2008;322:590–594.
68. Yang YD, Cho H, Koo JY, Tak MH, Cho Y, Shim WS, Park SP, Lee J, Lee B, Kim BM, Raouf R, Shin YK, Oh U. Tmem16a confers receptor-activated calcium-dependent chloride conductance. *Nature* 2008;455:1210–1215.
69. Huang F, Zhang H, Wu M, Yang H, Kudo M, Peters CJ, Woodruff PG, Solberg OD, Donne ML, Huang X, Sheppard D, Fahy JV, Wolters PJ, Hogan BL, Finkbeiner WE, Li M, Jan YN, Jan LY, Rock JR. Calcium-activated chloride channel tmem16a modulates mucin secretion and airway smooth muscle contraction. *Proceedings of the National Academy of Sciences of the United States of America* 2012;109:16354–16359.
70. Kobayashi S, Bannister ML, Gangopadhyay JP, Hamada T, Parness J, Ikemoto N. Dantrolene stabilizes domain interactions within the ryanodine receptor. *The Journal of biological chemistry* 2005;280:6580–6587.
71. Lopez JR, Cordovez G, Linares N, Allen PD. Cyclic adp-ribose induces a larger than normal calcium release in malignant hyperthermia-susceptible skeletal muscle fibers. *Pflugers Archiv : European journal of physiology* 2000;440:236–242.
72. Gyorke S, Carnes C. Dysregulated sarcoplasmic reticulum calcium release: Potential pharmacological target in cardiac disease. *Pharmacology & therapeutics* 2008;119:340–354.
73. Munch G, Bolck B, Sugaru A, Brixius K, Bloch W, Schwinger RH. Increased expression of isoform 1 of the sarcoplasmic reticulum Ca^{2+}-release channel in failing human heart. *Circulation* 2001;103:2739–2744.
74. Zhang CH, Li Y, Zhao W, Lifshitz LM, Li H, Harfe BD, Zhu MS, Zhuge R. The transmembrane protein 16a Ca^{2+}-activated Cl^{-} channel in airway smooth muscle contributes to airway hyperresponsiveness. *American journal of respiratory and critical care medicine* 2013;187:374–381.

Regulation of Airway Smooth Muscle Contraction by Ca^{2+} Signaling: Physiology Revealed by Microscopy Studies of Lung Slices

Michael J. Sanderson

Contents

Abstract An examination of the physiology of airway smooth muscle cells (SMCs) in lung slices from mice, rats, and humans with laser scanning microscopy reveals that agonist-induced airway SMC contraction is driven by Ca^{2+} oscillations. An increase in the frequency of Ca^{2+} oscillations correlates with an increase in airway contraction. In addition, contractile agonists simultaneously increase airway SMC Ca^{2+} sensitivity to enhance the action of the Ca^{2+} oscillations. These Ca^{2+} oscillations are primarily mediated by the activity of the inositol trisphosphate receptor (IP_3R). Although airway SMCs contract in response to membrane

M.J. Sanderson (✉)
Department of Microbiology and Physiological Systems, University of Massachusetts Medical School, 55 Lake Ave North, Worcester, MA 01655, USA
e-mail: Michael.Sanderson@umassmed.edu

Y.-X. Wang (ed.), *Calcium Signaling In Airway Smooth Muscle Cells*,
DOI 10.1007/978-3-319-01312-1_7,

depolarization, Ca^{2+} influx via voltage-gated Ca^{2+} channels does not appear to have a major role in agonist-induced airway SMC contraction. Airway relaxation induced by β_2-adrenergic receptor agonists involves both a reduction in the Ca^{2+} oscillation frequency and Ca^{2+} sensitivity.

Keywords Confocal microscopy • Inositol trisphosphate receptor • Contraction • Ca^{2+} oscillation frequency

1 Introduction

Asthma is a common lung disease characterized by airway hyperresponsiveness, a response mediated by the excessive contraction of airway smooth muscle cells (SMCs) [1]. Consequently, the goal of our research is to understand the basic mechanism controlling SMC contractility to determine how changes in this regulation might be responsible for aberrant SMC behavior. While it has been known for some time that SMC contraction is regulated by $[Ca^{2+}]_i$, the correlation between $[Ca^{2+}]_i$ dynamics, SMC contraction, and the caliber of intrapulmonary airways has only recently been appreciated.

A summary of the basic mechanisms regulating airway SMC contraction is shown in Fig. 1. Force development is regulated by the phosphorylation of the regulatory myosin light chain (MLC). When MLC is phosphorylated, myosin can undergo cross-bridge cycling with actin to generate filament sliding and force. Conversely, dephosphorylation of MLC prevents myosin-actin interaction and results in airway SMC relaxation. MLC phosphorylation is mediated by Ca^{2+}/calmodulin activation of MLC kinase (MLCK). The dynamics and regulation of the Ca^{2+} changes leading to MLCK activation will be addressed. MLC dephosphorylation is mediated by MLC phosphatase, which in turn is regulated by a number of other phosphorylation pathways. The role of these pathways in augmenting Ca^{2+}-dependent MLC phosphorylation will also be considered.

Originally, and out of necessity to address the behavior of otherwise inaccessible small airway SMCs, many experiments were conducted with isolated or cultured individual SMCs or with muscle strips obtained from the large airways (trachea or bronchi). Although these isolated preparations were very informative, they could not be used to determine if the physiology of the SMCs from the upper airways represented that of SMC in the smaller airways, how SMC behavior modified airway size, or how SMC activity was influenced by surrounding cells, the extracellular matrix, and lung parenchyma. An additional problem has always been that SMCs often alter their phenotype when in culture, a response complicating the extrapolation of data to the intact animal.

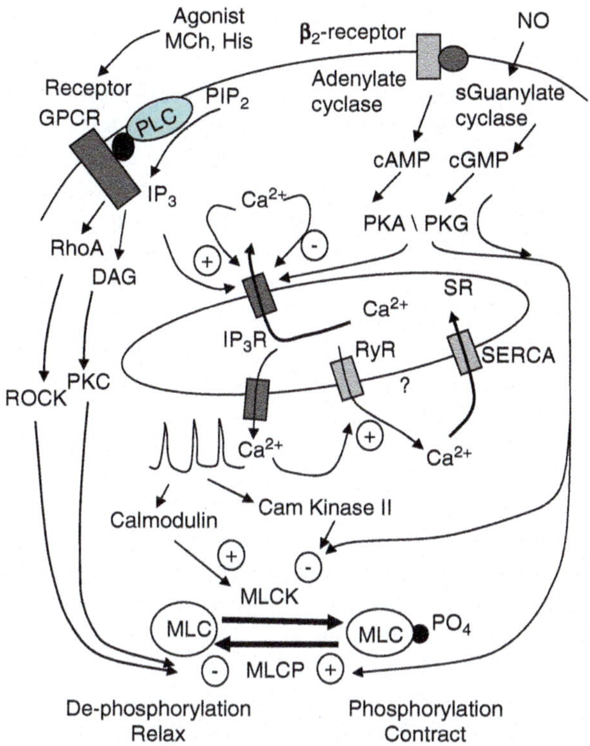

Fig. 1 *The regulation of airway smooth muscle cell (SMC) contraction:* agonists stimulate G-protein-coupled receptors (GPCRs) and phospholipase C (PLC) to synthesize inositol trisphosphate (IP_3) from membrane phosphatidylinositol 4,5-bisphosphate (PIP_2). IP_3 releases Ca^{2+} from the sarcoplasmic reticulum (SR) via the IP_3 receptor (IP_3R), Ca^{2+} is returned to the SR by sarcoplasmic/endoplasmic reticulum Ca^{2+} ATPase (SERCA) pumps; Ca^{2+} release and uptake lead to Ca^{2+} oscillations. The ryanodine receptor (RyR) mediates KCl-induced Ca^{2+} oscillations and may contribute to initial agonist-induced Ca^{2+} oscillations. Ca^{2+} oscillations stimulate myosin light chain (MLC) kinase (MLCK) via calmodulin to phosphorylate MLC and initiate contraction. Agonists inactivate MLC phosphatase (MLCP) via receptors, protein kinase C (PKC), or Rho-activated kinase (ROCK) to decrease MLC dephosphorylation and enhance contraction. Airway SMC relaxation is induced by reducing Ca^{2+} oscillation frequency by increasing cAMP/cGMP, which results in the inhibition of the IP_3R or by stimulating MLCP activity via cAMP. MLCK activity can also be modulated by phosphorylation

More recently, progress has been made toward understanding airway SMC physiology by the development of advanced microscopy techniques such as confocal and two-photon microscopy and our further development of living lung slices from mice, rats, and humans that maintain inherent SMC function. Consequently, the objective of this chapter is to emphasize what we have learned from the Ca^{2+} imaging studies of lung slices in relation to airway contraction and how this relates to earlier results from isolated SMCs. Extensive reviews of Ca^{2+} signaling in isolated or cultured cells are available elsewhere [29, 33, 42].

2 Preparation of Lung Slices

The rationale for using lung slices is that airways and their SMCs in lung slices better reflect the in vivo state of SMCs than do isolated SMCs growing in plastic dishes. Therefore, the method of lung slice preparation is relatively important. The techniques used with mouse, rat, and human lungs are essentially the same [41]. Briefly, lungs are inflated with warm agarose (approximately 2 %) to a volume that is below total lung capacity (to avoid barotrauma). Some air is introduced into the lung after the agarose to clear the agarose from the airways into the alveoli. The tissue is cooled to solidify the agarose and stiffen the lungs for slicing. Multiple lung slices (approximately 170–250 μm) are cut with a vibratome and suspended in culture medium without serum in a humidified incubator at 37 °C with 10 % CO_2 for up to 3 days until use. Individual lung slices are mounted in perfusion chambers for microscopy observation. Only lung slices with actively beating airway cilia are used for the experiments.

In comparison to isolated SMCs, lung slices have not been exposed to enzymes that degrade extracellular matrix and cell interactions. Thus, it is likely that SMCs within lung slices have all their membrane receptors relatively intact. Furthermore, the normal recoil forces opposing SMC contraction, mediated by the extracellular matrix, remain, and these can be exploited to study airway relaxation. All other key cell types that might influence SMC activity, such as epithelial cells or mast cells, are present in their correct relative proportions and positions [27, 41]. One caveat of the lung slice preparation might be that cell trauma has occurred at the slice edge. However, it is unreasonable to suggest that this damage is worse than that experienced by isolated cells. On the contrary, most cells of the lung slice are undamaged. Clearly, drug application to a lung slice does not mimic drug inhalation, and drugs have ready access to the SMCs. But, again this is no different from isolated cell preparations. A major difference between whole lungs and lung slices is that the alveoli of lung slices are filled with agarose rather than air [27]. This is essential to maintain the inflated state of the lung slice, but it may alter the recoil forces acting on the SMCs. On the other hand, cultured SMCs are adhered to rigid plastic while muscle strips are commonly observed with isometric tension; these are conditions that must be accepted if experiments are to be attempted.

3 Morphology of Airways and SMCs in Lung Slices

Within the lung slice, airways and adjacent arterioles are easily identified by their relatively large circular shape (in transverse cross section) that contrasts with finer mesh work of the alveoli tissue (Fig. 2). The airway usually has a larger diameter than the arteriole. Although not a focus of this chapter, this anatomical association of the airway and arteriole provides a unique ability to perform comparative physiology of airway and arteriole SMCs [35, 36]. The airways are

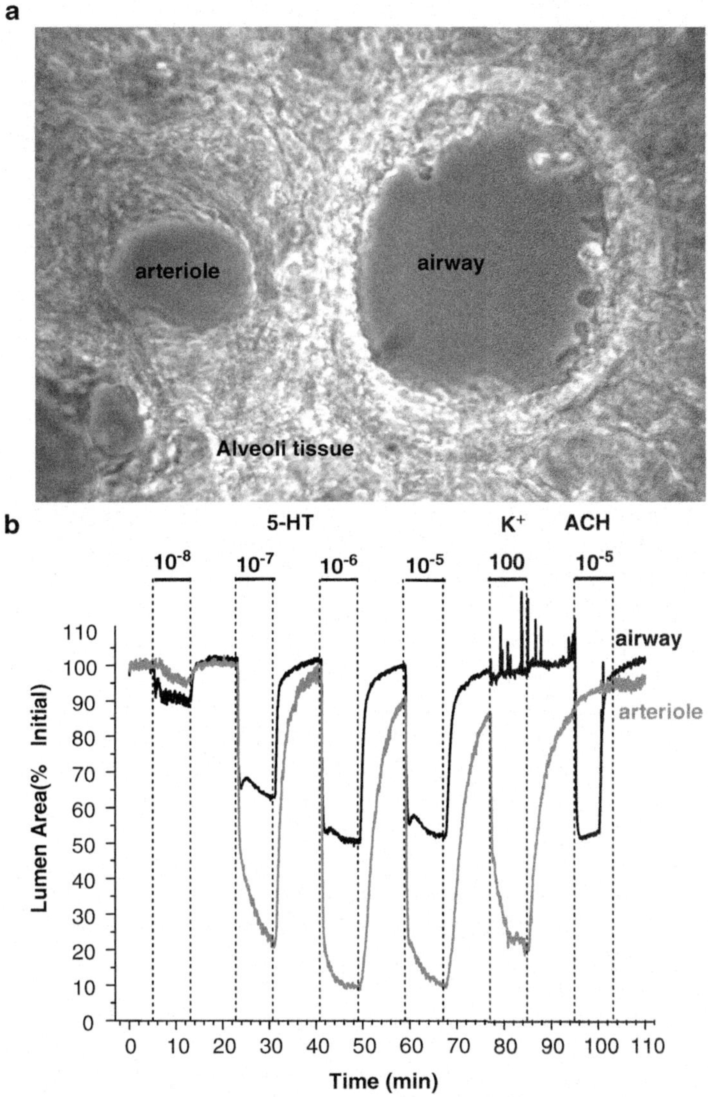

Fig. 2 *Contraction of airways and arterioles in a lung slice* (**a**) Phase-contrast image of a mouse lung slice showing larger airway with adjacent arteriole surrounded by alveoli parenchyma. (**b**) Contraction of airway (black trace) and arteriole (gray trace) in response to increasing concentrations of 5HT (10^{-8}–10^{-5} M), 100 mM KCl, and 10^{-5} acetylcholine (ACh). Contraction of the airway and arteriole increases and saturates with increasing agonist concentration. In response to KCl, the airway shows twitchy contractions with little change in lumen size, whereas the arteriole displays a substantial decrease in lumen size. The airway contracts in response to ACh, whereas the arteriole is unresponsive

characteristically lined with a ciliated, cuboidal epithelium. In healthy lung slices, the cilia are highly active, and lung slices can serve as an ideal preparation to observe ciliary activity [19]. The orientation of the lung slice can also be manipulated to be longitudinal to the airways to view ciliary activity in profile. However, a transverse cut is better for SMCs in order to minimize the numbers of circular SMCs being sliced. Staining for SMC α-actin localizes the SMCs directly below the epithelium [36]. Pulmonary veins are also found within lung slices, but these are identified by their solitary or isolated location.

4 Contraction of Airways in Lung Slices

Airway contraction and relaxation is routinely observed in lung slices with phase-contrast optics and time-lapse recordings in response to a variety of contractile agonists (Fig. 2). In mice and rats, we have observed a rapidly developing, prolonged airway contraction induced by acetylcholine (ACh), methacholine (MCh), carbachol, serotonin (5HT), ATP, and endothelin [5, 9, 10, 34–36]. In addition, histamine and leukotriene D_4 induced contraction in human lung slices [39]. However, there are species-dependent differences as well as muscle-type differences in agonist stimulation; mouse airways are responsive to 5HT but not histamine. Similarly, mouse arteriole SMCs are responsive to 5HT but not MCh. Following the removal of most agonists, airway relaxation occurs and airways regain their original size. Contraction induced with endothelin [34] or LTD_4 [39] occurs at low concentrations (approximately 10 nM) and is slow to reverse.

SMC contraction is commonly believed to be strongly associated with membrane depolarization; hence, KCl is often used as a contractile stimulant (Fig. 2). While in rat and human airways KCl contractions were obvious, they were associated with an underlying "twitching" response of individual SMCs. In mouse airway SMCs, the uncoordinated SMC twitching induced by KCl usually did not generate substantial sustained airway contraction [2, 5, 35, 36, 39]. Thus, for airway SMCs, it appears that membrane depolarization is less important or potent than agonist-receptor stimulation for sustained airway contraction.

5 Agonist-Induced Ca^{2+} Signaling Airway SMCs

It is well recognized that Ca^{2+} increases stimulate airway SMC contraction, but our understanding of the nature of this Ca^{2+} signal has dramatically changed. Traditionally, it was thought that Ca^{2+} increased in a biphasic manner, an initial spike in Ca^{2+} correlating with the onset of contraction followed by a decreased Ca^{2+} plateau associated with sustained contraction. With the advent of improved Ca^{2+} reporters and Ca^{2+} imaging systems with high temporal and spatial resolution, agonist-induced Ca^{2+} changes in isolated SMCs have been observed to occur as oscillations rather than changes in steady-state levels [29, 38]. The frequency of these Ca^{2+} oscillations increased with increased agonist concentration. Because it is difficult to extrapolate findings with isolated cells to airway mechanics, we sought to determine if such changes occurred in airway SMCs in lung slices. For these studies, it is essential to use confocal or two-photon microscopy to overcome light scattering and poor image contrast induced by the increased thickness of lung slice tissues. We constructed a custom-built microscopy system [40] especially for this purpose, but commercial systems may also be used if they have sufficient temporal resolution.

Agonist-induced Ca^{2+} oscillations: it is important to emphasize that all the contractile agonists we have investigated with mice, rat, or human lung slices have induced Ca^{2+} oscillations in airway SMCs (Fig. 3). It is equally interesting that the agonists 5HT and endothelin also induced Ca^{2+} oscillations in intrapulmonary blood vessel SMCs [34, 35]. These Ca^{2+} oscillations occurred as local increases in Ca^{2+} that propagate along the cells as Ca^{2+} waves, a behavior observed by line-scan analysis (Fig. 4a). The initiation site for each Ca^{2+} wave is often located toward one end of the cell. However, Ca^{2+} waves do not always propagate the full length of the cell (Fig. 4a). When a Ca^{2+} wave terminates prematurely, it is common to observe a new Ca^{2+} wave initiating from the opposite end of the cell. Subsequently, the initial propagation direction resumes as this appears to be the dominant process, which suggests that Ca^{2+} oscillations are driven by pacemaker activity. It is not clear why the initiation sites are organized toward one end of the cell, but it may reflect a higher density of IP_3Rs [45].

Frequency of Ca^{2+} oscillations: a second important property of agonist-induced Ca^{2+} oscillations is that their frequency increases with agonist concentration and that this increased frequency correlates with increased airway (and arteriole) contraction (Fig. 3). This correlation has led to the idea that the extent of airway contraction is a function of frequency-modulated (FM) regulation [12]. The advantage of FM signaling over analog signaling is a better discrimination of the control signal over background noise.

Although airway contraction appears to be FM regulated, the exact FM relationship between the Ca^{2+} oscillation frequency and airway contraction appears to be dependent on the agonist, tissue, and species (Fig. 3). For example, 5HT (at equal concentrations) induces relatively fast Ca^{2+} oscillations in airway SMCs but slower Ca^{2+} oscillations in arteriole SMCs, yet arteriole SMCs display a greater contraction. Likewise, human airway SMCs show lower Ca^{2+} oscillation frequencies in response to histamine than mouse airway SMCs in response to MCh, but human airways show greater contraction. These observations indicate that while the frequency of Ca^{2+} oscillations is important in determining relative airway contraction, the extent of airway contraction is modulated by additional mechanisms (i.e., Ca^{2+} sensitivity, considered later). The frequency of the Ca^{2+} oscillations is temperature dependent; in mouse, the frequency of the Ca^{2+} oscillations increased from approximately 17 to approximately 67 min^{-1} (stimulated with 200 nM MCh) as the temperature increased from 20 °C to

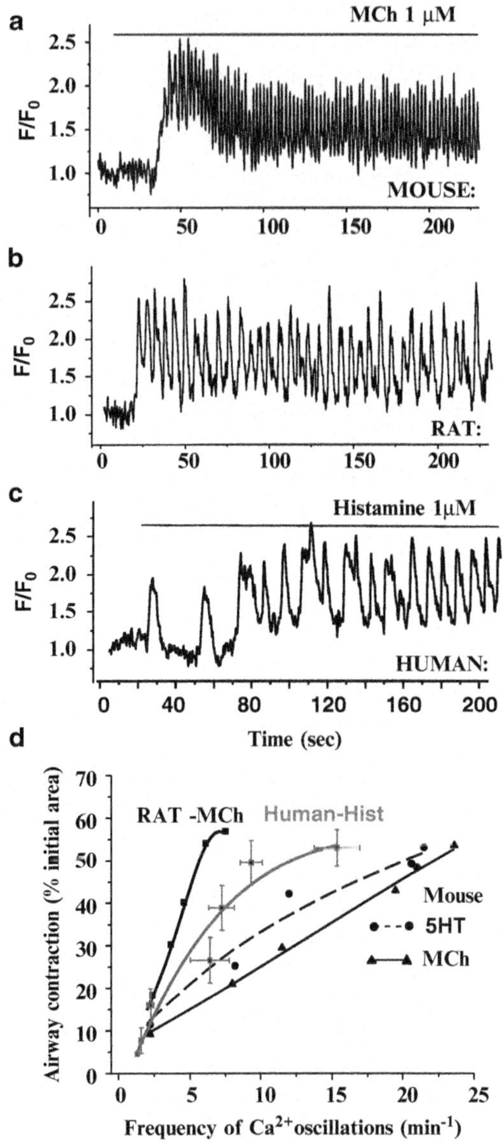

Fig. 3 *Agonist-induced Ca^{2+} oscillations in airway smooth muscle cells (SMCs):* airway SMCs in (**a**) mouse, (**b**) rat, and (**c**) human lung slices display Ca^{2+} oscillations in response to methlcholine (MCh) and histamine. (**d**) Airway contraction as a function of SMC Ca^{2+} oscillation frequency for rat, human, and mouse airways

37 °C [5]. Surprisingly, airway contraction did not show a correlative increase; in fact, airway contraction appeared independent of temperature. This, again, underscores the idea that the FM regulation of airway contraction involves additional mechanisms. In this case, we believe the opposing relaxation process of SMCs, mediated by MLCP (Fig. 1), is also temperature dependent and is increased in a compensating manner to offset the increased activity of MLKC such that MLC phosphorylation is not significantly altered.

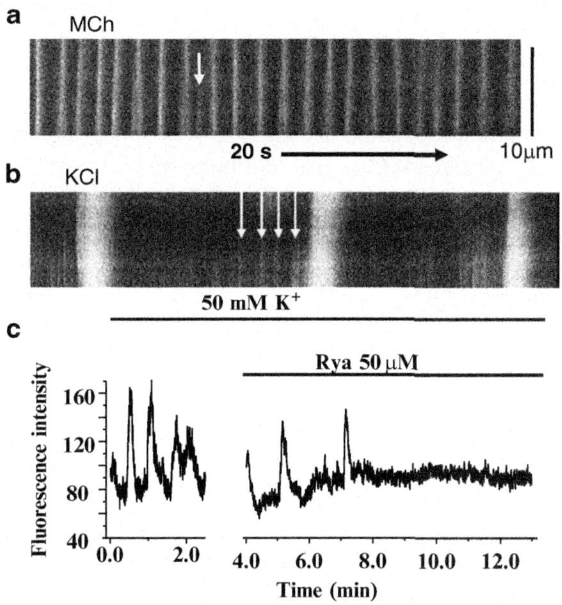

Fig. 4 *Ca^{2+} oscillations induced by KCl* (**a**) Line scan of methalcholine-induced Ca^{2+} oscillations in mouse SMC. *Horizontal axis*: time; *vertical axis*: distance along SMC; slightly sloping *white lines*: Ca^{2+} oscillations propagating along cell. Most lines are oriented toward *top right*, and the gradient of the slope represents wave propagation velocity. *White arrow*: premature termination of Ca^{2+} wave. The next wave has a slope oriented toward *bottom right*, indicating a reversed direction. (**b**) KCl-induced Ca^{2+} oscillations appear as broad bands and occur with a lower frequency. Smaller Ca^{2+} signals (*white arrows*) occur prior to the initiation of the major Ca^{2+} wave. (**c**) KCl-induced Ca^{2+} oscillations in mouse SMCs are inhibited by ryanodine while basal Ca^{2+} is elevated

From our mathematical modeling (see chapter by Sneyd et al.), it appears that a major factor contributing to airway SMC force generation is the shape of the Ca^{2+} oscillation itself [45]. Basically, spikelike Ca^{2+} oscillations generate more force than flatter domelike Ca^{2+} oscillations. Because the shape of Ca^{2+} oscillations changes with their frequency, it is likely that the Ca^{2+} oscillation shape influences force generation more than a simple frequency component.

KCl-induced Ca^{2+} oscillations: although the weak and uncoordinated contractile responses of mouse airways to KCl were unexpected, these responses can be explained by the perhaps equally surprising Ca^{2+} changes induced by KCl. The opening of voltage-gated Ca^{2+} channels (VGCC) is often associated with the depolarization of membrane potential. As a result, a sustained increase in Ca^{2+} was expected to accompany KCl exposure. Instead of a rapid increase in Ca^{2+}, mouse airway SMCs displayed a slow increase in Ca^{2+} that culminated in the stimulation of Ca^{2+} oscillations from an elevated Ca^{2+} baseline with a low frequency ($1–2\ min^{-1}$) but long duration [36] (Fig. 4). Each Ca^{2+} oscillation correlated with a twitch contraction of a SMC, an uncoordinated transient and localized contraction of only part of the airway mediated by individual SMC. Similarly, low-frequency Ca^{2+} oscillations were induced by KCl in rat and human airway SMCs, as well as mouse intrapulmonary arteriole SMCs [5, 35, 39]. KCl-induced Ca^{2+} oscillations also propagated as Ca^{2+} waves along the cell. Prior to the propagation of KCl-induced Ca^{2+} oscillations, smaller, localized increases in Ca^{2+} were observed to occur in the region of wave initiation (Fig. 4). The size and frequency of these Ca^{2+} elemental events increased until a Ca^{2+} wave was initiated. After the passage of a Ca^{2+} wave, these small Ca^{2+} events were abolished but resumed again with time.

6 Variation in Airway Contraction

A key advantage of using lung slices to evaluate airway contraction is the ability to investigate small airway responses. However, the diameter of small airways within the lung slice varies, decreasing toward the periphery. By collecting serial lung sections and analyzing the responses of the same airway that runs through these serial sections, we were able to characterize the response of airway SMCs with respect to their location along the respiratory tract [6]. The more centrally located regions of the airways were the most responsive to contractile agonists. This contractile response was matched by a maximal Ca^{2+} oscillation frequency in the central region. It is not clear why the central region SMCs might be more sensitive to agonist stimulation. However, an important implication of these results for experimentation is that care should be taken to compare the physiology of similar sized airways. While this is a simple issue when using lung slices from the same species, it is more complicated when comparing responses between species. Clearly, the airway diameter within a central location is different between mice and humans. The question of whether it is best to compare airways with a similar size or similar airway generation remains a matter for debate with respect to a comparison of either structural or functional similarity.

In addition to variation in SMC contraction along the airway, SMC contraction can also vary between mouse strains. In mice, the strains A/J and C3H/HeJ were found to be hypo- and hyperresponsive to an inhaled MCh challenge, respectively [28]. Our comparison of the responses of airways in lung slices to MCh from these mice resulted in an opposite rating: C3H/HeJ > BALB/C > A/J [11]. This different order may be the result of direct access to the SMCs in lung slices or the use of halothane in the earlier in vivo study; we found that this anesthetic inhibited SMC contraction. Irrespective of this contractile order, we were surprised to find that the frequencies of the Ca^{2+} oscillations were similar in each strain. The implication of this is that airway responsiveness does not solely reflect Ca^{2+} dynamics. As will be discussed, the mechanism of Ca^{2+} sensitivity is likely to play a key role in determining the contractile response.

Because asthma is characterized by airway hyperresponsiveness, it is an obvious step to examine the responses of airways in lung slices from a mouse model of asthma. Unfortunately, our attempts to observe airway hyperresponsiveness in mouse lung slices have not met with success. Enhanced contraction of airways in lung slices from ovalbumin-sensitized mice and T-bet KO mice [23] was not observed (unpublished data). Similarly, increased airway reactivity has not been observed in mouse lung slices after treatment with a variety of inflammatory factors (unpublished data). The reasons for this remain unclear. However, others have reported contractile changes in such lung slices [8].

7 Mechanisms Underlying Agonist-Induced Ca^{2+} Oscillations and Waves

Involvement of IP_3R: in numerous other nonexcitable cell types, Ca^{2+} oscillations have been found to be mediated by the opening activity of the inositol trisphosphate receptor (IP_3R) of the sarcoplasmic reticulum (SR) [13, 14]. All of our experiments have indicated that the IP_3R also plays a dominant role in airway SMC Ca^{2+} oscillations [2]. In the absence of extracellular Ca^{2+} or presence of Ca^{2+} channel blockers (e.g., nifedipine), Ca^{2+} oscillations induced in mouse SMCs with MCh or 5HT, rat SMCs with MCh, or human SMCs with histamine continued for multiple cycles, indicating a primary reliance on SR Ca^{2+} release rather than Ca^{2+} influx. However, the rundown of the Ca^{2+} oscillations indicates there is some need for Ca^{2+} influx to maintain Ca^{2+} oscillations for extended periods [36]. The inhibition of Ca^{2+} oscillations with the IP_3R antagonist 2-APB and the stimulation of Ca^{2+} oscillations with the photolytic release of IP_3 from inactive caged IP_3 more directly implicate the importance of the IP_3R. The stimulation of Ca^{2+} increases by flash photolysis of caged IP_3 was also inhibited by 2-APB [2]. These results are consistent with the fact that most of the agonist receptors we studied are linked to G-protein receptor, $G_{q/11}$, and the activation of PLCβ and IP_3 production [31].

Involvement of RyR: in cardiac and striated muscle cells, as well as systemic vascular SMCs, the ryanodine receptor (RyR) of the SR plays a significant role in Ca^{2+} release and contraction [22]. While some studies reported a role for RyR in airway SMCs [16, 24, 37], our results, by contrast, indicated that the role for the RyR in Ca^{2+} oscillations of normal mouse, rat, and human airway SMC in lung slices was minimal [2, 39]. The agonist ryanodine is commonly used to inactivate the RyR, but exposure of airway SMCs displaying ongoing agonist-induced Ca^{2+} oscillations to ryanodine had no effect on either the frequency of the Ca^{2+} oscillation or Ca^{2+} wave propagation. Although the RyR is usually stimulated to open by increases in Ca^{2+}, a failure of ryanodine to alter the Ca^{2+} signaling, which requires an open RyR to be effective (use dependence), indicates that agonist-induced Ca^{2+} increases did not evoke any RyR activity. This failure is underscored by the fact that during the several minutes of observation, cytosolic Ca^{2+} increased more than 20 times per minute.

Other RyR inhibitors, such as tetracaine and procaine, appear to inhibit human airway SMC Ca^{2+} signaling [16, 17]. While these compounds also relax airways and inhibit Ca^{2+} signaling in mouse lung slices, these compounds appear to act nonspecifically and inhibit IP_3 production instead of inhibiting RyRs [2]. This is consistent with the known anesthetic properties of tetracaine and procaine. Support for this idea is provided by the fact that photolytic release of IP_3 in the presence of ryanodine or tetracaine still resulted in a Ca^{2+} increase, indicating the IP_3R was functional.

8 Mechanisms Underlying KCl-Induced Ca^{2+} Oscillations

The characteristics of KCl-induced Ca^{2+} oscillations are significantly different from those of agonist-induced Ca^{2+} oscillations and have a different underlying mechanism. KCl-induced Ca^{2+} oscillations are dependent on Ca^{2+} influx and are abolished by both Ca^{2+}-free conditions and Ca^{2+} channel blockers [36]. The occurrence of smaller elemental Ca^{2+} events before the peak of the Ca^{2+} oscillations reflects the activity of RyR in other cells, in particular cardiac cells, responding to overloading of the SR store with Ca^{2+}. Importantly, ryanodine abolished KCl-induced Ca^{2+} oscillations (Fig. 4). The elevated baseline in cytosolic Ca^{2+} is also consistent with the inhibition of the RyR in a partially open state. These results suggest that KCl-induced Ca^{2+} waves are the result of Ca^{2+}-induced Ca^{2+} release (CICR) via the RyR. Support for this hypothesis is provided by the photolytic release of caged Ca^{2+}. In normal conditions, a photolytic local increase in Ca^{2+} has a minimal effect on airway SMC. However, in the presence of KCl, where the SR Ca^{2+} content is hypothesized to be elevated, a similar photolytic release of Ca^{2+} initiates a single Ca^{2+} wave [3]. The nonspecific actions of tetracaine can be reexamined with KCl-induced Ca^{2+} oscillations. While tetracaine inhibited KCl-induced Ca^{2+} oscillations, it did so without elevating baseline Ca^{2+} and was reversible, results differing from RyR inhibition by ryanodine [2].

9 Interplay Between IP_3R and RyR

With the insight from KCl experiments showing that RyRs are present and functional in airway SMCs, the conundrum as to why ryanodine has little effect on agonist-induced signaling must be addressed, especially since both receptors are interconnected via the SR. Our initial approach to this problem was to construct a mathematical model that includes both the IP_3R and RyR [44, 45]. The model successfully simulates the experimental data describing Ca^{2+} oscillations mediated by the IP_3R and RyR in response to agonist and KCl [45].

The power of the modeling approach is its ability to predict experimental behavior and cellular activity that are not observable in our experiments. For example, we measure changes in cytosolic Ca^{2+}, but these changes are a complex function of the activities of many components including the IP_3R, RyR, SR Ca^{2+} content, and membrane Ca^{2+} pumps. Because the model can isolate the contribution of each component to agonist-induced Ca^{2+} oscillations, we were alerted to the possibility of an early involvement of the RyR in these Ca^{2+} oscillations when the SR Ca^{2+} content was high, a condition under which the RyR is more responsive to cytosolic Ca^{2+} [45]. At the initiation of agonist-induced Ca^{2+} oscillations, the SR Ca^{2+} content is high and the RyR appears to contribute to the first few Ca^{2+} transients; the small amount of Ca^{2+} released by the IP_3R, in response to IP_3, is proposed to trigger the opening of the sensitized RyR. However, the SR content becomes substantially reduced with time during agonist-induced Ca^{2+} oscillations operating via the IP_3R, and this renders the RyR insensitive to cytosolic Ca^{2+}. This is consistent with our findings that the application of ryanodine during ongoing agonist-induced Ca^{2+} oscillations has no effect. However, a major model prediction was that ryanodine needed to be added prior to the agonist if it is to be available to act on the early Ca^{2+} transients. Our preliminary data with this experimental approach indicate that while a Ca^{2+} increase still occurs (Ca^{2+} flows out of the SR through the IP_3R), ryanodine can inhibit Ca^{2+} oscillations. The importance of this result is that it emphasizes the role of the SR in regulating the Ca^{2+} oscillations. While ryanodine may inhibit agonist-induced Ca^{2+} oscillations, these Ca^{2+} oscillations do not take place via the RyR. The underlying explanation is that a decreased SR Ca^{2+} content, due to partially open RyR, results in an alteration of the activity of the IP_3R. Details of the experimental results and model predictions are described in the chapter by Sneyd et al.

10 Ca^{2+} Sensitivity: The Other Mechanism Regulating Airway SMC Contraction

While increases in Ca^{2+} are important to drive SMC contraction, the complementary mechanisms that determine the effectiveness of the $[Ca^{2+}]_i$ change, collectively known as Ca^{2+} sensitivity, are equally important [43]. Airway SMC contraction is proportional to the degree of MLC phosphorylation (Fig. 1). While MLC phosphorylation is primarily increased by Ca^{2+}-dependent MLCK activity, the efficacy of this reaction is influenced by the rate of MLC dephosphorylation by MLCP. Thus, for a given Ca^{2+} oscillation frequency or Ca^{2+} level, the activity of MLCP modulates the extent of contraction. However, the activity of MLCP is controlled by multiple mechanisms; phosphorylation of MLCP or accessory proteins by Rho kinase or PKC generally reduces MLCP activity to enhance MLC phosphorylation and contraction (increased Ca^{2+} sensitivity) (Fig. 1).

To study the relationship between contraction and Ca^{2+} sensitivity, it is necessary to clamp the cytosolic Ca^{2+} of airway SMCs to a fixed concentration by disabling or bypassing the normal Ca^{2+} signaling mechanisms. A common approach has been to permeabilize the cell membranes to Ca^{2+} with detergent or toxins. However, this can be disruptive and result in the loss of other cell constituents. Consequently, we developed a method of Ca^{2+} permeabilization that exploits the cells' own inherent influx Ca^{2+} channels (normally used to refill SR), and this involves treatment of the lung slice with caffeine (20 mM) and ryanodine (50 μM) [4, 5, 39]. The fact that this approach works underscores our early perception that functional RyRs must be present in airway SMCs. Caffeine stimulates the opening of RyR by increasing their sensitivity to Ca^{2+} [25]. Thus, in the presence of ryanodine, the open RyR is locked (inhibited) in a partially open state, and this allows Ca^{2+} to leak from the SR. This results in the emptying of the SR, which is believed to stimulate the opening of membrane store-operated Ca^{2+} (SOC) channels with an influx of Ca^{2+} [15, 30]. Although caffeine and ryanodine are washed away before further experimentation, the effect of ryanodine is irreversible. Consequently, the cell has a constant Ca^{2+} influx, which leads to a constant elevated cytosolic Ca^{2+}; the cytosolic Ca^{2+} is proportional to the external Ca^{2+} concentration.

Although this permeabilization method is very simple, the initial responses of mouse SMCs to this treatment appear confusing but are in fact very informative. The expected response of the airway SMCs to a sustained increase in Ca^{2+} would be sustained contraction. However, exposure to caffeine and ryanodine initiates a transient contraction, resulting in airway relaxation [4]. The explanation for this has two parts: first, caffeine, in addition to its ability to release Ca^{2+}, serves as a potent but readily reversible relaxant of agonist-contracted airway SMCs. This was first described by [4]. Second, high sustained Ca^{2+} in mouse airway SMCs also induces SMC relaxation (Fig. 5). Thus, in the presence of caffeine and ryanodine, the mouse airway SMCs are fully relaxed. Even when caffeine is washed away, the airway remains relaxed due to the high Ca^{2+}.

The best experimental way to use Ca^{2+}-permeabilized airway SMCs is to wash away excess ryanodine and caffeine with Ca^{2+}-free saline (Fig. 5). This provides a starting point of low Ca^{2+} within the SMCs with the corresponding relaxed state. Upon addition of extracellular Ca^{2+}, the cytosolic Ca^{2+} increases and the airway SMCs contract as expected. However, with sustained increased Ca^{2+}, the airway SMCs subsequently relax (Fig. 5) [4, 5]. This response is proposed to result from a Ca^{2+}-dependent decrease in Ca^{2+} sensitivity (perhaps by increased MLCP activity), but the details of this mechanism are not clear. Importantly, if the lung slice is now exposed to a contractile agonist, a contractile response (proportional to the agonist concentration) is once again observed but without a change in cytosolic Ca^{2+} (Fig. 5); this represents an agonist-dependent increase in Ca^{2+} sensitivity (decreased MLCP activity).

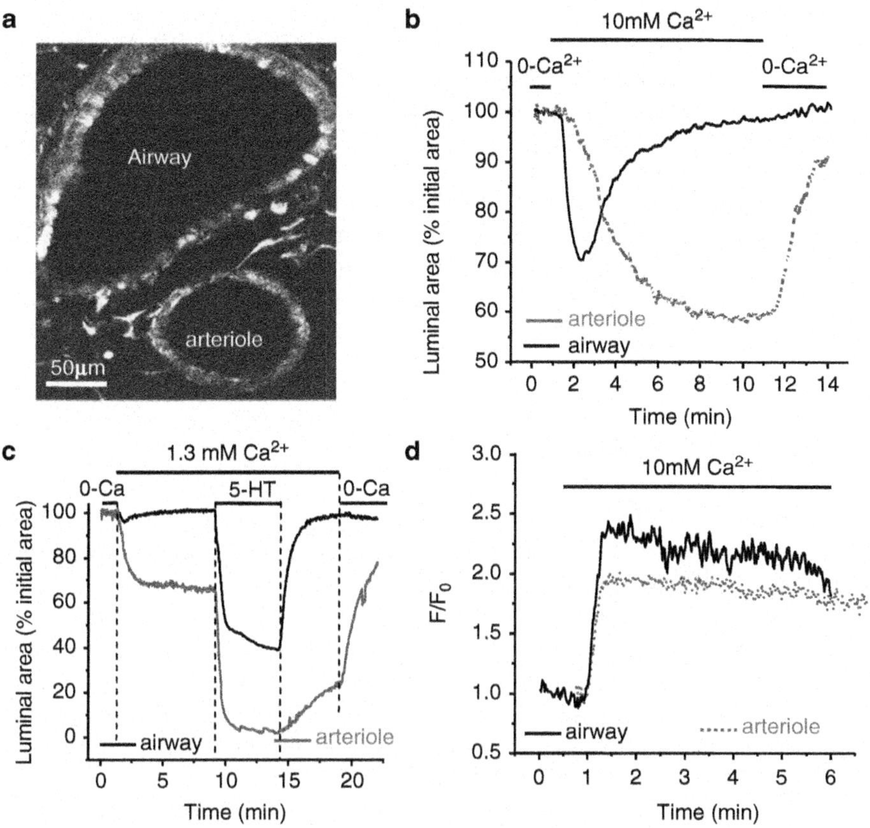

Fig. 5 *Ca^{2+}-permeabilized lung slices* (**a**) Airway and arteriole imaged with laser scanning microcopy after permeabilization with caffeine and ryanodine. (**b**) Addition of extracellular Ca^{2+} increases cytosolic Ca^{2+} to a steady state (**d**) but induces a transient contraction of airway and sustained contraction of arteriole. Removal of Ca^{2+} results in arteriole relaxation. (**c**) After addition of Ca^{2+}, the relaxed airway and partially contracted arteriole are induced to contract further in response to an agonist (5HT). Removal of agonist, followed by removal of Ca^{2+}, result in airway relaxation. During agonist addition there is no change in cytosolic Ca^{2+}

11 Comparative Ca^{2+} Sensitivity

Like variations in the Ca^{2+} oscillation frequency, the level of Ca^{2+} sensitivity varies between SMC types. This is clearly observed in lung slices when comparing the responses of an arteriole that is adjacent to the airway (Fig. 5). The airway and arteriole are identically exposed to caffeine and ryanodine and both show a sustained Ca^{2+} increase. However, with high Ca^{2+} the airway becomes relaxed, whereas the arteriole is partially contracted [4]. The addition of a contractile agonist induces the airway to contract and induces the arteriole to contract further (Fig. 5). Thus, the arteriole appears to have an inherent higher Ca^{2+} sensitivity than the airway.

Interestingly, the airways of both rats and humans behave more like that of mouse arterioles and have a higher intrinsic Ca^{2+} sensitivity [5, 39]. This increased Ca^{2+} sensitivity probably explains why rat and human airways display greater contraction in response to a lower rate of Ca^{2+} oscillation (Fig. 3). Similarly, increased Ca^{2+} sensitivity can also explain why the low rate of Ca^{2+} oscillation induced by KCl results in a stronger contraction of rat or human airways and arterioles as compared to only a "twitch" response in mouse airway SMCs.

In mouse, rat, and human airway SMCs, agonist stimulation substantially increases Ca^{2+} sensitivity. Thus, an agonist acts through two mechanisms to develop contraction: increased MLCK activity mediated by Ca^{2+} oscillations and decreased MLCP activity mediated by ROCK and PKC (Fig. 1).

12 Airway SMC Relaxation

A key therapy for asthma is the relaxation of contracted airway SMCs, and this has traditionally been achieved with β_2-adrenegeric receptor agonists. While these agonists are known to increase cAMP levels, their mode of action in bringing about SMC relaxation is less clear. A major hypothesis for their action was that cAMP increases resulted in the phosphorylation and activation of membrane K^+ channels [26]. The resulting membrane hyperpolarization was proposed to decrease the activity of VGCC and thereby decrease cytosolic Ca^{2+}. However, as discussed earlier, many agonists induce airway SMC contraction by stimulating Ca^{2+} oscillations that recycle internal Ca^{2+} and do not directly rely on Ca^{2+} influx through VGCCs.

On the other hand, cAMP-dependent activation of K^+ channels could counter KCl-induced depolarization and mediate relaxation of SMCs contracted with KCl. An alternative mechanism for inducing membrane hyperpolarization that is potentially available in SMCs is by a localized Ca^{2+} release via a cluster of RyRs (a Ca^{2+} spark) and the nearby activation of Ca^{2+}-activated K^+ channels; a Ca^{2+} spark results in a brief outward K^+ current. Ca^{2+} sparks are common in vascular and cardiac cells and also occur in tracheal SMCs [7, 46]. However, Ca^{2+} sparks generally occur spontaneously and often in cells that are subject to Ca^{2+} overload. The increased Ca^{2+} content of the SR leads to RyR sensitization and spontaneous release events. It is highly feasible that Ca^{2+} sparks help cells counter Ca^{2+} overload (excess Ca^{2+} influx), but the activation of Ca^{2+} sparks by β_2-adrenergic compounds in agonist-contracted airway SMCs has not been observed. Furthermore, based on our characterization of agonist- and KCl-induced Ca^{2+} signaling in airway SMCs, it is likely that the SR Ca^{2+} content is inadequate for RyR sensitization during ongoing Ca^{2+} oscillations [45]. Importantly, we have never observed Ca^{2+} sparks in airway SMCs in lung slices under resting or contracted conditions. However, our imaging systems can clearly visualize Ca^{2+} sparks in HeLa cells or atrial myocytes (unpublished data).

Although Ca^{2+} sparks do not appear to play a major role in the relaxation of agonist-contracted airway SMCs, this conclusion emphasizes the caveat that it is extremely important to understand the mechanism stimulating airway SMC contraction, if the mechanism of relaxation is also to be understood and exploited for therapy. Unfortunately, it is not known which agonist or stimulus is directly responsible for inducing human airway SMC contraction during an asthmatic episode. While the range of contractile stimulants is broad, including neurotransmitters, cytokines, leukotrienes, growth factors, and hormones, we have found that sustained SMC contraction is often associated with the presence of Ca^{2+} oscillations. Consequently, we further characterized the mechanism of action of bronchodilators on agonist-induced airway SMC Ca^{2+} oscillations.

Exposure of mouse, rat, and human airway SMCs, contracted with MCh, 5HT, or histamine, to short- (albuterol) or long-acting (formoterol) β_2-receptor agonists induced airway relaxation that was correlated with a reduction in the frequency of Ca^{2+} oscillations [3, 18, 20, 21, 39] (Fig. 6). The efficacy of formoterol was substantially greater than albuterol at all concentrations examined; a near full airway relaxation was achieved with 5 nM formoterol compared to 1 μM albuterol. The β-receptor agonist isoproterenol also induced airway relaxation, but this response was transient in comparison to albuterol or formoterol, most likely because of receptor desensitization or degradation [3].

While albuterol and formoterol both influenced Ca^{2+} oscillations, formoterol was more likely than albuterol to abolish Ca^{2+} oscillations. Although albuterol could transiently abolish Ca^{2+} oscillations, Ca^{2+} oscillations would commonly reappear, albeit at a lower frequency. While this reduction in Ca^{2+} oscillation activity correlates with the extent of airway relaxation, formoterol was also able to substantially relax contracted airway SMCs at very low concentrations (<5 nM, Fig. 6) with minimal effect on the Ca^{2+} oscillations. This suggested that β_2-adrenergic compounds might also modulate Ca^{2+} sensitivity. This mechanism of action was confirmed using Ca^{2+}-permeabilized lung slices. Both albuterol and formoterol relaxed agonist-contracted airways that had sustained high Ca^{2+}, but formoterol again had a greater effect.

The concept of short- and long-acting agonists is also reflected in the lung slice. Airway relaxation induced by albuterol persisted while albuterol was present, but airway contraction quickly returned when albuterol was removed. By contrast, the action of formoterol persisted for some time after removing it from human lung slices (Fig. 6). The ability of formoterol to act at low concentrations suggests it has a high affinity binding site and therefore might be harder to wash out of the preparation. Long-acting β-agonists have also been thought to have slower onset times, but we found there was little difference in the time taken to initiate relaxation between albuterol and formoterol. The difference in the duration of action of formoterol also suggests a different mode of action in mice and humans. The idea that the action of a long acting b2 agonist is the result of an increased accumulation (solubility) of the agent in the SMC membrane does not appear to be supported because it seems unlikely that this would be significantly different between mice and humans.

NO-induced airway relaxation: a second potent mechanism that can relax SMCs is the stimulation of cGMP production by nitric oxide (NO). Accordingly, agonist-

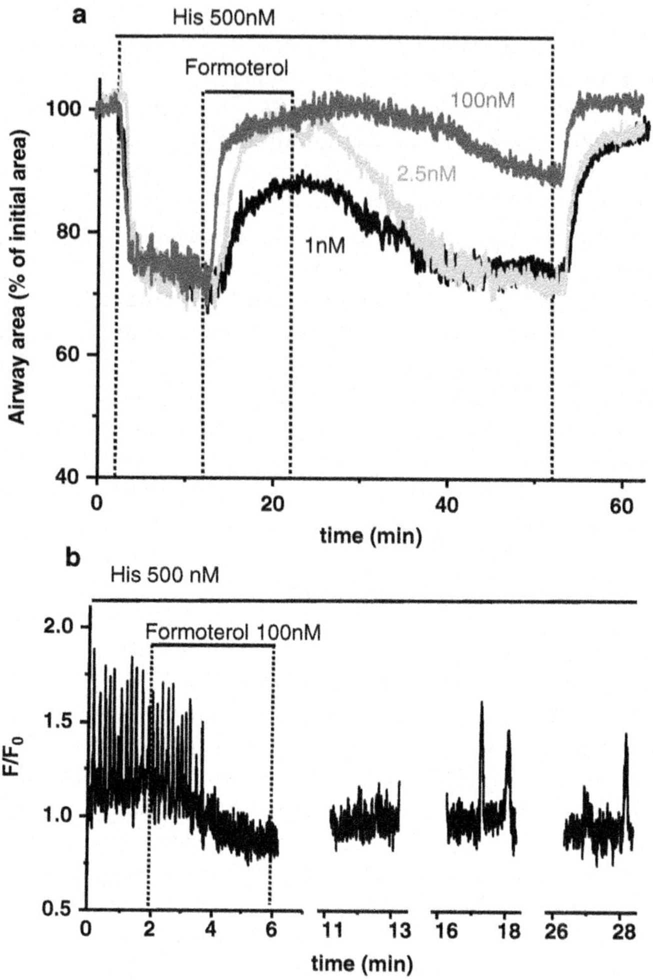

Fig. 6 *Formoterol-induced relaxation of human airways* (**a**) Airways were contracted with histamine and relaxed with different concentrations of formoterol (1, 2.5, and 100 nM). Greater and faster relaxation was obtained with increasing formoterol concentration. The time taken for the airway to recontract after formoterol removal was increased with concentration. (**b**) Formoterol (100 nM) abolished histamine-induced Ca^{2+} oscillations. After removal of formoterol, Ca^{2+} oscillations were not reestablished within recording time of 28 min, which is consistent with the prolonged relaxation induced by 100 nM formoterol

contracted airways relaxed and displayed a decreased Ca^{2+} oscillation frequency in response to NO donors [32]. The action of NO was enhanced with phosphodiesterase inhibitors and prevented by the inhibition of soluble guanylyl cyclase.

Mechanism of action of β_2-adrenergic receptor agonists: to determine whether a reduction in the Ca^{2+} oscillation frequency involved the IP_3R, we examined the effect of IP_3 photolysis in the presence of β_2-adrenergic receptor agonists and forskolin (an adenylyl cyclase activator used to increase cAMP). In all cases, we found that after the Ca^{2+} oscillations had slowed or stopped, the addition of IP_3 (by photolysis) reinitiated the Ca^{2+} oscillations [3, 21, 39]. Similar results were obtained with NO [32]. These results indicate that the sensitivity of the IP_3R for IP_3 was diminished by increased cAMP or cGMP. An alternative explanation for the slowing of the Ca^{2+} oscillations is that there was an insufficient amount of Ca^{2+} available in the SR due to decreased Ca^{2+} influx. However, if the Ca^{2+} influx was increased in the presence of forskolin, the low-frequency Ca^{2+} oscillations were further slowed or abolished. Thus, it appears that the IP_3R also became more sensitive to inhibition by Ca^{2+} [3]. It is also possible that the production of IP_3 was decreased by the inhibition of PLC. It is not yet clear how formoterol decreases Ca^{2+} sensitivity.

13 Summary

The lung slice, in conjunction with scanning laser microcopy, has proved to be a highly valuable and robust preparation with which to study small airway SMC physiology. The SMCs in the lung slice appear to be representative of SMCs in vivo and display dynamic Ca^{2+} signaling that correlates with airway contraction and relaxation. Agonist-induced sustained airway contraction is mediated by high-frequency Ca^{2+} oscillations, whereas KCl-induced contraction can be irregular and is mediated by low-frequency Ca^{2+} oscillations. Agonist-induced Ca^{2+} oscillations are mediated primarily through IP_3R activity, whereas KCl-induced Ca^{2+} oscillations involve the RyR. However, the activity of each receptor appears interlinked by SR Ca^{2+} content. In parallel, changes in Ca^{2+} sensitivity can modulate the extent of airway contraction induced by Ca^{2+} changes. Airway relaxation is achieved by reducing the frequency of the Ca^{2+} oscillations by inhibiting IP_3R activity and by reducing Ca^{2+} sensitivity. Clearly, the lung slice has greatly facilitated our understanding of normal SMC physiology, and its use in further exploring airway disease is certain.

Acknowledgments This work was supported by National Institutes of Health Grant HL 103405. I would also like to thank Albrecht Bergner, Jose Perez-Zogbhi, Yan Bai, Philippe Delmotte, and Anna Ressmeyer, who conducted most of the experimental work addressed in this chapter.

References

1. An SS, Bai TR, Bates JH, Black JL, Brown RH, Brusasco V, Chitano P, Deng L, Dowell M, Eidelman DH et al. 2007. Airway smooth muscle dynamics: a common pathway of airway obstruction in asthma. *Eur Respir J* **29:** 834–860.
2. Bai Y, Edelmann M, Sanderson MJ 2009. The contribution of inositol 1,4,5-trisphosphate and ryanodine receptors to agonist-induced Ca^{2+} signaling of airway smooth muscle cells. *Am J Physiol Lung Cell Mol Physiol* **297:** L347–361.

3. Bai Y, Sanderson MJ. 2006a. Airway smooth muscle relaxation results from a reduction in the frequency of Ca^{2+} oscillations induced by a cAMP-mediated inhibition of the IP_3 receptor. *Respir Res* **7:** 34.
4. Bai Y, Sanderson MJ. 2006b. Modulation of the Ca^{2+} sensitivity of airway smooth muscle cells in murine lung slices. *Am J Physiol Lung Cell Mol Physiol* **291:** L208–221.
5. Bai Y, Sanderson MJ. 2009. The contribution of Ca^{2+} signaling and Ca^{2+} sensitivity to the regulation of airway smooth muscle contraction is different in rats and mice. *Am J Physiol Lung Cell Mol Physiol* **296:** L947–958.
6. Bai Y, Zhang M, Sanderson MJ. 2007. Contractility and Ca^{2+} signaling of smooth muscle cells in different generations of mouse airways. *Am J Respir Cell Mol Biol* **36:** 122–130.
7. Bao R, Lifshitz LM, Tuft RA, Bellve K, Fogarty KE, ZhuGe R. 2008. A close association of RyRs with highly dense clusters of Ca^{2+}-activated Cl^- channels underlies the activation of STICs by Ca^{2+} sparks in mouse airway smooth muscle. *J Gen Physiol* **132:** 145–160.
8. Bergner A, Kellner J, Silva AK, Gamarra F, Huber RM. 2006. Ca^{2+}-signaling in airway smooth muscle cells is altered in T-bet knock-out mice. *Respir Res* **7:** 33.
9. Bergner A, Sanderson MJ. 2002a. Acetylcholine-induced Calcium Signaling and Contraction of Airway Smooth Muscle Cells in Lung Slices. *J Gen Physiol* **119:** 187–198.
10. Bergner A, Sanderson MJ. 2002b. ATP stimulates Ca^{2+} oscillations and contraction in airway smooth muscle cells of mouse lung slices. *Am J Physiol Lung Cell Mol Physiol* **283:** L1271–1279.
11. Bergner A, Sanderson MJ. 2003. Airway contractility and smooth muscle Ca^{2+} signaling in lung slices from different mouse strains. *J Appl Physiol* **95:** 1325–1332; discussion 1314.
12. Berridge MJ. 1997. The AM and FM of calcium signalling [news; comment]. *Nature* **386:** 759–760.
13. Berridge MJ. 2008. Smooth muscle cell calcium activation mechanisms. *J Physiol* **586:** 5047–5061.
14. Berridge MJ, Bootman MD, Roderick HL. 2003. Calcium signalling: dynamics, homeostasis and remodelling. *Nat Rev Mol Cell Biol* **4:** 517–529.
15. Cahalan MD. 2009. STIMulating store-operated Ca^{2+} entry. *Nat Cell Biol* **11:** 669–677.
16. Dai JM, Kuo KH, Leo JM, Pare PD, van Breemen C, Lee CH. 2007. Acetylcholine-induced asynchronous calcium waves in intact human bronchial muscle bundle. *Am J Respir Cell Mol Biol* **36:** 600–608.
17. Dai JM, Kuo KH, Leo JM, van Breemen C, Lee CH. 2006. Mechanism of ACh-induced asynchronous calcium waves and tonic contraction in porcine tracheal muscle bundle. *Am J Physiol Lung Cell Mol Physiol* **290:** L459–469.
18. Delmotte P, Ressmeyer AR, Bai Y, Sanderson MJ. 2010. Mechanisms of airway smooth muscle relaxation induced by beta2-adrenergic agonists. *Front Biosci* **15:** 750–764.
19. Delmotte P, Sanderson MJ. 2006. Ciliary beat frequency is maintained at a maximal rate in the small airways of mouse lung slices. *Am J Respir Cell Mol Biol* **35:** 110–117.
20. Delmotte P, Sanderson MJ. 2008. Effects of albuterol isomers on the contraction and Ca^{2+} signaling of small airways in mouse lung slices. *Am J Respir Cell Mol Biol* **38:** 524–531.
21. Delmotte P, Sanderson MJ. 2010. Effects of formoterol on contraction and Ca^{2+} signaling of mouse airway smooth muscle cells. *Am J Respir Cell Mol Biol* **42:** 373–381.
22. Fill M, Copello JA. 2002. Ryanodine receptor calcium release channels. *Physiol Rev* **82:** 893–922.
23. Finotto S, Neurath MF, Glickman JN, Qin S, Lehr HA, Green FH, Ackerman K, Haley K, Galle PR, Szabo SJ et al. 2002. Development of spontaneous airway changes consistent with human asthma in mice lacking T-bet. *Science* **295:** 336–338.
24. Kannan MS, Prakash YS, Brenner T, Mickelson JR, Sieck GC. 1997. Role of ryanodine receptor channels in Ca^{2+} oscillations of porcine tracheal smooth muscle. *Am J Physiol* **272:** L659–664.
25. Kong H, Jones PP, Koop A, Zhang L, Duff HJ, Chen SR. 2008. Caffeine induces Ca^{2+} release by reducing the threshold for luminal Ca^{2+} activation of the ryanodine receptor. *Biochem J* **414:** 441–452.

26. Kotlikoff MI, Kamm KE. 1996. Molecular mechanisms of beta-adrenergic relaxation of airway smooth muscle. *Annu Rev Physiol* **58:** 115–141.
27. Lauzon AM, Bates JH, Donovan G, Tawhai M, Sneyd J, Sanderson MJ. 2012. A multi-scale approach to airway hyperresponsiveness: from molecule to organ. *Front Physiol* **3:** 191.
28. Levitt RC, Mitzner W. 1989. Autosomal recessive inheritance of airway hyperreactivity to 5- hydroxytryptamine. *J Appl Physiol* **67:** 1125–1132.
29. Pabelick CM, Sieck GC, Prakash YS. 2001. Invited Review: Significance of spatial and temporal heterogeneity of calcium transients in smooth muscle. *J Appl Physiol* **91:** 488–496.
30. Peel SE, Liu B, Hall IP. 2008. ORAI and store-operated calcium influx in human airway smooth muscle cells. *Am J Respir Cell Mol Biol* **38:** 744–749.
31. Penn RB, Benovic JL. 2008. Regulation of heterotrimeric G protein signaling in airway smooth muscle. *Proc Am Thorac Soc* **5:** 47–57.
32. Perez-Zoghbi JF, Bai Y, Sanderson MJ. 2010. Nitric oxide induces airway smooth muscle cell relaxation by decreasing the frequency of agonist-induced Ca^{2+} oscillations. *J Gen Physiol* **135:** 247–259.
33. Perez-Zoghbi JF, Karner C, Ito S, Shepherd M, Alrashdan Y, Sanderson MJ. 2008. Ion channel regulation of intracellular calcium and airway smooth muscle function. *Pulm Pharmacol Ther.*
34. Perez-Zoghbi JF, Sanderson MJ. 2007. Endothelin-induced contraction of bronchiole and pulmonary arteriole smooth muscle cells is regulated by intracellular Ca^{2+} oscillations and Ca^{2+} sensitization. *Am J Physiol Lung Cell Mol Physiol* **293:** L1000–1011.
35. Perez JF, Sanderson MJ. 2005a. The contraction of smooth muscle cells of intrapulmonary arterioles is determined by the frequency of Ca^{2+} oscillations induced by 5-HT and KCl. *J Gen Physiol* **125:** 555–567.
36. Perez JF, Sanderson MJ. 2005b. The frequency of calcium oscillations induced by 5-HT, ACH, and KCl determine the contraction of smooth muscle cells of intrapulmonary bronchioles. *J Gen Physiol* **125:** 535–553.
37. Prakash YS, Kannan MS, Walseth TF, Sieck GC. 1998. Role of cyclic ADP-ribose in the regulation of $[Ca^{2+}]_i$ in porcine tracheal smooth muscle. *Am J Physiol* **274:** C1653–1660.
38. Prakash YS, Pabelick CM, Kannan MS, Sieck GC. 2000. Spatial and temporal aspects of ACh-induced $[Ca^{2+}]_i$ oscillations in porcine tracheal smooth muscle. *Cell Calcium* **27:** 153–162.
39. Ressmeyer AR, Bai Y, Delmotte P, Uy KF, Thistlethwaite P, Fraire A, Sato O, Ikebe M, Sanderson MJ. 2010. Human airway contraction and formoterol-induced relaxation is determined by Ca^{2+} oscillations and Ca^{2+} sensitivity. *Am J Respir Cell Mol Biol* **43:** 179–191.
40. Sanderson MJ. 2004. Acquisition of multiple real-time images for laser scanning microscopy. *Microscopy and analysis* **18:** 17–23.
41. Sanderson MJ. 2011. Exploring lung physiology in health and disease with lung slices. *Pulmonary Pharmacology and Therapeutics*: In Press.
42. Sanderson MJ, Delmotte P, Bai Y, Perez-Zogbhi JF. 2008. Regulation of airway smooth muscle cell contractility by Ca^{2+} signaling and sensitivity. *Proc Am Thorac Soc* **5:** 23–31.
43. Somlyo AP, Somlyo AV. 2003. Ca^{2+} sensitivity of smooth muscle and nonmuscle myosin II: modulated by G proteins, kinases, and myosin phosphatase. *Physiol Rev* **83:** 1325–1358.
44. Wang I, Politi AZ, Tania N, Bai Y, Sanderson MJ, Sneyd J. 2008. A mathematical model of airway and pulmonary arteriole smooth muscle. *Biophys J* **94:** 2053–2064.
45. Wang IY, Bai Y, Sanderson MJ, Sneyd J. 2010. A mathematical analysis of agonist- and KCl-induced Ca^{2+} oscillations in mouse airway smooth muscle cells. *Biophys J* **98:** 1170–1181.
46. ZhuGe R, Sims SM, Tuft RA, Fogarty KE, Walsh JV, Jr. 1998. Ca^{2+} sparks activate K^+ and Cl^- channels, resulting in spontaneous transient currents in guinea-pig tracheal myocytes. *J Physiol (Lond)* **513:** 711–718.

Temporal Aspects of Ca^{2+} Signaling in Airway Myocytes

Etienne Roux

Contents

Abstract This chapter discusses how variation in intracytoplasmic Ca^{2+} concentration ($[Ca^{2+}]_i$) over time can be considered as a signal for the contractile machinery. It presents an overview of the literature that describes the Ca^{2+} response pattern to different contractile or relaxant agonists in several types of airway smooth muscle cells (ASMCs) depending on the species and location in the airway tree and recording methods, insisting on the temporal aspects of this response. Since the dynamics of the Ca^{2+} signal depends on the dynamics of the mechanisms

E. Roux (✉)
Adaptation cardiovasculaire à l'ischémie INSERM U 1034, University of Bordeaux,
1, avenue Magellan U1034, F-33600 Pessac, France
e-mail: etienne.roux@u-bordeaux2.fr

Y.-X. Wang (ed.), *Calcium Signaling In Airway Smooth Muscle Cells*,
DOI 10.1007/978-3-319-01312-1_8,

responsible for this signal, the chapter presents an overview of the main mechanisms responsible for Ca^{2+} homeodynamics in ASMCs. By analyzing some examples, it shows how the kinetics of these mechanisms determine the pattern of the Ca^{2+} signal. The consequence of cell-to-cell variations in the Ca^{2+} signal is also discussed, with special attention to oscillatory versus nonoscillatory responses. The last part of the chapter presents the relationship between the parameter of the Ca^{2+} signal and the pattern of the contractile response. The mechanisms of the contractile apparatus itself will not be detailed, this question being beyond the scope of the chapter, but the temporal relationship between the Ca^{2+} signal and the subsequent contraction is analyzed.

Keywords Calcium dynamics • Oscillations • Kinetics • Model • Contraction

1 $[Ca^{2+}]_i$ Variations as a Signal

1.1 What Makes Ca^{2+} Variation a Signal?

A Ca^{2+} signal is, by definition, a variation over time or space of the intracellular Ca^{2+} concentration that is detected by a given intracellular machinery, so that the behavior of the cell is altered according to the variation in Ca^{2+} concentration. In this chapter, we consider only temporal variation of cytosolic Ca^{2+} concentration ($[Ca^{2+}]_i$), which is a major determinant of airway smooth muscle cells (ASMC) contraction. Though the Ca^{2+} signal may encode for several cell functions, such as contraction, proliferation, and secretion, this chapter will consider only the contractile response of ASMCs as a cell function.

Time-dependent variations in $[Ca^{2+}]_i$ is considered a signal insofar as they carry information that the intracellular machinery is able to decipher, i.e., is able to demonstrate a behavior specifically dependent on these $[Ca^{2+}]_i$ variations, and that is why biologists study temporal variation in $[Ca^{2+}]_i$. Hence, what is of interest for biologists is what makes sense for the cell. This determines which parameters should be used to characterize Ca^{2+} signals and the relevant correlations that can be established between the variation in the intensity of stimulation and the $[Ca^{2+}]_i$ response on the one hand, and between the parameters of the Ca^{2+} signal and the cellular functional response on the other hand. The scheme of the signaling pathway can be summarized as follows:

$$[\text{agonist}]\ (1) \rightarrow Ca^{2+}\text{signal}\ (2) \rightarrow \text{contractile response,}$$

where the first step (1) corresponds to the encoding of the Ca^{2+} signal and the second step (2) to the decoding of the Ca^{2+} signal. A signal can be basically defined as the oriented relationship through a communication channel between an issuing system (source) and a recipient system (receptor). The information is what modifies the state or the evolution of the system, and the message is the expression of this information. A given Ca^{2+} signal may carry several distinct messages, insofar as the Ca^{2+} signal can be deciphered by distinct recipient systems. For example, a Ca^{2+} signal may induce contraction, proliferation, or secretion, for example, each behavior being determined by a given effector system capable of deciphering the Ca^{2+} signal. This chapter focuses on the contractile apparatus.

According to this scheme, in ASMCs, the source of the contractile information is the agonist/agonist receptor complex and the recipient system or final receptor is the contractile apparatus. A time-dependent variation in $[Ca^{2+}]_i$ is the signal that carries the message, i.e., the contractile information. Hence, the temporal aspects of Ca^{2+} signals are at the heart of the notion of Ca^{2+} signal.

1.2 Signal and Message

The signal is the change in time or space of the cytosolic Ca^{2+} concentration. The message is the contractile information encoded in the signal. Hence, it is important to determine what makes a time-dependent variation in the concentration of free Ca^{2+} in the cytosol a message for the contractile machinery. Whatever the complexity of the mechanisms activated by an increase in $[Ca^{2+}]_i$, Ca^{2+} ions by themselves carry a message only if they bind to a target molecule. Hence, variation in $[Ca^{2+}]_i$ is a message only if it induces a correlated change in Ca^{2+} binding to a transmitter molecule, a Ca^{2+}-binding signaling protein. This corresponds to the schematic reaction

$$Ca^{2+} + M \underset{k_-}{\overset{k_+}{\rightleftharpoons}} Ca^{2+}\text{-}M$$

where M is the signaling molecule, and k_+ and k_- are the kinetic constants of the reaction. Accordingly, any change in $[Ca^{2+}]_i$ induced by an increase or decrease in the total amount of Ca^{2+} in the cytosol will induce a change in the concentration of the Ca^{2+}-M complex. Hence, variations in $[Ca^{2+}]_i$ can be viewed as carrying a message. However, this does not mean that the temporal pattern of the Ca^{2+}-M complex, i.e., the message, is similar to that of the Ca^{2+} signal. This is true only if k_+ and k_- are such that the reaction can be considered as instantaneous. Additionally, it should be noted that a given Ca^{2+} signal can carry distinct messages, not only in the sense that it activates different cell systems but also by the fact that the temporal sensitivity of the signaling molecules of the different recipient systems may differ.

1.3 Deciphering the Code

A signal is an encoded message. Deciphering the code entails determining how the cell converts the spatiotemporal change in $[Ca^{2+}]_i$ into a given change in the contractile status of the ASMCs. More precisely, the identification of the mechanisms responsible for the changes in $[Ca^{2+}]_i$ upon ASMC stimulation corresponds to the identification of the encoding mechanisms used by the Ca^{2+} homeodynamic system to convert the stimulation induced by the agonist into a Ca^{2+} signal, and the identification of the mechanisms responsible for the Ca^{2+}-contraction coupling means decrypting the code used by the cell to carry contractile information.

About the decryption of the Ca^{2+} code, it should be noted that the choice in the parameters used to describe the Ca^{2+} signal determines the underlying hypotheses on the nature of the coding system. Indeed, decryption of the code is done by establishing correlations between the contractile response and parameters of the Ca^{2+} signal. Since correlation cannot be established with unstudied parameters, exclusion of parameters should be based on explicit and relevant hypotheses on the possible encoding significance of such parameters. What may appear as a signal for the investigator may not be a message for the cell. For example, several studies use average $[Ca^{2+}]_i$ recording on tissue containing several ASMCs. This average $[Ca^{2+}]_i$ response can provide information on cell physiology. However, this $[Ca^{2+}]_i$ response is the average of the $[Ca^{2+}]_i$ signal of each ASMC but cannot be considered as a signal, even as a signal of whole tissue contraction, because it carries no information for each ASMC contractile apparatus. In addition, the area of the Ca^{2+} response, i.e., the integral of the $[Ca^{2+}]_i$ curve, in some studies called the Ca^{2+} index [1], gives no information on the dynamics of the Ca^{2+} response. Hence, considering it as a parameter of the Ca^{2+} signal for the cell requires the underlying hypothesis that the information carried by the Ca^{2+} signal is independent of the dynamics of the signal. Change in the Ca^{2+} index can be viewed as a sign that the Ca^{2+} signal is altered but not as a signal of a message in itself if the contractile apparatus is sensitive to the dynamics of $[Ca^{2+}]_i$ variations.

How can we decipher the contractile code? Description of the signal is not the same thing as understanding the message. Identification of the quantitative relationship between the parameters of the Ca^{2+} signal is a way of deciphering the contractile code. However, identification of correlation does not mean identification of the causal relationship between Ca^{2+} and contraction. Mathematical modeling is a useful tool for investigating the causal relationships between the pattern of the Ca^{2+} signal and that of the subsequent contraction, which is the decryption of the contractile code.

1.4 Different Temporal Aspects of Ca^{2+} Signal

The temporal dimension is included in the concept of a Ca^{2+} signal itself. The temporal aspect of a Ca^{2+} signal appears at different stages of the stimulation-response pathway. First, insofar as $[Ca^{2+}]_i$ change is an intracellular message induced by the stimulation of the cell by an extracellular agonist, and is hence a consequence of the activation of a network of intracellular mechanisms that determine the pattern of $[Ca^{2+}]_i$ variations, the temporal dimension of these interactions is critical for the shape of the Ca^{2+} signal and, hence, its informative content. This represents the encoding of the Ca^{2+} signal. Second, in as much as $[Ca^{2+}]_i$ variations represent a signal for the contractile machinery, in the sense that they determine the contractile behavior of ASMCs, both in amplitude and time, the time-dependent parameters that describe the Ca^{2+} signal should be considered information on contraction, i.e., information that can be detected by the contractile system. This represents the decoding of the Ca^{2+} signal, which depends on the kinetics of the functional interactions in the contractile system.

Hence, the temporal aspects of the Ca^{2+} signal include several distinct issues: (1) the temporal aspects of the Ca^{2+} signal pattern itself; (2) upstream of the Ca^{2+} signal, the temporal aspects of Ca^{2+} signal encoding, i.e., the dynamics of the mechanisms involved in the Ca^{2+} response to cell stimulation; and (3) downstream of the Ca^{2+} signal, the dynamics of the decoding machinery associated with the contractile apparatus that determines the pattern of the contractile response to Ca^{2+} signaling.

2 Overview of ASMC Ca^{2+} Handling

2.1 General Scheme: Bow-Tie Architecture

Free cytosolic Ca^{2+} concentration is the consequence of the balance between the ON mechanisms that tend to increase $[Ca^{2+}]_i$ and the OFF mechanisms that tend to decrease it. Signaling pathways that induce a Ca^{2+} signal can be viewed as acting on the ON/OFF balance and, by doing so, inducing a spatiotemporal change in $[Ca^{2+}]_i$ corresponding to the encoding of the Ca^{2+} signal. This Ca^{2+} signal can be "read" by the contractile machinery, and the resulting change in contraction corresponds to the decoding of the Ca^{2+} signal. This decoding process can be viewed also as a combination of ON mechanisms that tend to increase contraction for a given Ca^{2+} signal and OFF mechanisms that tend to decrease contraction. The Ca^{2+} signal is hence at the node of encoding and decoding processes, acting upstream and downstream of the Ca^{2+} signal, respectively, on which cell stimulation acts, and excitation-contraction coupling can be analyzed in the conceptual framework of this general scheme [2]. Such a scheme, called a bow-tie architecture, initially used to analyze engineered systems, is also useful for analyzing the information network that commands cell behavior such as excitation-contraction coupling [3–6]. This scheme is presented in Fig. 1a. According to this scheme, the structure involved in Ca^{2+} homeodynamics is the ON and OFF encoding mechanisms. They are summarized in Fig. 1b.

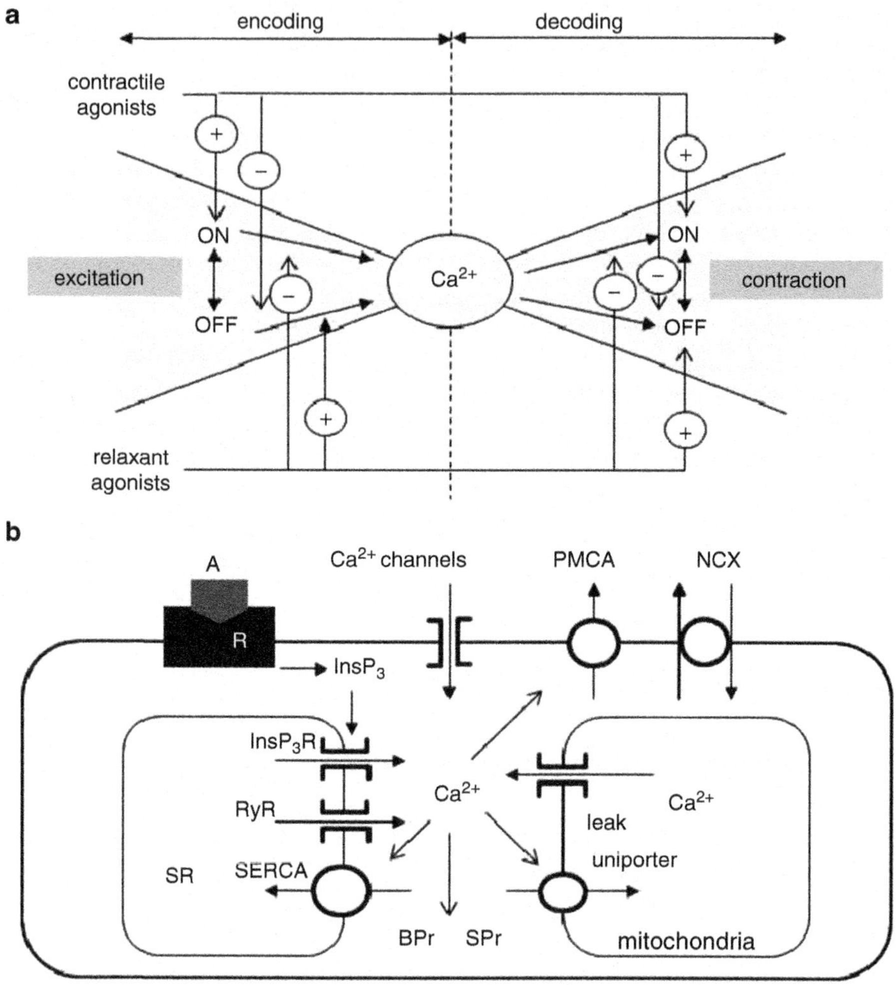

Fig. 1 General mechanisms of Ca^{2+} handling in ASMCs. (**a**) Representation of bow-tie architecture applied to ASMC excitation-contraction coupling. The Ca^{2+} signal pattern is determined by the temporal balance between ON and OFF mechanisms. The Ca^{2+} signal is also decoded by ON and OFF mechanisms on which depends the contraction. Contractant and relaxant agonists may act on the ON and OFF mechanisms of both encoding and decoding of the Ca^{2+} signal. (**b**) General scheme of main ON and OFF mechanisms of Ca^{2+} handling. The main ON mechanisms are (1) Ca^{2+} release from the SR through $InsP_3$ receptors ($InsP_3R$) activated by agonists (*A*) acting on membrane receptors coupled to PLC (*R*) or through ryanodine receptors (RyRs) and (2) extracellular Ca^{2+} influx through several Ca^{2+} channels (VOC, ROC, and SOC). The main OFF mechanisms are (1) Ca^{2+} extrusion through PMCA and NCX, (2) Ca^{2+} pumping into the SR by SERCA, (3) Ca^{2+} uptake by mitochondrial Ca^{2+} uniporter, and (4) Ca^{2+} binding to fast kinetics–low affinity cytosolic proteins (SPr) and slow kinetics–high affinity cytosolic proteins (BPr)

2.2 *Encoding Mechanisms of Ca^{2+} Signal*

2.2.1 ON Mechanisms

The ON mechanisms, by which $[Ca^{2+}]_i$ is increased upon ASMC stimulation, are basically Ca^{2+} release in the cytosol from the sarcoplasmic reticulum (SR) and extracellular Ca^{2+} influx through the plasma membrane. The major contractile agonists, such as acetylcholine and histamine, act primarily via Inositol 1,4,5-trisphosphate ($InsP_3$) production and Ca^{2+} release from the cytosol via $InsP_3$ receptors ($InsP_3R$) located in the sarcolemma [7–11]. Ca^{2+} release from the SR can also be due to the opening of the ryanodine receptors (RyR), which can be activated by $[Ca^{2+}]_i$ increase, the so-called Ca^{2+}-induced Ca^{2+} release (CICR), and cyclic ADP-ribose [12–14]. External Ca^{2+} influx can occur via different types of Ca^{2+} channel. L-type voltage-operated Ca^{2+} channels (VOCs) are present in ASMCs and are activated by membrane depolarization [15, 16]. Receptor-operated channels (ROCs) are activated by binding of the agonist on membrane receptors [17]. Both ROCs and VOCs can be physiologically involved in the Ca^{2+} response to airway smooth muscle stimulation [18]. For example, extracellular adenosine triphosphate (ATP) acts directly on P2X receptors, whose opening induces Ca^{2+} and Na^+ influx, which in turn depolarizes the plasma membrane, hence indirectly activating VOCs and subsequent additional Ca^{2+} influx [19]. Store-operated Ca^{2+} channels (SOCs) are activated by the emptying of intracellular Ca^{2+} stores [20, 21]. Though SOCs have been detected in ASMCs and contribute to general Ca^{2+} homeodynamics, their direct contribution to the physiological Ca^{2+} signal upon cell stimulation remains unclear [21–23].

2.2.2 OFF Mechanisms

A decrease in $[Ca^{2+}]_i$ results from Ca^{2+} extrusion out of the cell, Ca^{2+} uptake by intracellular organelles, or Ca^{2+} buffering by cytosolic Ca^{2+}-binding proteins. Ca^{2+} extrusion against its electrochemical gradient in ASMCs is mainly due to the activity of the plasma membrane Ca^{2+}-ATPase (PMCA) and the Na^+-Ca^{2+} exchanger (NCX), though the contribution of the latter to the Ca^{2+} signal is debated [24], contributing to either Ca^{2+} extrusion but also, in reverse mode, to Ca^{2+} influx [25–28]. Ca^{2+} storage in intracellular compartments is due to Ca^{2+} pumping into the SR by sarco-/endoplasmic Ca^{2+}-ATPase (SERCA) [24, 29] and to Ca^{2+} uptake into the mitochondria by the mitochondrial Ca^{2+} uniporter [2, 30, 31]. Another mechanism that should not be underestimated is free Ca^{2+} buffering by cytosolic proteins [32]. Two main types of Ca^{2+}-binding proteins have been identified: fast kinetics–low affinity Ca^{2+}-binding proteins, so-called signaling proteins, and slow kinetics–high affinity Ca^{2+}-binding proteins, so-called buffering proteins. Actually, these two types of Ca^{2+}-binding sites can coexist on the same protein [6]. Though the Ca^{2+} bound to these proteins remains in the cytosol, these Ca^{2+}-binding proteins can be functionally viewed as a cell compartment of Ca^{2+} sequestration since they remove free Ca^{2+} from the cytosol [2, 31].

3 Temporal Characterization of Ca^{2+} Signal in ASMCs

3.1 Overview of Biological Material and Methods

For the last two decades, a large number of studies have been published on the characterization of the cytosolic Ca^{2+} response to several agonists in ASMCs. The heterogeneity of these studies concerns the biological material used for these investigations, the type and mode of cell stimulation, and the methodology used for $[Ca^{2+}]_i$ recordings. Investigations were performed on ASMCs for different species, mainly rat [2, 8, 9, 19, 33–36], mouse [7, 33, 37–43], guinea-pig [25, 29, 44, 45], pig [12–14, 21, 22, 46, 47], horse [48, 49], cow [21, 50, 51], dog [24, 52–54], and human ASMCs [55–58], and interspecies differences have been demonstrated. Experiments were performed on either cultured cells or freshly isolated cells, and in some cases single cells in situ, for example in lung slides. These different conditions may be responsible for differences in the Ca^{2+} response pattern since it has been shown in rat tracheal cells that Ca^{2+} homeodynamics can be critically altered by cell culture, even for a short period (48 h) [19], and in situ recordings maintain cell-to-cell interactions that are disrupted in isolated myocytes. In most of the studies cited earlier, $[Ca^{2+}]_i$ recordings were done using intracellular Ca^{2+} probes such as indo-1 or fura-2 or similar fluorescent dyes, for which $[Ca^{2+}]_i$ variation is estimated by changes in fluorescence intensity or intensity ratio. These molecules allow real-time measurement of $[Ca^{2+}]_i$ variations, but, insofar as they are Ca^{2+}-binding molecules, their presence may alter the amount of free Ca^{2+} in the cytosol and, hence, the pattern of the Ca^{2+} response. Some studies have used an alternative electrophysiological method based on whole-cell patch clamp recording of Ca^{2+}-activated Cl^- currents (I_{ClCa}) [11, 59, 60]. Comparison of the Ca^{2+} signal profile obtained with both techniques indicates that fluorescent probes, at the usual concentrations, have no detectable effect on the $[Ca^{2+}]_i$ response pattern. For example, in freshly isolated rat trachea myocytes, similar $[Ca^{2+}]_i$ response patterns were obtained using indo-1 fluorescent probe or electrophysiological recording of I_{ClCa} [60].

Some differences may be due to the location along the airway tree. Studies have been performed either on tracheal or intrapulmonary bronchial myocytes, but few studies have compared the responses of these different cells in similar conditions. Since the histological organization of airways differs between trachea, extralobular, and intralobular bronchi, differences in ASMC $[Ca^{2+}]_i$ response depending on their location cannot be excluded. In rat airways, differences in contraction patterns and stimulation-contraction mechanisms have been evidenced in response to purinergic stimulation [19]. In mouse airways, a difference in Ca^{2+} signaling has been evidenced depending on bronchial generations [40].

3.2 General Pattern of $[Ca^{2+}]_i$ Response

Taken together, notwithstanding the differences observed in the Ca^{2+} signal depending on species, agonists, location along the airway tree, and recording methods, the results obtained from the literature show that a general Ca^{2+} pattern can be identified. Examples of original Ca^{2+} traces of ASMCs from different species in response to different agonists are presented in Fig. 2. Recordings on single cells show that the response is characterized by an initial fast and transient $[Ca^{2+}]_i$ rise. The first $[Ca^{2+}]_i$ peak can be followed either by a progressive decay to a more or less steady-state value below the maximum of the $[Ca^{2+}]_i$ peak but above the resting value, the so-called plateau, or by a succession of transient $[Ca^{2+}]_i$ peaks, the so-called Ca^{2+} oscillations. These Ca^{2+} oscillations can be superimposed on the plateau phase, i.e., $[Ca^{2+}]_i$ does not return to baseline between two oscillations. Fluorescent probes give an estimate of the variations in $[Ca^{2+}]_i$, but the absolute $[Ca^{2+}]_i$ value requires calibration, which was not done in all studies that investigated the Ca^{2+} response. When performed, the calculations give a resting $[Ca^{2+}]_i$ of around 100–200 nM, and the amplitude of the Ca^{2+} peak usually ranges up to 1 μM above the resting value. The time-to-peak, i.e., the time needed to reach the maximum $[Ca^{2+}]_i$, is usually not given, indicating that the authors pay more attention to the amplitude of the peak than to its timing. However, considering the few studies that have measured this value [2], and the estimation that can be made from the original traces shown in the others, the top of the Ca^{2+} peak is obtained within a few seconds.

The dynamics of $[Ca^{2+}]_i$ decay has been quantified by some studies [2, 24, 29]. They show that this decay follows a more or less exponential profile, the decay being faster when the stimulation is stopped before the decay begins, but the general pattern remains similar whether cell stimulation goes on or is stopped.

In the case of nonoscillatory response, the slow decay reaches a plateau, which is part of the Ca^{2+} signal, that is above the resting value. However, several studies have focused on the peak and on the oscillations, paying little attention to the plateau. Its significance in terms of contraction response is discussed later. Usually this plateau, when observed, persists as long as the stimulation continues and drops within seconds to baseline when it stops. Observations of original traces show that the plateau is not strictly constant over time and exhibits irregular variations. This can be attributable to stochastic variations in Ca^{2+} release and uptake, whose consequences for the overall Ca^{2+} signal is exacerbated by a clustering of SR Ca^{2+} channels [61]. In some cases, it slowly decreases even upon stimulation, for example upon ATP stimulation in rat tracheal myocytes [19]. In other cases, the Ca^{2+} signal may persist for several seconds after the end of the stimulation, as

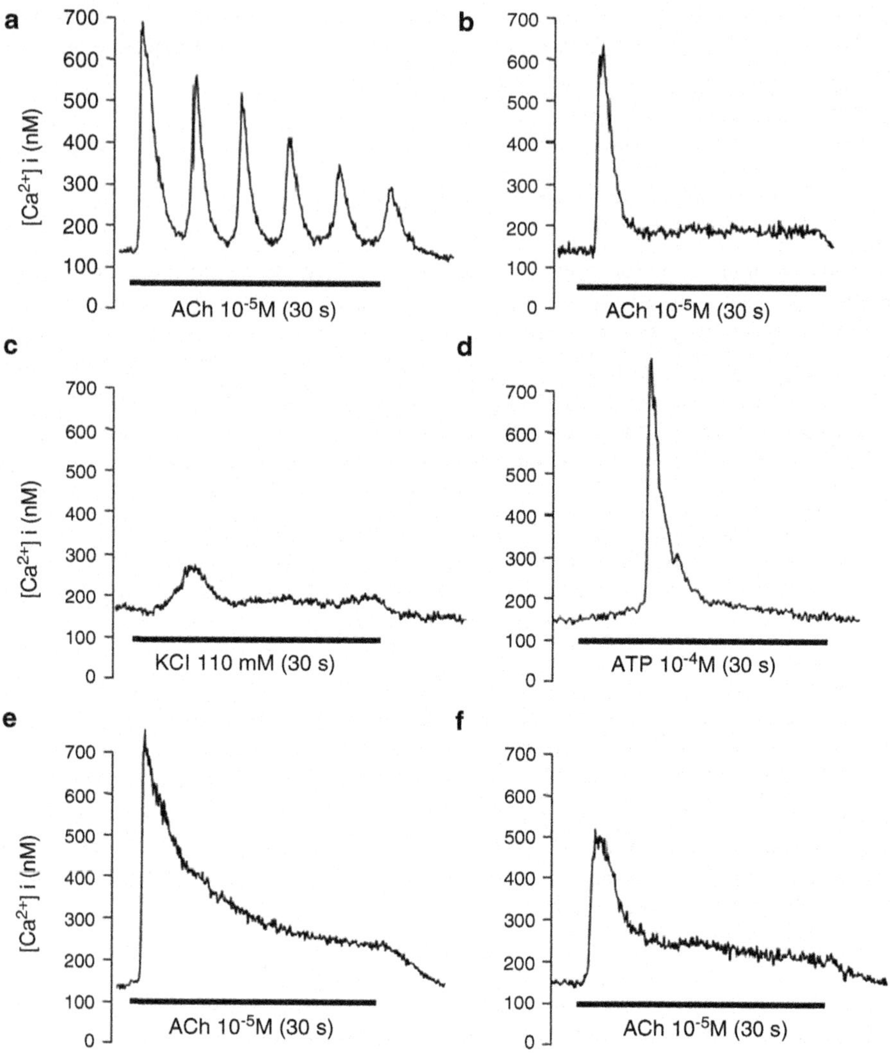

Fig. 2 Original typical traces of Ca^{2+} signals in freshly isolated ASMCs from several species: oscillatory (**a**) and nonoscillatory (**b**) Ca^{2+} response of rat tracheal myocyte to 10^{-5} M ACh. (**c**) Ca^{2+} response of rat tracheal myocyte to 110 mM KCl. (**d**) Ca^{2+} response of rat tracheal myocytes to 10^{-4} M ATP. (**e**) Ca^{2+} response of horse bronchial myocyte to 10^{-5} M ACh. (**f**) Ca^{2+} response of human bronchial myocyte to 10^{-5} M ACh. Cells were obtained from fresh tissues by enzymatic dissociation followed by mechanical dispersion and used within 1 day. $[Ca^{2+}]_i$ was measured by microspectrofluorimetry using the fluorescent dye Indo 1 (for the methods, see [19])

noticed in the same cell type with endothelin-1 [8]. In the case of oscillatory responses, the oscillations may be damped, with a progressive decay in their amplitude, or sustained as long as stimulation persists. Oscillatory responses have been evidenced in ASMCs from several species, such as mouse [42], guinea pig [29], pig [47, 62], and rat (with nonoscillatory responses as well) [9], whereas nonoscillatory responses have been observed in cow [50] and dog ASMCs [24]. In human airways, nonoscillatory Ca^{2+} responses have been described by some studies on isolated ASMCs [55, 57], whereas a recent in situ study describes an oscillatory Ca^{2+} signal [58].

The frequency of oscillations depends on species, agonists, and agonist concentration. In physiological conditions, oscillation frequency ranges between 2 and 3 per minute and 25–30 oscillations per minute. Some pathological conditions like exposure to oxidizing pollutants increase oscillation frequency [36]. Usually, oscillation frequency correlates, when recorded, with the intensity of contraction, and it is usually acknowledged that oscillation frequency encodes for the amplitude of contraction [34, 36–39, 42, 58, 63]. The mechanisms responsible for Ca^{2+} oscillations in ASMCs and their significance in contraction encoding are discussed later in the chapter.

3.3 *Mechanisms Responsible for Ca^{2+} Peak and Plateau*

3.3.1 ON Mechanisms

As shown earlier, the first step for the Ca^{2+} peak and plateau phases, for a large majority of agonists, is a fast Ca^{2+} increase followed by progressive decay that can usually be fitted by an exponential equation. This Ca^{2+} peak is observed with a quite similar profile whatever the source of Ca^{2+}, i.e., extracellular Ca^{2+} influx or Ca^{2+} release from the SR [42, 64]. In both cases, the elevation of $[Ca^{2+}]_i$ is due to a sudden increase in the amount of Ca^{2+} in the cytosol caused by the opening of Ca^{2+} channels, either located in the plasmalemma or in the sarcolemma. This influx is rapid because, when opened, these channels allow exergonic Ca^{2+} movement and the electrochemical gradient highly favors cytosolic Ca^{2+} influx. Mathematical modeling of Ca^{2+} influx through voltage-operated Ca^{2+} channels and $InsP_3R$ or RyR, based on experimental characterization of these channels, predicts a fast increase in $[Ca^{2+}]_i$, as observed in Pulmonary artery smooth muscle cell (PASMC) studies [2, 31, 65]. In the case of $InsP_3$-induced Ca^{2+} release, the kinetics of the Ca^{2+} response depend on the kinetics of the enzymatic reactions of $InsP_3$ production catalyzed by PLC. Hence, at the molecular level, the mechanisms responsible for $[Ca^{2+}]_i$ increase due to $InsP_3R$ stimulation are slower than that due to VOC activation. The velocity of $[Ca^{2+}]_i$ increase, however, depends not only on the kinetics of the molecular reactions that govern the opening and closure of Ca^{2+} channels but also on the ratio between the transmembrane fluxes and the volume of the cytosol. Experiments show that the kinetics of the Ca^{2+} peak induced by $InsP_3$-mediated agonist are almost similar to direct $InsP_3$ stimulation in both cases in ASMCs, indicating that, in this cell type, the enzymatic kinetics of $InsP_3$ production are not a limiting factor for $[Ca^{2+}]_i$ increase velocity [8].

3.3.2 OFF Mechanisms

The Ca^{2+} peak involves not only a rise in $[Ca^{2+}]_i$ but also the decrease in $[Ca^{2+}]_i$ that follows. Then, the OFF mechanisms responsible for cytosolic Ca^{2+} clearance are as important as the ON mechanisms responsible for $[Ca^{2+}]_i$ increase. Experimental results on the respective roles of SR Ca^{2+} uptake, mitochondria, and cytosolic buffering capacity are presented in Fig. 3. Intuitively, Ca^{2+} decay is linked to the closure, at least partial, of Ca^{2+} channels and subsequent activation of mechanisms that remove Ca^{2+} from the cytosol, in particular Ca^{2+} pumping into the SR. Studies of the dynamics of Ca^{2+} handling in ASMCs have proven this intuitive representation to be largely wrong.

First, as far as the stimulation persists, the Ca^{2+} channels remain open, at least in the first step of the stimulation. The end of the Ca^{2+} increase and the beginning of the Ca^{2+} decrease do not correspond to the closure of the Ca^{2+} channels. It has been established, for example, that the $[Ca^{2+}]_i$ peak, including the decay phase, induced by caffeine stimulation is not due to fast opening and closure of RyRs but to persistent RyR opening associated with simultaneous mechanisms of cytosolic Ca^{2+} clearance [2]. In the case of RyR closure, the exponential decay is faster than when RyRs remain open, but the profile is similar in both cases, as shown in Fig. 3a. Both experimental and theoretical studies have shown that the profile of the $[Ca^{2+}]_i$ peak is a consequence of the balance between ON and OFF mechanisms [2, 24, 29, 31, 66]. In this balance, the kinetics of the buffering capacity of OFF mechanisms are a critical determinant of the Ca^{2+} peak profile.

Role of Ca^{2+} Pumping in SR by SERCA

In the buffering system, Ca^{2+} pumping back into the cytosol is usually considered the most important aspect [67]. This is actually true in the sense that the SR as a Ca^{2+} store is important for cell physiology, but Ca^{2+} pumping back by SERCA is not the major determinant of the Ca^{2+} decay. Actually, if Ca^{2+} efflux from the cytosol is activated, then Ca^{2+} pumping back by SERCA cannot in principle account for $[Ca^{2+}]_i$ decreases since as soon as Ca^{2+} is pumped back into the cytosol it leaks through the opened Ca^{2+} channels. Accumulation of Ca^{2+} would require that the velocity of Ca^{2+} pumping against the concentration gradient would be higher than that of a passive Ca^{2+} leak, which is not the case [66]. Even when the Ca^{2+} channels are closed during the decay phase, as occurs for short (1 s) stimulations, inhibition of SERCA does not significantly modify the dynamics of the $[Ca^{2+}]_i$ clearance, as shown in Fig. 3b.

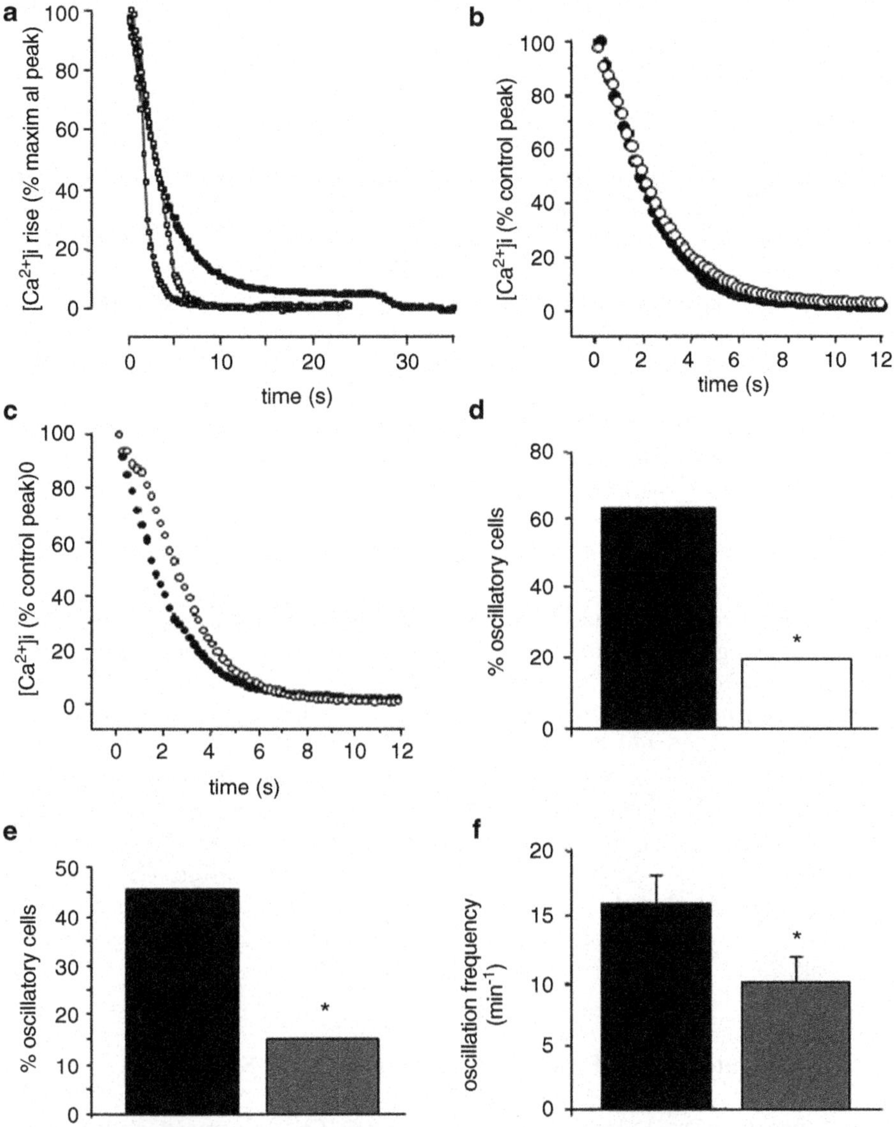

Fig. 3 Influence of OFF mechanisms on kinetics of Ca^{2+} signal. (**a**) Kinetics of $[Ca^{2+}]_i$ decay following Ca^{2+} peak induced by 5 mM caffeine stimulation for 1 s (*left trace*, *light gray squares*), 5 s (*dark gray squares*, *middle trace*), and 30 s (*right trace*, *black square*). (**b**) Influence of SERCA inhibition by cyclopiazonic acid (CPA) on the kinetics of $[Ca^{2+}]_i$ decay following 5 mM caffeine stimulation (1 s). *Black circles*: control; *open circles*: 10^{-5} M CPA. (**c, d**) Influence of mitochondrial Ca^{2+} uptake inhibition by 5.10^{-6} M FCCP on (**c**) the kinetics of $[Ca^{2+}]_i$ decay following 5 mM caffeine stimulation (1 s) (*black circles*: control; *open circles*: FCCP) and (**d**) the percentage of oscillating Ca^{2+} responses to 10^{-5} M ACh stimulation. (**e, f**) Influence of increased cytosolic Ca^{2+} buffering by 5.10^{-5} M BAPTA-AM on (**e**) the percentage of oscillating Ca^{2+} responses and (**f**) the frequency of Ca^{2+} oscillations to 10^{-5} M ACh stimulation. Cells were obtained from fresh tissues by enzymatic dissociation followed by mechanical dispersion and used within 1 day. $[Ca^{2+}]_i$ was measured by microspectrofluorimetry using the fluorescent dye Indo 1 (for the methods, see Roux and Marhl [2]). $^*P < 0.05$ (Student's *t*-test)

Role of Ca^{2+} Uptake in Mitochondria by Ca^{2+} Uniporter

It has been shown that SERCA is neither the only nor the fastest mechanism of free cytosolic Ca^{2+} clearance. When $[Ca^{2+}]_i$ increases, Ca^{2+} uptake into mitochondria by the mitochondrial Ca^{2+} uniporter plays a more important role than Ca^{2+} uptake by SERCA, even when SR Ca^{2+} channels are closed, because Ca^{2+} uptake by mitochondria occurs faster than that by the SR [30, 68–70]. As shown experimentally, for short caffeine stimulations, inhibition of SERCA does not significantly modify the slope of the $[Ca^{2+}]_i$ decay, whereas inhibition of the mitochondrial uniporter by Trifluorocarbonylcyanide phenylhydrazone (FCCP) does, as illustrated in Fig. 3c. Mitochondria seem also to be involved in the occurrence of oscillations since FCCP decreases the percentage of oscillatory response to ACh stimulation in freshly isolated rat tracheal myocytes (personal data) (Fig. 3d).

Role of Ca^{2+} Buffering by Cytosolic Ca^{2+}-Binding Proteins

Though significant, the effect of mitochondrial inhibition on the shape of the $[Ca^{2+}]_i$ decay following caffeine stimulation is modest, and theoretical analysis has shown that the shape of the $[Ca^{2+}]_i$ decay, as well as the amplitude of the Ca^{2+} peak, critically depends on Ca^{2+} binding to cytosolic proteins capable of buffering free cytosolic Ca^{2+} [2, 65, 66]. Though few experimental studies have been dedicated to investigating the role of cytosolic buffering capacity on the pattern of the Ca^{2+} signal, manipulation of the buffering capacity of the cytosol by BAPTA, a Ca^{2+} chelator, shows that an increase in the buffering capacity of the cytosol can alter Ca^{2+} signaling [71]. As shown in Fig. 3e, f, increased cytosolic Ca^{2+} buffering capacity by the Ca^{2+} chelator BAPTA decreases both the percentage of oscillating cells and the frequency of oscillations in response to cholinergic stimulation (personal data).

3.3.3 Integrated View of Ca^{2+} Handling upon ASMC Stimulation

Hence, besides $[Ca^{2+}]_i$ decay following an initial $[Ca^{2+}]_i$ increase, multiple Ca^{2+} buffering mechanisms act simultaneously, but not instantaneously. According to the theoretical analysis of Ca^{2+} handling, a fast $[Ca^{2+}]_i$ increase is followed by free cytosolic Ca^{2+} buffering by cytosolic proteins and Ca^{2+} uptake by the mitochondria prior to Ca^{2+} uptake in the cytosol, even in the case of cell stimulation and Ca^{2+} release into the cytosol [66]. This has several consequences for our understanding of Ca^{2+} signaling.

First, the temporal aspects of the multiple mechanisms involved in the profile of the Ca^{2+} signal are critical determinants of its timing and amplitude. Indeed, the amplitude of the Ca^{2+} peak depends on the buffering capacity of the so-called signaling proteins (fast kinetics Ca^{2+}-binding proteins). The sequential description of the mechanisms of the Ca^{2+} signal according to which Ca^{2+} is *first* released in the

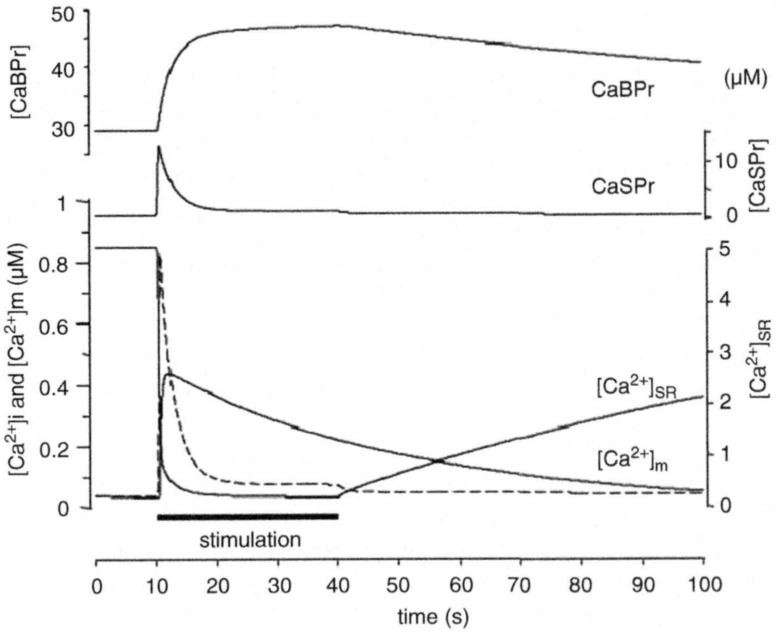

Fig. 4 Model prediction of Ca^{2+} movements upon 30 s caffeine stimulation. Simulated application of caffeine began at time 10 s. *Dashed line*: $[Ca^{2+}]_i$; *full lines*: concentrations of free Ca^{2+} in SR ($[Ca^{2+}]_{SR}$), in mitochondria ($[Ca^{2+}]_m$), and bound to signaling (CaSPr) and buffering (CaBPr) proteins. Though $[Ca^{2+}]_i$ returns to baseline at the end of caffeine stimulation (time 40 s), a return to resting Ca^{2+} concentration in the various cell compartments requires much more time

cytosol, *then* buffered by cytosolic protein, *later* taken up by the mitochondria, and *eventually* pumped back into the SR is only a metaphorical way to express the fact that these mechanisms have different kinetics that should not hide the fact that they occur simultaneously. Cell stimulation disrupts the dynamic equilibrium responsible for the $[Ca^{2+}]_i$ resting value, and the Ca^{2+} peak is the result of the subsequent transitory nonequilibrium state. The Ca^{2+} plateau is the consequence of the new dynamics of Ca^{2+} handling when the new equilibrium state is reached. Fast kinetics mechanisms are critical for the peak shape, whereas the plateau value is dependent on the relative buffering capacity of each cell compartment, even if the kinetics of these buffering reactions are slow. It should be noted that even if the plateau value seems slightly different from the resting $[Ca^{2+}]_i$ value, the equilibrium state may be quite different in both cases, with great differences in the amount of Ca^{2+} in the various intracellular compartments at rest versus upon cell stimulation.

The second consequence is that, although $[Ca^{2+}]_i$ values return to baseline within seconds after the end of cell stimulation, this does not mean that the cell is already back to its initial Ca^{2+} homeodynamic state. Theoretical studies, in correlation with experiments, show that complete Ca^{2+} refilling of the SR requires several minutes [2], as illustrated in Fig. 4. As a consequence, two successive cell stimulations triggering Ca^{2+} release from the SR will not result in similar Ca^{2+} response as long as Ca^{2+} handling is not back at its initial state, and the time needed to recover to this initial state corresponds to a relative refractory period. The delay needed for the return to resting equilibrium, which is determined by the temporal characteristics of the mechanisms of cytosolic Ca^{2+} clearance, has a quite opposite consequence for cell homeostasis depending on the source of Ca^{2+}. In the case of Ca^{2+} release from internal Ca^{2+} stores, repeated stimulations result in progressive depletion of the SR and subsequent decrease in stimulus responsiveness, whereas in the case of external Ca^{2+} influx, they would result in progressive Ca^{2+} overloading if the mechanisms of Ca^{2+} extrusion were saturated [2, 72].

In conclusion, it should be emphasized that the peak and the plateau involve similar mechanisms responsible for Ca^{2+} homeodynamics. The difference between the peak and the plateau profiles corresponds to two different phases of the transition between two equilibrium states of the Ca^{2+} handling, at rest and upon stimulation. The peak is a consequence of the transient nonequilibrium phase of the activated state, whereas the plateau corresponds to the state of the system when equilibrium is reached. When stimulation ends, return to the initial resting state is longer than the return of $[Ca^{2+}]_i$ to baseline, due to silent Ca^{2+} fluxes between the Ca^{2+} buffering elements that occur at constant free cytosolic Ca^{2+} concentration.

3.4 *Mechanisms of Ca^{2+} Oscillations*

As indicated earlier, Ca^{2+} oscillations have been reported in a variety of ASMCs, though it should not be forgotten that in some species, including humans, the Ca^{2+} signal can be a nonoscillatory one. In ASMCs, Ca^{2+} oscillations have been mainly observed in response to contractile agonists acting on G-protein-coupled receptors and $InsP_3$ production, such as ACh, 5-HT, endotheline-1, and ATP [9, 11, 13, 19, 42, 43, 73]. Additionally, experiments using direct stimulation by $InsP_3$ have shown that nonoscillatory $InsP_3$ concentration was able to generate oscillatory Ca^{2+} responses [8]. However, Ca^{2+} oscillations have also been noticed in response to stimulation acting via external Ca^{2+} influx [42]. Figure 5a, b compares Ca^{2+} oscillations induced by cholinergic stimulation and by direct $InsP_3$ exposure.

Basically, Ca^{2+} oscillations can be explained by asynchronous coupling between ON and OFF mechanisms. Since oscillations are observed in response to $InsP_3$, much attention has been paid to the kinetic properties of $InsP_3R$. Several isoforms of $InsP_3R$ have been identified, types 1, 2, and 3. These isoforms are sensitive to both cytosolic $InsP_3$ and Ca^{2+}. Both types 2 and 3 exhibit a sigmoid-shaped sensitivity curve to Ca^{2+} concentration, whereas type 1 has a bell-shaped sensitivity curve: for low $[Ca^{2+}]_i$, sensitivity to Ca^{2+} increases with $[Ca^{2+}]_i$, whereas for high $[Ca^{2+}]_i$, an increase in $[Ca^{2+}]_i$ decreases $InsP_3P$ opening [74–77]. Moreover, this loss in sensitivity for high $[Ca^{2+}]_i$ is delayed. $InsP_3R$ by itself seems to be an asynchronous ON/OFF switch capable of generating oscillations, an idea supported by the fact that $InsP_3R$ inhibition abolishes Ca^{2+} oscillations [7]. Several theoretical models

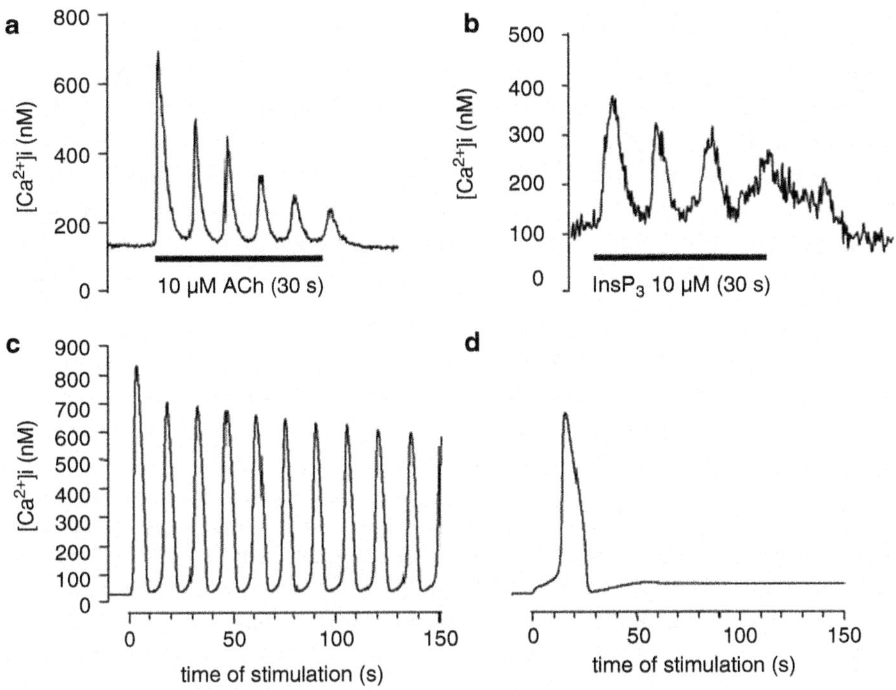

Fig. 5 Ca^{2+} oscillations. Original traces of Ca^{2+} response to ACh stimulation (**a**) and $InsP_3$ stimulation (**b**). Cells were obtained from fresh tissues by enzymatic dissociation followed by mechanical dispersion and used within 1 day. $[Ca^{2+}]_i$ was measured by microspectrofluorimetry using fluorescent dye indo-1. $InsP_3$ was applied on β-escine-permeabilized cells [8]. (**c**, **d**) Model prediction to 1 μM $InsP_3$ stimulation for different $InsP_3R$ isoform ratio. (**c**) 75 % IP_3R subtype 1 and 25 % subtype 3; (**d**) 25 % IP_3R subtype 1 and 75 % subtype 3. The model used is that of Haberichter et al. [65]

have been developed to explain how $InsP_3R$ can be a molecular oscillator [78]. However, the mechanisms of Ca^{2+} oscillations have appeared to be more complex and could not be attributed to the properties of the $InsP_3R$ type 1 alone. First, experimental studies have shown that type 1 $InsP_3R$ expression can be associated with nonoscillatory Ca^{2+} signals [57]. On the other end, studies in vascular SMCs showed intercellular variation in $InsP_3R$ isoform expression and that $InsP_3R$ alone cannot generate Ca^{2+} oscillations, which requires a mixture of type 1 and type 2 isoform expression in the same cell [79]. In rat ASMCs, an immunofluorescence study suggested that such a heterogeneous expression is observed, with 100 % of ASMCs expressing type 1 isoform, 27 % type 2, and 40 % type 3 (personal data). Theoretical modeling of the $InsP_3$-induced Ca^{2+} response indicates that both oscillatory and nonoscillatory Ca^{2+} signals can be predicted in cells expressing a mixture of type 1 and type 3 $InsP_3R$, depending on the total amount of $InsP_3R$ and the type1/type3 ratio, and the presence of cytosolic Ca^{2+}-binding proteins [65]. The role of the isoform ratio in the occurrence of Ca^{2+} oscillations is illustrated in Fig. 5c, d.

RyRs have also been shown to contribute to ACh-induced as well as KCl-induced Ca^{2+} oscillations [12, 13, 72].

It appears, then, that Ca^{2+} oscillations in ASMCs are not due to the intrinsic property of one specific molecular oscillator whose presence determines their occurrence. Rather, the occurrence of Ca^{2+} oscillations is an emerging property of the machinery of Ca^{2+} handling, and oscillations may appear when asynchronous ON/OFF mechanisms are simultaneously active. As shown by theoretical investigations, one cell may occupy different possible positions in the phase space, at one moment being in an oscillatory phase space and at another moment in a nonoscillatory space phase, depending on the intensity of cell stimulation [65, 72]. In a cell population, interindividual variations in $InsP_3R$ expression and in the ratio of isoform expression generate the distribution of the cells in the different phases of the space, so that heterogeneous cell behavior, oscillatory and nonoscillatory Ca^{2+} signals, may be the consequence of a homogeneous cell type associated with inherent cell-to-cell variations [65].

Oscillatory Ca^{2+} responses have been reported in some cases in response to KCl stimulation [42]. In the case of forced membrane depolarization, the mechanisms responsible for the Ca^{2+} signal, i.e., Ca^{2+} influx through VOCC and CICR through RyRs, are different from that activated by agonists acting on G-coupled membrane receptors like muscarinic receptors. This emphasizes the fact that oscillations may occur as a consequence of the activation of distinct pathways. However, in the case of KCl, Ca^{2+} oscillations require simultaneous involvement of ON/OFF mechanisms, unlike with $InsP_3$. Oscillations have also been reported in vascular SMCs, and theoretical modeling, including voltage-dependent Ca^{2+} influx and RyR activation, accounts for this oscillatory profile [80]. Though these models have been developed for vascular SMCs, the mechanisms they describe also account for what is observed in ASMCs, as was confirmed by recent modeling of KCl-induced oscillations in ASMCs [72].

Taken together, these results indicate that the pattern of the Ca^{2+} signal is critically determined by the kinetic constants of each element involved in ASMC Ca^{2+} handling. Perturbation of the functional properties of any of these structures may lead to loss of the oscillatory response, which does not mean that this structure is *the* element that determines the Ca^{2+} profile, but it is one element, among others, where functional properties contribute to shape the Ca^{2+} signal.

4 Decoding of Ca^{2+} Signals

4.1 Contractile System

How is a message carried by the Ca^{2+} signal? In other words, how does the Ca^{2+} signal determine the behavior of the contractile apparatus? What is, in the time-dependent variations of $[Ca^{2+}]_i$ that occur upon cell stimulation, information for the

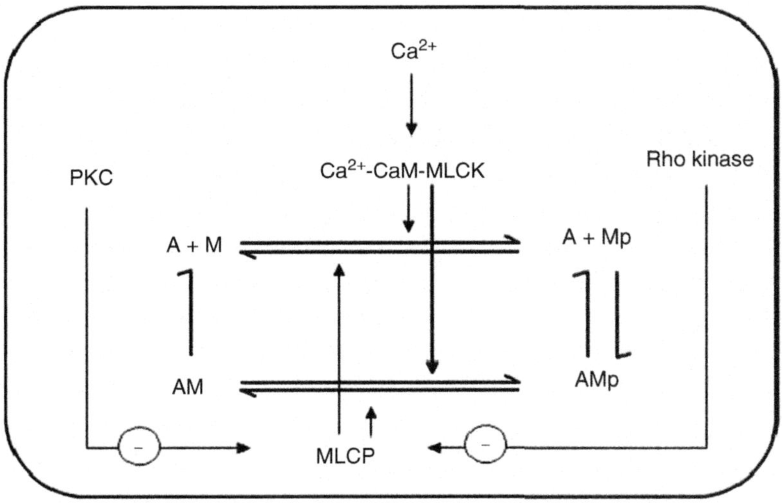

Fig. 6 General mechanisms of Ca^{2+}-contraction coupling in ASMCs. Ca^{2+} binds to calmodulin (*CaM*) and MLCK. The Ca^{2+}-CaM-MLCK complex is the active form of MLCK, which catalyzes MLC_{20} (*M*) phosphorylation. Phosphorylated MLC_{20} (*Mp*) binds to actin (*A*) to form the actin-myosin bridge (*AMp*) that can later be dephosphorylated (*AM*). Dephosphorylation is achieved by MLCP, whose activity is modulated by several kinases, mainly PKC and RhoK

contractile machinery, and what type of information is it? The object of this part of the chapter is not to detail the contractile apparatus and its regulation pathways but to present an overview of the "semantics" of the Ca^{2+} signal in terms of contraction. Basically, a rise in $[Ca^{2+}]_i$ activates the contractile apparatus by the following pathways, summarized in Fig. 6. Ca^{2+} binds to the cytosolic protein calmodulin (CaM), which can bind to the myosin light chain kinase (MLCK), and the Ca^{2+}-CaM-MLCK complex is the active form of MLCK. When activated, MLCK phosphorylates the 20 kDa regulatory myosin light chain (MLC_{20}), and this phosphorylation is required for the formation of the actin-myosin cross bridge and, hence, contraction. Dephosphorylation of MLC_{20} is ensured by myosin light chain phosphatase (MLCP). Both MLCK and MLCP activity can be up- and downregulated by several kinase- and phosphatase-dependent pathways [81]. The canonical way that controls contraction in ASMC is the Ca^{2+}-CaM-MLCK pathway [34, 82]. The other regulatory mechanisms are grouped into the so-called concept of regulation of Ca^{2+} sensitivity since they modify the contractile response for a given Ca^{2+} signal. Though useful, this distinction between the Ca^{2+}-induced contraction pathway on the one hand and the regulation of Ca^{2+} sensitivity on the other hand is somewhat artificial since regulation of Ca^{2+} sensitivity can itself be Ca^{2+}-dependent [83].

4.2 *Contractile Significance of Ca^{2+} Signal*

The semantics of the Ca^{2+} signal have long been investigated by correlating contraction with the Ca^{2+} signal parameters, mainly the Ca^{2+} peak, the plateau, and, when present, the oscillations [9, 19, 35–37, 42, 64]. Regarding the contraction itself, several distinct parameters are used to describe it. Usually, the intensity of the steady-state response is used to characterize the contractile response, measured by either isometric recording [19, 34, 36, 45, 51, 83–85] or changes in the trans-sectional area of the bronchial lumen [33, 40, 43], but other parameters are also used to describe the temporal aspects of the contractile response, such as shortening velocity in isotonic measurements [86, 87] or half-time maximal contraction (the time needed to achieve 50 % of the maximal contraction) in isotonic recording [34, 83, 86, 87].

Peak, plateau, and oscillation frequency all vary with the intensity of agonist stimulation, as does the amplitude of contraction. It is hence difficult to determine experimentally the specific message carried by each Ca^{2+} signal parameter because of the difficulty of varying one parameter independently of the others. Theoretical modeling is a way to investigate the specific significance of the distinct parameters of the Ca^{2+} signal since it is possible, in contrast to experimentation, to vary them independently [34, 83, 87–89].

4.2.1 Contractile Message of Initial Ca^{2+} Peak

Realistic computational modeling of Ca^{2+}-contraction coupling, including both the canonical MLCK pathway and MLCP regulation by RhoK and PKC, indicates that the amplitude of the Ca^{2+} peak determines the velocity of the early phase of the contraction [34, 83]. This may explain why ASM contraction develops slower for low than for high levels of agonist concentration. However, according to the model, the contractile apparatus is much less sensitive to a transient $[Ca^{2+}]_i$ rise than to a sustained one, and the peak does not determine the amplitude of contraction. This is basically due to the inertia of the MLCK-dependent activation of the contractile apparatus compared with the velocity of the Ca^{2+} peak, both in its increase and decrease, so that during the Ca^{2+} peak, the MLC phosphorylation rate decreases before phosphorylated MLC has reached its maximal value.

4.2.2 Contractile Message of Ca^{2+} Plateau

The amplitude of the plateau is, according to the model, the major determinant of the amplitude of contraction [34, 83]. Since the plateau is a stable $[Ca^{2+}]_i$ value, the system can reach its equilibrium. Usually, the amplitude of the plateau is considered small compared with the maximal peak value. However, both theoretical and experimental studies indicate that the level of phosphorylated MLC is not linearly related to the steady-state $[Ca^{2+}]_i$ value, and this explains the fact that the amplitude of the plateau is a critical determinant of the amplitude of the contraction [34].

4.2.3 Contractile Message of Ca^{2+} Oscillations

Experimental studies have demonstrated statistical correlations between the frequency of Ca^{2+} oscillations and the amplitude of contraction, suggesting that the Ca^{2+} oscillation frequency determines the amplitude of contraction [9, 19, 33, 35, 39, 43, 63]. However, such correlations do not establish an unequivocal causal relationship since, as stated earlier, other parameters simultaneously correlate with contraction intensity [9, 35]. Moreover, the meaning of frequency encoding should be clarified. Indeed, Ca^{2+} oscillations induce an average $[Ca^{2+}]_i$ increase that may induce contraction whether this increase is oscillating or not. Thus, if we consider oscillations as carrying a specific message, this means that the contractile apparatus is specifically sensitive to the dynamics of Ca^{2+} oscillations. In other words, the contractile system is oscillation-frequency-sensitive only if the contractile response to an oscillatory signal differs from that of a nonoscillatory signal with an identical average $[Ca^{2+}]_i$ value, i.e., for which the surface area under an oscillating trace is equal to that under a nonoscillating trace. It is hard, if not impossible, to design an experimental protocol to compare these two conditions. Additionally, since airway smooth muscle contraction is the response at the tissue level, the question arises as to whether the observed nonoscillatory contractile response at the tissue level is the consequence of the summation of asynchronous oscillatory responses at the cellular level or results from the integration at the cellular level of an oscillatory Ca^{2+} message into a nonoscillatory contractile response. For both questions, mathematical modeling is a useful tool to test these hypotheses. Computational modeling with realistic kinetic constants indicates that oscillation frequency within the range observed in experiments (5–30 oscillations min^{-1}) generates a nonoscillatory contractile response [34, 83]. Hence, it seems that the oscillatory message is integrated at the cellular level by the contractile machinery. This integration occurs at different steps of Ca^{2+}-contraction coupling, illustrated in Fig. 7. Model prediction indicates that the temporal pattern of MLCK activity is oscillating and follows more or less that of the Ca^{2+} signal, though MLCK activity decay might be slightly slower than Ca^{2+} decay (Fig. 7b). In contrast, the temporal pattern of MLC phosphorylation, though oscillatory, exhibits smoothed oscillations, the level of phosphorylated MLC remaining high between two oscillations (Fig. 7c). This is due to the inertia of the system, i.e., physiological MLCP activity is not sufficient to trigger an immediate drop in phosphorylated MLC when MLCK activity decreases between two Ca^{2+} oscillations. The second, and most important, step of the integration of the Ca^{2+} signal is the formation of actin-myosin bridges. Computational modeling predicts that the amount of actin-myosin bridges, which corresponds to isometric contraction, remains high between two Ca^{2+} peaks even if the amount of phosphorylated MLC tends to decrease. This is due to the fact that actin-myosin

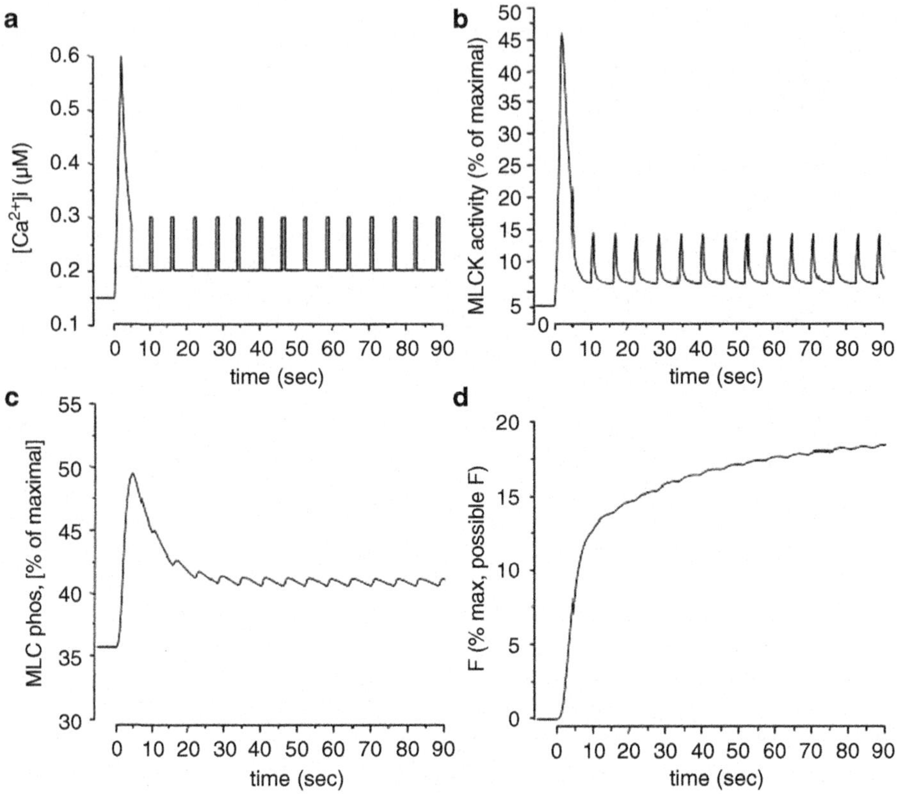

Fig. 7 Temporal integration of oscillatory Ca^{2+} signal. Model prediction of MLCK activation (**b**), MLC phosphorylation (**c**), and contraction (**d**) upon oscillating Ca^{2+} signal (**a**), according to the model of Mbikou et al. [83]

binding can be maintained even if MLC is dephosphorylated because of the low kinetic constant for the rupture of the dephosphorylated myosin-actin bridge, the so-called latch bridge. When the next Ca^{2+} peak and its subsequent rise in MLC phosphorylation occur before a significant drop in the actin-myosin bridge, which depends on the balance between oscillation frequency and the kinetics of actin-myosin bridge cycling, actin-myosin bridges tend to accumulate until an equilibrium is reached, and this corresponds to a slow additional increase in contraction (Fig. 7d). Changing the oscillation frequency within the physiological range in the model predicts an oscillation-frequency-dependent increase in contraction. It can hence be concluded that the oscillations actually encode for the intensity of ASM contraction since ASMC modeling predicts that the contractile system is specifically sensitive to oscillation frequency independent of the average $[Ca^{2+}]_i$ value. However, if we compare, as shown in Fig. 8, the sensitivity of the contractile machinery to the peak value (or transient $[Ca^{2+}]_i$ increase), the plateau value (or the average $[Ca^{2+}]_i$ value), and the oscillation frequency, the model prediction

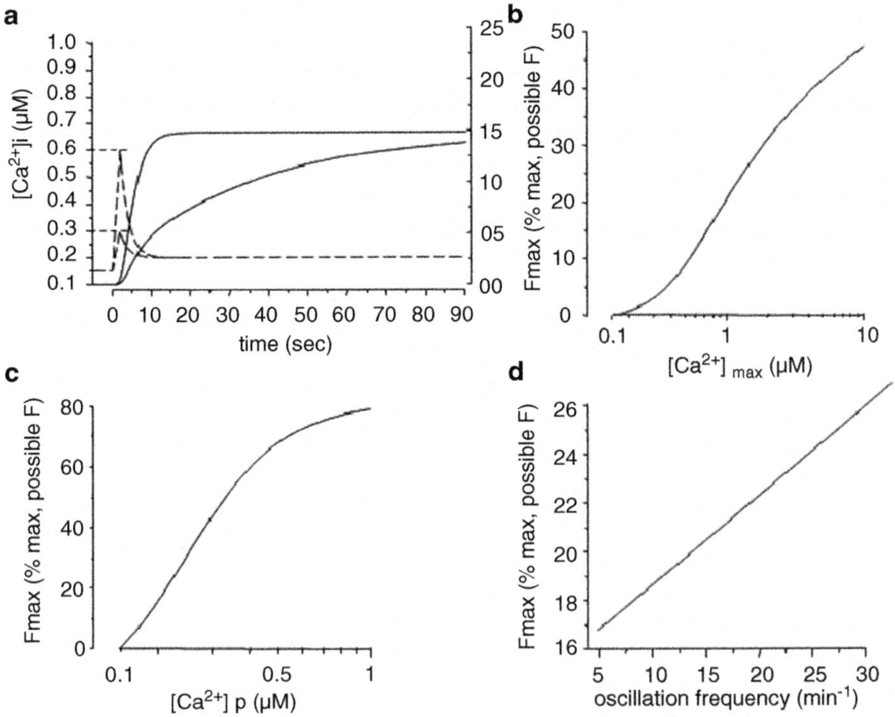

Fig. 8 Contraction encoding by peak, plateau, and frequency of oscillations. Model predictions are done using the model of Mbikou et al. [83]. (**a**) Predicted time course of force for two peak amplitudes (0.3 and 0.6 μM) for fixed resting $[Ca^{2+}]_i$ and plateau. (**b**) Predicted relationship between amplitude of transient $[Ca^{2+}]_i$ increase and force. (**c**) Predicted relationship between amplitude of sustained $[Ca^{2+}]_i$ and force. (**d**) Predicted relationship between frequency of oscillations and force. The simulated Ca^{2+} signal used is similar to that of Fig. 7, with variation in oscillation frequency

is that the contractile apparatus will be much more sensitive to the plateau value than to the frequency of Ca^{2+} oscillation per se. Hence, the importance of the plateau phase or, in the case of oscillatory Ca^{2+} signals, of the average $[Ca^{2+}]_i$ value over time independently of Ca^{2+} oscillations should not be underestimated. The sensitivity to a transient $[Ca^{2+}]_i$ rise is quite low, as shown in Fig. 8b, which explains the fact that, for a standard biphasic Ca^{2+} response, the amplitude of the Ca^{2+} peak does not determine the amplitude of the contraction, which depends on the plateau, but not its velocity (Fig. 8a).

It is tempting to speculate about the advantage of frequency encoding versus amplitude encoding in cell physiology. For example, it has been hypothesized that oscillation encoding may be more precise than amplitude encoding, hence ensuring more accurate functional response to cell stimulation, especially for low agonist concentrations [90]. Though this may be true in principle, one should be cautious, however, when speculating about the putative advantage of oscillatory versus nonoscillatory encoding, and at least in ASMCs, there is no clear evidence that oscillatory signaling may be advantageous. The main argument against such a speculation is that, in various species, ASMCs exhibit nonoscillatory Ca^{2+} signals. Additionally, analysis of cell-to-cell variability of the Ca^{2+} signal has shown, as a general property, high interindividual variations in the Ca^{2+} response to the same stimulation [91]. The Ca^{2+} signal appears, then, to be highly noisy, and it is hard to imagine the specific evolutionary advantage of highly precise decoding of a highly noisy signal. Robustness to the noise of the signal would, in principle, be more advantageous.

Functional analysis shows that Ca^{2+} oscillations can appear as an emerging property of the system responsible for intracellular Ca^{2+} handing and that these oscillations have functional consequences for the contractile apparatus. Whether the occurrence of an oscillating Ca^{2+} signal is a byproduct of the functional constraint of the system controlling Ca^{2+} homeodynamics [92] or a consequence of a selective pressure that has designed this system for the production of frequency-encoded Ca^{2+} signal is purely speculative and has no real support.

5 Conclusion

In conclusion, the temporal aspects of Ca^{2+} signals are inherent in the concept of Ca^{2+} signal, which is, by definition, a spatiotemporal change in $[Ca^{2+}]_i$. Temporal variations upon contractile stimulation exhibit a general pattern characterized by a transient Ca^{2+} peak followed either by a steady-state phase or Ca^{2+} oscillations. This pattern is the consequence of simultaneous, but not instantaneous, interactions between all the elements involved in Ca^{2+} handling, both ON and OFF mechanisms. Among the latter, Ca^{2+} pumping back into the SR is not the only important mechanism, and Ca^{2+} uptake by mitochondria and Ca^{2+} binding to cytosolic proteins are critical determinants of the Ca^{2+} signal. The transient peak is the consequence of a transitory nonequilibrium phase between the resting and the activated states of Ca^{2+} homeodynamics. If an equilibrium is reached, then $[Ca^{2+}]_i$ stabilizes with respect to the plateau, whereas if asynchronous interactions between ON and OFF occur, Ca^{2+} oscillations appear. These different parameters of the Ca^{2+} signal are contractile messages. According to theoretical analysis, the amplitude of the Ca^{2+} peak encodes for the velocity of contraction, whereas the plateau determines the amplitude of the contraction. When they exist, oscillatory signals seem to be integrated at the cell level, leading to nonoscillatory contractions due to the temporal inertia of the MLC dephosphorylation process and dephosphorylated myosin-actin bridges. Oscillation frequency seems to encode specifically for the amplitude of contraction, but the contractile system is less sensitive to Ca^{2+} oscillation frequency per se than to the average Ca^{2+} value over time, whether or not the Ca^{2+} signal is oscillatory.

References

1. Ikeda, M., K. Kurokawa, and Y. Maruyama, *Cyclic nucleotide-dependent regulation of agonist-induced calcium increases in mouse megakaryocytes.* J Physiol, 1992. **447**: p. 711–28.
2. Roux, E. and M. Marhl, *Role of sarcoplasmic reticulum and mitochondria in ca(2+) removal in airway myocytes.* Biophys J, 2004. **86**(4): p. 2583–95.
3. Csete, M. and J. Doyle, *Bow ties, metabolism and disease.* Trends in Biotechnology, 2004. **22** (9): p. 446–450.
4. Ma, H.W. and A.P. Zeng, *The connectivity structure, giant strong component and centrality of metabolic networks.* Bioinformatics, 2003. **19**(11): p. 1423–1430.
5. Marhl, M., M. Perc, and S. Schuster, *Selective regulation of cellular processes via protein cascades acting as band-pass filters for time-limited oscillations.* FEBS Lett., 2005. **579**(25): p. 5461–5465.
6. Schuster, S., B. Knoke, and M. Marhl, *Differential regulation of proteins by bursting calcium oscillations - a theoretical study.* Biosystems, 2005. **81**(1): p. 49–63.
7. Bai, Y., M. Edelmann, and M.J. Sanderson, *The contribution of inositol 1,4,5-trisphosphate and ryanodine receptors to agonist-induced Ca(2+) signaling of airway smooth muscle cells.* Am J Physiol Lung Cell Mol Physiol, 2009. **297**(2): p. L347–61.
8. Hyvelin, J.M., et al., *Cellular mechanisms of acrolein-induced alteration in calcium signaling in airway smooth muscle.* Toxicol Appl Pharmacol, 2000. **164**(2): p. 176–83.
9. Roux, E., et al., *$[Ca^{2+}]_i$ oscillations induced by muscarinic stimulation in airway smooth muscle cells: receptor subtypes and correlation with the mechanical activity.* Br J Pharmacol, 1997. **120**: p. 1294–1301.
10. Roux, E., et al., *Muscarinic stimulation of airway smooth muscle cells.* Gen Pharmacol, 1998. **31**(3): p. 349–56.
11. Liu, X. and J.M. Farley, *Frequency modulation of acetylcholine-induced Ca(++)-dependent Cl- current oscillations are mediated by 1, 4, 5-trisphosphate in tracheal myocytes.* J Pharmacol Exp Ther, 1996. **277**: p. 796–804.
12. Kannan, M.S., et al., *Role of ryanodine receptor channels in Ca2+ oscillations of porcine tracheal smooth muscle.* Am J Physiol, 1997. **272**(4 Pt 1): p. L659–64.
13. Prakash, Y.S., M.S. Kannan, and G.C. Sieck, *Regulation of intracellular calcium oscillations in porcine tracheal smooth muscle cells.* Am J Physiol, 1997. **272**(3 Pt 1): p. C966–75.
14. Prakash, Y.S., et al., *Role of cyclic ADP-ribose in the regulation of [Ca2+]i in porcine tracheal smooth muscle.* Am J Physiol, 1998. **274**(6 Pt 1): p. C1653–60.
15. Marthan, R., et al., *Calcium channel currents in isolated smooth muscle cells from human bronchus.* J Appl Physiol, 1989. **66**: p. 1706–1714.
16. Kotlikoff, L., *Calcium currents in isolated canine airway smooth muscle cells.* Am J Physiol, 1988. **254**: p. C793–C801.
17. Rodger, I.W., *Voltage-dependent and receptor-operated calcium channels*, in *Airways smooth muscle: biochemical control of contraction and relaxation*, D. Raeburn and M.A. Giembycz, Editors. 1994, Birkhäuser Verlag: Basel. p. 155–168.
18. McFadzean, I. and A. Gibson, *The developing relationship between receptor-operated and store-operated calcium channels in smooth muscle.* Br J Pharmacol, 2002. **135**(1): p. 1–13.
19. Mounkaila, B., R. Marthan, and E. Roux, *Biphasic effect of extracellular ATP on human and rat airways is due to multiple P2 purinoceptor activation.* Respir Res, 2005. **6**(1): p. 143.
20. Sanders, K.M., *Invited review: mechanisms of calcium handling in smooth muscles.* J Appl Physiol, 2001. **91**(3): p. 1438–49.
21. Helli, P.B., E. Pertens, and L.J. Janssen, *Cyclopiazonic acid activates a Ca2+−permeable, nonselective cation conductance in porcine and bovine tracheal smooth muscle.* J Appl Physiol, 2005. **99**(5): p. 1759–68.
22. Ay, B., et al., *Store-operated Ca2+ entry in porcine airway smooth muscle.* Am J Physiol Lung Cell Mol Physiol, 2004. **286**(5): p. L909–17.

23. Marthan, R., *Store-operated calcium entry and intracellular calcium release channels in airway smooth muscle.* Am J Physiol Lung Cell Mol Physiol, 2004. **286**(5): p. L907–8.
24. Janssen, L.J., D.K. Walters, and J. Wattie, *Regulation of [Ca2+]i in canine airway smooth muscle by Ca(2+)-ATPase and Na+/Ca2+ exchange mechanisms.* Am J Physiol, 1997. **273** (2 Pt 1): p. L322–30.
25. Flores-Soto, E., et al., *In airways ATP refills sarcoplasmic reticulum via P2X smooth muscle receptors and induces contraction through P2Y epithelial receptors.* Pflugers Arch, 2011. **461** (2): p. 261–75.
26. Liu, B., et al., *Reverse mode Na+/Ca2+ exchange mediated by STIM1 contributes to Ca2+ influx in airway smooth muscle following agonist stimulation.* Respir Res, 2010. **11**: p. 168.
27. Iwamoto, T. and M. Shigekawa, *Differential inhibition of Na+/Ca2+ exchanger isoforms by divalent cations and isothiourea derivative.* Am J Physiol, 1998. **275**(2 Pt 1): p. C423–30.
28. Pitt, A. and A.J. Knox, *Molecular characterization of the human airway smooth muscle Na+/Ca2+ exchanger.* Am J Respir Cell Mol Biol, 1996. **15**(6): p. 726–30.
29. Sims, S.M., Y. Jiao, and Z.G. Zheng, *Intracellular calcium stores in isolated tracheal smooth muscle cells.* Am J Physiol, 1996. **271**(2 Pt 1): p. L300–9.
30. Roux, E., J.-P. Mazat, and M. Marhl, *Role of mitochondria in calcium homeostasis and contraction of smooth muscle cells*, in *Mitochondria: Structure, Functions and Dysfunctions*, O.L. Svensson, Editor. 2009, Nova Publisher: Hauppauge NY.
31. Roux, E., et al., *Modelling of calcium handling in airway myocytes.* Prog Biophys Mol Biol, 2006. **90**(1–3): p. 64–87.
32. Marhl, M., et al., *Modelling oscillations of calcium and endoplasmic reticulum transmembrane potential; role of the signalling and buffering proteins and of the size of the Ca2+ sequestering ER subcompartments.* Bioelectrochem Bioenerg, 1998. **46**: p. 79–90.
33. Bai, Y. and M.J. Sanderson, *The contribution of Ca2+ signaling and Ca2+ sensitivity to the regulation of airway smooth muscle contraction is different in rats and mice.* Am J Physiol Lung Cell Mol Physiol, 2009. **296**(6): p. L947–58.
34. Mbikou, P., et al., *Theoretical and experimental investigation of calcium-contraction coupling in airway smooth muscle.* Cell Biochem Biophys, 2006. **46**(3): p. 233–52.
35. Roux, E., M. Duvert, and R. Marthan, *Combined effect of chronic hypoxia and in vitro exposure to gas pollutants on airway reactivity.* Am J Physiol Lung Cell Mol Physiol, 2002. **283**(3): p. L628–35.
36. Roux, E., et al., *Calcium signaling in airway smooth muscle cells is altered by in vitro exposure to the aldehyde acrolein.* Am J Respir Cell Mol Biol, 1998. **19**(3): p. 437–44.
37. Perez-Zoghbi, J.F., Y. Bai, and M.J. Sanderson, *Nitric oxide induces airway smooth muscle cell relaxation by decreasing the frequency of agonist-induced Ca2+ oscillations.* J Gen Physiol, 2010. **135**(3): p. 247–59.
38. Jiang, H., et al., *Phosphoinositide 3-kinase gamma regulates airway smooth muscle contraction by modulating calcium oscillations.* J Pharmacol Exp Ther, 2010. **334**(3): p. 703–9.
39. Delmotte, P. and M.J. Sanderson, *Effects of formoterol on contraction and Ca2+ signaling of mouse airway smooth muscle cells.* Am J Respir Cell Mol Biol, 2010. **42**(3): p. 373–81.
40. Bai, Y., M. Zhang, and M.J. Sanderson, *Contractility and Ca2+ signaling of smooth muscle cells in different generations of mouse airways.* Am J Respir Cell Mol Biol, 2007. **36**(1): p. 122–30.
41. Bergner, A., et al., *Ca2+−signaling in airway smooth muscle cells is altered in T-bet knock-out mice.* Respir Res, 2006. **7**: p. 33.
42. Perez, J.F. and M.J. Sanderson, *The frequency of calcium oscillations induced by 5-HT, ACH, and KCl determine the contraction of smooth muscle cells of intrapulmonary bronchioles.* J Gen Physiol, 2005. **125**(6): p. 535–53.
43. Bergner, A. and M.J. Sanderson, *Acetylcholine-induced calcium signaling and contraction of airway smooth muscle cells in lung slices.* J Gen Physiol, 2002. **119**(2): p. 187–98.

44. Campos-Bedolla, P., et al., *Airway smooth muscle relaxation induced by 5-HT(2A) receptors: role of Na(+)/K(+)-ATPase pump and Ca(2+)-activated K(+) channels.* Life Sci, 2008. **83** (11–12): p. 438–46.
45. Oguma, T., et al., *Roles of P2X receptors and Ca2+ sensitization in extracellular adenosine triphosphate-induced hyperresponsiveness in airway smooth muscle.* Clin Exp Allergy, 2007. **37**(6): p. 893–900.
46. Janssen, L.J., T. Tazzeo, and J. Zuo, *Enhanced myosin phosphatase and Ca(2+)-uptake mediate adrenergic relaxation of airway smooth muscle.* Am J Respir Cell Mol Biol, 2004. **30**(4): p. 548–54.
47. Prakash, Y.S., et al., *Spatial and temporal aspects of ACh-induced [Ca2+]i oscillations in porcine tracheal smooth muscle.* Cell Calcium, 2000. **27**(3): p. 153–62.
48. Fleischmann, B.K., Y.X. Wang, and M.I. Kotlikoff, *Muscarinic activation and calcium permeation of nonselective cation currents in airway myocytes.* Am J Physiol, 1997. **272** (1 Pt 1): p. C341-9.
49. Wang, Y.X., B.K. Fleischmann, and M.I. Kotlikoff, *M2 receptor activation of nonselective cation channels in smooth muscle cells: calcium and Gi/G(o) requirements.* Am J Physiol, 1997. **273**(2 Pt 1): p. C500–8.
50. Kajita, J. and H. Yamaguchi, *Calcium mobilization by muscarinic cholinergic stimulation in bovine single airway smooth muscle.* Am J Physiol, 1993. **264**: p. L496–L503.
51. Schaafsma, D., et al., *Differential Rho-kinase dependency of full and partial muscarinic receptor agonists in airway smooth muscle contraction.* Br J Pharmacol, 2006. **147**(7): p. 737–43.
52. Janssen, L.J. and S.M. Sims, *Ca(2+)-dependent Cl- current in canine tracheal smooth muscle cells.* Am J Physiol, 1995. **269**(1 Pt 1): p. C163–9.
53. McCann, J.D. and M.J. Welsh, *Calcium-activated potassium channels in canine airway smooth muscle.* J Physiol, 1986. **372**: p. 113–27.
54. Wade, G.R. and S.M. Sims, *Muscarinic stimulation of tracheal smooth muscle cells activates large-conductance Ca(2+)-dependent K+ channel.* Am J Physiol, 1993. **265**(3 Pt 1): p. C658–65.
55. Amrani, Y., et al., *Ca2+ increase and Ca(2+)-influx in human tracheal smooth muscle cells: role of Ca2+ pools controlled by sarco-endoplasmic reticulum Ca(2+)-ATPase 2 isoform.* Br J Pharmacol, 1995. **115**(7): p. 1204–10.
56. Govindaraju, V., et al., *The effects of extracellular purines and pyrimidines on human airway smooth muscle cells.* J Pharmacol Exp Ther, 2005. **315**(2): p. 941–8.
57. Hyvelin, J.M., et al., *Human isolated bronchial smooth muscle contains functional ryanodine/caffeine-sensitive Ca-release channels.* Am J Respir Crit Care Med, 2000. **162**(2 Pt 1): p. 687–94.
58. Ressmeyer, A.R., et al., *Human airway contraction and formoterol-induced relaxation is determined by Ca2+ oscillations and Ca2+ sensitivity.* Am J Respir Cell Mol Biol, 2010. **43** (2): p. 179–91.
59. Liu, X. and J.M. Farley, *Acetylcholine-induced Ca++−dependent chloride current oscillations are mediated by inositol 1,4,5-trisphosphate in tracheal myocytes.* J Pharmacol Exp Ther, 1996. **277**(2): p. 796–804.
60. Roux, E., et al., *Modelling of Ca2+−activated chloride current in tracheal smooth muscle cells.* Acta Biotheoretica, 2001. **49**(4): p. 291–300.
61. Marhl, M., et al., *Complex calcium oscillations and the role of mitochondria and cytosolic proteins.* Biosystems, 2000. **57**(2): p. 75–86.
62. Liu, X. and J.M. Farley, *Acetylcholine-induced chloride current oscillations in swine tracheal smooth muscle cells.* J Pharmacol Exp Ther, 1996. **276**(1): p. 178–86.
63. Bai, Y. and M.J. Sanderson, *Airway smooth muscle relaxation results from a reduction in the frequency of Ca2+ oscillations induced by a cAMP-mediated inhibition of the IP3 receptor.* Respir Res, 2006. **7**: p. 34.

64. Ouedraogo, N., et al., *Effects of intravenous anesthetics on normal and passively sensitized human isolated airway smooth muscle.* Anesthesiology, 1998. **88**(2): p. 317–26.
65. Haberichter, T., et al., *The influence of different InsP(3) receptor isoforms on Ca(2+) signaling in tracheal smooth muscle cells.* Bioelectrochemistry, 2002. **57**(2): p. 129.
66. Marhl, M., D. Noble, and E. Roux, *Modeling of molecular and cellular mechanisms involved in Ca2+ signal encoding in airway myocytes.* Cell Biochem Biophys, 2006. **46**(3): p. 285–302.
67. Berridge, M.J., P. Lipp, and M.D. Bootman, *The versatility and universality of calcium signalling.* Nat Rev Mol Cell Biol, 2000. **1**(1): p. 11–21.
68. Babcock, D.F., et al., *Mitochondrial participation in the intracellular Ca2+ network.* J Cell Biol, 1997. **136**(4): p. 833–44.
69. Drummond, R.M. and F.S. Fay, *Mitochondria contribute to Ca2+ removal in smooth muscle cells.* Pflugers Arch, 1996. **431**(4): p. 473–82.
70. Fall, C.P. and J.E. Keizer, *Mitochondrial modulation of intracellular Ca(2+) signaling.* J Theor Biol, 2001. **210**(2): p. 151–65.
71. Wylam, M.E., A. Xue, and G.C. Sieck, *Mechanisms of intrinsic force in small human airways.* Respir Physiol Neurobiol, 2012. **181**(1): p. 99–108.
72. Wang, I.Y., et al., *A mathematical analysis of agonist- and KCl-induced Ca(2+) oscillations in mouse airway smooth muscle cells.* Biophys J, 2010. **98**(7): p. 1170–81.
73. Bergner, A. and M.J. Sanderson, *ATP stimulates Ca2+ oscillations and contraction in airway smooth muscle cells of mouse lung slices.* Am J Physiol Lung Cell Mol Physiol, 2002. **283**(6): p. L1271–9.
74. Hagar, R.E. and B.E. Ehrlich, *Regulation of the type III InsP(3) receptor by InsP(3) and ATP.* Biophysical Journal, 2000. **79**(1): p. 271–278.
75. Moraru, II, et al., *Regulation of type 1 inositol 1,4,5-trisphosphate-gated calcium channels by InsP3 and calcium: Simulation of single channel kinetics based on ligand binding and electrophysiological analysis.* J Gen Physiol, 1999. **113**(6): p. 837–49.
76. Ramos-Franco, J., et al., *Single channel function of recombinant type-1 inositol 1,4,5-trisphosphate receptor ligand binding domain splice variants.* Biophys J, 1998. **75**(6): p. 2783–93.
77. Ramos-Franco, J., M. Fill, and G.A. Mignery, *Isoform-specific function of single inositol 1,4,5-trisphosphate receptor channels.* Biophys J, 1998. **75**(2): p. 834–9.
78. Dupont, G. and A. Goldbeter, *One-pool model for Ca2+ oscillations involving Ca2+ and inositol 1,4,5-trisphosphate as co-agonists for Ca2+ release.* Cell Calcium, 1993. **14**(4): p. 311–22.
79. Morel, J.L., et al., *Crucial role of type 2 inositol 1,4,5-trisphosphate receptors for acetylcholine-induced Ca2+ oscillations in vascular myocytes.* Arterioscler Thromb Vasc Biol, 2003. **23**(9): p. 1567–75.
80. Parthimos, D., D.H. Edwards, and T.M. Griffith, *Minimal model of arterial chaos generated by coupled intracellular and membrane Ca2+ oscillators.* Am J Physiol, 1999. **277**(3 Pt 2): p. H1119–44.
81. Roux, E., P. Mbikou, and A. Fajmut, *Role of Protein Kinase Network in Excitation-Contraction Coupling in Smooth Muscle Cell*, in *Protein Kinases*, G. Da Silva Xavier, Editor. 2012, InTech: Rijeka. p. 287–320.
82. Mbikou, P., E. Roux, and A. Fajmut, Couplage excitation-contraction du muscle lisse des voies aeriennes. 2010, Sarrebruck: Éditions universitaires européennes. 144.
83. Mbikou, P., et al., *Contribution of Rho kinase to the early phase of the calcium-contraction coupling in airway smooth muscle.* Exp Physiol, 2011. **96**(2): p. 240–58.
84. Hai, C.M. and B. Szeto, *Agonist-induced myosin phosphorylation during isometric contraction and unloaded shortening in airway smooth muscle.* Am J Physiol, 1992. **262**(1 Pt 1): p. L53–62.
85. Ouedraogo, N., R. Marthan, and E. Roux, *The effects of propofol and etomidate on airway contractility in chronically hypoxic rats.* Anesth Analg, 2003. **96**(4): p. 1035–41.

86. Hai, C.M. and R.A. Murphy, *Regulation of shortening velocity by cross-bridge phosphorylation in smooth muscle.* Am J Physiol Cell Physiol, 1988. **255**(1): p. C86–94.
87. Hai, C.-M. and H.R. Kim, *An expanded latch-bridge model of protein kinase C-mediated smooth muscle contraction.* J Appl Physiol, 2005. **98**(4): p. 1356–1365.
88. Hai, C.M. and R.A. Murphy, *Cross-bridge phosphorylation and regulation of latch state in smooth muscle.* Am. J. Physiol. Cell Physiol., 1988. **254**(1): p. C99–106.
89. Lauzon, A.M., et al., *A multi-scale approach to airway hyperresponsiveness: from molecule to organ.* Front Physiol, 2012. **3**: p. 191.
90. Berridge, M.J., P.H. Cobbold, and K.S. Cuthbertson, *Spatial and temporal aspects of cell signalling.* Philos Trans R Soc Lond B Biol Sci, 1988. **320**(1199): p. 325–43.
91. Marhl, M., et al., *Importance of cell variability for calcium signaling in rat airway myocytes.* Biophys Chem, 2010. **148**: p. 42–50.
92. Burdakov, D. and A. Verkhratsky, *Biophysical re-equilibration of Ca2+ fluxes as a simple biologically plausible explanation for complex intracellular Ca2+ release patterns.* FEBS Lett, 2006. **580**(2): p. 463–8.

Mechanisms Underlying Ca^{2+} Store Refilling in Airway Smooth Muscle

Charlotte K. Billington, Ian P. Hall, and Carl P. Nelson

Contents

Abstract In airway smooth muscle (ASM) cells, contractile tone is tightly coupled to the intracellular calcium (Ca^{2+}) concentration, with Ca^{2+} release from intracellular stores, followed by influx through plasma membrane ion channels providing the signature Ca^{2+} signal leading to contraction. One such Ca^{2+} influx pathway, via store-operated calcium (SOC) channels, is particularly prominent in ASM cells. SOC entry not only supplies Ca^{2+} to replenish the depleted intracellular stores but also provides a sustained Ca^{2+} signal to maintain contraction and potentially initiate downstream cellular responses. The molecular determinants of store-operated Ca^{2+} entry were poorly defined until the relatively recent discovery of the central roles of the stromal interaction molecules and Orai proteins in this pathway. In this chapter, we describe the current understanding of the mechanism of SOC entry and its functional significance in ASM cells.

C.K. Billington, Ph.D. (✉) • I.P. Hall • C.P. Nelson
Division of Respiratory Medicine, The University of Nottingham,
Nottingham NG7 2UH, UK
e-mail: charlotte.billington@nottingham.ac.uk

Y.-X. Wang (ed.), *Calcium Signaling In Airway Smooth Muscle Cells*,
DOI 10.1007/978-3-319-01312-1_9,

Keywords Store-operated calcium channels • Stromal interaction molecules • Orai • Airway smooth muscle • Calcium

1 Introduction

Airway smooth muscle (ASM) tone is the critical determinant of bronchial diameter and, thus, bronchial airflow. At rest, ASM is relatively quiescent, but activation of ASM cells by bronchoconstrictors (e.g. acetylcholine, histamine) can lead, in asthmatics, to marked airway constriction and bronchospasm. In asthma, bronchial hyperresponsiveness is a major cause of respiratory dysfunction, and many common asthma therapies target ASM to directly regulate airway tone and ameliorate these symptoms.

In ASM cells, contractile tone is predominantly determined by the intracellular calcium (Ca^{2+}) concentration. Many bronchoconstrictors act through G_q protein-coupled receptors to elevate Ca^{2+} through phospholipase C (PLC) activation and inositol trisphosphate (IP_3)-dependent Ca^{2+} release from intracellular stores [2]. The tight regulation of intracellular Ca^{2+} levels is critical for the regulation of myriad cellular functions and, ultimately, cell viability [43]. Accordingly, in ASM cells, Ca^{2+} modulates such key cellular events as contraction, proliferation, migration and cytokine synthesis. Dysfunction in one or a combination of these events may initiate, orchestrate and perpetuate the permanent structural changes to the airways observed in people with chronic asthma. The molecular components involved in regulating intracellular Ca^{2+} levels in ASM therefore constitute attractive therapeutic targets in the treatment of asthma (and other airway diseases) and, as such, have attracted considerable research efforts.

A variety of Ca^{2+}-gating pathways, including voltage-dependent calcium channels (VDCCs), receptor-operated calcium channels (ROCs) and sodium/calcium exchangers, may be found in the plasma membrane of ASM. However, the dominant mechanism by which intracellular Ca^{2+} is modulated in ASM cells is via store-operated calcium (SOC) entry (SOCE), and it is this mechanism which primarily drives ASM contraction [61]. SOCE [formerly referred to as capacitative calcium entry (CCE)] was first postulated in 1986 by James Putney and is the process by which a decrease in endoplasmic reticulum (ER) Ca^{2+} content signals to induce an influx of extracellular Ca^{2+} via plasma membrane Ca^{2+} channels [51, 55]. The current mediating this Ca^{2+} entry was first identified by Hoth and Penner ([28]), who coined the term I_{CRAC} (for calcium release-activated calcium current).

Experimentally, Sarcoplasmic reticulum (SR) Ca^{2+} stores may be depleted by a variety of manipulations, often in combination with the removal of extracellular Ca^{2+}, including inhibition of the sarcoplasmic/endoplasmic reticulum (SR/ER) Ca^{2+}-ATPase (SERCA) pump (e.g. with thapsigargin [46]); emptying of the SR by ryanodine or caffeine treatment [3, 4, 10]; exposure to the Ca^{2+} chelator ethylene glycol-*bis*(β-aminoethyl)-N,N,N′,N′-tetraacetic acid (EGTA); or the activation of G-protein-coupled receptors (GPCRs), predominantly in response to contractile agents and coupled to the G_q family of G proteins [8, 24]. In ASM cells, bradykinin and histamine (acting via the B2 and H1 G_q-protein-coupled receptors, respectively) are commonly used as ligands due to their pathophysiological and clinical relevance.

Upon extracellular exposure to a contractile agonist such as bradykinin or histamine, an initial rise in intracellular Ca^{2+} is observed, which is a direct result of Ca^{2+} being released from intracellular SR stores (via the IP_3R) into the cytoplasm. This rise in intracellular Ca^{2+} causes the ASM cell to contract. The subsequent influx of Ca^{2+} from the extracellular milieu is largely via SOCE, and this process is central to the maintenance of contraction [61], although the contribution of L-type VDCC, ROC and reverse-mode Na^+/Ca^{2+} exchange to Ca^{2+} homeostasis/contraction in ASM cells has also been demonstrated [15, 16]. SOCE is also crucial in enabling the cell to re-establish its levels of Ca^{2+} to those appropriate for each intracellular location and, hence, allows the cell to return to a basal state.

Refilling of intracellular stores is clearly a crucial stage of Ca^{2+} homeostasis, and in the past decade understanding of the molecules and mechanisms involved has advanced rapidly with the discovery of the STIM and Orai families of proteins and increased understanding of the roles of transient receptor potential (TRP) channels. In this chapter we will describe the current understanding of STIM, Orai and canonical TRP (TRPC) channels and how they orchestrate store refilling. We will also address some of the emerging evidence for the involvement of SOCE in ASM cell physiology and pathophysiology. Although some general concepts may be drawn from studies in non-ASM cells, where possible we will focus on what is known specifically in ASM cells.

2 I_{CRAC} and I_{SOC}

In the absence of a molecular definition of the mechanism of SOCE, considerable efforts went into biophysically characterizing the currents underlying Ca^{2+} entry, with the I_{CRAC} current representing the best characterized SOCE current. I_{CRAC} is defined as a small-conductance (0.02 pS), non-voltage-gated, inwardly rectifying and highly Ca^{2+}-selective (approx. 1,000-fold selective for Ca^{2+} over Na^+) current [44]. In ASM cells, a small, transient inward current, consistent with I_{CRAC} (i.e. a positive reversal potential and limited outward rectification) was identified [46]. However, in common with many other cell types, including vascular smooth muscle [1], larger, less Ca^{2+}-selective SOC currents (I_{SOC}) have been observed in ASM cells. Indeed, store depletion with cyclopiazonic acid (CPA) has been shown to elicit a larger, more outwardly rectifying cation current in both bovine [25] and human [45, 46, 61] ASM cells. These findings highlight that ASM cells can utilize multiple pathways to gate Ca^{2+} in response to store depletion, adding further complexity to the task of dissecting the molecular determinants of SOCE.

3 STIM

Stromal interaction molecules (STIM) STIM1 and STIM2 were first reported in 1996 [42]; however, a role for STIM in Ca^{2+} signalling was not suggested until 2005, when two RNAi screens highlighted STIM as a modulator of Ca^{2+} [32, 52]. The field has since advanced rapidly, and STIM is now recognized as the key molecule alerting cells to increase the flow of Ca^{2+} across the plasma membrane following depletion of intracellular stores.

The domain architecture of STIM1 and STIM2 has been carefully researched and is shown in Fig. 1. Apart from a small proportion of STIM1 molecules which have been observed on the plasma membrane, STIMs consist of one transmembrane region with the N-terminus located in the lumen of the ER store and the soluble C-terminus projecting into the cytosol [59]. Systematic mutagenesis experiments have revealed the protein domains which are crucial for Ca^{2+} sensing, subcellular reorganization, punctae formation and facilitation of a CRAC current. The step-by-step role of STIM in the regulation of SOCE was recently reviewed by Michael Cahalan [11] and is summarized schematically in Fig. 2.

Under basal conditions, STIMs likely exist as dimers with Ca^{2+} ions bound to the cEF-hand domain in the N-terminus. The combined sterile-α motif (SAM) and EF-hand domains function as highly sensitive Ca^{2+} sensors, and when ER Ca^{2+} stores are depleted, these Ca^{2+} ions are released from STIMs, resulting in a conformational change. When the exodus of Ca^{2+} from ER stores leads to an ER luminal concentration of less than approx. 300 μM, the "empty-handed" STIMs

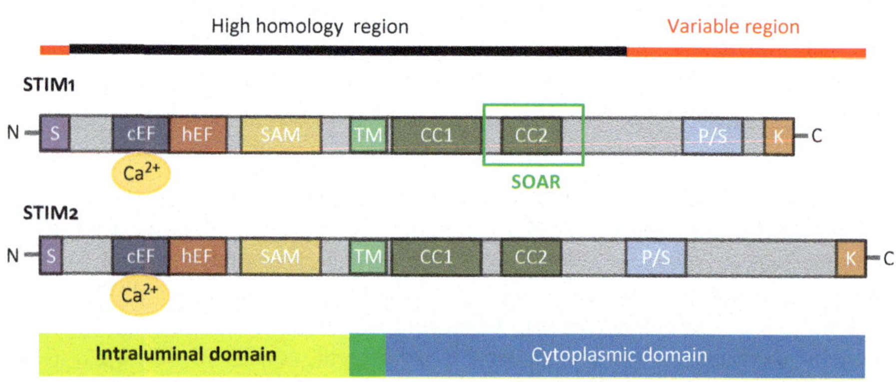

Fig. 1 Schematic illustration of molecular domains of STIM1 and STIM2. Both proteins consist of a single transmembrane domain (*TM*), separating cytoplasmic and intraluminal portions of the proteins. The N-terminal signal peptides (*S*) are cleaved during translation. The intraluminal domains contain canonical Ca^{2+}-binding EF-hand domains (*cEF*), non-Ca^{2+}-binding hidden EF-hand domains (*hEF*) and a sterile α-motif (*SAM*). The cytoplasmic domains contain two coiled-coil domains (*CC1* and *CC2*), a proline/serine-rich domain (*P/S*) and a C-terminal lysine-rich domain (*K*), involved in phospholipid interactions. The STIM-Orai-activating region (*SOAR*) within STIM1 is also indicated

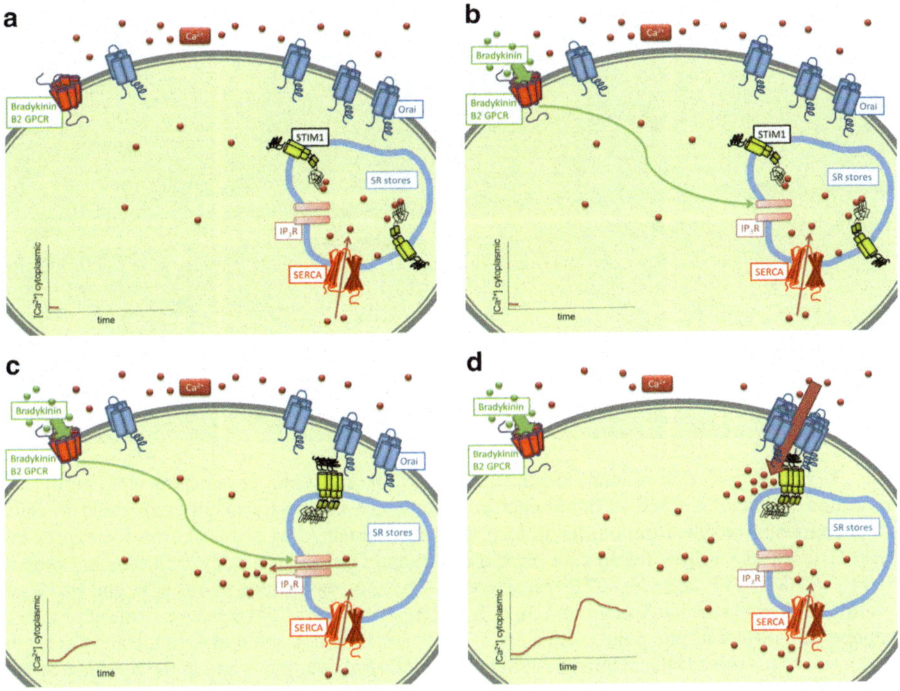

Fig. 2 The role of STIM1 and Orai in Ca^{2+} store refilling in ASM. The four images (**a–d**) illustrate the sequential process of intracellular Ca^{2+} release, store-operated Ca^{2+} entry and Ca^{2+} store refilling in response to activation of the B2 bradykinin receptor in ASM cells. In each case, the graph in the *bottom left-hand corner* demonstrates the relative Ca^{2+} concentration in the cytoplasm occurring under the conditions depicted in the corresponding figure. (**a**) Under basal conditions the vast majority of STIM1 molecules are found spanning the membrane of intracellular stores with the SOAR-containing C-terminus located in the cytoplasm and the SAM- and EF-domain-containing N-terminus located inside the intracellular store. At resting levels of Ca^{2+} within the ER store (~400 μM), STIM1 is believed to exist as a dimer. Ca^{2+} ions bound to the EF-hand regions of STIM1 maintain the localization and conformation observed under basal conditions. SERCA pumps maintain basal Ca^{2+} levels within the ER stores by actively pumping Ca^{2+} ions into the store. (**b**) Exposure of the ASM cell to an agonist such as bradykinin promotes activation of the bradykinin B2 GPCR and of its downstream signaling pathway [via $G_{\alpha q}$ and PLC-induced inositol 1,4,5-trisphosphate (IP_3) generation], resulting in activation of the IP_3 receptor (IP_3R). (**c**) Activation of IP_3R drives the release of Ca^{2+} ions from the ER stores into the cytoplasm, resulting in a net drop in ER store Ca^{2+} concentration. When the concentration reaches ~300 μM, the lost contact between Ca^{2+} ions and the EF-hand domains of the STIM1 molecules prompts STIM1 to undergo conformational changes, exposing the SOAR domain. STIM1 also rapidly oligomerizes and translocates along the ER membrane to regions in close proximity to the plasma membrane. (**d**) The exposure of the SOAR domain allows Orai molecules on the plasma membrane to be harnessed, resulting in punctae containing STIM1 and Orai. They are believed to exist in a 1:1 ratio of STIM1 dimers to Orai subunits, i.e. eight STIM1 molecules interacting with one tetrameric Orai molecule. As a result of these interactions, extracellular Ca^{2+} enters the cell via the Orai channel, increasing intracellular Ca^{2+} and allowing the refilling of ER stores via SERCA activity. As intracellular Ca^{2+} returns to resting levels, the localization and conformation of STIM1 and Orai also return to the basal conditions depicted in Fig. 2a

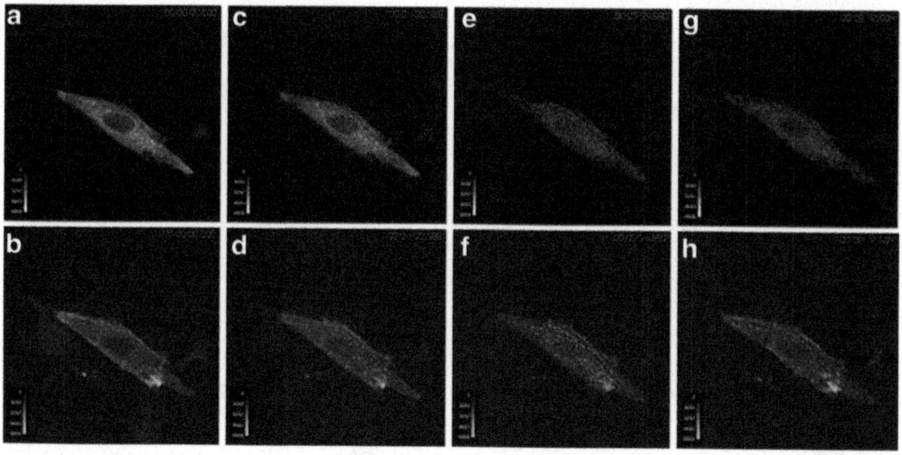

Fig. 3 To observe the cellular localization of STIM and Orai, researchers commonly use recombinant versions fused to fluorescent proteins which can be transiently expressed in cells. In this series of images from our laboratory, the same human ASM cell is shown overexpressing STIM1 fused to a yellow fluorescent protein (YFP) and Orai fused to a cyan fluorescent protein (CFP) (STIM1-YFP and Orai-CFP respectively). The *top row* of images (**a**, **c**, **e**, and **g**) shows STIM1-YFP, whereas the *bottom row* (**b**, **d**, **f**, and **h**) shows Orai-CFP. Each vertical set of panels represents a different time point, with **a** and **b** imaged at time 0, **c** and **d** at 1 min, **e** and **f** at 6 min and **g** and **h** at 8 min. Thus, in human ASM cells under basal conditions and in the presence of low (0.1 mM) extracellular Ca^{2+}, STIM1-YFP exhibits a cytoplasmic distribution (panel **a**), whereas Orai-CFP localizes more to the plasma membrane (panel **b**) [7]. Following store depletion induced by combined exposure to 1 μM bradykinin (acting via the mechanisms depicted in Fig. 2) and the SERCA pump inhibitor 1 μM thapsigargin, both STIM1-YFP and Orai-CFP alter their cellular locations and co-localize in punctae (panels **c–f**). Removal of bradykinin and thapsigargin from the extracellular environment, combined with perfusion of buffer containing 2 mM extracellular Ca^{2+} into the system, causes the punctae to disperse, and both STIM1-YFP and Orai-CFP begin to return to their initial cellular locations (**g** and **h**)

undergo a conformational change exposing the STIM-Orai activating region (SOAR) domain. These "empty handed" STIMs are driven to oligomerize and then translocate along the ER membrane to regions where the ER membrane is in close proximity (one study suggests 10–25 nm [65]) to the plasma membrane (shown schematically in Fig. 2 and in human ASM cells in Fig. 3). In addition to conformational changes in STIMs, the mechanism by which STIMs translocate along the ER membrane following store depletion also involves remodeling of the ER, and the details of this process were revealed by elegant experimentation and systematically reviewed in 2011 by Shen et al. [54].

As the name suggests, the SOAR domain of STIMs is that responsible for the interaction between STIMs and Orai (Orai is described below in Sect. 4). As observed and described in Figs. 2 and 3, aggregates of STIM and Orai molecules form junctional assemblies, resulting in an influx of extracellular Ca^{2+} into the cytoplasm. Although not yet confirmed in ASM cells, in other cell systems a number of binding and regulatory proteins have been identified which can also modulate this process. These include junctate, calmodulin, CRAC regulatory protein 2A, Golli and SOCE-associated regulatory factor; their putative roles have recently been reviewed [56, 58]. Roles for these proteins have been demonstrated in both the formation and dissociation of junctional assemblies. Whilst Orai channels appear to be turned off by a calcium-dependent inactivation (CDI) process involving the STIM1 cytoplasmic inhibitory domain region, STIM1 itself returns to its resting conformation and location following the increase in Ca^{2+} concentration within the ER stores [56].

Peel et al. confirmed the expression of STIM1 and STIM2 in human cultured ASM cells in 2006 [45]. Using siRNA specifically targeted against STIM1 or STIM2, a functional role was found for STIM1 (but not STIM2), with STIM1 knockdown leading to decreased store-operated Ca^{2+} influx following store depletion by CPA. Interestingly, the functional importance of STIM1 varied according to the mechanism by which intracellular stores were depleted, with Ca^{2+} re-entry following histamine-mediated store depletion being much more susceptible to STIM1 knockdown than that following exposure to bradykinin [45].

4 Orai

Following on from the identification of the STIM proteins as key mediators in the SOCE process, a further significant advance was made with the discovery of the Orai family of proteins [20]. Named after the sisters guarding the gates of heaven in Greek mythology, three Orai genes have been identified (Orai1–3) from genome-wide RNAi screens, each encoding a protein containing four transmembrane domains, with both N- and C-termini located intracellularly (Fig. 4) [20, 63, 69]. This topology did not immediately suggest that the Orai proteins were ion channels, yet Orai1 (also known as CRACM) appears to constitute the pore-forming subunits of I_{CRAC} channels. A modified linkage analysis identified a single point mutation in Orai1 (R91W), which leads to severe combined immunodeficiency (SCID) syndrome through the disruption of Ca^{2+} entry in T-cells [20]. The executive role of Orai1 in SOCE is further supported by the findings that I_{CRAC} activity is rescued in SCID patients' T-cells by expression of wild-type Orai1 [20] and that the R91W mutation in Orai1 ablates I_{CRAC} in cells [20, 63, 69]. Finally, when recombinant Orai1 is co-expressed with STIM1, large I_{CRAC} currents can be detected with properties very similar to those of native I_{CRAC} [36, 47, 57, 69].

The functional roles of Orai2 and Orai3 remain somewhat unclear, but both can couple with STIM1 to generate I_{CRAC}-like channels, with properties slightly different from those of the Orai1-containing channels [17, 33]. In addition, functional Orai channels appear to exist as tetramers [29, 37], with hetero-oligomerization between Orai isoforms likely to generate further diversity [33]. How these multiple potential channels relate to the variety of SOC channels observed in vivo remains to be established.

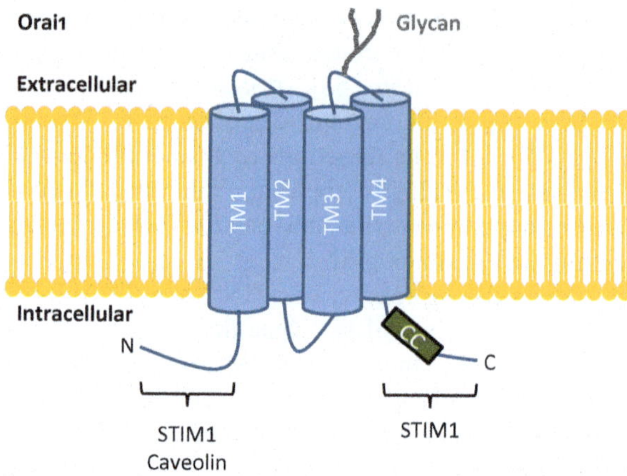

Fig. 4 Schematic illustration of Orai1. Orai1 consists of four transmembrane domains (*TM1–TM4*), with intracellular N- and C-termini. The second extracellular domain (between *TM3* and *TM4*) is heavily glycosylated, and the C-terminus contains a coiled-coil (*CC*) domain. The C-terminus of Orai1 (in particular the coiled-coil domain) plays a critical role in STIM1 binding, but the role of the N-terminus in association with STIM1 is less clear (see [19] for review)

Using an approach identical to that adopted for investigations into the role of STIMs in human cultured ASM cells, Peel et al. uncovered the contribution of the Orai proteins to store refilling [46]. Expression studies confirmed that all three Orai isoforms were present at the level of mRNA in ASM. However, due to a lack of availability of antibodies against Orai2 or Orai3 at that time, only the presence of Orai1 could be confirmed at the protein level. Subsequent siRNA-based experiments revealed a role for Orai1 but not Orai2 in both CPA- and thapsigargin-mediated Ca^{2+} influx. The role of Orai1 in ASM cell SOCE (in response to CPA) was recently confirmed by Sathish et al. [53] using siRNA targeting Orai1. Interestingly, Peel et al. [46] also found that Orai3 knockdown resulted in a decrease in both CPA-mediated store release and subsequent Ca^{2+} influx, suggesting that this isoform could regulate basal Ca^{2+} levels or Ca^{2+} release from stores.

Electrophysiologically, cultured human ASM cells exhibited a two-component, store-operated current following exposure to the SERCA inhibitor CPA [46]. The first component was a small, transient inward current with characteristics typical of I_{CRAC} (very positive reversal potential with limited outward rectification), while a subsequent component was more similar to the I_{SOC} currents previously reported in human bronchial smooth muscle cells [61]. The initial I_{CRAC}-like current was inhibited by Orai1 depletion, whilst the I_{SOC}-like current was relatively unaffected [46]. These findings suggest a key role for Orai1 in the I_{CRAC}-like current, consistent with findings in other systems [20, 63, 69], but the molecular nature of the larger I_{SOC}-like current remains unclear. It is likely that one or more of the five TRPC homologues identified in human ASM cells (TRPC1, 3, 4, 5 and 6) [14, 40, 64] might contribute to this current (see Sect. 5 for further discussion).

In addition to STIM1, numerous other Orai1 binding partners have been identified [see [58] for a review]. For instance, the N-terminus of Orai1 has been demonstrated to interact with caveolin-1 to regulate dynamin-dependent internalization of the channel in oocytes [67]. An interaction between caveolin-1 and Orai1 may be important in regulating SOCE in ASM cells, as it was recently reported that caveolin-1 over-expression enhanced Orai1 levels and elevated SOCE, whilst siRNA-mediated knockdown of caveolin-1 blunted SOCE and reduced Orai1 expression [53]. The mechanism of this regulation by caveolin-1 has not been investigated, but given the findings in oocytes [67], it would be interesting to examine whether manipulations in caveolin-1 expression altered the internalization of the channel in ASM cells.

5 TRPC

The TRPC family consists of seven genes (TRPC1–7), which have been proposed as key molecular components of the non-selective cation currents (NSCCs) found in a variety of excitable and non-excitable cells [38]. In human ASM cells, five members of the TRPC family (TRPC1, 3, 4, 5 and 6) have been identified at the mRNA level, with TRPC1, 4, 5 and 6 confirmed at the protein level [14, 40, 64]. However, the functions of these TRPC channels in ASM are less well defined.

In human ASM cells, Corteling et al. [14] observed a histamine-induced Ca^{2+} entry pathway with characteristics similar to those of TRPC-dependent entry, while the large I_{SOC} currents identified in ASM cells [46, 61] were also consistent with TRPC channel activity. Indeed, Sweeney et al. found that TRPC1 mRNA was elevated under conditions in which I_{SOC} was enhanced (in proliferating cells), suggesting a role for TRPC1 in mediating at least part of the observed I_{SOC} current. However, a causal link between TRPC1 mRNA levels and I_{SOC} was not firmly established, so this remains speculative. In contrast, TNFα treatment selectively increased TRPC3 expression in human ASM cells, which was demonstrated to be responsible for an elevated basal intracellular Ca^{2+} and enhanced SOCE, suggesting a pathophysiological role for TRPC3 in inflammatory airway disease [64].

But which TRPC isoforms are involved in gating Ca^{2+} in healthy ASM cells? The role of TRPC6 was investigated in guinea pig ASM cells, using siRNA to knock down the channel [23]. The researchers found that TRPC6 depletion had no effect on oleyl acetyl glycerol (OAG)-mediated Ca^{2+} entry, suggesting that TRPC6 is not a primary Ca^{2+}-conducting channel in ASM cells. However, the authors suggested that the channel could still be contributing to NSCCs without directly gating Ca^{2+} if Na^{+} ions were the main conductance through TRPC6, as has been demonstrated in human embryonic kidney (HEK) cells [18].

In contrast, TRPC3 was reported to be crucial in generating NSCCs (whose constitutive activity at rest contributes to the relatively depolarized resting membrane potential) in normal mouse ASM cells, with an emerging role for TRPC1 in an asthmatic mouse model [66]. This differs somewhat from the findings of White et al. [64], who only found a significant expression/function of TRPC3 following TNFα treatment, but this may reflect the different species used in the two studies. The role of TRPC isoforms in mediating Ca^{2+} entry in ASM cells therefore remains unresolved. Indeed, it has been argued that the major cellular role of TRPC channels is not in SOCE but in receptor-operated Ca^{2+} entry, as a result of their sensitivity to diacylglycerol, generated predominantly by PLC activity (see [50] and references therein).

The potential for TRPC channels to interact with STIM and Orai proteins adds a further layer of complexity to the study of their role in Ca^{2+} homeostasis and may partly explain the difficulties in clearly identifying their functions. It is possible that STIM1 provides a common mechanism for sensing store depletion and stimulating the activity of an array of store-operated currents, since it has been shown to interact with TRPC1-6, in addition to the Orai family (see [11] and [68] for a review). For instance, TRPC1 and TRPC4 have recently been demonstrated to interact with STIM1 and form functional store-operated channels in murine and human lung endothelial cells [60]. Furthermore, Orai1 has been reported to regulate or contribute to TRPC currents in a variety of systems [30, 31, 41]. So what is the nature of the relationship between STIM1, Orai1 and TRPC channels and how does it relate to the range of store-operated currents and Ca^{2+} signals observed in cells?

It has been established in a number of cell backgrounds that while the interaction between STIM1 and Orai1 generates an I_{CRAC} current, the I_{SOC} current requires the expression of STIM1, Orai1 and TRPC1 [12, 13, 30]. As a result, in Jurkat cells, where I_{CRAC} is the predominant store-operated current, only Orai1 (and STIM1) is required to mediate SOCE [30]. However, in other cells (such as HEK-293 cells), SOCE also requires TRPC1 expression as a result of a dependence on I_{SOC} for Ca^{2+} entry [12, 30]. Kim et al. [30] found that Orai1 and TRPC1 were mutually dependent upon each other for their interaction with STIM1 and concluded that they were present within the same complex in HEK-293 cells. However, Cheng et al. [13] presented an alternative model in human salivary gland cells, in which STIM1 complexes with TRPC1 and Orai1 separately (to generate I_{CRAC} and NSCC currents respectively). They propose that Ca^{2+} entry through I_{CRAC} is necessary for the insertion of TRPC1 into the plasma membrane, generating a subsequent NSCC through TRPC1/STIM1 complexes. Intriguingly, they reported that I_{CRAC} and the NSCC performed distinct functions within the cell, with I_{CRAC} leading to nuclear factor of activated T-cell (NFAT) stimulation and the NSCC being responsible for NFκB stimulation and sustained K_{Ca} channel activity [13]. The potential for generating multiple intracellular signals via distinct SOCE pathways would be worthy of investigation in ASM cells, where two components of store-operated current have been isolated [46].

6 Plasma Membrane Sodium/Calcium Exchanger

The plasma membrane sodium/calcium exchanger (NCX) is an ion transporter protein which imports three sodium ions into a cell in exchange for one calcium ion via an electrochemical gradient. However, an excess of sodium within the cell can force the NCX into reverse mode and hence play a role in Ca^{2+} influx into the cell. While this system has been extensively researched in other cells, most notably cardiomyocytes, only a few studies have researched the role of NCX transporters in ASM. Studies using KB-R7943, a relatively selective antagonist of the reverse mode of NCX, reveal a role for this transporter in store refilling as exposure to KB-R7943 decreased ASM contraction induced by acetylcholine, histamine, 5-hydroxytryptamine or caffeine [15, 26, 27, 34]. However, it should be noted that KB-R7943 has recently been shown to block L-type VDCCs in addition to NCX [21]. In the light of the potential role of L-type VDCCs in ASM Ca^{2+} homeostasis, e.g. [15], results obtained using KB-R7943 should perhaps be interpreted with caution.

The NCX transporter is encoded for by three genes (NCX1, 2 and 3), with NCX1 occurring as diverse splice variants. In human ASM cells only NCX1 has been identified and, unlike cardiomyocytes in which NCX1.1 is the predominant functional variant, it is variant NCX1.3 which is present in human ASM [34, 48]. Liu et al. observed that NCX1.3 was functional in human ASM cells in both forward and reverse mode, and these currents could be inhibited by both KB-R7943 and siRNA-mediated knockdown [34]. The similar effects of KB-R7943 and NCX1.3 siRNA suggest that, at least in these experiments, KB-R7943 was acting through NCX inhibition. Interestingly, siRNA-mediated knockdown of STIM1 also inhibited the observed outwardly rectifying current, providing the first potential link between SOC and NCX1.

7 Functional Significance of SOCE in ASM Cells

As discussed earlier, ASM intracellular Ca^{2+} levels impact such physiologically critical functions as contraction, migration, proliferation, cytokine secretion and cell adhesion. While the roles of STIM, Orai and SOCE have yet to be dissected for many of these aspects of cell biology, significant evidence has accrued on their role in ASM proliferation. Zou et al. utilized a short hairpin RNA (shRNA) approach to specifically knock down STIM1 and Orai1 and found this to attenuate SOCE, as expected, but also to inhibit serum- and PDGF-BB-induced ASM cell proliferation [70]. Further evidence for the involvement of SOCE in ASM proliferation was provided by Gao et al. [22], who found that IL-13 increased thapsigargin-stimulated SOCE in rat bronchial smooth muscle cells. This effect on SOCE partially contributed to the pro-proliferative effects of IL-13 on ASM cells, suggesting that enhanced SOCE could be implicated in inflammatory remodeling of the airways, such as occur in asthma [22].

Interestingly, SOCE was observed to be higher in proliferating ASM cells when compared with quiescent cells, as was expression of Orai1 [70]. Sweeney et al. [61] also reported that SOCE was enhanced in proliferating porcine ASM cells, while similar results have been obtained in vascular smooth muscle cells [5]. This seems to contrast with the well-established notion that SOCE is substantially attenuated during mitosis [49, 62]. It is possible that this discrepancy indicates a difference in Ca^{2+} handling during cell division in smooth muscle. However, it is worth noting that studies in ASM cells to date have relied upon the relatively crude approach of comparing SOCE in serum-starved versus serum-replete cells, whereas a more detailed dissection of SOCE/I_{CRAC} during different phases of the cell cycle (e.g. [62]) has led to the consensus in other cell types that SOCE is lost in cells undergoing mitosis.

Direct evidence for the role of SOCE in ASM contraction is surprisingly limited, but Sweeney et al. [61] investigated the contribution of SOCE to acetylcholine (ACh)-induced constriction in rat bronchial rings. They found that in the absence of extracellular Ca^{2+}, only a weak, transient constriction was observed in response to ACh. When extracellular Ca^{2+} was restored, however, a 2.8-fold greater (and more sustained) contraction was obtained, highlighting the significant contribution of SOCE to cholinergic bronchoconstriction [61]. In addition, Ohga et al. reported that the novel SOCE blocker YM-58483/BTP-2 attenuated ovalbumin-mediated bronchoconstriction in guinea pigs, illustrating that SOCE may also be important for in vivo bronchoconstriction [39].

8 Ca^{2+} Oscillations in ASM

Physiological levels of bronchoconstrictors often trigger sustained oscillations in intracellular Ca^{2+} levels, rather than large, global elevations [6]. While it is generally accepted that these oscillations predominantly result from Ca^{2+} store release, they require Ca^{2+} influx to support them, and it is believed that this influx is via SOCE [9, 50]. Indeed, Putney has recently proposed that SOCE may do more than simply replenish the pool of Ca^{2+} available for oscillatory release and that in some cases Ca^{2+} entry through Orai channels may directly generate Ca^{2+} oscillations and drive downstream signaling events [50].

In human ASM cells, asynchronous Ca^{2+} waves can occur through repetitive cycles of SR Ca^{2+} release and reuptake by SERCA [15]. These Ca^{2+} waves are important in generating contraction in intact human bronchial smooth muscle bundles and were shown to rely upon ROC/SOC entry (as well as Na^{+}/Ca^{2+} exchanger and, to a lesser extent, L-type VDCCs), as they could be abolished by the ROC/SOC channel blocker SKF96365 [15]. A similar profile was observed in porcine tracheal smooth muscle bundles [16]. In addition, inhibition of SOCE ablated LTD4-induced Ca^{2+} oscillations in human ASM cells [35]. However, these experiments were performed on cells pretreated with IL-13, an inflammatory mediator known to enhance SOCE [22]. It may, therefore, be that the role of SOCE in Ca^{2+} oscillations was exaggerated in this instance, but since the IL-13 pretreatment could (to some extent) mimic the inflammatory environment found in asthmatic airways, this study might provide an indication of the contribution of SOCE to asthmatic ASM hyper-contractility [35]. Whether SOCE-derived Ca^{2+} acts in an executive (as proposed by Putney) or facilitatory (as commonly believed) manner in ASM cells remains to be established. The pharmacological tools utilized in the aforementioned studies are relatively poorly selective, so the use of more selective molecular techniques (e.g. siRNA targeting STIM and Orai family members) might provide stronger evidence for the role of SOCE in Ca^{2+} signaling/contraction, as well as greater mechanistic insight into the link between Ca^{2+} entry and downstream cellular function.

9 Summary

In this chapter we have discussed the mechanisms underlying Ca^{2+} store refilling in ASM. SOCE-mediated Ca^{2+} homeostasis appears to be the most functionally important Ca^{2+}-handling process in ASM, controlling key physiological endpoints including contraction, proliferation and cytokine release. Although some key participants in SOCE have been identified and their role assessed in ASM (e.g. STIM, Orai1), there remains a significant amount of mechanistic information which has only been demonstrated in other cell systems. Whether the process of SOCE, so elegantly dissected in cell lines, occurs in the same manner in ASM cells is as yet largely unknown. Unanswered questions include the following: (1) In what combination do STIM1, Orai1 and the variety of TRPC channels expressed in ASM cells assemble to generate the observed SOCE signals? (2) Do distinct Ca^{2+} entry pathways generate discrete intracellular signals, each coupled to the regulation of a different subset of downstream effectors? (3) What role do Orai2 and Orai3 play in ASM cells? (4) What is the contribution of SOCE to pathophysiological changes (e.g. in contractile hyperresponsiveness) in ASM cells? These (and many other) questions provide the challenge for future research in the field of ASM Ca^{2+} signaling.

References

1. Albert AP, Saleh SN, Large WA (2008). Inhibition of native TRPC6 channel activity by phosphatidylinositol 4,5-bisphosphate in mesenteric artery myocytes. *J Physiol* **586**(13): 3087–3095.
2. Barnes PJ (1998). Pharmacology of airway smooth muscle. *Am J Respir Crit Care Med* **158** (5 Pt 3): S123-132.
3. Bazan-Perkins B, Flores-Soto E, Barajas-Lopez C, Montano LM (2003). Role of sarcoplasmic reticulum Ca2+ content in Ca2+ entry of bovine airway smooth muscle cells. *Naunyn Schmiedebergs Arch Pharmacol* **368**(4): 277–283.

4. Bazan-Perkins B, Sanchez-Guerrero E, Carbajal V, Barajas-Lopez C, Montano LM (2000). Sarcoplasmic reticulum Ca2+ depletion by caffeine and changes of [Ca2+](i) during refilling in bovine airway smooth muscle cells. *Archives of medical research* **31**(6): 558–563.
5. Berra-Romani R, Mazzocco-Spezzia A, Pulina MV, Golovina VA (2008). Ca2+ handling is altered when arterial myocytes progress from a contractile to a proliferative phenotype in culture. *Am J Physiol Cell Physiol* **295**(3): C779–790.
6. Berridge MJ (2008). Smooth muscle cell calcium activation mechanisms. *J Physiol* **586** (Pt 21): 5047–5061.
7. Billington CK, Peel SE, Swan C, Liu X, Liu B, Hall IP (2010). Real Time Confocal Imaging and FRET-based Studies of Dynamic Stim1 and Orai Interactions in Human Airway Smooth Muscle cells during Store-Operated Calcium Influx. *Am J Respir Crit Care Med* **181:** A5306.
8. Billington CK, Penn RB (2003). Signaling and regulation of G protein-coupled receptors in airway smooth muscle. *Respir Res* **4:** 2.
9. Bird GS, Putney JW, Jr. (2005). Capacitative calcium entry supports calcium oscillations in human embryonic kidney cells. *J Physiol* **562**(Pt 3): 697–706.
10. Bourreau JP, Abela AP, Kwan CY, Daniel EE (1991). Acetylcholine Ca2+ stores refilling directly involves a dihydropyridine-sensitive channel in dog trachea. *Am J Physiol* **261**(3 Pt 1): C497-505.
11. Cahalan MD (2009). STIMulating store-operated Ca(2+) entry. *Nat Cell Biol* **11**(6): 669–677.
12. Cheng KT, Liu X, Ong HL, Ambudkar IS (2008). Functional requirement for Orai1 in store-operated TRPC1-STIM1 channels. *J Biol Chem* **283**(19): 12935–12940.
13. Cheng KT, Liu X, Ong HL, Swaim W, Ambudkar IS (2011). Local Ca(2)+ entry via Orai1 regulates plasma membrane recruitment of TRPC1 and controls cytosolic Ca(2)+ signals required for specific cell functions. *PLoS biology* **9**(3): e1001025.
14. Corteling RL, Li S, Giddings J, Westwick J, Poll C, Hall IP (2004). Expression of transient receptor potential C6 and related transient receptor potential family members in human airway smooth muscle and lung tissue. *Am J Respir Cell Mol Biol* **30**(2): 145–154.
15. Dai JM, Kuo KH, Leo JM, Pare PD, van Breemen C, Lee CH (2007). Acetylcholine-induced asynchronous calcium waves in intact human bronchial muscle bundle. *Am J Respir Cell Mol Biol* **36**(5): 600–608.
16. Dai JM, Kuo KH, Leo JM, van Breemen C, Lee CH (2006). Mechanism of ACh-induced asynchronous calcium waves and tonic contraction in porcine tracheal muscle bundle. *Am J Physiol Lung Cell Mol Physiol* **290**(3): L459-469.
17. DeHaven WI, Smyth JT, Boyles RR, Putney JW, Jr. (2007). Calcium inhibition and calcium potentiation of Orai1, Orai2, and Orai3 calcium release-activated calcium channels. *J Biol Chem* **282**(24): 17548–17556.
18. Estacion M, Sinkins WG, Jones SW, Applegate MA, Schilling WP (2006). Human TRPC6 expressed in HEK 293 cells forms non-selective cation channels with limited Ca2+ permeability. *J Physiol* **572**(Pt 2): 359–377.
19. Fahrner M, Muik M, Derler I, Schindl R, Fritsch R, Frischauf I, et al. (2009). Mechanistic view on domains mediating STIM1-Orai coupling. *Immunological reviews* **231**(1): 99–112.
20. Feske S, Gwack Y, Prakriya M, Srikanth S, Puppel SH, Tanasa B, et al. (2006). A mutation in Orai1 causes immune deficiency by abrogating CRAC channel function. *Nature* **441**(7090): 179–185.
21. Flores-Soto E, Reyes-Garcia J, Sommer B, Chavez J, Barajas-Lopez C, Montano LM (2012). PPADS, a P2X receptor antagonist, as a novel inhibitor of the reverse mode of the Na(+)/Ca(2)(+) exchanger in guinea pig airway smooth muscle. *Eur J Pharmacol* **674**(2–3): 439–444.
22. Gao YD, Zou JJ, Zheng JW, Shang M, Chen X, Geng S, et al. (2010). Promoting effects of IL-13 on Ca2+ release and store-operated Ca2+ entry in airway smooth muscle cells. *Pulm Pharmacol Ther* **23**(3): 182–189.
23. Godin N, Rousseau E (2007). TRPC6 silencing in primary airway smooth muscle cells inhibits protein expression without affecting OAG-induced calcium entry. *Molecular and cellular biochemistry* **296**(1–2): 193–201.

24. Gosens R, Roscioni SS, Dekkers BG, Pera T, Schmidt M, Schaafsma D, et al. (2008). Pharmacology of airway smooth muscle proliferation. *Eur J Pharmacol* **585**(2–3): 385–397.
25. Helli PB, Janssen LJ (2008). Properties of a store-operated nonselective cation channel in airway smooth muscle. *Eur Respir J* **32**(6): 1529–1539.
26. Hirota S, Janssen LJ (2007a). Store-refilling involves both L-type calcium channels and reverse-mode sodium-calcium exchange in airway smooth muscle. *Eur Respir J* **30**(2): 269–278.
27. Hirota S, Pertens E, Janssen LJ (2007b). The reverse mode of the Na(+)/Ca(2+) exchanger provides a source of Ca(2+) for store refilling following agonist-induced Ca(2+) mobilization. *Am J Physiol Lung Cell Mol Physiol* **292**(2): L438-447.
28. Hoth M, Penner R (1992). Depletion of intracellular calcium stores activates a calcium current in mast cells. *Nature* **355**(6358): 353–356.
29. Ji W, Xu P, Li Z, Lu J, Liu L, Zhan Y, et al. (2008). Functional stoichiometry of the unitary calcium-release-activated calcium channel. *Proc Natl Acad Sci U S A* **105**(36): 13668–13673.
30. Kim MS, Zeng W, Yuan JP, Shin DM, Worley PF, Muallem S (2009). Native Store-operated Ca2+ Influx Requires the Channel Function of Orai1 and TRPC1. *J Biol Chem* **284**(15): 9733–9741.
31. Liao Y, Erxleben C, Yildirim E, Abramowitz J, Armstrong DL, Birnbaumer L (2007). Orai proteins interact with TRPC channels and confer responsiveness to store depletion. *Proc Natl Acad Sci U S A* **104**(11): 4682–4687.
32. Liou J, Kim ML, Heo WD, Jones JT, Myers JW, Ferrell JE, Jr., et al. (2005). STIM is a Ca2+ sensor essential for Ca2+–store-depletion-triggered Ca2+ influx. *Curr Biol* **15**(13): 1235–1241.
33. Lis A, Peinelt C, Beck A, Parvez S, Monteilh-Zoller M, Fleig A, et al. (2007). CRACM1, CRACM2, and CRACM3 are store-operated Ca2+ channels with distinct functional properties. *Curr Biol* **17**(9): 794–800.
34. Liu B, Peel SE, Fox J, Hall IP (2010). Reverse mode Na+/Ca2+ exchange mediated by STIM1 contributes to Ca2+ influx in airway smooth muscle following agonist stimulation. *Respir Res* **11**: 168.
35. Matsumoto H, Hirata Y, Otsuka K, Iwata T, Inazumi A, Niimi A, et al. (2012). Interleukin-13 enhanced Ca2+ oscillations in airway smooth muscle cells. *Cytokine* **57**(1): 19–24.
36. Mercer JC, Dehaven WI, Smyth JT, Wedel B, Boyles RR, Bird GS, et al. (2006). Large store-operated calcium selective currents due to co-expression of Orai1 or Orai2 with the intracellular calcium sensor, Stim1. *J Biol Chem* **281**(34): 24979–24990.
37. Mignen O, Thompson JL, Shuttleworth TJ (2008). Orai1 subunit stoichiometry of the mammalian CRAC channel pore. *J Physiol* **586**(2): 419–425.
38. Montell C, Birnbaumer L, Flockerzi V (2002). The TRP channels, a remarkably functional family. *Cell* **108**(5): 595–598.
39. Ohga K, Takezawa R, Yoshino T, Yamada T, Shimizu Y, Ishikawa J (2008). The suppressive effects of YM-58483/BTP-2, a store-operated Ca2+ entry blocker, on inflammatory mediator release in vitro and airway responses in vivo. *Pulm Pharmacol Ther* **21**(2): 360–369.
40. Ong HL, Chen J, Chataway T, Brereton H, Zhang L, Downs T, et al. (2002). Specific detection of the endogenous transient receptor potential (TRP)-1 protein in liver and airway smooth muscle cells using immunoprecipitation and Western-blot analysis. *Biochem J* **364**(Pt 3): 641–648.
41. Ong HL, Cheng KT, Liu X, Bandyopadhyay BC, Paria BC, Soboloff J, et al. (2007). Dynamic assembly of TRPC1-STIM1-Orai1 ternary complex is involved in store-operated calcium influx. Evidence for similarities in store-operated and calcium release-activated calcium channel components. *J Biol Chem* **282**(12): 9105–9116.
42. Oritani K, Kincade PW (1996). Identification of stromal cell products that interact with pre-B cells. *J Cell Biol* **134**(3): 771–782.
43. Orrenius S, Zhivotovsky B, Nicotera P (2003). Regulation of cell death: the calcium-apoptosis link. *Nat Rev Mol Cell Biol* **4**(7): 552–565.

44. Parekh AB, Putney JW, Jr. (2005). Store-operated calcium channels. *Physiol Rev* **85**(2): 757–810.
45. Peel SE, Liu B, Hall IP (2006). A key role for STIM1 in store operated calcium channel activation in airway smooth muscle. *Respir Res* **7**: 119.
46. Peel SE, Liu B, Hall IP (2008). ORAI and store-operated calcium influx in human airway smooth muscle cells. *Am J Respir Cell Mol Biol* **38**(6): 744–749.
47. Peinelt C, Vig M, Koomoa DL, Beck A, Nadler MJ, Koblan-Huberson M, et al. (2006). Amplification of CRAC current by STIM1 and CRACM1 (Orai1). *Nat Cell Biol* **8**(7): 771–773.
48. Pitt A, Knox AJ (1996). Molecular characterization of the human airway smooth muscle Na+/Ca2+ exchanger. *Am J Respir Cell Mol Biol* **15**(6): 726–730.
49. Preston SF, Sha'afi RI, Berlin RD (1991). Regulation of Ca2+ influx during mitosis: Ca2+ influx and depletion of intracellular Ca2+ stores are coupled in interphase but not mitosis. *Cell regulation* **2**(11): 915–925.
50. Putney JW (2011). The physiological function of store-operated calcium entry. *Neurochem Res* **36**(7): 1157–1165.
51. Putney JW, Jr. (1986). A model for receptor-regulated calcium entry. *Cell Calcium* **7**(1): 1–12.
52. Roos J, DiGregorio PJ, Yeromin AV, Ohlsen K, Lioudyno M, Zhang S, et al. (2005). STIM1, an essential and conserved component of store-operated Ca2+ channel function. *J Cell Biol* **169**(3): 435–445.
53. Sathish V, Abcejo AJ, Thompson MA, Sieck GC, Prakash YS, Pabelick CM (2012). Caveolin-1 regulation of store-operated Ca2+ influx in human airway smooth muscle. *Eur Respir J* **40**(2): 470–478.
54. Shen WW, Frieden M, Demaurex N (2011). Remodelling of the endoplasmic reticulum during store-operated calcium entry. *Biology of the cell / under the auspices of the European Cell Biology Organization* **103**(8): 365–380.
55. Smyth JT, Hwang SY, Tomita T, DeHaven WI, Mercer JC, Putney JW (2010). Activation and regulation of store-operated calcium entry. *J Cell Mol Med* **14**(10): 2337–2349.
56. Soboloff J, Rothberg BS, Madesh M, Gill DL (2012). STIM proteins: dynamic calcium signal transducers. *Nat Rev Mol Cell Biol* **13**(9): 549–565.
57. Soboloff J, Spassova MA, Tang XD, Hewavitharana T, Xu W, Gill DL (2006). Orai1 and STIM reconstitute store-operated calcium channel function. *J Biol Chem* **281**(30): 20661–20665.
58. Srikanth S, Gwack Y (2012). Orai1, STIM1, and their associating partners. *J Physiol* **590**(Pt 17): 4169–4177.
59. Stathopulos PB, Ikura M (2010). Partial unfolding and oligomerization of stromal interaction molecules as an initiation mechanism of store operated calcium entry. *Biochemistry and cell biology = Biochimie et biologie cellulaire* **88**(2): 175–183.
60. Sundivakkam PC, Freichel M, Singh V, Yuan JP, Vogel SM, Flockerzi V, et al. (2012). The Ca (2+) sensor stromal interaction molecule 1 (STIM1) is necessary and sufficient for the store-operated Ca(2+) entry function of transient receptor potential canonical (TRPC) 1 and 4 channels in endothelial cells. *Mol Pharmacol* **81**(4): 510–526.
61. Sweeney M, McDaniel SS, Platoshyn O, Zhang S, Yu Y, Lapp BR, et al. (2002). Role of capacitative Ca2+ entry in bronchial contraction and remodeling. *J Appl Physiol* **92**(4): 1594–1602.
62. Tani D, Monteilh-Zoller MK, Fleig A, Penner R (2007). Cell cycle-dependent regulation of store-operated I(CRAC) and Mg2+−nucleotide-regulated MagNuM (TRPM7) currents. *Cell Calcium* **41**(3): 249–260.
63. Vig M, Peinelt C, Beck A, Koomoa DL, Rabah D, Koblan-Huberson M, et al. (2006). CRACM1 is a plasma membrane protein essential for store-operated Ca2+ entry. *Science* **312**(5777): 1220–1223.

64. White TA, Xue A, Chini EN, Thompson M, Sieck GC, Wylam ME (2006). Role of transient receptor potential C3 in TNF-alpha-enhanced calcium influx in human airway myocytes. *Am J Respir Cell Mol Biol* **35**(2): 243–251.
65. Wu MM, Buchanan J, Luik RM, Lewis RS (2006). Ca2+ store depletion causes STIM1 to accumulate in ER regions closely associated with the plasma membrane. *J Cell Biol* **174**(6): 803–813.
66. Xiao JH, Zheng YM, Liao B, Wang YX (2010). Functional role of canonical transient receptor potential 1 and canonical transient receptor potential 3 in normal and asthmatic airway smooth muscle cells. *Am J Respir Cell Mol Biol* **43**(1): 17–25.
67. Yu F, Sun L, Machaca K (2010). Constitutive recycling of the store-operated Ca2+ channel Orai1 and its internalization during meiosis. *J Cell Biol* **191**(3): 523–535.
68. Yuan JP, Kim MS, Zeng W, Shin DM, Huang G, Worley PF, et al. (2009). TRPC channels as STIM1-regulated SOCs. *Channels (Austin)* **3**(4): 221–225.
69. Zhang SL, Yeromin AV, Zhang XH, Yu Y, Safrina O, Penna A, et al. (2006). Genome-wide RNAi screen of Ca(2+) influx identifies genes that regulate Ca(2+) release-activated Ca(2+) channel activity. *Proc Natl Acad Sci U S A* **103**(24): 9357–9362.
70. Zou JJ, Gao YD, Geng S, Yang J (2011). Role of STIM1/Orai1-mediated store-operated Ca(2)(+) entry in airway smooth muscle cell proliferation. *J Appl Physiol* **110**(5): 1256–1263.

Novel Mechanisms in Ca^{2+} Homeostasis and Internal Store Refilling of Airway Smooth Muscle

Luke J. Janssen

Contents

Abstract Calcium is known to play a key role in excitation–contraction coupling of airway smooth muscle (ASM) and may also be important in other cellular responses, such as gene expression, migration, proliferation, and apoptosis. The sarcoplasmic reticulum acts as an agonist-releasable store of Ca^{2+} and as a sink to buffer changes in cytosolic $[Ca^{2+}]_i$. ASM also expresses, in great abundance, other Ca^{2+}-mobilizing effectors such as voltage-dependent Ca^{2+} channels (Ca_v) and sodium/calcium exchangers (NCX) on the plasmalemma, as well as ryanodine receptors (RyRs) on the SR membrane. These three had long been held to be important in mediating electromechanical coupling (Ca_v), extrusion of cytosolic Ca^{2+} (NCX) and Ca^{2+}-induced Ca^{2+} release (RyR), respectively. However, more recent data and careful consideration have challenged those associations. In this chapter, we explore the novel hypothesis that all three contribute to refilling of the SR, perhaps orchestrated or powered by electrical slow waves (which are also found in ASM of all species studied to date).

L.J. Janssen (✉)
Firestone Institute for Respiratory Health, St. Joseph's Hospital and the Department of Medicine, McMaster University, L-314, St. Joseph's Hospital, 50 Charlton Ave. East, Hamilton, ON L8N 4A6, Canada
e-mail: janssenl@mcmaster.ca

Y.-X. Wang (ed.), *Calcium Signaling In Airway Smooth Muscle Cells*,
DOI 10.1007/978-3-319-01312-1_10, © Springer International Publishing Switzerland 2014

Keywords Excitation–contraction coupling • Ca^{2+} handling • Airway smooth muscle

Abbreviations

$[Ca^{2+}]_i$	Cytosolic concentrations of calcium
ASM	Airway smooth muscle
Ca_v	Voltage-dependent calcium channel
CICR	Ca^{2+}-induced Ca^{2+} release
EM	Electromechanical
MLC	Myosin light chain
MLCK	Myosin light chain kinase
NCX	Na^+/Ca^{2+} exchange
RyR	Ryanodine receptor
SERCA	Sarcoplasmic/endoplasmic reticulum Ca^{2+}-ATPase
SR	Sarcoplasmic reticulum

1 Introduction

Despite decades of development of asthma therapies, there is still a need for a better bronchodilator. Generally, asthma is well controlled by the use of inhaled steroids, but all asthmatics will on occasion need treatment for acute symptoms (at times life-threatening or lethal). When rapid relief of asthma symptoms is required, inhaled β-agonists remain the most widely used. These exert their beneficial effect by rapidly improving airflow obstruction predominately by relaxing constricted airway smooth muscle (ASM). β-agonists are effective in most circumstances but are not fully sufficient: severe exacerbations continue to occur (even in mild asthmatics on average every 1–2 years [86]) and cause considerable morbidity and mortality. Also, there are growing concerns about the safety of frequent or regular β-agonist use [100, 101].

A number of other classes of bronchodilators were introduced several decades ago, including anti-cholinergics, anti-leukotrienes, and phosphodiesterase inhibitors. Anti-cholinergics are routinely added to inhaled β-agonists in the treatment of very severe acute asthma in hospital setting [68, 102], but only inhaled β-agonists are recommended for the treatment of airflow obstruction in the day-to-day management of asthma. No substantially new pharmacological agent targeting ASM has appeared in the past several decades. Any truly new advances have been aimed at controlling airway inflammation, which is also important but should not eclipse efforts aimed at controlling bronchoconstriction directly. Bronchial thermoplasty has recently been introduced to ablate the ASM itself using thermal energy [20, 21, 82, 89], but it is relatively invasive and currently can only be applied to the first few generations of airways. Others have proposed chemical agents which specifically target and reduce ASM mass in all airway generations (e.g., silencing RNA, promoters of programmed cell death), but nothing is yet on the horizon. Altogether, then, we have made very little progress over the past several decades with respect to finding novel forms of bronchodilator therapy. This impasse is largely due to the limitations of our current knowledge of ASM physiology. Further progress will require a better understanding of the mechanisms underlying ASM contraction and the identification of new therapeutic targets.

Ultimately, contraction of ASM is triggered by elevation of $[Ca^{2+}]_i$, which leads to phosphorylation of myosin light chain (MLC) via activation of Ca^{2+}/calmodulin-dependent MLC kinase (MLCK) [6, 47]. Ca^{2+} can be stored within an intracellular organelle and released in a carefully regulated fashion (described in Sect. 3). Alternatively, Ca^{2+} can enter cells from the extracellular space but must first cross the plasmalemma via some form of ion channel or exchanger (described in more detail in Sect. 4). In what follows, we describe these various Ca^{2+}-mobilization pathways in more detail. In the course of doing so, we will call particular attention to four different entities which are present in ASM from all species studied to date: electrical slow waves, L-type voltage-dependent Ca^{2+} channels (Ca_v), ryanodine receptors (RyRs), and the sodium/calcium exchanger (NCX). Their ubiquity in ASM from all species would suggest that they play some key roles in ASM physiology. However, as we shall show below, all four are still very poorly understood. In the course of presenting these four entities in the context of Ca^{2+} handling in ASM physiology, we will lay out a hypothesis in which they work together to maintain the filling state of the internal Ca^{2+} store (which has more recently been recognized as the prime trigger for ASM contraction).

2 Membrane Potential and Electrical Slow Waves

In cardiac, skeletal, gastrointestinal, lymphatic, and vascular smooth muscles, resting membrane potential is well below the threshold required for Ca_v activation. In those tissues, excitatory stimuli trigger action potentials which depolarize the membrane into the positive range, resulting in Ca_v activation as well as a more slowly developing inactivation (note: these two are *different* processes, *not opposite* processes). Within tens of milliseconds, though, the action potential disappears, the membrane repolarizes, inactivation reverses, and the membrane is able to generate another action potential.

In ASM, on the other hand, membrane potential rarely depolarizes into the positive range, even during excitation. Instead, in those preparations in which this has been studied – human [52], guinea pig [1, 4, 14, 18, 39, 51, 104, 106], dog [57, 58, 60, 70], cow [16, 69, 106], ferret [84], and mouse (Fig. 1) – ASM exhibits quasi-sinusoidal oscillations centred around −35 mV, with amplitudes of up to 25 mV and frequency of less than 10 Hz [26, 57–61]. The depolarizing and hyperpolarizing phases comprise Cl- and K^+ currents, respectively [60, 63]; interestingly, L-type Ca^{2+}

Fig. 1 *Electrophysiological recording in mouse bronchial smooth muscle*. Membrane potential and electrical slow waves in murine bronchial smooth muscle recorded using an intracellular microelectrode before and after 20 mM Tetraethylammonium (TEA)

channel blockers and Ca^{2+}-free media also seem to disrupt slow wave activity [39, 40, 60, 69, 103]. In in vivo recordings, their frequency is increased by vagal stimulation and decreased by sympathetic stimulation [8, 70, 78]. In isolated tissues, their frequency is increased by submaximal concentrations of acetylcholine or histamine (supramaximal concentrations depolarize the membrane so much that slow waves disappear [1, 14, 16, 18, 69, 103]) and decreased by β-adrenergic stimulation [3, 51, 103], agents which act through adenylate cyclase [51, 103] or agents which directly hyperpolarize the membrane [4, 51, 104].

The presence of electrical slow waves in the ASM of all species studied to date suggests that they play an important role in the unique aspects of ASM physiology and might therefore be exploited in developing effective airway-specific agents to modulate bronchoconstriction. However, their role in ASM physiology has not yet been adequately explained [74]. In what follows, we propose novel roles for Ca_v and NCX, which are critically dependent on membrane potential and electrical slow wave activity.

3 Release of Internally Sequestered Ca^{2+}

The sarcoplasmic reticulum (SR) plays a key role in Ca^{2+} handling, acting as a sink to buffer cytosolic $[Ca^{2+}]_i$ and an agonist-releasable store of Ca^{2+} to trigger contractions [105] (Fig. 2). Bronchoconstrictors cause generation of IP_3 which opens Ca^{2+} channels on the SR, leading to recurring Ca^{2+} waves [41, 42, 62]. It is the frequency of those Ca^{2+} waves, rather than their amplitude, which determines the degree of contraction, being increased by bronchoconstrictors [9, 10, 72, 85, 90, 91, 93, 98] and decreased by bronchodilators [7]; this regulatory mechanism is elaborated more fully within this compendium by Dr. M. Sanderson in the chapter "Novel Mechanisms in Ca^{2+} Homeostasis and Internal Store Refilling of Airway Smooth Muscle."

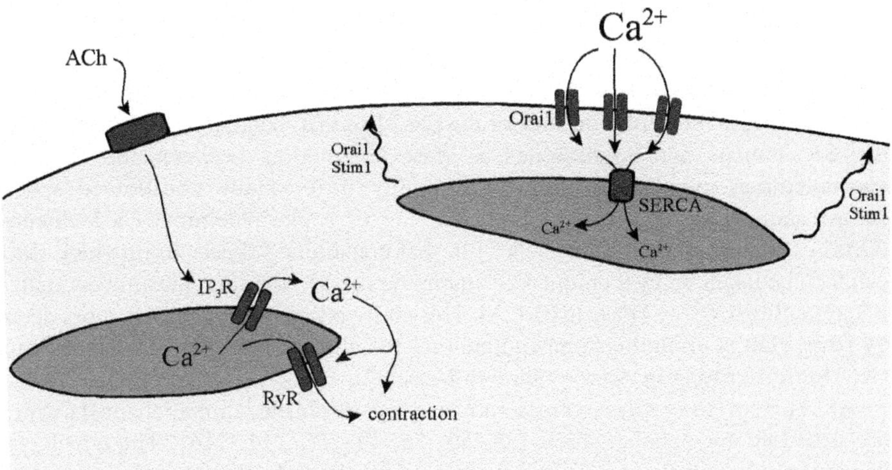

Fig. 2 *Ca^{2+} release and uptake by sarcoplasmic reticulum (SR)*. Ca^{2+} can be released from the SR via IP_3-gated Ca^{2+} channels (activated by agonist-induced phosphoinositide cascade) or from RyRs (which in turn might be triggered by IP_3-mediated Ca^{2+} release). STIM1 and Orai1 then orchestrate a tight apposition and interaction between the depleted SR and the plasmalemma, forming a tight, diffusionally constrained space. Ca^{2+} entry via Orai1 and possibly other sources (Fig. 3) provide a source of Ca^{2+} which can be taken up by SERCA

SR membranes also express a second group of Ca^{2+}-release channels, known as ryanodine receptors (RyRs) for their pharmacological sensitivity to that plant alkaloid. These are activated by cytosolic Ca^{2+}, a process often referred to as Ca^{2+}-induced Ca^{2+} release (CICR). RyRs are also activated by caffeine, which enhances the Ca^{2+} sensitivity of RyRs such that even basal $[Ca^{2+}]_i$ is sufficient to trigger CICR. Finally, cyclic ADP ribose can also activate RyRs, prompting some researchers to propose the latter as the endogenous ligand for these channels [29, 92, 113, 115], just as IP_3 is the endogenous ligand for the other Ca^{2+}-release pathway. Once again, the pervasive presence of RyRs in ASM of all species would suggest an important role in ASM physiology, but that role is still unclear and highly disputed: some claim they mediate contraction [24, 25, 30, 31, 45], while others refute this [53, 112] or even claim a role in relaxant mechanisms [32, 107, 117]. We have presented evidence [107] which suggests that RyRs are involved in discharging Ca^{2+} from an overly filled SR but preferentially direct that discharge toward the plasmalemma (perhaps because RyRs are preferentially expressed on those portions of the SR juxtaposed to the plasmalemma), where it can be extruded by the plasmalemmal Ca^{2+}-ATPase or Na^+/Ca^{2+} exchange, without triggering a mechanical response. Conversely, in the following section, we speculate on the possibility that RyRs might work in concert with Ca_v and electrical slow waves to mediate refilling of the SR.

4 Influx of External Ca^{2+} via Plasmalemmal Ion Channels

Once the SR releases stored Ca^{2+}, it must be refilled. A protein in the SR membrane (STIM1) senses the depletion state of the SR, via an EF-hand domain extended into the SR lumen, and orchestrates a close interaction between the SR and plasmalemmal membranes, forming a highly diffusionally constrained space. At the same time, it also stimulates opening of a plasmalemmal Ca^{2+} channel (Orai1) which markedly raises [Ca^{2+}] in that constrained space, from which Ca^{2+} can then be taken up by a unique Ca^{2+} pump on the SR: the sarcoplasmic/endoplasmic reticulum Ca^{2+}-ATPase (SERCA). This entire process is outlined in more detail by Dr. I. Hall et al. in this compendium (see the chapter "Mechanisms Underlying Ca^{2+} Store Refilling in Airway Smooth Muscle").

ASMs from *all* species express another class of plasmalemmal channels which are activated by depolarization [38, 50, 55, 71, 77, 114, 116]. These voltage-dependent Ca^{2+} channels (Ca_v) are identical to those found in cardiac, skeletal, vascular, lymphatic, and gastrointestinal smooth muscles. In all those other smooth muscle cell types, Ca_v play a central role in electromechanical (EM) coupling, as was mentioned earlier in the context of electrical slow waves (Sect. 2). As such, Ca^{2+} channel blockers and K^+ channel openers are invaluable in controlling tone in many disease states in which there is unwanted smooth muscle contraction, including, for example, cardiovascular hypertension, stroke, myocardial infarction, and gastrointestinal motility disorders [33, 36, 95]. The facts that ASM cells also express those same channels in abundance and that excitation of ASM is often accompanied by some degree of membrane depolarization (caused by opening of Ca^{2+}-dependent Cl channels and suppression of K^+ conductances [54, 65]) might suggest that a similar EM coupling mechanism operates in ASM. However, there is a tremendous body of evidence suggesting otherwise (reviewed in detail elsewhere [56]), including the facts that (1) membrane potential rarely reaches into the range of voltages needed to substantially activate Ca_v [55, 60], (2) in vitro contractions of ASM preparations are seemingly unaffected by Ca_v blockers [12, 13, 28, 34, 67] or by clamping the membrane potential to very negative values [66] and (3) muscarinic agonists can actually suppress Ca_v currents [109, 118]. Most importantly, there is mixed opinion regarding the value of Ca^{2+} channel blockers in asthma; some find them to be useful [15, 17, 22, 23, 35, 79–81, 87, 88, 96], while others do not [6, 19, 43, 44]. Altogether, it seems that EM coupling and voltage-dependent Ca^{2+} influx through Ca_v is of minor importance in ASM; we feel it is time to consider other roles for these channels in ASM physiology.

Ca_v may contribute to SR refilling [12, 13, 27, 48, 66, 75, 94, 119] during extreme levels of excitation since submaximal stimulation elicits little or no membrane depolarization. For example, we have shown that voltage pulses allowed SR refilling in ASM pretreated with the SERCA inhibitor cyclopiazonic acid [66] at a concentration found elsewhere to maximally inhibit that Ca^{2+} pump [5, 108, 120]. This was quite surprising since innumerable studies of Ca^{2+} uptake in a wide variety of cell types have all shown SERCA to be the only type of Ca^{2+}

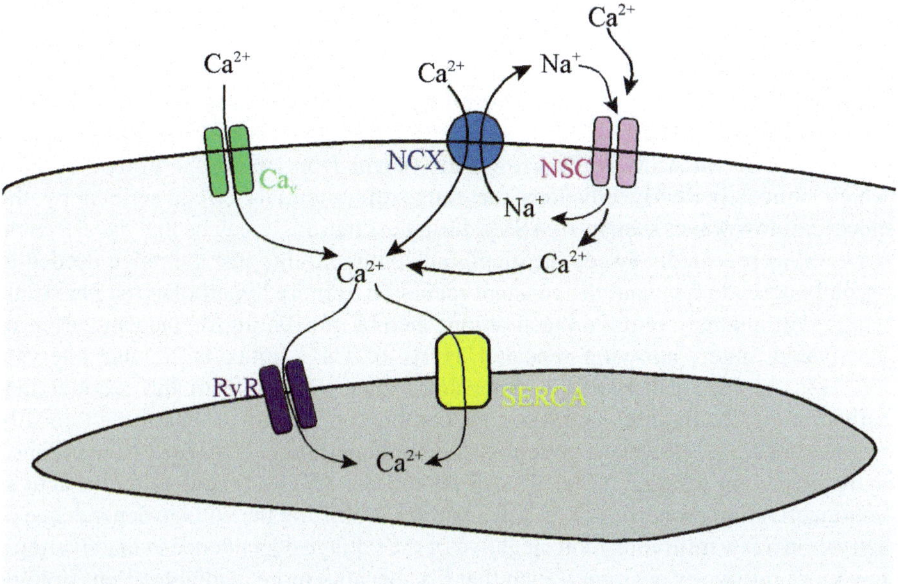

Fig. 3 *Novel roles for Ca_v, NCX, RyR in Ca^{2+} store refilling of airway smooth muscle (ASM).* Excitation of ASM leads to several important changes, including (1) activation of plasmalemmal non-selective cation channels (*NSC*), with consequent Na^+ influx; (2) membrane depolarization; (3) orchestration by STIM1 and Orai1 of a tight apposition and interaction between the depleted sarcoplasmic reticulum (SR) and the plasmalemma, forming a tight, diffusionally constrained space. Ca^{2+} entry into this space – via Ca_v and reverse-mode NCX, which are in turn coordinated by electrical slow waves – raises $[Ca^{2+}]_i$ well above that within the depleted SR. With the concentration gradient for Ca^{2+} across the SR membrane now reversed, Ca^{2+} can enter the SR via passive diffusion through RyRs (which have been opened by the elevated subplasmalemmal $[Ca^{2+}]_i$) or via active uptake by SERCA

pump on the SR, and no study has identified a SERCA which is insensitive to cyclopiazonic acid. As such, it seemed that Ca^{2+} somehow transitioned from the extracellular space directly into the SR, bypassing SERCA altogether, through some unknown SR entry pathway. Based on recent developments in this field, we have hypothesized that this pathway into the SR might comprise RyRs [46, 64]. That is, entry of external Ca^{2+} into the diffusionally constrained space formed between the SR and plasmalemma (see Sect. 3 and Fig. 3) would raise $[Ca^{2+}]_i$ to a level well above that within the depleted SR. With the Ca^{2+} concentration gradient across the *SR* membrane reversed in this way, opening of any Ca^{2+}-permeable channels on the SR would provide a pathway for that subplasmalemmal Ca^{2+} to *enter* the depleted SR *in retrograde fashion*. RyRs are activated by cytosolic Ca^{2+} and highly conductive to Ca^{2+} (tenfold more so than IP_3-gated ones). Du et al. have provided *indirect* evidence for a functional interaction between Ca_v and RyR in rat ASM [30], although they did not use direct electrophysiological approaches to characterize this. Also, STIM1 can interact directly with $Ca_v1.2$ in vascular smooth muscle [110].

Electrophysiological studies of voltage-dependent ion currents will sometimes point to a very narrow range of voltages in which the current of interest can persist presumably indefinitely (i.e., absent other regulatory mechanisms such as phosphorylation of the channel). This range is just above those needed to activate the channels and yet just below those at which the channels become inactivated. In the case of Ca_v in ASM, this narrow band ranges from approx. −40 to + 10 mV, which almost perfectly envelops the range of potentials circumscribed by the electrical slow waves seen in all ASMs during excitation. It may be that the electrical slow waves repeatedly sweep the membrane potential into just the range needed to partially activate Ca_v, and the constant recursion to more hyperpolarized potentials would continuously remove inactivation, thereby preventing the accumulation of inactivated states, allowing repeated spurts of Ca^{2+} influx at regular intervals (~1 Hz). At face value, this hypothesis is also vulnerable to the rebuttal that activation of Ca_v requires voltages considerably more positive than those typically recorded in ASM. However, protein–protein interactions such as the two mentioned in the preceding paragraph (i.e., Ca_v−RyR and Ca_v−STIM1) could alter the activation/inactivation properties of the Ca_v channels – shifting the voltage dependence of activation a few millivolts more negative or the voltage dependence in inactivation a few more millivolts positive – such that Ca_v became more available at physiologically relevant membrane voltages.

5 Influx of Ca^{2+} via NCX

A decade or more ago, NCX was seen to contribute primarily toward extrusion of internal Ca^{2+}, using the energy liberated from allowing sodium to flow down its concentration gradient to drive Ca^{2+} "uphill" against its concentration gradient. The plasmalemmal Ca^{2+}-ATPase, on the other hand, accomplishes this same Ca^{2+} flux using the energy liberated by hydrolysis of ATP. The turnover rate of NCX is approx. tenfold higher than that of the plasmalemmal Ca^{2+} pump (100–150 s^{-1} vs. 5,000 s^{-1}, respectively [11, 99]), suggesting that NCX is far more suited to rapid and massive bulk movements of Ca^{2+} out of the cell (and, as we will see below, *into* the cell). However, the affinity of the plasmalemmal (and SR) Ca^{2+} pump for Ca^{2+} is approx. tenfold higher than that of NCX (approx. 0.3:M versus approx. 6:M, respectively [11, 99]), making those pumps much more suited to fine-tuning $[Ca^{2+}]_i$. The molecular biology and pharmacology of NCX has been outlined in detailed elsewhere [11] and will not be recapitulated here.

However, in the context of ASM physiology, membrane potential, and electrical slow waves, it is well worth considering carefully the electrophysiological properties of NCX. NCX exchanges Na^+ ions on one side of a membrane for Ca^{2+} ions on the other side and can mediate this exchange equally well in either direction. Moreover, this process is electrogenic: three Na^+ ions are exchanged for each Ca^{2+} ion. As such, Ca^{2+} extrusion is favoured at very negative membrane potentials, but Ca^{2+} *influx* occurs at more depolarized potentials or when cytosolic $[Na^+]$ rises

[46]. In the past, Ca^{2+} influx through "reverse-mode" NCX was seen to be simply an interesting phenomenon attainable only under highly non-physiologically relevant experimental conditions (e.g., patch-clamp electrophysiological studies). However, it came to be realized that there could be marked compartmentalization within cells, produced in part by the arrangement of organelles and placement of homeostatic enzymes, including the diffusionally constrained compartment referred to previously. During excitation by muscarinic or other agonists which activate non-selective cation currents [37, 65, 111], Na^+ entering into this space could accumulate to remarkably high levels, at least briefly. Concurrently, excitation also brings on a degree of membrane depolarization. When the reversal potential for NCX is calculated for a physiologically relevant range of voltage, [Na^+] and [Ca^{2+}], it can be seen that the electrical slow waves seen in ASM sweep the membrane potential back and forth across the reversal potential for NCX [46, 49, 97], allowing the latter to alternate between the forward and reverse modes and to be exquisitely modulated by changes in cytosolic [Na^+] and [Ca^{2+}]. Our group and others have examined the effects of selective inhibitors of reverse-mode NCX activity – particularly KRB7943 and SEA0400 – on contractile responses in intact tissues and have concluded that Ca^{2+} influx via NCX works in concert with SERCA to mediate SR refilling [2, 48, 49]: we would now propose that retrograde flux into the SR via RyRs might contribute to this refilling as well (as proposed in Sect. 4 above). Interestingly, NCX expression is increased in a murine model of asthma [73]; this may reflect a compensatory change for the decreased expression of SERCA which was found in asthmatic ASM [76].

6 Conclusion

Although some debate whether ASM plays any important role at all in *normal* airway physiology [83], very few would contest that ASM is a major problem in asthma. Whether that problem is due to an altered function of individual ASM cells or hyperplasia of the whole airway wall tissue, bronchodilation (or inhibition of contraction) is the singular goal of rescue asthma therapy. Ca^{2+} homeostasis is critical to both ASM contraction and relaxation. The current understanding of ASM Ca^{2+} handling and excitation–contraction coupling are presently inadequate to fully explain many of the data already available. ASM expresses Ca_v, RyR, and NCX in abundance, and these three had long been held to be important in mediating EM coupling (Ca_v), CICR-mediated contraction (RyR), and extrusion of cytosolic Ca^{2+} (NCX), respectively. However, more recent data and careful consideration have challenged these associations. Instead, all three can now be proposed to contribute to refilling of the SR, perhaps orchestrated or powered by electrical slow waves. Further studies are crucial to bridge the gaps and will undoubtedly provide the basis for the development of new approaches to relax ASM and provide symptomatic relief for asthmatic patients.

References

1. **Ahmed F, Foster RW, Small RC and Weston AH**. Some features of the spasmogenic actions of acetylcholine and histamine in guinea-pig isolated trachealis. *Br J Pharmacol* 83: 227–233, 1984.
2. **Algara-Suarez P, Mejia-Elizondo R, Sims SM, Saavedra-Alanis VM and Espinosa-Tanguma R**. The 1.3 isoform of Na^+-Ca^{2+} exchanger expressed in guinea pig tracheal smooth muscle is less sensitive to KB-R7943. *J Physiol Biochem* 66: 117–125, 2010.
3. **Allen SL, Beech DJ, Foster RW, Morgan GP and Small RC**. Electrophysiological and other aspects of the relaxant action of isoprenaline in guinea-pig isolated trachealis. *Br J Pharmacol* 86: 843–854, 1985.
4. **Allen SL, Boyle JP, Cortijo J, Foster RW, Morgan GP and Small RC**. Electrical and mechanical effects of BRL34915 in guinea-pig isolated trachealis. *Br J Pharmacol* 89: 395–405, 1986.
5. **Amoako D, Qian Y, Kwan CY and Bourreau JP**. Probing excitation-contraction coupling in trachealis smooth muscle with the mycotoxin cyclopiazonic acid. *Clin Exp Pharmacol Physiol* 23: 733–737, 1996.
6. **Ann TM, Harman E, Chesrown S and Hendeles L**. Efficacy of calcium channel blockers as maintenance therapy for asthma. *Br J Clin Pharmacol* 53: 243–249, 2002.
7. **Bai Y and Sanderson MJ**. Airway smooth muscle relaxation results from a reduction in the frequency of Ca^{2+} oscillations induced by a cAMP-mediated inhibition of the IP_3 receptor. *Respir Res* 7: 34, 2006.
8. **Beinfield WH and Seifter J**. Spontaneous mechanical activities of dog trachealis muscle in vivo. *J Appl Physiol* 48: 320–328, 1980.
9. **Bergner A and Sanderson MJ**. Acetylcholine-induced calcium signaling and contraction of airway smooth muscle cells in lung slices. *J Gen Physiol* 119: 187–198, 2002.
10. **Bergner A and Sanderson MJ**. ATP stimulates Ca2+ oscillations and contraction in airway smooth muscle cells of mouse lung slices. *Am J Physiol Lung Cell Mol Physiol* 283: L1271–L1279, 2002.
11. **Blaustein MP and Lederer WJ**. Sodium/calcium exchange: its physiological implications. *Physiol Rev* 79: 763–854, 1999.
12. **Bourreau JP, Abela AP, Kwan CY and Daniel EE**. Acetylcholine Ca^{2+} stores refilling directly involves a dihydropyridine- sensitive channel in dog trachea. *Am J Physiol* 261: C497–C505, 1991.
13. **Bourreau JP, Kwan CY and Daniel EE**. Distinct pathways to refill ACh-sensitive internal Ca^{2+} stores in canine airway smooth muscle. *Am J Physiol* 265: C28–C35, 1993.
14. **Boyle JP, Davies JM, Foster RW, Good DM, Kennedy I and Small RC**. Spasmogen action in guinea-pig isolated trachealis: involvement of membrane K^+-channels and the consequences of K^+-channel blockade. *Br J Pharmacol* 93: 319–330, 1988.
15. **Burghuber OC, Kneussl M, Harmuth P, Silberbauer K, Sinzinger H and Haber P**. Inhibition of acetylcholine-induced bronchoconstriction in asthmatics by nifedipine. *Respiration* 50: 265–272, 1986.
16. **Cameron AR, Johnston CF, Kirkpatrick CT and Kirkpatrick MC**. The quest for the inhibitory neurotransmitter in bovine tracheal smooth muscle. *Q J Exp Physiol* 68: 413–426, 1983.
17. **Cerrina J, Denjean A, Alexandre G, Lockhart A and Duroux P**. Inhibition of exercise-induced asthma by a calcium antagonist, nifedipine. *Am Rev Respir Dis* 123: 156–160, 1981.
18. **Coburn RF**. Stretch-induced membrane depolarization in ferret trachealis smooth muscle cells. *J Appl Physiol* 62: 2320–2325, 1987.
19. **Corris PA, Nariman S and Gibson GJ**. Nifedipine in the prevention of asthma induced by exercise and histamine. *Am Rev Respir Dis* 128: 991–992, 1983.
20. **Cox G, Miller JD, McWilliams A, Fitzgerald JM and Lam S**. Bronchial thermoplasty for asthma. *Am J Respir Crit Care Med* 173: 965–969, 2006.

21. **Cox G, Thomson NC, Rubin AS, Niven RM, Corris PA, Siersted HC, Olivenstein R, Pavord ID, McCormack D, Chaudhuri R, Miller JD and Laviolette M**. Asthma control during the year after bronchial thermoplasty. *N Engl J Med* 356: 1327–1337, 2007.
22. **Crimi N, Palermo F, Sorace R, Gibellino F and Mistretta A**. Effect of a calcium antagonist, nifedipine, in exercise-induced asthma. *Respiration* 45: 262–264, 1984.
23. **Cuss FM and Barnes PJ**. The effect of inhaled nifedipine on bronchial reactivity to histamine in man. *J Allergy Clin Immunol* 76: 718–723, 1985.
24. **Dai JM, Kuo KH, Leo JM, Pare PD, van Breemen C and Lee CH**. Acetylcholine-induced asynchronous calcium waves in intact human bronchial muscle bundle. *Am J Respir Cell Mol Biol* 36: 600–608, 2007.
25. **Dai JM, Kuo KH, Leo JM, van Breemen C and Lee CH**. Mechanism of ACh-induced asynchronous calcium waves and tonic contraction in porcine tracheal muscle bundle. *Am J Physiol Lung Cell Mol Physiol* 290: L459–L469, 2006.
26. **Daniel EE, Abela AP, Janssen LJ and O'Byrne PM**. Effects of inflammatory mediators on canine airway neuromuscular function. *Can J Physiol Pharmacol* 70: 624–634, 1992.
27. **Daniel EE, Bourreau JP, Abela A and Jury J**. The internal calcium store in airway muscle: emptying, refilling and chloride. Possible new directions for drug development. *Biochem Pharmacol* 43: 29–37, 1992.
28. **Daniel EE, Jury J, Serio R and Jager LP**. Role of depolarization and calcium in contractions of canine trachealis from endogenous or exogenous acetylcholine. *Can J Physiol Pharmacol* 69: 518–525, 1991.
29. **Deshpande DA, Walseth TF, Panettieri RA and Kannan MS**. CD38-cyclic ADP-ribose-mediated Ca^{2+} signaling contributes to airway smooth muscle hyperresponsiveness. *FASEB J* 2003.
30. **Du W, McMahon TJ, Zhang ZS, Stiber JA, Meissner G and Eu JP**. Excitation-contraction coupling in airway smooth muscle. *J Biol Chem* 281: 30143–30151, 2006.
31. **Du W, Stiber JA, Rosenberg PB, Meissner G and Eu JP**. Ryanodine receptors in muscarinic receptor-mediated bronchoconstriction. *J Biol Chem* 280: 26287–26294, 2005.
32. **Ethier MF and Madison JM**. IL-4 inhibits calcium transients in bovine trachealis cells by a ryanodine receptor-dependent mechanism. *FASEB J* 20: 154–156, 2006.
33. **Faraci FM and Heistad DD**. Regulation of the cerebral circulation: role of endothelium and potassium channels. *Physiological Reviews* 78: 53–97, 1998.
34. **Farley JM and Miles PR**. The sources of calcium for acetylcholine-induced contractions of dog tracheal smooth muscle. *J Pharmacol Exp Ther* 207: 340–346, 1978.
35. **Fish JE and Norman PS**. Effects of the calcium channel blocker, verapamil, on asthmatic airway responses to muscarinic, histaminergic, and allergenic stimuli. *Am Rev Respir Dis* 133: 730–734, 1986.
36. **Fleckenstein A, Frey M and Fleckenstein-Grun G**. Antihypertensive and arterial anticalcinotic effects of calcium antagonists. *American Journal of Cardiology* 57: 1D–10D, 1986.
37. **Fleischmann BK, Wang YX and Kotlikoff MI**. Muscarinic activation and calcium permeation of nonselective cation currents in airway myocytes. *Am J Physiol* 272: C341–C349, 1997.
38. **Fleischmann BK, Wang YX, Pring M and Kotlikoff MI**. Voltage-dependent calcium currents and cytosolic calcium in equine airway myocytes. *J Physiol* 492 (Pt 2): 347–358, 1996.
39. **Foster RW, Okpalugo BI and Small RC**. Antagonism of Ca^{2+} and other actions of verapamil in guinea-pig isolated trachealis. *Br J Pharmacol* 81: 499–507, 1984.
40. **Foster RW, Small RC and Weston AH**. Evidence that the spasmogenic action of tetraethylammonium in guinea- pig trachealis is both direct and dependent on the cellular influx of calcium ion. *Br J Pharmacol* 79: 255–263, 1983.

41. **Ghayur MN, Gilani AH and Janssen LJ**. Ginger attenuates acetylcholine-induced contraction and Ca^{2+} signalling in murine airway smooth muscle cells. *Can J Physiol Pharmacol* 86: 264–271, 2008.
42. **Ghayur MN, Gilani AH, Rasheed H, Khan A, Iqbal Z, Ismail M, Saeed SA and Janssen LJ**. Cardiovascular and airway relaxant activities of peony root extract. *Can J Physiol Pharmacol* 86: 793–803, 2008.
43. **Gordon EH, Wong SC and Klaustermeyer WB**. Comparison of nifedipine with a new calcium channel blocker, flordipine, in exercise-induced asthma. *J Asthma* 24: 261–265, 1987.
44. **Hartmann V and Magnussen H**. Effect of diltiazem on histamine- and carbachol-induced bronchospasm in normal and asthmatic subjects. *Chest* 87: 174–179, 1985.
45. **Hay DW, Luttmann MA, Muccitelli RM and Goldie RG**. Endothelin receptors and calcium translocation pathways in human airways. *Naunyn Schmiedebergs Arch Pharmacol* 359: 404–410, 1999.
46. **Hirota S, Helli P and Janssen LJ**. Ionic mechanisms and Ca^{2+} handling in airway smooth muscle. *Eur Respir J* 30: 1400–1419, 2007.
47. **Hirota S, Helli PB, Catalli A, Chew A and Janssen LJ**. Airway smooth muscle excitation-contraction coupling and airway hyperresponsiveness. *Can J Physiol Pharmacol* 83: 725–732, 2005.
48. **Hirota S and Janssen LJ**. Store-refilling involves both L-type calcium channels and reverse-mode sodium-calcium exchange in airway smooth muscle. *Eur Respir J* 30: 269–278, 2007.
49. **Hirota S, Pertens E and Janssen LJ**. The reverse mode of the Na^{+}/Ca^{2+} exchanger provides a source of Ca^{2+} for store refilling following agonist-induced Ca^{2+} mobilization. *Am J Physiol Lung Cell Mol Physiol* 292: L438–L447, 2007.
50. **Hisada T, Kurachi Y and Sugimoto T**. Properties of membrane currents in isolated smooth muscle cells from guinea-pig trachea. *Pflugers Arch* 416: 151–161, 1990.
51. **Honda K, Satake T, Takagi K and Tomita T**. Effects of relaxants on electrical and mechanical activities in the guinea-pig tracheal muscle. *Br J Pharmacol* 87: 665–671, 1986.
52. **Honda K and Tomita T**. Electrical activity in isolated human tracheal muscle. *Jpn J Physiol* 37: 333–336, 1987.
53. **Hyvelin JM, Martin C, Roux E, Marthan R and Savineau JP**. Human isolated bronchial smooth muscle contains functional ryanodine/caffeine-sensitive Ca-release channels. *Am J Respir Crit Care Med* 162: 687–694, 2000.
54. **Janssen LJ**. Acetylcholine and caffeine activate Cl- and suppress K^{+} conductances in human bronchial smooth muscle. *Am J Physiol* 270: L772–L781, 1996.
55. **Janssen LJ**. T-type and L-type Ca^{2+} currents in canine bronchial smooth muscle: characterization and physiological roles. *Am J Physiol* 272: C1757–C1765, 1997.
56. **Janssen LJ**. Ionic mechanisms and Ca^{2+} regulation in airway smooth muscle contraction: do the data contradict dogma? *Am J Physiol Lung Cell Mol Physiol* 282: L1161–L1178, 2002.
57. **Janssen LJ and Daniel EE**. Characterization of the prejunctional beta adrenoceptors in canine bronchial smooth muscle. *J Pharmacol Exp Ther* 254: 741–749, 1990.
58. **Janssen LJ and Daniel EE**. Pre- and postjunctional muscarinic receptors in canine bronchi. *Am J Physiol* 259: L304–L314, 1990.
59. **Janssen LJ and Daniel EE**. Classification of postjunctional beta adrenoceptors mediating relaxation of canine bronchi. *J Pharmacol Exp Ther* 256: 670–676, 1991.
60. **Janssen LJ and Daniel EE**. Depolarizing agents induce oscillations in canine bronchial smooth muscle membrane potential: possible mechanisms. *J Pharmacol Exp Ther* 259: 110–117, 1991.
61. **Janssen LJ and Daniel EE**. Pre- and postjunctional effects of a thromboxane mimetic in canine bronchi. *Am J Physiol* 261: L271–L276, 1991.

62. **Janssen LJ, Farkas L, Rahman T and Kolb MR**. ATP stimulates Ca^{2+}-waves and gene expression in cultured human pulmonary fibroblasts. *Int J Biochem Cell Biol* 41: 2477–2484, 2009.
63. **Janssen LJ, Hague C and Nana R**. Ionic mechanisms underlying electrical slow waves in canine airway smooth muscle. *Am J Physiol* 275: L516–L523, 1998.
64. **Janssen LJ and Killian K**. Airway smooth muscle as a target of asthma therapy: history and new directions. *Respir Res* 7: 123, 2006.
65. **Janssen LJ and Sims SM**. Acetylcholine activates non-selective cation and chloride conductances in canine and guinea-pig tracheal myocytes. *J Physiol* 453: 197–218, 1992.
66. **Janssen LJ and Sims SM**. Emptying and refilling of Ca^{2+} store in tracheal myocytes as indicated by ACh-evoked currents and contraction. *Am J Physiol* 265: C877–C886, 1993.
67. **Janssen LJ, Wattie J, Lu-Chao H and Tazzeo T**. Muscarinic excitation-contraction coupling mechanisms in tracheal and bronchial smooth muscles. *J Appl Physiol* 91: 1142–1151, 2001.
68. **Kaliner MA**. Evolution of asthma treatments. *Ann Allergy* 71: 300–305, 1993.
69. **Kirkpatrick CT**. Excitation and contraction in bovine tracheal smooth muscle. *J Physiol* 244: 263–281, 1975.
70. **Kondo T, Tamura K, Onoe K, Takahira H, Ohta Y and Yamabayashi H**. In vivo recording of electrical activity of canine tracheal smooth muscle. *J Appl Physiol* 72: 135–142, 1992.
71. **Kotlikoff MI**. Calcium currents in isolated canine airway smooth muscle cells. *Am J Physiol* 254: C793–C801, 1988.
72. **Kuo KH, Dai J, Seow CY, Lee CH and van Breemen C**. Relationship between asynchronous Ca^{2+} waves and force development in intact smooth muscle bundles of the porcine trachea. *Am J Physiol Lung Cell Mol Physiol* 285: L1345–L1353, 2003.
73. **Liu B, Peel SE, Fox J and Hall IP**. Reverse mode Na^{+}/Ca^{2+} exchange mediated by STIM1 contributes to Ca^{2+} influx in airway smooth muscle following agonist stimulation. *Respir Res* 11: 168, 2010.
74. **Liu C, Zuo J, Pertens E, Helli PB and Janssen LJ**. Regulation of Rho/ROCK signaling in airway smooth muscle by membrane potential and $[Ca^{2+}]_i$. *Am J Physiol Lung Cell Mol Physiol* 289: L574–L582, 2005.
75. **Liu X and Farley JM**. Depletion and refilling of acetylcholine- and caffeine-sensitive Ca^{2+} stores in tracheal myocytes. *J Pharmacol Exp Ther* 277: 789–795, 1996.
76. **Mahn K, Hirst SJ, Ying S, Holt MR, Lavender P, Ojo OO, Siew L, Simcock DE, McVicker CG, Kanabar V, Snetkov VA, O'connor BJ, Karner C, Cousins DJ, Macedo P, Chung KF, Corrigan CJ, Ward JP and Lee TH**. Diminished sarco/endoplasmic reticulum Ca^{2+} ATPase (SERCA) expression contributes to airway remodelling in bronchial asthma. *Proc Natl Acad Sci U S A* 106: 10775–10780, 2009.
77. **Marthan R, Martin C, Amedee T and Mironneau J**. Calcium channel currents in isolated smooth muscle cells from human bronchus. *J Appl Physiol* 66: 1706–1714, 1989.
78. **McCaig DJ**. Electrophysiology of neuroeffector transmission in the isolated, innervated trachea of the guinea-pig. *Br J Pharmacol* 89: 793–801, 1986.
79. **Middleton E Jr**. Airway smooth muscle, asthma, and calcium ions. *J Allergy Clin Immunol* 73: 643–650, 1984.
80. **Middleton E Jr**. Calcium antagonists and asthma. *J Allergy Clin Immunol* 76: 341–346, 1985.
81. **Middleton E Jr**. The treatment of asthma–beyond bronchodilators. *N Engl Reg Allergy Proc* 6: 235–237, 1985.
82. **Miller JD, Cox G, Vincic L, Lombard CM, Loomas BE and Danek CJ**. A prospective feasibility study of bronchial thermoplasty in the human airway. *Chest* 127: 1999–2006, 2005.
83. **Mitzner W**. Airway smooth muscle: the appendix of the lung. *Am J Respir Crit Care Med* 169: 787–790, 2004.

84. **Murlas CG and Doupnik CA**. Electromechanical coupling of ferret airway smooth muscle in response to leukotriene C4. *J Appl Physiol* 66: 2533–2538, 1989.
85. **Nuttle LC and Farley JM**. Frequency modulation of acetylcholine-induced oscillations in Ca^{2+} and Ca^{2+}-activated Cl- current by cAMP in tracheal smooth muscle. *J Pharmacol Exp Ther* 277: 753–760, 1996.
86. **O'Byrne PM, Barnes PJ, Rodriguez-Roisin R, Runnerstrom E, Sandstrom T, Svensson K and Tattersfield A**. Low dose inhaled budesonide and formoterol in mild persistent asthma: the OPTIMA randomized trial. *Am J Respir Crit Care Med* 164: 1392–1397, 2001.
87. **Patel KR**. Calcium antagonists in exercise-induced asthma. *Br Med J (Clin Res Ed)* 282: 932–933, 1981.
88. **Patel KR**. The effect of calcium antagonist, nifedipine in exercise-induced asthma. *Clin Allergy* 11: 429–432, 1981.
89. **Pavord ID, Cox G, Thomson NC, Rubin AS, Corris PA, Niven RM, Chung KF and Laviolette M**. Safety and efficacy of bronchial thermoplasty in symptomatic, severe asthma. *Am J Respir Crit Care Med* 176: 1185–1191, 2007.
90. **Perez JF and Sanderson MJ**. The frequency of calcium oscillations induced by 5-HT, ACH, and KCl determine the contraction of smooth muscle cells of intrapulmonary bronchioles. *J Gen Physiol* 125: 535–553, 2005.
91. **Perez-Zoghbi JF and Sanderson MJ**. Endothelin-induced contraction of bronchiole and pulmonary arteriole smooth muscle cells is regulated by intracellular Ca^{2+} oscillations and Ca^{2+} sensitization. *Am J Physiol Lung Cell Mol Physiol* 293: L1000–L1011, 2007.
92. **Prakash YS, Kannan MS, Walseth TF and Sieck GC**. Role of cyclic ADP-ribose in the regulation of $[Ca^{2+}]i$ in porcine tracheal smooth muscle. *Am J Physiol* 274: C1653–C1660, 1998.
93. **Prakash YS, Pabelick CM, Kannan MS and Sieck GC**. Spatial and temporal aspects of ACh-induced $[Ca^{2+}]_i$ oscillations in porcine tracheal smooth muscle. *Cell Calcium* 27: 153–162, 2000.
94. **Qian Y and Bourreau JP**. Two distinct pathways for refilling Ca^{2+} stores in permeabilized bovine trachealis muscle. *Life Sci* 64: 2049–2059, 1999.
95. **Quayle JM, Nelson MT and Standen NB**. ATP-sensitive and inwardly rectifying potassium channels in smooth muscle. *Physiol Rev* 77: 1165–1232, 1997.
96. **Rafferty P, Varley JG, Edwards JS and Holgate ST**. Inhibition of exercise-induced asthma by nifedipine: a dose–response study. *Br J Clin Pharmacol* 24: 479–484, 1987.
97. **Rahman M, Inman M, Kiss L and Janssen LJ**. Reverse-mode NCX-current in mouse airway smooth muscle: Na^+- and voltage-dependence, contributions to Ca^{2+}-influx and contraction, and altered expression in a model of allergen-induced hyperresponsiveness. *Acta Physiol (Oxf)* 2011.
98. **Roux E, Guibert C, Savineau JP and Marthan R**. $[Ca^{2+}]_i$ oscillations induced by muscarinic stimulation in airway smooth muscle cells: receptor subtypes and correlation with the mechanical activity. *Br J Pharmacol* 120: 1294–1301, 1997.
99. **Roux E, Noble PJ, Noble D and Marhl M**. Modelling of calcium handling in airway myocytes. *Prog Biophys Mol Biol* 90: 64–87, 2006.
100. **Sears MR**. The evolution of beta2-agonists. *Respir Med* 95 Suppl B: S2–S6, 2001
101. **Sears MR, Taylor DR, Print CG, Lake DC, Li QQ, Flannery EM, Yates DM, Lucas MK and Herbison GP**. Regular inhaled beta-agonist treatment in bronchial asthma. *Lancet* 336: 1391–1396, 1990.
102. **SILBERT NE**. Treatment of bronchial conditions with a combination of anticholinergic, adrenergic, and antihistaminic drugs; preliminary report. *Ann Allergy* 10: 469–472, 1952.
103. **Small RC**. Electrical slow waves and tone of guinea-pig isolated trachealis muscle: effects of drugs and temperature changes. *Br J Pharmacol* 77: 45–54, 1982.
104. **Small RC, Berry JL, Foster RW, Blarer S and Quast U**. Analysis of the relaxant action of SDZ PCO 400 in airway smooth muscle from the ox and guinea-pig. *Eur J Pharmacol* 219: 81–88, 1992.

105. **Somlyo AP and Somlyo AV**. From pharmacomechanical coupling to G-proteins and myosin phosphatase. *Acta Physiol Scand* 164: 437–448, 1998.
106. **Souhrada M, Souhrada JF and Cherniack RM**. Evidence for a sodium electrogenic pump in airway smooth muscle. *J Appl Physiol* 51: 346–352, 1981.
107. **Tazzeo T, Zhang Y, Keshavjee S and Janssen LJ**. Ryanodine receptors decant internal Ca^{2+} store in human and bovine airway smooth muscle. *Eur Respir J* 2008.
108. **Uyama Y, Imaizumi Y and Watanabe M**. Effects of cyclopiazonic acid, a novel Ca^{2+}-ATPase inhibitor, on contractile responses in skinned ileal smooth muscle. *Br J Pharmacol* 106: 208–214, 1992.
109. **Wade GR, Barbera J and Sims SM**. Cholinergic inhibition of Ca^{2+} current in guinea-pig gastric and tracheal smooth muscle cells. *J Physiol* 491 (Pt 2): 307–319, 1996.
110. **Wang Y, Deng X, Mancarella S, Hendron E, Eguchi S, Soboloff J, Tang XD and Gill DL**. The calcium store sensor, STIM1, reciprocally controls Orai and CaV1.2 channels. *Science* 330: 105–109, 2010.
111. **Wang YX, Fleischmann BK and Kotlikoff MI**. M2 receptor activation of nonselective cation channels in smooth muscle cells: calcium and Gi/G(o) requirements. *Am J Physiol* 273: C500–C508, 1997.
112. **Wang YX and Kotlikoff MI**. Muscarinic signaling pathway for calcium release and calcium-activated chloride current in smooth muscle. *Am J Physiol* 273: C509–C519, 1997.
113. **Wang YX, Zheng YM, Mei QB, Wang QS, Collier ML, Fleischer S, Xin HB and Kotlikoff MI**. FKBP12.6 and cADPR regulation of Ca^{2+} release in smooth muscle cells. *Am J Physiol Cell Physiol* 286: C538–C546, 2004.
114. **Welling A, Felbel J, Peper K and Hofmann F**. Hormonal regulation of calcium current in freshly isolated airway smooth muscle cells. *Am J Physiol* 262: L351–L359, 1992.
115. **White TA, Johnson S, Walseth TF, Lee HC, Graeff RM, Munshi CB, Prakash YS, Sieck GC and Kannan MS**. Subcellular localization of cyclic ADP-ribosyl cyclase and cyclic ADP- ribose hydrolase activities in porcine airway smooth muscle. *Biochim Biophys Acta* 1498: 64–71, 2000.
116. **Worley JF and Kotlikoff MI**. Dihydropyridine-sensitive single calcium channels in airway smooth muscle cells. *Am J Physiol* 259: L468–L480, 1990.
117. **Yamaguchi H, Kajita J and Madison JM**. Isoproterenol increases peripheral $[Ca^{2+}]_i$ and decreases inner $[Ca^{2+}]_i$ in single airway smooth muscle cells. *Am J Physiol* 268: C771–C779, 1995.
118. **Yamakage M, Hirshman CA and Croxton TL**. Cholinergic regulation of voltage-dependent Ca^{2+} channels in porcine tracheal smooth muscle cells. *Am J Physiol* 269: L776–L782, 1995.
119. **Yang CM, Yo YL and Wang YY**. Intracellular calcium in canine cultured tracheal smooth muscle cells is regulated by M3 muscarinic receptors. *Br J Pharmacol* 110: 983–988, 1993.
120. **Zheng XF, Guan YY and Kwan CY**. Cyclopiazonic acid causes endothelium-dependent relaxation in rat aorta. *Chung-Kuo Yao Li Hsueh Pao - Acta Pharmacologica Sinica* 14: 21–26, 1993.

The Role of Mitochondria in Calcium Regulation in Airway Smooth Muscle

Philippe Delmotte, Li Jia, and Gary C. Sieck

Contents

Abstract Excitation–contraction coupling represents a cascade of events that connects the initiating signal, an elevation of cytosolic Ca^{2+} concentration ($[Ca^{2+}]_{cyt}$), with the ensuing cross-bridge recruitment and cycling that underlies ATP-consuming mechanical work. Mitochondria are the power plant of the cell. An elevation in $[Ca^{2+}]_{cyt}$ is followed by an increase in mitochondrial Ca^{2+} ($[Ca^{2+}]_{mito}$). Moreover, an increase in $[Ca^{2+}]_{mito}$ leads to an increase in ATP production. Therefore, the relationship between elevations in $[Ca^{2+}]_{cyt}$ and $[Ca^{2+}]_{mito}$ reflects a coupling between energy demand and energy supply. Accordingly, mitochondrial Ca^{2+} regulation mediates excitation–energy

P. Delmotte • L. Jia • G.C. Sieck, Ph.D. (✉)
Department of Physiology and Biomedical Engineering, Mayo Clinic College of Medicine, 4-184 W Joseph SMH, 200 First Street SW, Rochester, MN 55905, USA
e-mail: Sieck.Gary@mayo.edu

Y.-X. Wang (ed.), *Calcium Signaling In Airway Smooth Muscle Cells*,
DOI 10.1007/978-3-319-01312-1_11, © Springer International Publishing Switzerland 2014

coupling in ASM. To this end, mitochondrial movement within ASM cells may help couple ATP production to localized differences in energy demand.

Keywords Excitation–Contraction Coupling • ATP • Metabolism • Mitochondrial Ca^{2+} • Cytosolic Ca^{2+}

1 Introduction

Airway smooth muscle (ASM) is recognized as being a major end effector of acute airway narrowing in asthma [42, 123, 124, 126]. The mechanisms that underlie the enhanced ASM contractility observed in inflamed airway in conditions such as asthma remain unknown [91, 108, 124, 126]. In this context, it is important to better understand the mechanisms underlying excitation–contraction coupling in ASM, which describes the physiological process of converting a stimulus into a mechanical response (i.e., contraction) [118, 123, 124].

Excitation–contraction coupling involves both an increase in cytosolic Ca^{2+} concentration ($[Ca^{2+}]_{cyt}$) and an increase in the demand for energy in the form of adenosine triphosphate (ATP). In this regard, mitochondria have a tremendous potential to exert a central role in excitation–contraction coupling.

It is well known that mitochondria produce most of the ATP used in mammalian cells. Mitochondrial pathways leading to ATP production [e.g., intramitochondrial dehydrogenases associated with the tricarboxylic acid (TCA) cycle] are Ca^{2+} dependent [35]. Accordingly, the relationship between cytosolic and mitochondrial Ca^{2+} has attracted considerable interest [13, 17, 33, 37, 38, 49, 51].

The regulation of $[Ca^{2+}]_{cyt}$ is a complex process that involves sarcoplasmic reticulum (SR) Ca^{2+} release and reuptake and Ca^{2+} plasma membrane fluxes [7, 8, 54, 62, 91, 103, 105, 108, 120, 124, 125, 127]. It has been shown in many cell types, including ASM, that an increase in $[Ca^{2+}]_{cyt}$ is followed by an increase in intramitochondrial Ca^{2+} concentration ($[Ca^{2+}]_{mito}$) [23, 32, 36, 38]. Significantly, an increase in $[Ca^{2+}]_{mito}$ plays an important role in the control of the rate of ATP production [13, 46, 50]. In summary, an increase in $[Ca^{2+}]_{cyt}$ leads to an increase of $[Ca^{2+}]_{mito}$, which stimulates ATP synthesis. In this regard, mitochondria exert an excitation–energy coupling, which mirrors excitation–contraction coupling with Ca^{2+} as a key regulator. Recent studies in ASM cells have also shown that changes in $[Ca^{2+}]_{mito}$ regulation can alter $[Ca^{2+}]_{cyt}$ [23, 32], suggesting a close relationship between both Ca^{2+} signals and potential feedback mechanisms [13]. Mitochondria also display dynamic tubular networks undergoing constant fission and fusion, which are essential to maintain normal morphology and mitochondrial function [22, 88]. Recent studies suggest that Ca^{2+} plays a primary role in modulating mitochondrial dynamics (fission and fusion events, movement). The dynamics of mitochondria are believed to couple ATP synthesis to localized differences in energy demand within the cell [13, 22, 50, 88].

2 Mechanisms Underlying $[Ca^{2+}]_{cyt}$ Response

An increase in $[Ca^{2+}]_{cyt}$ is the main initiating step in excitation–contraction coupling, leading to cross-bridge formation and muscle contraction in response to agonist stimulation. Agonists induce a transient increase in $[Ca^{2+}]_{cyt}$ via multiple mechanisms, including SR Ca^{2+} release and plasma membrane Ca^{2+} influx. The subsequent decrease in $[Ca^{2+}]_{cyt}$ is achieved by SR Ca^{2+} reuptake and plasma membrane Ca^{2+} efflux.

2.1 SR Ca^{2+} Release

Agonists binding to G_q/G_{11} protein- or tyrosine-kinase-coupled receptors activate phospholipase C (PLC) that cleaves phosphatidylinositol 4,5-biphosphate (PIP_2) into diacylglycerol (DAG) and inositol 1,4,5-trisphosphate (IP_3). IP_3 is soluble and diffuses through the cytosol, while DAG remains bound to the membrane. Activation of $G_{\alpha q}$ and $G_{\alpha i}$-type G-protein-coupled receptors leads to the production of cyclic ADP-ribose (cADPR) synthesized from NAD^+ by the bifunctional ectoenzymes of the CD38 family (ADP-ribosyl cyclases) (Fig. 1). IP_3 and cADPR in turn trigger release of Ca^{2+} from the SR through IP_3 receptor (IP_3R) and ryanodine receptor (RyR) channels, respectively [3, 69, 105, 106, 127]. The initial release of Ca^{2+} from the SR activates nearby clusters of IP_3R and RyR channels to amplify the elevation in $[Ca^{2+}]_{cyt}$, an amplification and propagation process called Ca^{2+}-induced Ca^{2+} release (CICR) [105, 124] (Fig. 1). The kinetics of IP_3R and RyR channel opening and closing are modulated by $[Ca^{2+}]_{cyt}$ levels where $[Ca^{2+}]_{cyt}$ has the potential for either positive or negative feedback – the relationship of channel open probability to $[Ca^{2+}]_{cyt}$ is represented by a bell-shaped curve. The open probability of the RyR channels can also be modulated by cADPR as well as ryanodine, caffeine, and FK 506-binding proteins [68, 106, 127, 140, 145].

The role of the CD38-cADPR-RyR pathway in SR Ca^{2+} release in ASM is not as well characterized as IP_3 [3, 69, 105, 106, 127]. It appears that agonists binding to $G_{\alpha i}$-protein-coupled receptors activate CD38 [67], which is a bifunctional ectoenzyme that mediates the synthesis of cADPR from nicotinamide adenine dinucleotide (NAD^+) via ADP-ribosyl cyclase activity [110]. CD38 is also responsible for the hydrolysis of cADPR to ADPR. cADPR is soluble and diffuses through the cytosol to interact with RyRs at the SR to increase their open probability. It is likely that this interaction with SR RyRs is mediated by removal of FKPB12.6, which normally binds to RyR2 [139]. Suppression of CD38 by siRNA decreases ADP-ribosyl cyclase activity, cADPR levels, and Ca^{2+} response to various agonists

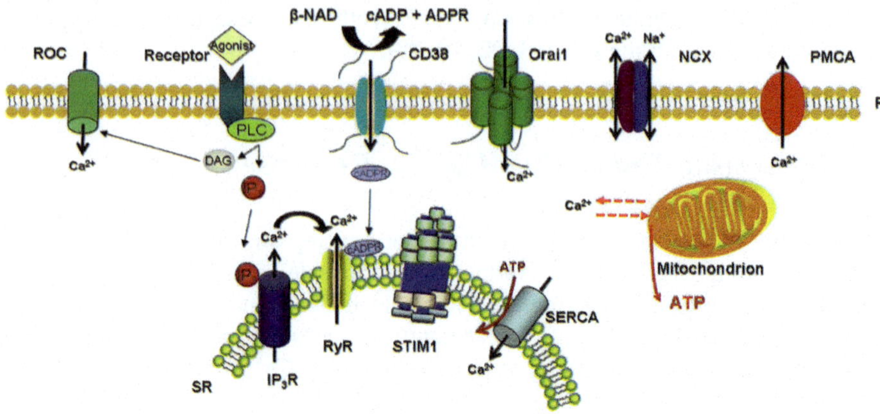

Fig. 1 *Schematic illustration of major mechanisms involved in agonist-induced Ca^{2+} increase.* Agonists binding to phospholipase C (*PLC*)-coupled receptor lead to production of inositol 1,4,5-trisphosphate (*IP_3*) and diacylglycerol (*DAG*). Agonist stimulation also leads to production of cyclic ADP-ribose (*cADPR*) by the ADP-ribosyl cyclase CD38. IP_3 and cADPR trigger release of Ca^{2+} from the sarcoplasmic reticulum (*SR*) through IP_3 receptor (*IP_3R*) channels and ryanodine receptor (*RyR*) channels, respectively. DAG may activate receptor-operated channels. The initial elevation in $[Ca^{2+}]_{cyt}$ consequently activates RyRs to amplify the elevation in $[Ca^{2+}]_{cyt}$, referred to as Ca^{2+}-induced Ca^{2+} release (*CICR*). The depletion of SR Ca^{2+} causes STIM1 aggregation and activates store-operated channel (*SOC*) Orai1, trigging store-operated Ca^{2+} entry. The increase in $[Na^{+}]$ due to the activation of SOC drives Na^{+}/Ca^{2+} exchange by NCX on the plasma membrane, resulting in Ca^{2+} influx. Increase in $[Ca^{2+}]_{cyto}$ leads to an increase in $[Ca^{2+}]_{mito}$ and stimulates ATP synthesis by the mitochondria. The SR Ca^{2+}-ATPases (SERCA), NCX, the plasma membrane Ca^{2+} ATPases (PMCA) and mitochondrial Ca^{2+} uptake are involved in reducing $[Ca^{2+}]_{cyto}$. Additionally, mitochondrial Ca^{2+} uptake and release near Ca^{2+} channels might affect the kinetics of these channels

such as acetylcholine (ACh), histamine, and bradykinin [127]. cADPR-induced elevation of $[Ca^{2+}]_{cyt}$ in human ASM is blocked by the cADPR receptor antagonist and by RyR channels blockers, but not by IP_3R channel blockers [106].

2.2 SR Ca^{2+} Reuptake

Following agonist stimulation, sarcoplasmic/endoplasmic reticulum Ca^{2+} ATPase (SERCA) is the major mechanism for replenishing the SR Ca^{2+} store, with two Ca^{2+} ions pumped into the SR for every ATP consumed against a gradient of approximately 1 mM in the SR lumen [79]. The mammalian SERCA family contains three tissue-specific members, SERCA1, SERCA2, and SERCA3, of which SERCA2 is expressed in muscle cells. The predominant isoform expressed in ASM is SERCA2B [76]. Regulation of SERCA activity controls the extent of SR Ca^{2+} reuptake and, thereby, the subsequent $[Ca^{2+}]_{cyt}$ response [121, 122]. SERCA activity is inhibited by phospholamban (PLB) in cardiac muscle [134] and in various smooth muscles [111, 112]. However, in human ASM, PLB protein is not expressed, although PLB mRNA is detected [122]. Recently it was shown that SERCA was phosphorylated via calmodulin-dependent protein kinase II (CaMKII) in human ASM and that this enhances SERCA activity [122]. When human ASM cells were exposed to the CaMKII antagonist KN-93, the rate of fall of ACh- or bradykinin-induced $[Ca^{2+}]_{cyt}$ transients was slowed, indicating CaMKII-mediated SERCA regulation. Exposure of human ASM to proinflammatory cytokines (TNFα and IL-13) decreases SERCA expression and slows the rate of fall of agonist-induced $[Ca^{2+}]_{cyt}$ transients. However, cytokine exposure does not appear to affect CaMKII-mediated SERCA phosphorylation [122].

2.3 *Plasma Membrane Ca^{2+} Fluxes*

In addition to SR Ca^{2+} release, Ca^{2+} influx occurs through voltage-gated [143] and receptor-gated [86] Ca^{2+} channels, as well as in response to agonist-induced SR Ca^{2+} depletion [i.e., store-operated Ca^{2+} entry (SOCE) or capacitative Ca^{2+} entry] [3, 89, 127].

After agonist-induced elevation of $[Ca^{2+}]_{cyt}$ the subsequent decline in $[Ca^{2+}]_{cyt}$ is primarily mediated by SERCA; however, plasma membrane Ca^{2+} ATPases and the bidirectional plasma membrane Na^+/Ca^{2+} exchanger (NCX) also contribute to the reduction in $[Ca^{2+}]_{cyt}$ [57, 62, 63, 74, 120, 121] (Fig. 1).

2.3.1 Receptor-Gated and Voltage-Gated Channels

Agonists binding to G-coupled receptors stimulate the production of DAG or its derivatives such as arachidonic acid to activate receptor-operated, nonselective cation channels (permeable to both Na^+ and Ca^{2+}). Plasma membrane Ca^{2+} influx through these channels may precede or accompany Ca^{2+} influx in response to SR Ca^{2+} depletion (SOCE) [54, 81, 85, 86]. Not all the molecular identities of all receptor-operated channels in ASM are fully characterized. Transient receptor potential (TRP) proteins appear to constitute some or part of these Ca^{2+} channels [114]. The TRP protein superfamily comprises multiple subfamily members. In ASM, three subfamilies (TRPC, TRPM, and TRPV) have been proposed to form agonist-activated, cation-permeable channels that participate in mediating Ca^{2+} influx [48]. The permeability to cations differs greatly across the various family members [114]. TRP proteins assemble into homo- or heteromeric tetramers to form channels [102] that are activated by a wide variety of stimuli and may be regulated by growth factors and cytokines [64, 71, 141].

Another major mechanism for Ca^{2+} influx across the plasma membrane are voltage-operated Ca^{2+} channels (L-type) [57, 120]. Activation of voltage-dependent Ca^{2+} influx alone does not trigger contraction in ASM. However, voltage-dependent Ca^{2+} influx very likely contributes to the initial elevation of $[Ca^{2+}]_{cyt}$ induced by agonists, and such voltage-dependent Ca^{2+} influx may trigger subsequent Ca^{2+}-induced SR Ca^{2+} release as well as contribute to SR Ca^{2+} refilling [7, 62].

2.3.2 Na^{+}/Ca^{2+} Exchanger

NCX is an ion transporter expressed in the plasma membrane that exchanges Na^{+} and Ca^{2+} in either Ca^{2+} efflux or Ca^{2+} influx modes, depending on the transmembrane ion gradients for Na^{+} and Ca^{2+} and membrane potential [9]. In the bidirectional exchange mode, NCX uses the energy within the trans-plasma membrane Na^{+} gradient to exchange one Ca^{2+} for three Na^{+} ions. NCX normally utilizes the Na^{+} concentration gradient across the cell membrane to move Ca^{2+} out of the cell. However, accumulation of Na^{+} on the inner surface of the plasma membrane with depolarization can reduce Ca^{2+} extrusion and even force the NCX into the reverse Ca^{2+} influx mode [57, 74, 120]. In ASM, SR Ca^{2+} store depletion activates SOCE channels. These channels are permeable to both Na^{+} and Ca^{2+}. Elevation of Na^{+} therefore causes a reverse mode of the NCX and Ca^{2+} entry [57]. Both inward exchange and outward exchange modes of NCX have been examined in human ASM and contribute to $[Ca^{2+}]_{cyt}$ responses to agonist stimulation [120]. Furthermore, NCX Ca^{2+} influx mode may be linked to SR Ca^{2+} store depletion via the regulatory protein STIM1 in ASM [74]. NCX has been found to contribute to enhanced $[Ca^{2+}]_{cyt}$ response to agonists in inflamed airways [120].

2.3.3 Store-Operated Calcium Entry

Contractile agonists induce SR Ca^{2+} release and depletion of SR Ca^{2+}, triggering subsequent Ca^{2+} influx via Ca^{2+} release-activated Ca^{2+} (CRAC) channels located in the plasma membrane and SOCE [3, 103, 127]. SOCE is generally slow, with delays ranging from tens to hundreds of seconds following IP_3-induced SR Ca^{2+} release [101]. The stromal interaction molecule (STIM) proteins were identified as an essential intracellular link between the SR and the plasma membrane [73, 115]. Two mammalian homologs, STIM1 and STIM2, have been identified [100, 142]. STIM proteins function as sensors for SR $[Ca^{2+}]$ and with SR Ca^{2+} depletion, STIM proteins aggregate (oligomerize) adjacent to the plasma, where they organize the highly Ca^{2+} selective Orai channels into clusters and activate these channels to initiate SOCE [16] (Fig. 2). Orai proteins form Ca^{2+}-selective channels across the plasma membrane and participate in activating SOCE [43, 130] (Fig. 2). Three Orai homologs – Orai1, Orai2, and Orai3 – have been identified in human ASM cells [103]. Orai1 was identified as being critical for the activation of SOCE in human ASM since siRNA suppression substantially inhibits SOCE, while suppression of Orai2 or Orai3 does not appear to affect SOCE [103]. Orai1 is a pore-forming subunit of CRAC channels [109]. Orai1 is predominantly expressed in the plasma membrane in human ASM [119] and interacts with STIM1 after store depletion [83].

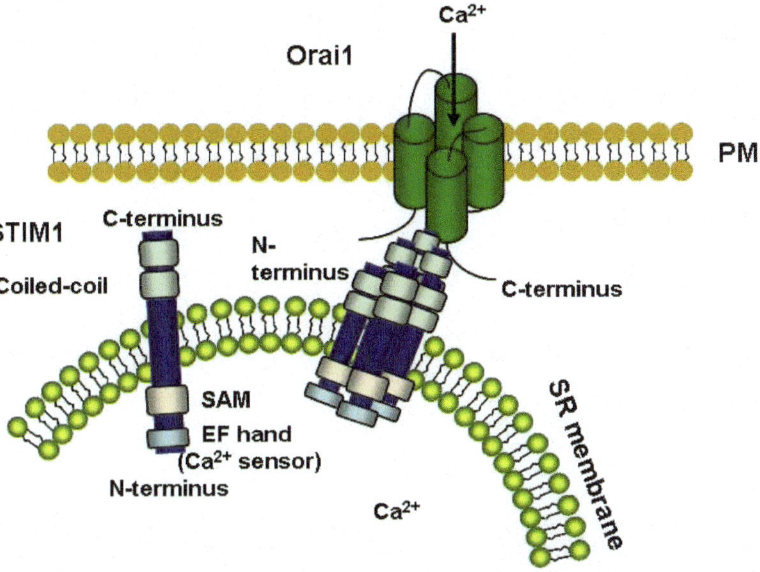

Fig. 2 *STIM1 and Orai1*. STIM1 is located in the sarcoplasmic reticulum (*SR*) membrane. The N-terminus of STIM1 toward the SR lumen contains a Ca^{2+}-binding EF-hand motif and SAM domains. Orai1 is a four-transmembrane segment protein localized in the plasma membrane (*PM*) with its N- and C-termini extended to the cytoplasm. STIM1 senses SR Ca^{2+} store depletion via EF-hand motif. The depletion of the SR Ca^{2+} store causes STIM1 oligomerization and interacts with Orai1 at the SR–plasma membrane junction, triggering store-operated Ca^{2+} entry

3 Relationship Between Cytosolic and Mitochondrial Ca^{2+} Responses

In ASM, as in many others cell types, an increase in $[Ca^{2+}]_{cyt}$ is temporally related to an increase in $[Ca^{2+}]_{mito}$ (~10 s) (Fig. 3) and is often referred to as mitochondrial Ca^{2+} buffering [23–25, 32, 34, 56, 61, 70, 117, 133]. However, the idea that mitochondrial Ca^{2+} uptake might serve to buffer $[Ca^{2+}]_{cyt}$ is controversial. Inhibition of mitochondrial Ca^{2+} uptake increases $[Ca^{2+}]_{cyt}$ in ASM [32]. Indeed, mitochondrial Ca^{2+} uptake or mitochondrial Ca^{2+} release in proximity to SR IP_3 or RyR Ca^{2+} channels may affect their open probability and thereby affect local $[Ca^{2+}]_{cyt}$ (referred as Ca^{2+} microdomains). Similarly, mitochondrial Ca^{2+} uptake or release may affect NCX at the plasma membrane.

More importantly, mitochondrial Ca^{2+} is a potent regulator of the metabolic pathways involved in ATP synthesis. Therefore, while an elevation of $[Ca^{2+}]_{cyt}$ is an essential component of excitation–contraction coupling, a corresponding increase in $[Ca^{2+}]_{mito}$ may be important in coupling excitation and its associated increase in ATP consumption to mitochondrial ATP production (i.e., excitation–energy coupling).

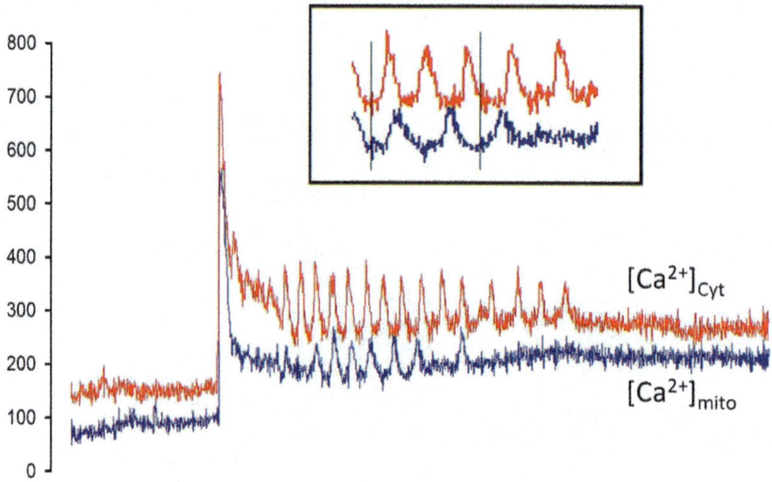

Fig. 3 *An increase in [Ca^{2+}]$_{cyt}$ is temporally related to an increase in [Ca^{2+}]$_{mito}$*. In cells loaded with both fluo-3 (cytosolic Ca^{2+}, [Ca^{2+}]$_{cyt}$, *red*) and rhod-2 (mitochondrial Ca^{2+}; [Ca^{2+}]$_{mito}$, *blue*), 1 μM ACh-induced [Ca^{2+}]$_{cyt}$ oscillations that were reflected by delayed (approx. 10 s), dampened [Ca^{2+}]$_{mito}$ oscillations

3.1 Mitochondrial Ca^{2+} Transport Mechanisms

Mitochondrial Ca^{2+} transport into mitochondria involves the movement of Ca^{2+} ions across two membranes, the outer and inner membranes. Influx of Ca^{2+} across the outer membrane of mitochondria into the intermembrane space is now thought to be regulated via voltage-dependent anion channels (VDAC or mitochondrial porin) at the outer membrane [58]. Three mechanisms or modes of Ca^{2+} influx and two Ca^{2+} efflux modes have been characterized in the inner membrane [58] (Fig. 4).

3.1.1 Mitochondrial Ca^{2+} Uptake or Influx

The main mechanism of mitochondrial Ca^{2+} influx across the inner membrane is the mitochondrial Ca^{2+} uniporter (MCU) [50, 113]. MCU facilitates passive transport of Ca^{2+} down its electrochemical gradient without coupling Ca^{2+} transport to another ion. Therefore, mitochondrial Ca^{2+} uptake is driven by membrane potential ($\Delta\Psi_m$) [50, 113]. The MCU, which is sensitive to ruthenium red, presents the electrophysiological characteristics of a low-affinity but highly selective Ca^{2+} channel. An oxygen-bridged dinuclear ruthenium amine complex, ruthenium 360 (Ru360), has been widely used to specifically inhibit MCU-mediated mitochondrial Ca^{2+} uptake [58, 80]. In human ASM, Ru360 has no effect on basal [Ca^{2+}]$_{mito}$ but significantly reduces the increase in [Ca^{2+}]$_{mito}$

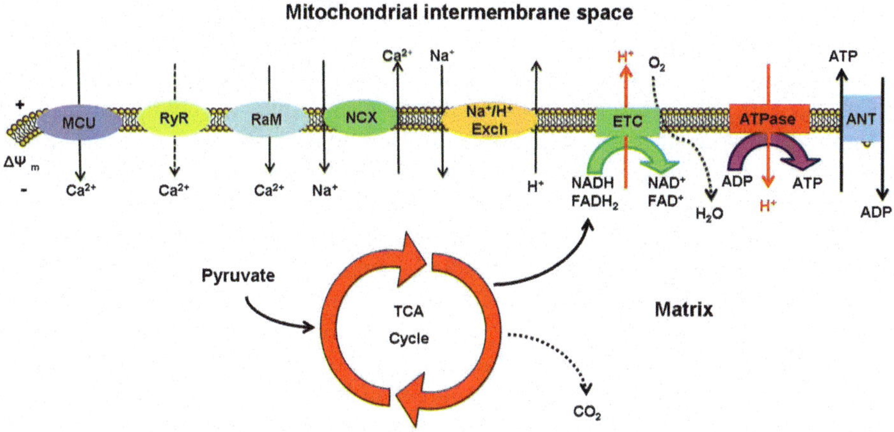

Fig. 4 *Mitochondrial Ca^{2+} transport mechanisms and ATP production in mitochondria. MCU* mitochondrial Ca^{2+} uniporter, *mRyR* mitochondrial ryanodine receptor, *RaM* rapid-uptake mode, *mNCX* mitochondrial sodium calcium exchanger, *TCA cycle* tricarboxylic acid cycle, *ETC* electron transport chain, $\Delta\Psi$ mitochondrial membrane potential, *ANT* adenine nucleotide transporter

induced by histamine [32]. Interestingly, Ru360 has a more pronounced effect on histamine-induced $[Ca^{2+}]_{mito}$ responses measured in mitochondria localized in the perinuclear area compared to mitochondria localized in more distal regions of cells [32]. The molecular identity of MCU was recently reported [5, 30], and this should lead to a better understanding of the regulation of mitochondrial Ca^{2+} uptake.

Studies using isolated mitochondria have suggested that MCU does not transport Ca^{2+} below a threshold $[Ca^{2+}]_{cyt}$ level in the 200–250 nM range [50]. Therefore, MCU-mediated mitochondrial Ca^{2+} influx is unlikely to be activated under resting or baseline $[Ca^{2+}]_{cyt}$ conditions (~100 nM). Maximal activation of MCU requires $[Ca^{2+}]_{cyt}$ levels that are much higher than normally achieved during physiological stimulation (Kd of 20–30 μM). However, agonist-induced $[Ca^{2+}]_{cyt}$ increase is followed by an increase in $[Ca^{2+}]_{mito}$ in many cell types, including human ASM [23–25, 32, 34, 56, 61, 70, 117, 133]. This apparent inconsistency was recently reconciled by the Ca^{2+} hotspot theory, where the $[Ca^{2+}]_{cyt}$ near the SR IP_3 or RyR Ca^{2+} channels and plasma membrane Ca^{2+} influx channels can transiently increase to much higher levels [17, 36, 45]. Different studies have since confirmed the existence of Ca^{2+} hotspots or microdomains where $[Ca^{2+}]_{cyt}$ was five to ten times higher than elsewhere in the cytosol and up to 100 μM [45]. Mitochondria are typically located in close proximity to the SR and plasma membrane, and thus the MCU are transiently exposed to much higher $[Ca^{2+}]_{cyt}$ levels sufficient for higher levels of MCU activation.

The second mechanism of Ca^{2+} influx across the inner mitochondrial membrane is termed rapid-uptake mode (RaM). The RaM mechanism has not been fully characterized, but it appears that RaM-mediated mitochondrial Ca^{2+} influx is faster (millisecond time scale) and accounts for more rapid changes in $[Ca^{2+}]_{mito}$ in response to changes in $[Ca^{2+}]_{cyt}$ [50, 58]. RaM-mediated mitochondrial Ca^{2+} influx is activated at lower $[Ca^{2+}]_{cyt}$ levels and responds to small, rapid changes in $[Ca^{2+}]_{cyt}$ [50, 58]. However, RaM-mediated Ca^{2+} influx is rapidly inactivated, thereby restricting the time course of mitochondrial Ca^{2+} entry. As such, RaM-mediated Ca^{2+} influx is well suited to respond to small-amplitude transient increases in $[Ca^{2+}]_{cyt}$ such as Ca^{2+} sparks [90, 91, 125].

The third mechanism of mitochondrial Ca^{2+} influx is the mitochondrial ryanodine receptor (mRyR). Antibodies against RyR channels identified a 600 kDa protein in the inner mitochondrial membrane. These mRyR channels were shown to bind $[^3H]$ryanodine with high affinity (Kd = 9.8 nM). The open probability of mRyR channels is increased at sub to low micromolar $[Ca^{2+}]$ levels and is inactivated at very high $[Ca^{2+}]$ (more than 50 μM), the range that also favors MCU activation [50]. As might be expected, the mRyR channels display a bell-shaped Ca^{2+} dependency of ryanodine binding with a half-activation concentration of approximately 2 μM [50]. In the microdomains encompassing SR and mitochondria, $[Ca^{2+}]_{cyt}$ may fluctuate between 1 and 100 μM during agonist stimulation [45, 50]; thus the Ca^{2+} sensitivity of mRyR is well within the physiological range. It has been speculated that the mRyR could be activated in the lower half of a projected microdomain range of $[Ca^{2+}]_{cyt}$ (less than 50 μM), while MCU would be inactivated or less effective [50, 58]. In the higher range (more than 50 μM), MCU would be activated and mRyR inactivated.

3.1.2 Mitochondrial Ca^{2+} Release or Efflux

Efflux of Ca^{2+} from mitochondria in excitable cells is mainly achieved by Ca^{2+} exchange for Na^+ (Li^+ may substitute for Na^+), with subsequent exchange of Na^+ out of the matrix for protons [50] (Fig. 4). As a result mitochondrial Ca^{2+} efflux changes mitochondrial membrane potential ($\Delta\Psi_m$) [50, 94]. The mitochondrial Na^+/Ca^{2+} exchanger (mNCX) was recently identified and is distinct from the plasma membrane Na^+/Ca^{2+} exchanger [94, 95]. It appears that substantial efflux of Ca^{2+} can be accomplished via the mNCX [26–28]. Mitochondrial efflux of Ca^{2+} via the mNCX has a sigmoidal dependence on $[Na^+]$ concentration, with an EC50 in the physiological range and a relatively high Vmax [26–28, 50].

A second mechanism of mitochondrial Ca^{2+} efflux is Na^+ independent but instead is linked to an exchange of protons. It appears that this Ca^{2+}/H^+ efflux mechanism exchanges two H^+ ions for a single Ca^{2+} ion [50, 58]. Most descriptions of this mitochondrial Ca^{2+} efflux mechanism are for nonexcitable cells such as liver and kidney, but there is little evidence for such a mechanism in ASM.

3.2 *Mitochondria and SR Ca^{2+} Release*

Depending on the cell type, mitochondria have been found to either inhibit or promote SR Ca^{2+} release [4, 18, 19, 24, 25, 56, 61]. Given that the open probability of IP_3R and RyR channels is influenced by $[Ca^{2+}]_{cyt}$ and that the relationships for both channels are represented by bell-shaped curves, an elevation of $[Ca^{2+}]_{cyt}$ has the potential for either positive or negative feedback. In smooth muscle other than ASM, inhibition of mitochondrial Ca^{2+} uptake has been reported to either prevent or not alter Ca^{2+} release via IP_3R channels [4, 18, 19, 25]. Inhibition of mitochondrial Ca^{2+} uptake also increases the frequency of Ca^{2+} sparks, suggesting that mitochondria normally exert an inhibitory effect on RyR channel activity via buffering $[Ca^{2+}]_{cyt}$ levels [60, 92]. RyRs are activated by 1–10 μM $[Ca^{2+}]_{cyt}$ [41], and it is possible that mitochondrial Ca^{2+} uptake reduces local $[Ca^{2+}]_{cyt}$ in SR/mitochondria microdomains below this level [60, 92]. It has also been shown that depolarization of mitochondrial membrane potential reduces the frequency of Ca^{2+} sparks in arterial smooth muscle cells [24]. Mitochondrial depolarization reduces Ca^{2+} efflux, and this might also reduce local $[Ca^{2+}]_{cyt}$, which in turn reduces the open probability of RyR channels [24].

Mitochondrial reactive oxygen species (ROS) generation, while thought to be damaging byproducts of cellular respiration, is also proposed to be important for a multitude of cell signaling processes [12]. An increase in $[Ca^{2+}]_{mito}$ is believed to stimulate ROS generation [13]. However, the exact mechanism of Ca^{2+}-induced ROS generation is unclear [12, 13]. Modulation of the mitochondrial bioenergetics (i.e., ATP synthesis) or mitochondrial biogenesis, exposure to cytokines, change in oxygen concentration (particularly hypoxia), or agonist such as histamine can all affect mitochondrial ROS generation [23, 104]. In cardiac or vascular smooth muscle cells, an increase in endogenous mitochondrial ROS leads to an elevation in $[Ca^{2+}]_{cyt}$ [23, 72, 104]. Inhibition of RyR channels blunts ROS-induced increases in $[Ca^{2+}]_{cyt}$ [72], suggesting that ROSs increase the open probability of RyR channels. Furthermore, RyR has a critical thiol residue involved in the gating of the Ca^{2+} channel that can be modified by ROSs [39].

3.3 *Mitochondria and SR Ca^{2+} Reuptake*

The physical proximity of mitochondria and SR has led several researchers to explore the possible interactions between mitochondria and SERCA [19, 56, 61, 116]. One hypothesis is that mitochondria alter SR Ca^{2+} refilling by competing with SERCA pumps for removal of cytosolic Ca^{2+} [96]. It has been proposed that mitochondrial Ca^{2+} uptake could generate microdomains of low $[Ca^{2+}]_{cyt}$ next to SERCA pumps [47, 78]. Changes in $[Ca^{2+}]_{mito}$ can also affect ATP synthesis and, thereby, SERCA activity. In endothelial cells, mitochondrial Ca^{2+} uptake was found to be essential for SR Ca^{2+} refilling [78]. In ASM, inhibition of mitochondrial

Ca^{2+} uptake was found to slightly alter the rate of decay of a caffeine-induced increase in $[Ca^{2+}]_{cyt}$ [116]. Although these investigators concluded that this effect was not mediated by a decrease in maximum SERCA activity, they did not exclude the possibility that diminished ATP availability was an underlying cause.

3.4 *Mitochondria and Plasma Membrane Ca^{2+} Fluxes*

In addition to SR Ca^{2+} release and reuptake, plasma membrane Ca^{2+} influx/efflux is a major component of $[Ca^{2+}]_{cyt}$ regulation. There is evidence in non-ASM tissues that mitochondria can influence SOCE [59, 82, 97]. The extent of Ca^{2+} influx via SOCE is not simply dependent on local plasma membrane Ca^{2+} gradients but is regulated by the state of SR repletion and even by other intracellular signaling cascades (e.g., DAG) [97]. Several studies have shown that mitochondrial Ca^{2+} uptake stimulates SOCE [84, 96, 97]. The main hypothesis is that by taking up Ca^{2+} released from SR, mitochondria can enhance store depletion and thereby sustain SOCE [84, 96, 97]. A recent study suggests that mitochondrial Ca^{2+}release is also important to SOCE regulation in endothelial cells [87]. In contrast, mitochondrial depolarization does not affect SOCE in several cell types [66, 87, 98, 135]. The physiological purpose of SOCE enhancement by mitochondria, however, is not clear. Recently, the relation between mitochondria and STIM1/Orai1 (the major components of SOCE) has been investigated. Inhibition of mitochondrial Ca^{2+} uptake was found to impair STIM1 aggregation and to reduce Orai1 activity [55, 87]. On the other hand, in cells where STIM1 protein is reduced, mitochondrial Ca^{2+} uptake is also reduced [55]. The tethering of mitochondria to the SR through the mitochondrial dynamin-related protein Mfn 2 was found to be essential for STIM1 aggregation [128]. In contrast, proximity of the mitochondria to the plasma membrane or mitochondria motility does not seem to affect SOCE or STIM1 aggregation [87]. Recent studies also suggest that mitochondrial ROS generation following agonist stimulation activates STIM1 aggregation [23, 55]. The cysteine 56 residue of STIM1 is proposed to act as a sensor for oxidant-dependent activation of STIM1.

4 Excitation–Contraction Coupling

In smooth muscle, contraction requires cyclic attachment and detachment of the myosin head to actin (i.e., cross-bridge cycling) and the hydrolysis of ATP by actin-activated myosin ATPase (actomyosin ATPase) [65, 123, 124]. When ASM cells are exposed to ACh or other agonists, the resulting transient increase in $[Ca^{2+}]_{cyt}$ leads to muscle contraction, with a delay of several hundred milliseconds [124]. Ca^{2+}-dependent regulation of cross-bridge recruitment requires phosphorylation of the regulatory domain of myosin (MLC_{20}) by myosin light chain kinase (MLCK).

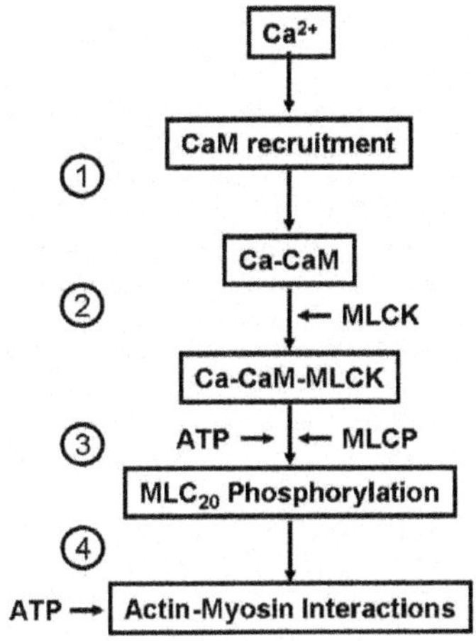

Fig. 5 *Schematic illustration of excitation–contraction coupling time frame and time delays between elevation in* $[Ca^{2+}]_{cyt}$ *and force generation.* Delay 1: mobilization of calmodulin (*CaM*). Delay 2: binding of Ca^{2+} to CaM and subsequent activation of myosin light chain kinase (*MLCK*). Delay 3: phosphorylation of regulatory myosin light chain MLC_{20}. Delay 4: cross-bridge recruitment. *MLCP* myosin light chain phosphatase

MLCK activity is stimulated by an increase in $[Ca^{2+}]_{cyt}$ through the binding of Ca^{2+} to calmodulin (CaM). The extent of MLC_{20} phosphorylation is also regulated by myosin light chain phosphatase (MLCP), which is not Ca^{2+} dependent but is regulated by the Rho kinase pathway [131]. As a consequence, the phosphorylation of MLC_{20} reflects the balance between MLCK and MLCP activities and, therefore, Ca^{2+}-dependent and independent (i.e., Ca^{2+} sensitivity) mechanisms.

The delay between the agonist-induced increase in $[Ca^{2+}]_{cyt}$ and contraction reflects the time required for a cascade of events: (1) mobilization of CaM, (2) binding of Ca^{2+} to CaM and subsequent activation of MLCK, (3) phosphorylation of regulatory myosin light chain MLC_{20}, and finally (4) cross-bridge recruitment (Fig. 5). The relative contribution of each of these four components to the overall delay in excitation–contraction coupling in ASM was quantified by measuring the response to rapid (virtually instantaneous) release of caged compounds by flash photolysis [124]. It was found that mobilization of CaM provides the most significant delay, while the delay due to phosphorylation of MLC_{20} and cross-bridge recruitment is relatively minor [124] (Fig. 5). Thus, following an increase in $[Ca^{2+}]_{cyt}$ the demand for ATP driven by cross-bridge cycling occurs within several hundred milliseconds (less than 1 s).

Given that an elevation in $[Ca^{2+}]_{cyt}$ leads to an increase in $[Ca^{2+}]_{mito}$, which in turn stimulates ATP synthesis, the spatial and temporal aspects of agonist-induced $[Ca^{2+}]_{cyt}$ increase will likely influence both excitation–contraction coupling and excitation–energy coupling. Agonist-induced increases in $[Ca^{2+}]_{cyt}$ are transient and also display other dynamic features such as propagating Ca^{2+} oscillations. Thus, simplification of contractile responses to global and homogeneous elevation of $[Ca^{2+}]_{cyt}$ across smooth muscle cells may not accurately reflect the dynamic nature of ASM Ca^{2+} signaling (Fig. 3) [105, 124]. The dynamic interplay between SR Ca^{2+} release via IP_3R or RyR channels and SR Ca^{2+} uptake via SERCA, in ASM cells, leads to transient changes in $[Ca^{2+}]_{cyt}$ in the form of propagating Ca^{2+} oscillations at a frequency that ranges from approx. 5 to 40 min^{-1} and an amplitude that ranges from approx. 100 to 800 nM of $[Ca^{2+}]_{cyt}$ [105, 107, 125]. Local $[Ca^{2+}]_{cyt}$ oscillations induced by agonist stimulation propagate to adjacent regions of the cell as a traveling wave of $[Ca^{2+}]_{cyt}$. The Ca^{2+} wave usually originates from a site within an ASM cell in which there is increased frequency of spontaneous Ca^{2+} sparks and maintains the same wave front direction across the cytosol [90, 91, 105, 107, 125]. These spatial and temporal aspects of cytosolic Ca^{2+} waves are likely to play an important role in regulating energy supply by the mitochondria. As mentioned, spontaneous yet spatially isolated $[Ca^{2+}]_{cyt}$ transients or Ca^{2+} sparks are observed at discrete locations within ASM cells. The Ca^{2+} spark is a stochastic event that represents the most elementary Ca^{2+} release event within a cell and is attributed to the activation kinetics of RyR channels [90, 91, 125].

5 Excitation–Energy Coupling

Several decades ago, it was proposed that an accumulation of Ca^{2+} by the mitochondria stimulates ATP synthesis [129, 138]. Since then, it has been shown that physiological elevation in $[Ca^{2+}]_{cyt}$ leads to an increase in $[Ca^{2+}]_{mito}$ that stimulates ATP synthesis [13, 17, 35, 36].

Mitochondria produce most of the ATP used in cells through oxidative phosphorylation. Nicotinamide Adenine Dinucleotide (H) (NADH) donates electrons to the electron transport chain within the inner membrane of the mitochondria (Fig. 4). As the electron transport chain catalyzes electron flow, H^+ ions are pumped against their concentration gradient out of the mitochondrial matrix. The final stage of the electron transport chain requires O_2 as an electron acceptor and produces H_2O (at complex IV). A small percentage of electrons directly leaks to oxygen, resulting in the formation of ROS. Electrons flowing through the electron transport chain result in O_2 consumption and the formation of a proton motive force (pmf), which has electrical ($\Delta\Psi$) and chemical (ΔpH) features. The pmf favors the flow of protons into the mitochondrial matrix. The electrochemical gradient across the membrane is used to drive ATP synthesis from ADP and inorganic Phosphate (Pi) by the ATP synthase (complex V). Synthesized ATP is then transported to the cytosol in exchange for cytosolic ADP by the adenine nucleotide transporter (ANT) (Fig. 4).

There is substantial evidence to support a role for Ca^{2+} as an important effector of mitochondrial ATP production (other effectors include ADP [20, 21], P_i [11], O_2, and substrate availability [15]). An emerging consensus is that an elevation of $[Ca^{2+}]_{mito}$ stimulates mitochondrial NADH production, which in turn stimulates ATP production. The mechanisms behind this response are not completely understood. Several key enzymes are involved in ATP synthesis, including intramitochondrial dehydrogenases [pyruvate dehydrogenase (PDH), NAD-isocitrate dehydrogenase (ICDH), and oxoglutarate dehydrogenase (OGDH)] as well as ANT. Activity of these enzymes is stimulated by an increase in $[Ca^{2+}]_{mito}$ [13, 35]. Finally, an increase in $[Ca^{2+}]_{mito}$ can also increase NADH levels in the mitochondria by stimulating the glycerol phosphate shuttle [35] and the aspartate/glutamate transporters [93].

6 Mitochondrial Dynamics

Mitochondria are spatially dynamic due to their ability to move along the cytoskeleton within the cytosol and through continuous fission and fusion, which leads to a wild spectrum of possible ultrastructures or shapes – from individual punctate structures to continuous, tubular networks (Fig. 6). Mitochondrial dynamics (fission and fusion events and movement) are essential to couple $[Ca^{2+}]_{cyt}$ and $[Ca^{2+}]_{mito}$, and therefore ATP synthesis, to the localized energy demand of local cytosolic regions. Changes in mitochondrial dynamics have been observed in a number of pathophysiological conditions, most notably in neurodegenerative diseases, indicating that a "normal" relationship between mitochondrial structure and function is essential for normal physiology [88]. While a unique relationship between fission/fusion, mitochondrial shape, and ATP production has not been clearly defined, it has been observed that changes in mitochondrial shape usually accompany changes in ATP production and vice versa [6, 44]. A growing number of novel proteins that regulate mitochondrial dynamics have been identified [88]. The balance between fission and fusion is notably regulated by a set of large GTPase proteins. Mitofusin (Mfn) 1, Mfn2, and optic atrophy 1 (OPA1) mediate mitochondrial fusion. Mutations in Mfn1 lead to small fragmented mitochondria, whereas mutations in Mfn2 lead to large fragmented mitochondria located near the nucleus [88]. The dynamin-related protein1 (Drp1) has been reported to be involved in

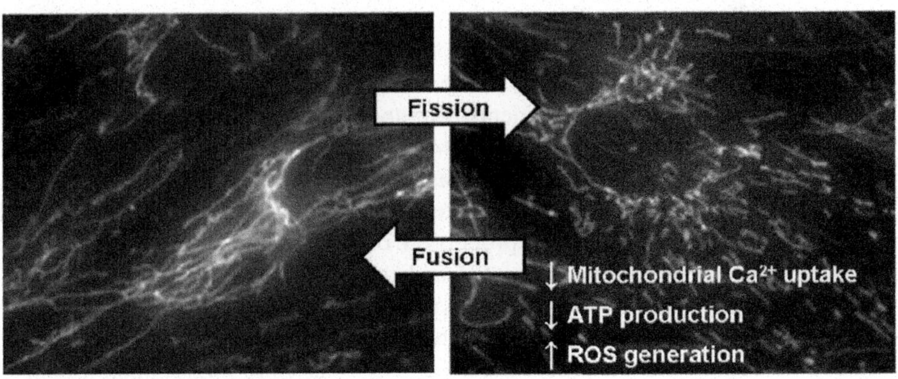

Fig. 6 *Mitochondrial fusion and fission in human ASM*. Fluorescent images of human ASM cells loaded with MitoTracker Green AM showing filamentous mitochondrial network corresponding to fusion events and fragmented mitochondrial network corresponding to fission events

fission events. Overexpression of Drp1 induces mitochondrial fragmentation and impairs mitochondrial Ca^{2+} uptake during agonist stimulation [132].

In addition, mitochondrial volume may affect mitochondrial functions. In liver cells, an increase in $[Ca^{2+}]_{mito}$ was shown to increase mitochondrial volume and stimulate ATP production [52, 53]. In asthma, severe mitochondrial ultrastructural changes such as loss of cristae and swelling have been observed [1, 75]. Little is known about mitochondrial movement and fission/fusion in ASM. An increase in mitochondrial mass has also been reported in asthma and associated with ASM cell proliferation and airway remodeling. A number of factors regulate mitochondrial mass, most notably peroxisome-proliferator-activated receptor γ coactivator-1α (PGC-1α), which has been found to be stimulated by increased levels of $[Ca^{2+}]_{cyt}$ [136].

Another important but often overlooked aspect of mitochondria is their motility, which plays an important role in fission/fusion events and allows mitochondria to accumulate where high amounts of energy are needed [31, 88, 137]. Mitochondrial movements can be categorized into two groups: random/wiggling motions and directed motion trajectories. Rapid and long-range movements require cytoskeletal tracks, particularly microtubules and microtubule-associated motors, such as kinesin and dynein. F-actin and intermediate filaments could also be involved [2, 10, 144]. Mitochondrial movement is believed to be correlated to $[Ca^{2+}]_{cyt}$. Several studies have shown that the pattern of mitochondrial movement follows the spatial and temporal pattern of transient $[Ca^{2+}]_{cyt}$ responses, which result in a redistribution of mitochondria to areas of high $[Ca^{2+}]_{cyt}$ [2, 10, 144]. In addition, mitochondrial movements within cells are inhibited by depleting local ATP levels. This suggests that $[Ca^{2+}]_{cyt}$ and ATP may coordinate recruitment of mitochondria [14]. In a study examining porcine ASM cells using electron microscopy, van Breemen and colleagues reported that the proportion of mitochondria near (less than 30 nm) the SR membrane decreases with agonist stimulation [29]. However, no data correlating these changes to mitochondrial Ca^{2+} uptake were provided. These results are in sharp contrast to findings in other cell types using techniques other than electron microscopy, where it has been reported that agonist stimulation leads to an increase in mitochondrial density near the SR [14, 77]. In human ASM, preliminary studies in our lab indicate that histamine stimulation increases the longer-range, directed movements of mitochondria from more distal compartments of the cell to the perinuclear region of the cell. In contrast, we observed a decrease in the random wiggle-type movements of mitochondria during histamine stimulation. This decrease was more pronounced in the perinuclear region versus distal regions, again suggesting heterogonous subpopulations of mitochondria.

Mitochondria are generally not homogeneously distributed within cells. In ASM and in other cells, a significant proportion of the mitochondrial population is observed in close proximity to the SR and the nucleus [29, 32]. Mitochondrial movements play an important role in dynamic mitochondrial redistribution. The main hypothesis is that mitochondria strategically situate near regions of high ATP demand or regions with the potential for Ca^{2+} signal sensing and regulation such as the SR, plasma membrane, or the nucleus [14, 32, 40, 99, 144]. A mitochondrial subpopulation very close to the SR (which is dynamically recruited from the total population) is likely employed in excitation–energy coupling; however, the influence of other factors in the context of ASM cells, including heterogeneity of the $[Ca^{2+}]_{cyt}$ response or cytoskeleton remodeling associated with ASM contraction, pose interesting questions for further study.

7 Summary and Conclusions

Smooth muscle appears to be optimized for maximum energy efficiency, whereas the skeletal muscle contractile response is optimized for maximum power output. Therefore, efficient matching of energy supply with energy demand may be a more significant governing factor for the ASM contractile response. The ASM cell is a physiological model of energy efficiency, especially in comparison to skeletal muscle. The dynamic nature of the ASM cell allows for adaptation to various stimuli while maintaining a stable time course of muscle force. Given evidence that a change in metabolite levels is not detectable in smooth muscle cells, credit must also be given to the processes that supply energy in just the right amount at the right time. The main feature of this process includes a linkage between an elevation of $[Ca^{2+}]_{cyt}$ and $[Ca^{2+}]_{mito}$ with subsequent decoding of $[Ca^{2+}]_{mito}$ into increased ATP production (Fig. 7).

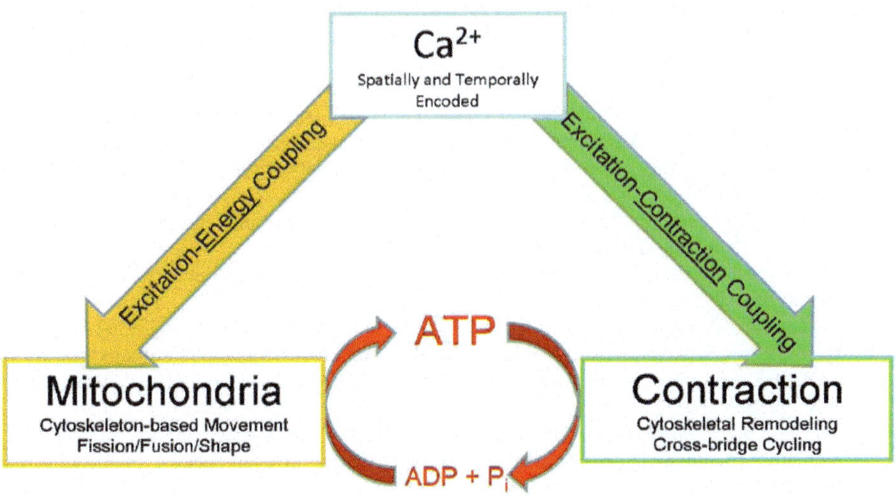

Fig. 7 *Excitation–contraction coupling is mirrored by an excitation–energy coupling performed by mitochondria.* Both processes are initiated by elevated $[Ca^{2+}]_{cyt}$ levels and result in the production of ATP during a contractile response. Spatial and temporal heterogeneity of the $[Ca^{2+}]_{cyt}$ response, contraction-associated cytoskeleton remodeling and cross-bridge cycling, and cytoskeleton-based mitochondrial motility and fission/fusion events might influence these relationships

The control of mitochondrial energetics in ASM is far from being established and is at this stage speculative. However, a few key observations support the idea that mitochondria perform an excitation–energy coupling function, thereby coupling energy demand with energy supply. An elevation of $[Ca^{2+}]_{cyt}$ signals muscle contraction and concomitant ATP hydrolysis. The coordinated elevation of $[Ca^{2+}]_{mito}$ leads to an increase in mitochondrial ATP production, thereby matching the increase in ATP consumption. Mitochondrial dynamics and redistribution likely determine the quality of mitochondrial Ca^{2+} sensing, and therefore mitochondrial compartmentalization might ensure appropriate and efficient ATP production to localized cytosolic regions.

References

1. Aguilera-Aguirre L, Bacsi A, Saavedra-Molina A, Kurosky A, Sur S, Boldogh I (2009) Mitochondrial dysfunction increases allergic airway inflammation. J Immunol 183: 5379–5387.
2. Anesti V, Scorrano L (2006) The relationship between mitochondrial shape and function and the cytoskeleton. Biochim Biophys Acta 1757:692–699.
3. Ay B, Prakash YS, Pabelick CM, Sieck GC (2004) Store-operated Ca^{2+} entry in porcine airway smooth muscle. Am J Physiol Lung Cell Mol Physiol 286:909–917.
4. Balemba OB, Bartoo AC, Nelson MT, Mawe GM (2008) Role of mitochondria in spontaneous rhythmic activity and intracellular calcium waves in the guinea pig gallbladder smooth muscle. Am J Physiol Gastrointest Liver Physiol 294:467–476.
5. Baughman JM, Perocchi F, Girgis HS, Plovanich M, Belcher-Timme CA, Sancak Y, Bao XR, Strittmatter L, Goldberger O, Bogorad RL, Koteliansky V, Mootha VK (2011) Integrative genomics identifies MCU as an essential component of the mitochondrial calcium uniporter. Nature 476:341–345.
6. Benard G, Bellance N, James D, Parrone P, Fernandez H, Letellier T, Rossignol R (2007) Mitochondrial bioenergetics and structural network organization. J Cell Sci 120:838–848.
7. Berridge MJ (2008) Smooth muscle cell calcium activation mechanisms. J Physiol 586:5047–5061.
8. Berridge MJ, Lipp P, Bootman MD (2000) The versatility and universality of calcium signalling. Nat Rev Mol Cell Biol 1:11–21.
9. Blaustein MP, Lederer WJ (1999) Sodium/calcium exchange: its physiological implications. Physiol Rev 79:763–854.
10. Boldogh IR, Pon LA (2007) Mitochondria on the move. Trends Cell Biol 17:502–510.
11. Bose S, French S, Evans FJ, Joubert F, Balaban RS (2003) Metabolic network control of oxidative phosphorylation: multiple roles of inorganic phosphate. J Biol Chem 278:39155–39165.
12. Brookes PS, Levonen AL, Shiva S, Sarti P, Darley-Usmar VM (2002) Mitochondria: regulators of signal transduction by reactive oxygen and nitrogen species. Free Radic Biol Med 33:755–764.
13. Brookes PS, Yoon Y, Robotham JL, Anders MW, Sheu SS (2004) Calcium, ATP, and ROS: a mitochondrial love-hate triangle. Am J Physiol Cell Physiol 287:817–833.
14. Brough D, Schell MJ, Irvine RF (2005) Agonist-induced regulation of mitochondrial and endoplasmic reticulum motility. Biochem J 392:291–297.
15. Brown GC (1992) Control of respiration and ATP synthesis in mammalian mitochondria and cells. Biochem J 284:1–13.
16. Cahalan MD (2009) STIMulating store-operated Ca^{2+} entry. Nat Cell Biol 11:669–677.

17. Carafoli E (2010) The fateful encounter of mitochondria with calcium: how did it happen? Biochim Biophys Acta 1797:595–606.
18. Chalmers S, McCarron JG (2008) The mitochondrial membrane potential and Ca^{2+} oscillations in smooth muscle. J Cell Sci 121:75–85.
19. Chalmers S, McCarron JG (2009) Inhibition of mitochondrial calcium uptake rather than efflux impedes calcium release by inositol-1,4,5-trisphosphate-sensitive receptors. Cell Calcium 46:107–113.
20. Chance B, Williams GR (1955) Respiratory enzymes in oxidative phosphorylation. I. Kinetics of oxygen utilization. J Biol Chem 217:383–393.
21. Chance B, Williams GR (1956) The respiratory chain and oxidative phosphorylation. Adv Enzymol Relat Subj Biochem 17:65–134.
22. Chen H, Chan DC (2005) Emerging functions of mammalian mitochondrial fusion and fission. Hum Mol Genet 14:283–289.
23. Chen T, Zhu L, Wang T, Ye H, Huang K, Hu Q (2010) Mitochondria depletion abolishes agonist-induced Ca^{2+} plateau in airway smooth muscle cells: potential role of H_2O_2. Am J Physiol Lung Cell Mol Physiol 298:178–188.
24. Cheranov SY, Jaggar JH (2004) Mitochondrial modulation of Ca^{2+} sparks and transient KCa currents in smooth muscle cells of rat cerebral arteries. J Physiol 556:755–771.
25. Collins TJ, Lipp P, Berridge MJ, Li W, Bootman MD (2000) Inositol 1,4,5-trisphosphate-induced Ca^{2+} release is inhibited by mitochondrial depolarization. Biochem J 347:593–600.
26. Crompton M, Capano M, Carafoli E (1976) Respiration-dependent efflux of magnesium ions from heart mitochondria. Biochem J 154:735–742.
27. Crompton M, Heid I (1978) The cycling of calcium, sodium, and protons across the inner membrane of cardiac mitochondria. Eur J Biochem 91:599–608.
28. Crompton M, Moser R, Ludi H, Carafoli E (1978) The interrelations between the transport of sodium and calcium in mitochondria of various mammalian tissues. Eur J Biochem 82:25–31.
29. Dai J, Kuo KH, Leo JM, van Breemen C, Lee CH (2005) Rearrangement of the close contact between the mitochondria and the sarcoplasmic reticulum in airway smooth muscle. Cell Calcium 37:333–340.
30. De Stefani D, Raffaello A, Teardo E, Szabo I, Rizzuto R (2011) A forty-kilodalton protein of the inner membrane is the mitochondrial calcium uniporter. Nature 476:336–340.
31. De Vos KJ, Allan VJ, Grierson AJ, Sheetz MP (2005) Mitochondrial function and actin regulate dynamin-related protein 1-dependent mitochondrial fission. Curr Biol 15:678–683.
32. Delmotte P, Yang B, Thompson MA, Pabelick CM, Prakash YS, Sieck GC (2012) Inflammation alters regional mitochondrial Ca^{2+} in human airway smooth muscle cells. Am J Physiol Cell Physiol 303:244–256.
33. Demaurex N, Poburko D (2009) Cell biology. A revolving door for calcium. Science 326:57–58.
34. Demaurex N, Poburko D, Frieden M (2009) Regulation of plasma membrane calcium fluxes by mitochondria. Biochim Biophys Acta 1787:1383–1394.
35. Denton RM (2009) Regulation of mitochondrial dehydrogenases by calcium ions. Biochim Biophys Acta 1787:1309–1316.
36. Drago I, Pizzo P, Pozzan T (2011) After half a century mitochondrial calcium in- and efflux machineries reveal themselves. EMBO J 30:4119–4125.
37. Duchen MR (2000a) Mitochondria and Ca^{2+} in cell physiology and pathophysiology. Cell Calcium 28:339–348.
38. Duchen MR (2000b) Mitochondria and calcium: from cell signalling to cell death. J Physiol 529:57–68.
39. Favero TG, Zable AC, Abramson JJ (1995) Hydrogen peroxide stimulates the Ca^{2+} release channel from skeletal muscle sarcoplasmic reticulum. J Biol Chem 270:25557–25563.
40. Filippin L, Magalhaes PJ, Di Benedetto G, Colella M, Pozzan T (2003) Stable interactions between mitochondria and endoplasmic reticulum allow rapid accumulation of calcium in a subpopulation of mitochondria. J Biol Chem 278:39224–39234.

41. Fill M, Copello JA (2002) Ryanodine receptor calcium release channels. Physiol Rev 82:893–922.
42. Fredberg JJ (2004) Bronchospasm and its biophysical basis in airway smooth muscle. Respir Res 5:2.
43. Frischauf I, Schindl R, Derler I, Bergsmann J, Fahrner M, Romanin C (2008) The STIM/Orai coupling machinery. Channels (Austin) 2:261–268.
44. Galloway CA, Yoon Y (2012) Perspectives on: SGP symposium on mitochondrial physiology and medicine: what comes first, misshape or dysfunction? The view from metabolic excess. J Gen Physiol 139:455–463.
45. Giacomello M, Drago I, Bortolozzi M, Scorzeto M, Gianelle A, Pizzo P, Pozzan T (2010) Ca^{2+} hot spots on the mitochondrial surface are generated by Ca^{2+} mobilization from stores, but not by activation of store-operated Ca^{2+} channels. Mol Cell 38:280–290.
46. Glancy B, Balaban RS (2012) Role of mitochondrial Ca^{2+} in the regulation of cellular energetics. Biochemistry 51:2959–2973.
47. Glitsch MD, Bakowski D, Parekh AB (2002) Store-operated Ca^{2+} entry depends on mitochondrial Ca^{2+} uptake. Embo J 21:6744–6754.
48. Gosling M, Poll C, Li S (2005) TRP channels in airway smooth muscle as therapeutic targets. Naunyn Schmiedebergs Arch Pharmacol 371:277–284.
49. Gunter TE, Pfeiffer DR (1990) Mechanisms by which mitochondria transport calcium. Am J Physiol 258:755–786.
50. Gunter TE, Sheu SS (2009) Characteristics and possible functions of mitochondrial Ca^{2+} transport mechanisms. Biochim Biophys Acta 1787:1291–1308.
51. Gunter TE, Yule DI, Gunter KK, Eliseev RA, Salter JD (2004) Calcium and mitochondria. FEBS Lett 567:96–102.
52. Halestrap AP (1989) The regulation of the matrix volume of mammalian mitochondria in vivo and in vitro and its role in the control of mitochondrial metabolism. Biochim Biophys Acta 973:355–382.
53. Halestrap AP, Quinlan PT, Whipps DE, Armston AE (1986) Regulation of the mitochondrial matrix volume in vivo and in vitro. The role of calcium. Biochem J 236:779–787.
54. Hall IP (2000) Second messengers, ion channels and pharmacology of airway smooth muscle. Eur Respir J 15:1120–1127.
55. Hawkins BJ, Irrinki KM, Mallilankaraman K, Lien YC, Wang Y, Bhanumathy CD, Subbiah R, Ritchie MF, Soboloff J, Baba Y, Kurosaki T, Joseph SK, Gill DL, Madesh M (2010) S-glutathionylation activates STIM1 and alters mitochondrial homeostasis. J Cell Biol 190:391–405.
56. Hernandez-SanMiguel E, Vay L, Santo-Domingo J, Lobaton CD, Moreno A, Montero M, Alvarez J (2006) The mitochondrial Na^+/Ca^{2+} exchanger plays a key role in the control of cytosolic Ca^{2+} oscillations. Cell Calcium 40:53–61.
57. Hirota S, Janssen LJ (2007) Store-refilling involves both L-type calcium channels and reverse-mode sodium-calcium exchange in airway smooth muscle. Eur Respir J 30:269–278.
58. Hoppe UC (2010) Mitochondrial calcium channels. FEBS Lett 584:1975–1981.
59. Hoth M, Button DC, Lewis RS (2000) Mitochondrial control of calcium-channel gating: a mechanism for sustained signaling and transcriptional activation in T lymphocytes. Proc Natl Acad Sci U S A 97:10607–10612.
60. Isaeva EV, Shirokova N (2003) Metabolic regulation of Ca^{2+} release in permeabilized mammalian skeletal muscle fibres. J Physiol 547:453–462.
61. Ishii K, Hirose K, Iino M (2006) Ca^{2+} shuttling between endoplasmic reticulum and mitochondria underlying Ca^{2+} oscillations. EMBO Rep 7:390–396.
62. Janssen LJ (2002) Ionic mechanisms and Ca^{2+} regulation in airway smooth muscle contraction: do the data contradict dogma? Am J Physiol Lung Cell Mol Physiol 282:1161–1178.
63. Janssen LJ, Walters DK, Wattie J (1997) Regulation of $[Ca^{2+}]_i$ in canine airway smooth muscle by Ca^{2+}-ATPase and Na^+/Ca^{2+} exchange mechanisms. Am J Physiol 273:322–330.

64. Ji RR, Samad TA, Jin SX, Schmoll R, Woolf CJ (2002) p38 MAPK activation by NGF in primary sensory neurons after inflammation increases TRPV1 levels and maintains heat hyperalgesia. Neuron 36:57–68.
65. Jones KA, Lorenz RR, Prakash YS, Sieck GC, Warner DO (1999) ATP hydrolysis during contraction of permeabilized airway smooth muscle. Am J Physiol 277:334–342.
66. Jousset H, Malli R, Girardin N, Graier WF, Demaurex N, Frieden M (2008) Evidence for a receptor-activated Ca^{2+} entry pathway independent from Ca^{2+} store depletion in endothelial cells. Cell Calcium 43:83–94.
67. Jude JA, Wylam ME, Walseth TF, Kannan MS (2008) Calcium signaling in airway smooth muscle. Proc Am Thorac Soc 5:15–22.
68. Kannan MS, Fenton AM, Prakash YS, Sieck GC (1996) Cyclic ADP-ribose stimulates sarcoplasmic reticulum calcium release in porcine coronary artery smooth muscle. Am J Physiol 270:801–806.
69. Kannan MS, Prakash YS, Brenner T, Mickelson JR, Sieck GC (1997) Role of ryanodine receptor channels in Ca^{2+} oscillations of porcine tracheal smooth muscle. Am J Physiol 272:659–664.
70. Kanwar YS, Sun L (2008) Shuttling of calcium between endoplasmic reticulum and mitochondria in the renal vasculature. Am J Physiol Renal Physiol 295:1301–1302.
71. Kanzaki M, Zhang YQ, Mashima H, Li L, Shibata H, Kojima I (1999) Translocation of a calcium-permeable cation channel induced by insulin-like growth factor-I. Nat Cell Biol 1:165–170.
72. Lin MJ, Yang XR, Cao YN, Sham JS (2007) Hydrogen peroxide-induced Ca^{2+} mobilization in pulmonary arterial smooth muscle cells. Am J Physiol Lung Cell Mol Physiol 292:1598–1608.
73. Liou J, Kim ML, Heo WD, Jones JT, Myers JW, Ferrell JE, Jr., Meyer T (2005) STIM is a Ca^{2+} sensor essential for Ca^{2+}-store-depletion-triggered Ca^{2+} influx. Curr Biol 15:1235–1241.
74. Liu B, Peel SE, Fox J, Hall IP (2010) Reverse mode Na^{+}/Ca^{2+} exchange mediated by STIM1 contributes to Ca^{2+} influx in airway smooth muscle following agonist stimulation. Respir Res 11:168.
75. Mabalirajan U, Dinda AK, Kumar S, Roshan R, Gupta P, Sharma SK, Ghosh B (2008) Mitochondrial structural changes and dysfunction are associated with experimental allergic asthma. J Immunol 181:3540–3548.
76. Mahn K, Ojo OO, Chadwick G, Aaronson PI, Ward JP, Lee TH (2010) Ca^{2+} homeostasis and structural and functional remodelling of airway smooth muscle in asthma. Thorax 65:547–552.
77. Malaiyandi LM, Honick AS, Rintoul GL, Wang QJ, Reynolds IJ (2005) Zn^{2+} inhibits mitochondrial movement in neurons by phosphatidylinositol 3-kinase activation. J Neurosci 25:9507–9514.
78. Malli R, Frieden M, Osibow K, Zoratti C, Mayer M, Demaurex N, Graier WF (2003) Sustained Ca^{2+} transfer across mitochondria is Essential for mitochondrial Ca^{2+} buffering, store-operated Ca^{2+} entry, and Ca^{2+} store refilling. J Biol Chem 278:44769–44779.
79. Marin J, Encabo A, Briones A, Garcia-Cohen EC, Alonso MJ (1999) Mechanisms involved in the cellular calcium homeostasis in vascular smooth muscle: calcium pumps. Life Sci 64:279–303.
80. Matlib MA, Zhou Z, Knight S, Ahmed S, Choi KM, Krause-Bauer J, Phillips R, Altschuld R, Katsube Y, Sperelakis N, Bers DM (1998) Oxygen-bridged dinuclear ruthenium amine complex specifically inhibits Ca^{2+} uptake into mitochondria in vitro and in situ in single cardiac myocytes. J Biol Chem 273:10223–10231.
81. Mignen O, Shuttleworth TJ (2000) I(ARC), a novel arachidonate-regulated, noncapacitative Ca^{2+} entry channel. J Biol Chem 275:9114–9119.
82. Montero M, Alonso MT, Carnicero E, Cuchillo-Ibanez I, Albillos A, Garcia AG, Garcia-Sancho J, Alvarez J (2000) Chromaffin-cell stimulation triggers fast millimolar mitochondrial Ca^{2+} transients that modulate secretion. Nat Cell Biol 2:57–61.

83. Muik M, Frischauf I, Derler I, Fahrner M, Bergsmann J, Eder P, Schindl R, Hesch C, Polzinger B, Fritsch R, Kahr H, Madl J, Gruber H, Groschner K, Romanin C (2008) Dynamic coupling of the putative coiled-coil domain of ORAI1 with STIM1 mediates ORAI1 channel activation. J Biol Chem 283:8014–8022.
84. Munoz E, Valero RA, Quintana A, Hoth M, Nunez L, Villalobos C (2011) Nonsteroidal anti-inflammatory drugs inhibit vascular smooth muscle cell proliferation by enabling the Ca^{2+}-dependent inactivation of calcium release-activated calcium/orai channels normally prevented by mitochondria. J Biol Chem 286:16186–16196.
85. Murray RK, Fleischmann BK, Kotlikoff MI (1993) Receptor-activated Ca influx in human airway smooth muscle: use of Ca imaging and perforated patch-clamp techniques. Am J Physiol 264:485–490.
86. Murray RK, Kotlikoff MI (1991) Receptor-activated calcium influx in human airway smooth muscle cells. J Physiol 435:123–144.
87. Naghdi S, Waldeck-Weiermair M, Fertschai I, Poteser M, Graier WF, Malli R (2010) Mitochondrial Ca^{2+} uptake and not mitochondrial motility is required for STIM1-Orai1-dependent store-operated Ca^{2+} entry. J Cell Sci 123:2553–2564.
88. Otera H, Mihara K (2011) Molecular mechanisms and physiologic functions of mitochondrial dynamics. J Biochem 149:241–251.
89. Pabelick CM, Ay B, Prakash YS, Sieck GC (2004) Effects of volatile anesthetics on store-operated Ca^{2+} influx in airway smooth muscle. Anesthesiology 101:373–380.
90. Pabelick CM, Prakash YS, Kannan MS, Sieck GC (1999) Spatial and temporal aspects of calcium sparks in porcine tracheal smooth muscle cells. Am J Physiol 277:1018–1025.
91. Pabelick CM, Sieck GC, Prakash YS (2001) Invited review: significance of spatial and temporal heterogeneity of calcium transients in smooth muscle. J Appl Physiol 91:488–496.
92. Pacher P, Thomas AP, Hajnoczky G (2002) Ca^{2+} marks: miniature calcium signals in single mitochondria driven by ryanodine receptors. Proc Natl Acad Sci U S A 99:2380–2385.
93. Palmieri L, Pardo B, Lasorsa FM, del Arco A, Kobayashi K, Iijima M, Runswick MJ, Walker JE, Saheki T, Satrustegui J, Palmieri F (2001) Citrin and aralar1 are Ca^{2+}-stimulated aspartate/glutamate transporters in mitochondria. EMBO J 20:5060–5069.
94. Palty R, Sekler I (2012) The mitochondrial Na^{+}/Ca^{2+} exchanger. Cell Calcium 52:9–15.
95. Palty R, Silverman WF, Hershfinkel M, Caporale T, Sensi SL, Parnis J, Nolte C, Fishman D, Shoshan-Barmatz V, Herrmann S, Khananshvili D, Sekler I (2010) NCLX is an essential component of mitochondrial Na^{+}/Ca^{2+} exchange. Proc Natl Acad Sci U S A 107:436–441.
96. Parekh AB (2003) Store-operated Ca^{2+} entry: dynamic interplay between endoplasmic reticulum, mitochondria and plasma membrane. J Physiol 547:333–348.
97. Parekh AB, Putney JW, Jr. (2005) Store-operated calcium channels. Physiol Rev 85:757–810.
98. Park KS, Jo I, Pak K, Bae SW, Rhim H, Suh SH, Park J, Zhu H, So I, Kim KW (2002) FCCP depolarizes plasma membrane potential by activating proton and Na^{+} currents in bovine aortic endothelial cells. Pflugers Arch 443:344–352.
99. Park MK, Ashby MC, Erdemli G, Petersen OH, Tepikin AV (2001) Perinuclear, perigranular and sub-plasmalemmal mitochondria have distinct functions in the regulation of cellular calcium transport. EMBO J 20:1863–1874.
100. Parker NJ, Begley CG, Smith PJ, Fox RM (1996) Molecular cloning of a novel human gene (D11S4896E) at chromosomal region 11p15.5. Genomics 37:253–256.
101. Patterson RL, van Rossum DB, Gill DL (1999) Store-operated Ca^{2+} entry: evidence for a secretion-like coupling model. Cell 98:487–499.
102. Pedersen SF, Owsianik G, Nilius B (2005) TRP channels: an overview. Cell Calcium 38:233–252.
103. Peel SE, Liu B, Hall IP (2008) ORAI and store-operated calcium influx in human airway smooth muscle cells. Am J Respir Cell Mol Biol 38:744–749.
104. Perez-Vizcaino F, Cogolludo A, Moreno L (2010) Reactive oxygen species signaling in pulmonary vascular smooth muscle. Respir Physiol Neurobiol 174:212–220.

105. Prakash YS, Kannan MS, Sieck GC (1997) Regulation of intracellular calcium oscillations in porcine tracheal smooth muscle cells. Am J Physiol 272:966–975.
106. Prakash YS, Kannan MS, Walseth TF, Sieck GC (1998) Role of cyclic ADP-ribose in the regulation of $[Ca^{2+}]_i$ in porcine tracheal smooth muscle. Am J Physiol 274:1653–1660.
107. Prakash YS, Pabelick CM, Kannan MS, Sieck GC (2000) Spatial and temporal aspects of ACh-induced $[Ca^{2+}]_i$ oscillations in porcine tracheal smooth muscle. Cell Calcium 27:153–162.
108. Prakash YS, Sathish V, Thompson MA, Pabelick CM, Sieck GC (2009) Asthma and sarcoplasmic reticulum Ca^{2+} reuptake in airway smooth muscle. Am J Physiol Lung Cell Mol Physiol 297:794.
109. Prakriya M, Feske S, Gwack Y, Srikanth S, Rao A, Hogan PG (2006) Orai1 is an essential pore subunit of the CRAC channel. Nature 443:230–233.
110. Prasad GS, McRee DE, Stura EA, Levitt DG, Lee HC, Stout CD (1996) Crystal structure of Aplysia ADP ribosyl cyclase, a homologue of the bifunctional ectozyme CD38. Nat Struct Biol 3:957–964.
111. Raeymaekers L, Hofmann F, Casteels R (1988) Cyclic GMP-dependent protein kinase phosphorylates phospholamban in isolated sarcoplasmic reticulum from cardiac and smooth muscle. Biochem J 252:269–273.
112. Raeymaekers L, Jones LR (1986) Evidence for the presence of phospholamban in the endoplasmic reticulum of smooth muscle. Biochim Biophys Acta 882:258–265.
113. Raffaello A, De Stefani D, Rizzuto R (2012) The mitochondrial Ca^{2+} uniporter. Cell Calcium 52:16–21.
114. Ramsey IS, Delling M, Clapham DE (2006) An introduction to TRP channels. Annu Rev Physiol 68:619–647.
115. Roos J, DiGregorio PJ, Yeromin AV, Ohlsen K, Lioudyno M, Zhang S, Safrina O, Kozak JA, Wagner SL, Cahalan MD, Velicelebi G, Stauderman KA (2005) STIM1, an essential and conserved component of store-operated Ca^{2+} channel function. J Cell Biol 169:435–445.
116. Roux E, Marhl M (2004) Role of sarcoplasmic reticulum and mitochondria in Ca^{2+} removal in airway myocytes. Biophys J 86:2583–2595.
117. Rudolf R, Mongillo M, Magalhaes PJ, Pozzan T (2004) In vivo monitoring of Ca^{2+} uptake into mitochondria of mouse skeletal muscle during contraction. J Cell Biol 166:527–536.
118. Sandow A (1952) Excitation-contraction coupling in muscular response. Yale J Biol Med 25:176–201.
119. Sathish V, Abcejo AJ, Thompson MA, Sieck GC, Prakash YS, Pabelick CM (2012) Caveolin-1 regulation of store-operated Ca^{2+} influx in human airway smooth muscle. Eur Respir J 40:470–478.
120. Sathish V, Delmotte PF, Thompson MA, Pabelick CM, Sieck GC, Prakash YS (2011) Sodium-calcium exchange in intracellular calcium handling of human airway smooth muscle. PLoS One 6:e23662.
121. Sathish V, Leblebici F, Kip SN, Thompson MA, Pabelick CM, Prakash YS, Sieck GC (2008) Regulation of sarcoplasmic reticulum Ca^{2+} reuptake in porcine airway smooth muscle. Am J Physiol Lung Cell Mol Physiol 294:787–796.
122. Sathish V, Thompson MA, Bailey JP, Pabelick CM, Prakash YS, Sieck GC (2009) Effect of proinflammatory cytokines on regulation of sarcoplasmic reticulum Ca^{2+} reuptake in human airway smooth muscle. Am J Physiol Lung Cell Mol Physiol 297:26–34.
123. Sieck GC, Gransee HM (2012) Respiratory muscles structure, function & regulation (Colloquium series on integrated systems physiology, from molecule to function to disease). Morgan and Claypool Life Sciences, San Rafael.
124. Sieck GC, Han YS, Pabelick CM, Prakash YS (2001) Temporal aspects of excitation-contraction coupling in airway smooth muscle. J Appl Physiol 91:2266–2274.
125. Sieck GC, Kannan MS, Prakash YS (1997) Heterogeneity in dynamic regulation of intracellular calcium in airway smooth muscle cells. Can J Physiol Pharmacol 75:878–888.

126. Sieck GC, Prakash YS, Han YS, Kannan MS (1999) Airway smooth muscle excitation-contraction coupling. Biophys J 76:A285–A285.
127. Sieck GC, White TA, Thompson MA, Pabelick CM, Wylam ME, Prakash YS (2008) Regulation of store-operated Ca^{2+} entry by CD38 in human airway smooth muscle. Am J Physiol Lung Cell Mol Physiol 294:378–385.
128. Singaravelu K, Nelson C, Bakowski D, de Brito OM, Ng SW, Di Capite J, Powell T, Scorrano L, Parekh AB (2011) Mitofusin 2 regulates STIM1 migration from the Ca^{2+} store to the plasma membrane in cells with depolarized mitochondria. J Biol Chem 286:12189–12201.
129. Slater EC, Cleland KW (1953) The effect of calcium on the respiratory and phosphorylative activities of heart-muscle sarcosomes. Biochem J 55:566–590.
130. Smyth JT, Hwang SY, Tomita T, DeHaven WI, Mercer JC, Putney JW (2010) Activation and regulation of store-operated calcium entry. J Cell Mol Med 14:2337–2349.
131. Somlyo AP, Somlyo AV (2000) Signal transduction by G-proteins, rho-kinase and protein phosphatase to smooth muscle and non-muscle myosin II. J Physiol 522:177–185.
132. Szabadkai G, Simoni AM, Chami M, Wieckowski MR, Youle RJ, Rizzuto R (2004) Drp-1-dependent division of the mitochondrial network blocks intraorganellar Ca^{2+} waves and protects against Ca^{2+}-mediated apoptosis. Mol Cell 16:59–68.
133. Szado T, Kuo KH, Bernard-Helary K, Poburko D, Lee CH, Seow C, Ruegg UT, van Breemen C (2003) Agonist-induced mitochondrial Ca^{2+} transients in smooth muscle. FASEB J 17:28–37.
134. Tada M, Katz AM (1982) Phosphorylation of the sarcoplasmic reticulum and sarcolemma. Annu Rev Physiol 44:401–423.
135. To MS, Aromataris EC, Castro J, Roberts ML, Barritt GJ, Rychkov GY (2010) Mitochondrial uncoupler FCCP activates proton conductance but does not block store-operated Ca^{2+} current in liver cells. Arch Biochem Biophys 495:152–158.
136. Trian T, Benard G, Begueret H, Rossignol R, Girodet PO, Ghosh D, Ousova O, Vernejoux JM, Marthan R, Tunon-de-Lara JM, Berger P (2007) Bronchial smooth muscle remodeling involves calcium-dependent enhanced mitochondrial biogenesis in asthma. J Exp Med 204:3173–3181.
137. Varadi A, Cirulli V, Rutter GA (2004) Mitochondrial localization as a determinant of capacitative Ca^{2+} entry in HeLa cells. Cell Calcium 36:499–508.
138. Vasington FD, Murphy JV (1962) Ca ion uptake by rat kidney mitochondria and its dependence on respiration and phosphorylation. J Biol Chem 237:2670–2677.
139. Wang YX, Zheng YM, Mei QB, Wang QS, Collier ML, Fleischer S, Xin HB, Kotlikoff MI (2004) FKBP12.6 and cADPR regulation of Ca^{2+} release in smooth muscle cells. Am J Physiol Cell Physiol 286:538–546.
140. White TA, Kannan MS, Walseth TF (2003) Intracellular calcium signaling through the cADPR pathway is agonist specific in porcine airway smooth muscle. FASEB J 17:482–484.
141. White TA, Xue AL, Chini EN, Thompson M, Sieck GC, Wylam ME (2006) Role of transient receptor potential C3 in TNF-alpha-enhanced calcium influx in human airway myocytes. Am J Resp Cell Mol 35:243–251.
142. Williams RT, Manji SS, Parker NJ, Hancock MS, Van Stekelenburg L, Eid JP, Senior PV, Kazenwadel JS, Shandala T, Saint R, Smith PJ, Dziadek MA (2001) Identification and characterization of the STIM (stromal interaction molecule) gene family: coding for a novel class of transmembrane proteins. Biochem J 357:673–685.
143. Worley JF, 3rd, Kotlikoff MI (1990) Dihydropyridine-sensitive single calcium channels in airway smooth muscle cells. Am J Physiol 259:468–480.
144. Yi M, Weaver D, Hajnoczky G (2004) Control of mitochondrial motility and distribution by the calcium signal: a homeostatic circuit. J Cell Biol 167:661–672.
145. Zalk R, Lehnart SE, Marks AR (2007) Modulation of the ryanodine receptor and intracellular calcium. Annu Rev Biochem 76:367–385.

Role of Caveolae in the Airway

Christina M. Pabelick, Brij B. Singh, and Y.S. Prakash

Contents

Abstract Caveolae are flask-shaped invaginations of the plasma membrane that are rich in lipids and serve as microdomains to facilitate interactions between proteins at the membrane as well as intracellular components, thus modulating signal transduction, protein and lipid transport, and other processes. Constituent caveolar proteins such as caveolins and cavins also have scaffolding domains that anchor and regulate protein function. There is now evidence that caveolae and their constituent proteins are present in airway smooth muscle in a variety of species. Caveolae in airway cells contain or interact with molecules such as receptors, ion channels, and regulatory proteins that are key to the roles of airway epithelium and

C.M. Pabelick, M.D. (✉) • Y.S. Prakash, M.D., Ph.D.
Department of Anesthesiology, Mayo Clinic, 4-184 W Joseph SMH, Rochester, MN 55905, USA

Department of Physiology and Biomedical Engineering, Mayo Clinic, 4-184 W Joseph SMH, Rochester, MN 55905, USA
e-mail: pabelick.christina@mayo.edu

B.B. Singh, Ph.D.
Department of Biochemistry and Molecular Biology, School of Medicine and Health Sciences, University of North Dakota, Grand Forks, ND, USA

Y.-X. Wang (ed.), *Calcium Signaling In Airway Smooth Muscle Cells*,
DOI 10.1007/978-3-319-01312-1_12,

smooth muscle in regulating airway structure and function. Furthermore, caveolar protein expression and regulation appear to be important in mediating and modulating the effects of inflammation on the airway, thereby contributing to the pathophysiology of diseases such as asthma.

Keywords Caveolae • Caveolin • Cavin • Airway

1 Caveolae, Caveolins, and Cavins

Lipid rafts are unique and distinct from nonraft phospholipid bilayers by being in a liquid-ordered state [1, 2], a result of enrichment of cholesterol and sphingolipids. Unlike planar membranes, caveolae, a subset of lipid rafts, are flask or omega-shaped membrane invaginations that can be viewed using microscopy techniques. The importance of caveolae lies in their recognized role as signaling organizers in a variety of cell types [3–5]. By virtue of the clustering of receptor, signaling molecules, ion channels, and other proteins, caveolae function as cell signaling modulators and additionally play a role in regulating vesicle transport kinetics and the movement of other plasma membrane proteins.

Caveolae contain caveolin proteins as a major structural element that makes them unique. Caveolin-1 (Cav1), caveolin-2 (Cav2), and caveolin-3 (Cav3) [6] have been identified, and higher-order oligomers of these isoforms within plasma membrane are critical for the invagination of caveolae. While Cav1 and Cav2 have been shown to be present in most cell types, Cav3 is expressed primarily in muscle, especially skeletal and cardiac muscle. Cav1 is the most studied isoform in terms of its regulation. Although how Cav1 is trafficked to the plasma membrane is not entirely understood, interactions with signaling intermediates such as ARF1 (in *Caenorhabditis elegans*) and proteins such as amyloid beta, insulin, and Na/K ATPase have been shown to regulate Cav1 trafficking and stabilization. Cav2 is thought to provide a "supporting" role in helping Cav1 form invaginations and in most cell types is coexpressed with Cav1. On the other hand, in striated muscle, Cav3 is localized to the sarcolemma and is critical for caveolar formation even without Cav1.

More recently, a new family of proteins called cavins (cavin1–4) has been identified as being key to caveolar formation and shape and caveolin function. The significance of caveolins and cavins in the formation of caveolae is underscored by the fact that genetic disruption of these proteins results in loss of caveolar formation [7–9]. Cavin1 (also known as PTRF – polymerase I transcript releasing factor, or Cav-p60) is thought to be an integral component of caveolar biogenesis [4, 8] in that sequestration of intracellular Cav1 into caveolae requires cavin1. Furthermore, loss of cavin1 increases the lateral mobility of Cav1 to noncaveolar rafts followed by accelerated lysosomal degradation. Cavin1 colocalizes with Cav1 within lipid rafts but does not directly interact with Cav1, although indirect interactions mediated by cytoskeletal proteins cannot be ruled out. Cavin2 [also known as serum deprivation protein response (SDPR)] is also enriched in caveolae and colocalizes with Cav1. However, unlike with cavin1, overexpression of cavin2 does not increase caveolae number per se but induces changes in caveolar morphology, resulting in extensively distended tubule formation. Conversely, cavin2 downregulation induces loss of Cav1 and cavin1 expression, indirectly leading to decreased caveolar formation. Similarly, cavin3 [sdr-related gene product that binds to c-kinase (SRBC)] has been shown to associate with Cav1 and facilitate intracellular trafficking of Cav1 and may thus be important for endocytosis of Cav1, especially in liver and brain. Cavin4 has been found to be associated with sarcolemmal caveolae in striated muscle. Reduced cavin4 expression has been reported in muscle diseases associated with Cav3 dysfunction, suggesting a functional role for this protein.

Caveolin isoform expression depends on the type of smooth muscle, with all three isoforms expressed in vascular smooth muscle [10], while human airway smooth muscle (ASM) expresses only Cav1 and Cav2 [11, 12]. Limited data in airway epithelium suggest expression of both Cav1 and Cav2 [13, 14]. There is currently very limited information on cavins in ASM or airways [15]. In ongoing studies, we have found substantial expression of PTRF/cavin1 and SRBC/cavin3, while only mRNA for SDPR/cavin2 can be detected, findings that we confirmed in mouse airways [16]. However, their role or interactions with Cav1 in modulating airway structure or function is not known. Furthermore, there is currently no information on cavin isoform expression in airway epithelium.

2 Regulation of Caveolin and Cavin Expression

In other cell types, activation of the extracellular signal-regulated kinases (ERK) pathway plays a critical role in Cav1 expression [17], potentially by inhibiting transcription of the *cav1* gene via phosphorylation of key transcription factors. Growth factors tend to inhibit Cav1 expression via the ERK pathway [18, 19]. Activation of Src family kinases and cyclic AMP (cAMP)-dependent protein kinase have also been implicated in reducing Cav1 [20, 21]. Lipid molecules can positively or negatively regulate Cav1 expression. For example, transcription is enhanced by free cholesterol [22, 23], while high-density lipoproteins decrease Cav1 expression [23, 24]. Cav2 expression is linked to Cav1, but, unlike transcriptional regulation of Cav1 expression, it is increased degradation of Cav2 in the absence of Cav1 that is responsible. Normal regulation of caveolin isoform expression in the airway has not been examined. In a previous study, we found that pharmacological inhibition of MAP kinases, especially ERK, as well as NFκB reduces Cav1 protein expression in human ASM cells [12].

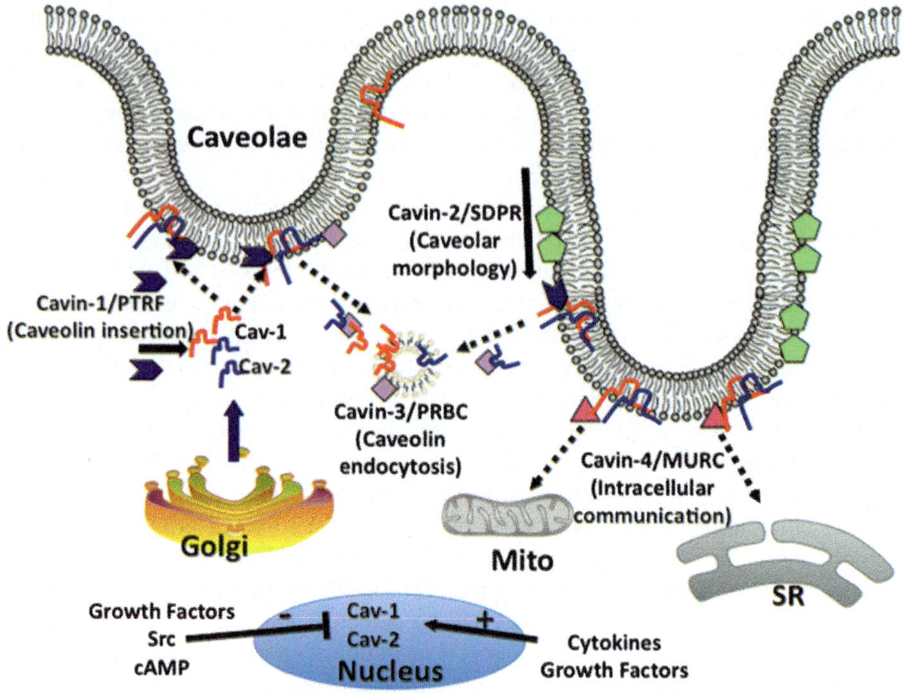

Fig. 1 Caveolins and cavins in airway smooth muscle (*ASM*). Unlike striated muscle, ASM largely expresses caveolin-1 and caveolin-2. Regulation of caveolin expression is complex and can occur at the level of gene and protein expression, as well as by protein insertion into the plasma membrane. In this regard, besides modulators such as cytokines and growth factors, recently identified cavin proteins appear to be important. Cavin1 regulates caveolin insertion, cavin2 the shape of caveolae, cavin3 internalization of caveolins, and cavin4 the interaction of caveolae with intracellular structures, particularly the sarcoplasmic reticulum

There are currently limited data on the regulation of cavin expression in any cell system. Importantly, oxidative stress upregulates cavin1 expression [25] through unknown mechanisms. Separately, cytoskeletal disruption and cholesterol depletion also decrease cavin1 expression [26], as can factors such as insulin, while isoproterenol increases expression [27]. Serum deprivation reduces cavin2 and 3; however, the mechanism that regulates their expression is not known. There is currently no information on cavin isoform expression in airway epithelium or ASM. Figure 1 provides a model of the role of cavins and Cav1 in caveolar morphology and function.

3 Functional Roles of Caveolins and Cavins

As invaginations harboring a variety of proteins, caveolae can facilitate interactions between the plasma membrane and intracellular organelles, such as the endoplasmic reticulum and mitochondria, as well as components such as the cytoskeleton. Although caveolins are relatively small (21–24 kDa), their distinct scaffolding region is an essential part of their ability to bind many signaling proteins [4, 7]. Similarly, domains on cavin proteins are important for protein–protein interactions. Caveolins bind and shuttle cholesterol and other proteins to/from the plasma membrane [5]. In smooth muscle cells, including those of the airway, caveolae contain proteins that regulate intracellular Ca^{2+} ($[Ca^{2+}]_i$) and contractility, as well as cellular proliferation [28–31]. Caveolae in canine ASM express binding proteins for L-type Ca^{2+} channels and plasma membrane Ca^{2+} ATPase [32], while human ASM caveolae harbor agonist receptors (e.g., muscarinic M3 and histaminergic H1), Ca^{2+} influx mechanisms (e.g., transient receptor potential canonical (TRPC) channels and Orai1) [11], and proteins such as RhoA that regulate force [11, 33]. In human ASM cells, we previously showed that caveolar muscarinic and histaminergic receptors are critical for $[Ca^{2+}]_i$ responses to these bronchoconstrictor agonists [11] such that the Ca^{2+} responses are substantially blunted in the presence of Cav1 siRNA or when Cav1 function is inhibited using a scaffolding domain inhibitory peptide. Cav1 also appears to be important for TRPC channel localization and function, and in human ASM, depletion of Cav1 inhibits store-operated calcium entry (SOCE) [34]. Furthermore, the SOCE regulatory protein (indeed, perhaps even the channel) Orai1 is expressed within caveolae in human ASM. Importantly, the sensor for sarcoplasmic reticulum (SR) Ca^{2+} levels, STIM1, interacts with Cav1 and Orai1 within caveolae to regulate SOCE. In addition to these roles in regulating $[Ca^{2+}]_i$, Cav1 is also important for the regulation of Ca^{2+} sensitivity for force generation. In porcine ASM, we demonstrated that caveolar depletion with cyclodextrin or Cav1 scaffolding domain inhibitor peptide decreased force responses to acetylcholine (ACh). Inhibiting RhoA with fasudil or Y27632, is more pronounced in the absence of Cav1 signaling.

In endothelial cells, nitric oxide synthase (NOS) is a classic caveolae-associated signaling enzyme. Direct interaction between Cav1 and endothelial nitric oxide synthase (eNOS) has been demonstrated, with targeting of eNOS to the caveolae dependent on posttranslational lipid modifications of eNOS [35, 36]. Importantly, under basal conditions, Cav1 inhibits eNOS activity [36], which is overcome by elevation of $[Ca^{2+}]_i$ (e.g., by inducers of nitric oxide such as ACh or ATP), recruitment of calmodulin, and involvement of phosphoinositide-3 (PI3)/Akt. While the role of epithelial NO as a bronchodilator is less clear, a similar role for Cav1 in regulating epithelial NOS has not been well studied. Using Cav1 knockout (KO) mice, we recently reported [16] the surprising finding that hypercontractility of the Cav1 KO airway (which contrasts with the in vitro human ASM work) may involve impaired Cav1/NOS interactions. However, further examination of Cav1 in epithelium is yet to be done.

In addition to G-protein-coupled receptors (GPCRs), receptor tyrosine kinase pathways, which are activated by factors such as epidermal growth factor (EGF), insulin-like growth factor (IGF), and platelet-derived growth factor (PDGF), are localized in caveolae and regulate cell proliferation and differentiation. Furthermore, regulation of kinase cascades such as PI3/Akt and mitogen-activated protein kinases (MAPK) are dependent on raft/caveolae integrity [5]. Cav1 inhibits Raf/MEK-1/ERK-2 signaling to the nucleus [17]. Studies in human ASM [29] have demonstrated a role for caveolae in ERK signaling. Although Gosens et al. showed that upon activation p42/p44 MAP kinase appears to be localized to caveolae-free membrane domains, some of the upstream molecules important for ERK signaling remained within caveolae. In contrast, we demonstrated in porcine ASM the importance of Cav1 in the regulation of ASM contractility [37] and did not find a role for ERK signaling. A dynamic association between protein kinase C (PKC) and caveolae has also been demonstrated in other cell types [8, 38] and may need to be explored in ASM. In this regard, similar to eNOS association, caveolin peptides that bind to the Cav1 scaffolding domain inhibit PKC activity [39]. Moreover, PKC-mediated vasoconstriction is increased in some arteries of Cav1-deficient mice [40]. Importantly, SDPR (cavin2) and SRBC (cavin3) bind to PKCα and PKCδ, and thus caveolar association of PKC may be mediated in part by cavins.

Of the four cavins identified, PTRF/cavin1 [8, 41], SDPR/cavin2 [42], and SRBC/cavin3 [43] appear to be important in caveolar formation and shape, while MURC/cavin4 [44, 45] helps enhance interactions with intracellular structures. Considering their novelty, there is currently very limited information on cavins in ASM or airways [46]. We demonstrated the presence of cavin1–3 in mouse lung [16]. However, their role or interactions with Cav1 in modulating airway contractility is not known. Furthermore, there is no information in regard to cavin expression and their role in airway epithelium.

4 Caveolae in Airway Inflammation

Airway inflammation leads to increased contractility and enhanced $[Ca^{2+}]_i$ responses of ASM. The effect of inflammation on caveolae and caveolins in the inflamed airway have only recently been examined and are thought to be important [47]. In a mostly biochemical study using human ASM, it has been shown that lipid rafts are important in tumor necrosis factor alpha (TNFα)-mediated effects [48]. In that study, tumor necrosis factor receptor 1 (TNFR1) was localized to both lipid raft and nonraft plasma membrane regions. Lipid rafts are important for TNFR1-mediated activation of RhoA but not NFκB and MAPK. However, this study did not examine the functional role of Cav1 during airway inflammation. In ovalbumin-sensitized/challenged mice, Cav1 is necessary for IL-4-mediated enhancement of transforming growth factor beta (TGFβ) signaling, but this study did not examine airway reactivity per se [49]. Cav1 can regulate NFκB and inflammatory responses

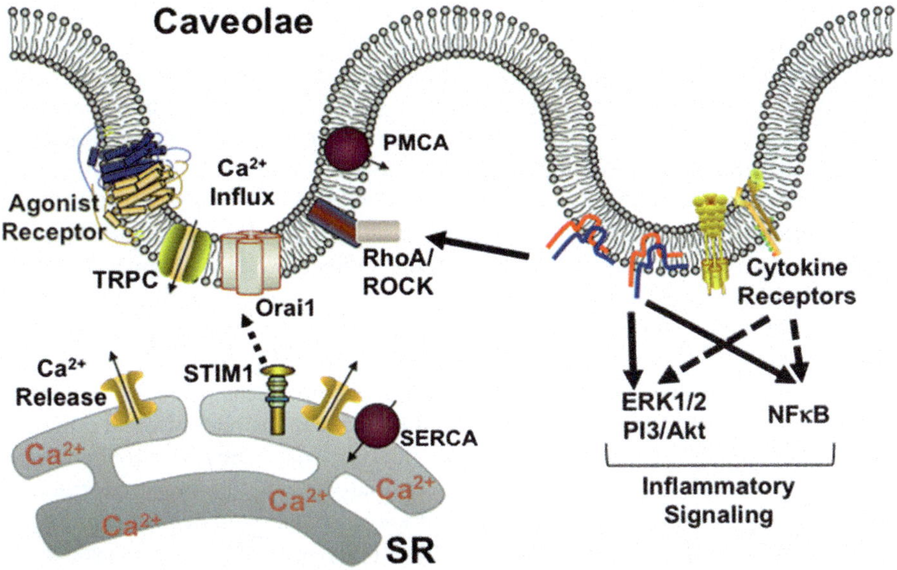

Fig. 2 Caveolins are important in airway smooth muscle signaling. By harboring a number of regulatory proteins including agonist receptors, Ca^{2+} influx channels (e.g., TRPC and store-operated channel Orai1) and the RhoA pathway for Ca^{2+} sensitization. In turn, caveolins can regulate intracellular signaling cascades such as ERK and NFkB, thereby modulating the response of inflammation, relevant to diseases such as asthma

to sepsis [47]. Previous studies have demonstrated a role for caveolins in regulation of p42/44 MAP kinase [29] and RhoA and Rho kinase signaling [50], thereby modulating cellular proliferation, relevant to diseases such as asthma. Cytokines such as TNFα can enhance muscarinic receptor activation as well as Ca^{2+} influx mechanisms [51–53]. We recently demonstrated that such cytokine effects involve Cav1 [12], where TNFα and IL-13 increase Cav1 expression in human ASM. Importantly, downregulation of Cav1 expression blunted cytokine-induced enhancement of $[Ca^{2+}]_i$ responses. Interestingly, the effects on IL-13 are not as profound as with TNFα. The reason for the relative smaller effect appears to be due to the fact that the IL-13 receptor is not expressed within caveolae of human ASM (in comparison to TNFα receptor). In a recent study, we also found that although both Orai1 and STIM1 play an important role in TNFα-induced enhancement of SOCE, caveolar Orai1 is particularly important in mediating cytokine effects. Figure 2 demonstrates the importance of caveolae and Cav1 in $[Ca^{2+}]_i$ regulation and airway inflammation.

In different smooth muscle types such as vascular, gallbladder, and intestine, as well as airway, caveolins play a role in force regulation [28, 30, 37, 54–56]. We further demonstrated recently that TNFα enhancement of ASM contractility is dependent on RhoA and involves Cav1, such that cyclodextrin or Cav1 siRNA blunts cytokine effects [37]. Previous studies examining cell proliferation have reported Cav1 regulation of p42/44 MAP kinase [29] as well as RhoA and Rho kinase signaling [50], which are important in mediating cytokine effects on ASM. While much of the work on caveolins in ASM has been done in vitro, what is interesting is that the data from in vivo models such as the Cav1 KO mouse actually demonstrate airway hyperresponsiveness in the absence of Cav1. In a recent study using ovalbumin-sensitized and challenged Cav1 KO mice, we showed increased airway reactivity to methacholine compared to wild-type controls [16] and further demonstrated a complex role for epithelial caveolins involving a balance between arginase and NOS in terms of L-arginine metabolism. Wild-type mice that were sensitized/challenged showed increased Cav1 in both ASM and epithelium, consistent with in vitro findings using inflammatory cytokines. A potential confounding factor that underlies the discrepancy between in vitro and in vivo findings may be a time- and context-dependent role for Cav1 that has not been explored. Short-term alterations in Cav1 expression as done in vitro may demonstrate different effects in comparison to the permanent absence of this important protein throughout life in vivo. Regardless, what does happen is that airway inflammation leads to upregulation of cavins, consistent with their known function in enhancing caveolar structure and function [8, 42, 43, 57]. Interestingly, Cav1 KO airways showed a dramatic decrease in the expression of cavins, suggesting a transcriptional modulation of these proteins by Cav1 and their potential role in contractile responses of the airway. However, these issues remain to be examined.

5 Summary and Conclusions

Caveolae are present in the airway, and a variety of caveolin and cavin isoforms are expressed. Cav1 appears to be important in harboring a number of key proteins, such as receptors and ion channels, that regulate $[Ca^{2+}]_i$ and force in the airway. Furthermore, Cav1 itself signals intracellularly to regulate contractility and cell proliferation. Importantly, caveolins are modulators and mediators of inflammatory signaling in the airway, which is relevant to diseases such as asthma. However, their role in the airway is certainly complex and remains to be completely elucidated.

References

1. Brown DA, London E (1998) Functions of lipid rafts in biological membranes. Annu Rev Cell Dev Biol 14: 111–136. DOI 10.1146/annurev.cellbio.14.1.111
2. Pike LJ, Han X, Chung KN, Gross RW (2002) Lipid rafts are enriched in arachidonic acid and plasmenylethanolamine and their composition is independent of caveolin-1 expression: a quantitative electrospray ionization/mass spectrometric analysis. Biochemistry 41: 2075–2088
3. Pike LJ (2005) Growth factor receptors, lipid rafts and caveolae: an evolving story. Biochimica et biophysica acta 1746: 260–273. DOI 10.1016/j.bbamcr.2005.05.005

4. Thomas CM, Smart EJ (2008) Caveolae structure and function. Journal of cellular and molecular medicine 12: 796–809. DOI 10.1111/j.1582-4934.2008.00295.x
5. Patel HH, Murray F, Insel PA (2008) Caveolae as organizers of pharmacologically relevant signal transduction molecules. Annu Rev Pharmacol Toxicol 48: 359–391. DOI 10.1146/annurev.pharmtox.48.121506.124841
6. Tang Z, Scherer PE, Okamoto T, Song K, Chu C, Kohtz DS, Nishimoto I, Lodish HF, Lisanti MP (1996) Molecular cloning of caveolin-3, a novel member of the caveolin gene family expressed predominantly in muscle. J Biol Chem 271: 2255–2261
7. Razani B, Combs TP, Wang XB, Frank PG, Park DS, Russell RG, Li M, Tang B, Jelicks LA, Scherer PE, Lisanti MP (2002) Caveolin-1-deficient mice are lean, resistant to diet-induced obesity, and show hypertriglyceridemia with adipocyte abnormalities. J Biol Chem 277: 8635–8647. DOI 10.1074/jbc.M110970200
8. Hill MM, Bastiani M, Luetterforst R, Kirkham M, Kirkham A, Nixon SJ, Walser P, Abankwa D, Oorschot VM, Martin S, Hancock JF, Parton RG (2008) PTRF-Cavin, a conserved cytoplasmic protein required for caveola formation and function. Cell 132: 113–124. DOI S0092-8674(07)01546-2 [pii] 10.1016/j.cell.2007.11.042
9. Liu L, Brown D, McKee M, Lebrasseur NK, Yang D, Albrecht KH, Ravid K, Pilch PF (2008) Deletion of Cavin/PTRF causes global loss of caveolae, dyslipidemia, and glucose intolerance. Cell Metab 8: 310–317. DOI S1550-4131(08)00242-8 [pii] 10.1016/j.cmet.2008.07.008
10. Cohen AW, Hnasko R, Schubert W, Lisanti MP (2004) Role of caveolae and caveolins in health and disease. Physiol Rev 84: 1341–1379. DOI 10.1152/physrev.00046.2003
11. Prakash YS, Thompson MA, Vaa B, Matabdin I, Peterson TE, He T, Pabelick CM (2007) Caveolins and intracellular calcium regulation in human airway smooth muscle. Am J Physiol Lung Cell Mol Physiol 293: L1118–1126. DOI 00136.2007 [pii] 10.1152/ajplung.00136.2007
12. Sathish V, Abcejo AJ, Vanoosten SK, Thompson MA, Prakash YS, Pabelick CM (2011) Caveolin-1 in Cytokine-Induced Enhancement of Intracellular Ca^{2+} in Human Airway Smooth Muscle. Am J Physiol Lung Cell Mol Physiol. DOI ajplung.00019.2011 [pii] 10.1152/ajplung.00019.2011
13. Byrne S, Cheent A, Dimond J, Fisher G, Ockleford CD (2001) Immunocytochemical localization of a caveolin-1 isoform in human term extra-embryonic membranes using confocal laser scanning microscopy: implications for the complexity of the materno-fetal junction. Placenta 22: 499–510. DOI 10.1053/plac.2001.0697S0143-4004(01)90697-6 [pii]
14. Silva WI, Maldonado HM, Lisanti MP, Devellis J, Chompre G, Mayol N, Ortiz M, Velazquez G, Maldonado A, Montalvo J (1999) Identification of caveolae and caveolin in C6 glioma cells. Int J Dev Neurosci 17: 705–714. DOI S0736-5748(99)00040-4 [pii]
15. Sharma P, Ghavami S, Stelmack GL, McNeill KD, Mutawe MM, Klonisch T, Unruh H, Halayko AJ (2010) beta-Dystroglycan binds caveolin-1 in smooth muscle: a functional role in caveolae distribution and Ca^{2+} release. Journal of cell science 123: 3061–3070. DOI 10.1242/jcs.066712
16. Aravamudan B, VanOosten SK, Meuchel LW, Vohra P, Thompson M, Sieck GC, Prakash YS, Pabelick CM (2012) Caveolin-1 knockout mice exhibit airway hyperreactivity. Am J Physiol Lung Cell Mol Physiol 303: L669–681. DOI 10.1152/ajplung.00018.2012
17. Engelman JA, Zhang XL, Razani B, Pestell RG, Lisanti MP (1999) p42/44 MAP kinase-dependent and -independent signaling pathways regulate caveolin-1 gene expression. Activation of Ras-MAP kinase and protein kinase a signaling cascades transcriptionally down-regulates caveolin-1 promoter activity. J Biol Chem 274: 32333–32341
18. Liu J, Razani B, Tang S, Terman BI, Ware JA, Lisanti MP (1999) Angiogenesis activators and inhibitors differentially regulate caveolin-1 expression and caveolae formation in vascular endothelial cells. Angiogenesis inhibitors block vascular endothelial growth factor-induced down-regulation of caveolin-1. J Biol Chem 274: 15781–15785
19. Park WY, Cho KA, Park JS, Kim DI, Park SC (2001) Attenuation of EGF signaling in senescent cells by caveolin. Ann N Y Acad Sci 928: 79–84

20. Razani B, Wang XB, Engelman JA, Battista M, Lagaud G, Zhang XL, Kneitz B, Hou H, Jr., Christ GJ, Edelmann W, Lisanti MP (2002) Caveolin-2-deficient mice show evidence of severe pulmonary dysfunction without disruption of caveolae. Molecular and cellular biology 22: 2329–2344
21. Smart EJ, Graf GA, McNiven MA, Sessa WC, Engelman JA, Scherer PE, Okamoto T, Lisanti MP (1999) Caveolins, liquid-ordered domains, and signal transduction. Molecular and cellular biology 19: 7289–7304
22. Bist A, Fielding PE, Fielding CJ (1997) Two sterol regulatory element-like sequences mediate up-regulation of caveolin gene transcription in response to low density lipoprotein free cholesterol. Proceedings of the National Academy of Sciences of the United States of America 94: 10693–10698
23. Fielding CJ, Bist A, Fielding PE (1997) Caveolin mRNA levels are up-regulated by free cholesterol and down-regulated by oxysterols in fibroblast monolayers. Proceedings of the National Academy of Sciences of the United States of America 94: 3753–3758
24. Frank PG, Galbiati F, Volonte D, Razani B, Cohen DE, Marcel YL, Lisanti MP (2001) Influence of caveolin-1 on cellular cholesterol efflux mediated by high-density lipoproteins. American journal of physiology Cell physiology 280: C1204–1214
25. Volonte D, Galbiati F (2011) Polymerase I and transcript release factor (PTRF)/cavin-1 is a novel regulator of stress-induced premature senescence. J Biol Chem 286: 28657–28661. DOI 10.1074/jbc.C111.235119
26. Liu L, Pilch PF (2008) A critical role of cavin (polymerase I and transcript release factor) in caveolae formation and organization. J Biol Chem 283: 4314–4322. DOI M707890200 [pii] 10.1074/jbc.M707890200
27. Aboulaich N, Chui PC, Asara JM, Flier JS, Maratos-Flier E (2011) Polymerase I and transcript release factor regulates lipolysis via a phosphorylation-dependent mechanism. Diabetes 60: 757–765. DOI 10.2337/db10-0744
28. Daniel EE, Eteraf T, Sommer B, Cho WJ, Elyazbi A (2009) The role of caveolae and caveolin 1 in calcium handling in pacing and contraction of mouse intestine. Journal of cellular and molecular medicine 13: 352–364
29. Gosens R, Stelmack GL, Dueck G, McNeill KD, Yamasaki A, Gerthoffer WT, Unruh H, Gounni AS, Zaagsma J, Halayko AJ (2006) Role of caveolin-1 in p42/p44 MAP kinase activation and proliferation of human airway smooth muscle. Am J Physiol Lung Cell Mol Physiol 291: L523–534. DOI 00013.2006 [pii] 10.1152/ajplung.00013.2006
30. Pojoga LH, Adamova Z, Kumar A, Stennett AK, Romero JR, Adler GK, Williams GH, Khalil RA (2010) Sensitivity of NOS-dependent vascular relaxation pathway to mineralocorticoid receptor blockade in caveolin-1-deficient mice. Am J Physiol Heart Circ Physiol 298: H1776–1788
31. Poljakovic M, Porter DW, Millecchia L, Kepka-Lenhart D, Beighley C, Wolfarth MG, Castranova V, Morris SM, Jr. (2007) Cell- and isoform-specific increases in arginase expression in acute silica-induced pulmonary inflammation. J Toxicol Environ Health A 70: 118–127. DOI 769176703 [pii] 10.1080/15287390600755075
32. Darby PJ, Kwan CY, Daniel EE (2000) Caveolae from canine airway smooth muscle contain the necessary components for a role in Ca^{2+} handling. Am J Physiol Lung Cell Mol Physiol 279: L1226–1235
33. Hunter I, Nixon GF (2006) Spatial compartmentalization of tumor necrosis factor (TNF) receptor 1-dependent signaling pathways in human airway smooth muscle cells. Lipid rafts are essential for TNF-alpha-mediated activation of RhoA but dispensable for the activation of the NF-kappaB and MAPK pathways. J Biol Chem 281: 34705–34715. DOI M605738200 [pii] 10.1074/jbc.M605738200
34. Sathish V, Abcejo AJ, Thompson MA, Sieck GC, Prakash YS, Pabelick CM (2012) Caveolin-1 regulation of store-operated Ca^{2+} influx in human airway smooth muscle. The European respiratory journal : official journal of the European Society for Clinical Respiratory Physiology 40: 470–478. DOI 10.1183/09031936.00090511

35. Garcia-Cardena G, Oh P, Liu J, Schnitzer JE, Sessa WC (1996) Targeting of nitric oxide synthase to endothelial cell caveolae via palmitoylation: implications for nitric oxide signaling. Proceedings of the National Academy of Sciences of the United States of America 93: 6448–6453
36. Ju H, Zou R, Venema VJ, Venema RC (1997) Direct interaction of endothelial nitric-oxide synthase and caveolin-1 inhibits synthase activity. J Biol Chem 272: 18522–18525
37. Sathish V, Yang B, Meuchel LW, VanOosten SK, Ryu AJ, Thompson MA, Prakash YS, Pabelick CM (2011) Caveolin-1 and force regulation in porcine airway smooth muscle. Am J Physiol Lung Cell Mol Physiol 300: L920–929. DOI ajplung.00322.2010 [pii] 10.1152/ajplung.00322.2010
38. Lisanti MP, Scherer PE, Tang Z, Sargiacomo M (1994) Caveolae, caveolin and caveolin-rich membrane domains: a signalling hypothesis. Trends Cell Biol 4: 231–235
39. Oka N, Yamamoto M, Schwencke C, Kawabe J, Ebina T, Ohno S, Couet J, Lisanti MP, Ishikawa Y (1997) Caveolin interaction with protein kinase C. Isoenzyme-dependent regulation of kinase activity by the caveolin scaffolding domain peptide. J Biol Chem 272: 33416–33421
40. Shakirova Y, Bonnevier J, Albinsson S, Adner M, Rippe B, Broman J, Arner A, Sward K (2006) Increased Rho activation and PKC-mediated smooth muscle contractility in the absence of caveolin-1. American journal of physiology Cell physiology 291: C1326–1335. DOI 10.1152/ajpcell.00046.2006
41. Vinten J, Johnsen AH, Roepstorff P, Harpoth J, Tranum-Jensen J (2005) Identification of a major protein on the cytosolic face of caveolae. Biochim Biophys Acta 1717: 34–40. DOI S0005-2736(05)00285-3 [pii] 10.1016/j.bbamem.2005.09.013
42. Hansen CG, Bright NA, Howard G, Nichols BJ (2009) SDPR induces membrane curvature and functions in the formation of caveolae. Nat Cell Biol 11: 807–814. DOI ncb1887 [pii]10.1038/ncb1887
43. McMahon KA, Zajicek H, Li WP, Peyton MJ, Minna JD, Hernandez VJ, Luby-Phelps K, Anderson RG (2009) SRBC/cavin-3 is a caveolin adapter protein that regulates caveolae function. EMBO J 28: 1001–1015. DOI emboj200946 [pii] 10.1038/emboj.2009.46
44. Bastiani M, Liu L, Hill MM, Jedrychowski MP, Nixon SJ, Lo HP, Abankwa D, Luetterforst R, Fernandez-Rojo M, Breen MR, Gygi SP, Vinten J, Walser PJ, North KN, Hancock JF, Pilch PF, Parton RG (2009) MURC/Cavin-4 and cavin family members form tissue-specific caveolar complexes. J Cell Biol 185: 1259–1273. DOI jcb.200903053 [pii] 10.1083/jcb.200903053
45. Ogata T, Ueyama T, Isodono K, Tagawa M, Takehara N, Kawashima T, Harada K, Takahashi T, Shioi T, Matsubara H, Oh H (2008) MURC, a muscle-restricted coiled-coil protein that modulates the Rho/ROCK pathway, induces cardiac dysfunction and conduction disturbance. Mol Cell Biol 28: 3424–3436. DOI MCB.02186-07 [pii] 10.1128/MCB.02186-07
46. Sharma P, Ghavami S, Stelmack GL, McNeill KD, Mutawe MM, Klonisch T, Unruh H, Halayko AJ (2010) beta-Dystroglycan binds caveolin-1 in smooth muscle: a functional role in caveolae distribution and Ca^{2+} release. J Cell Sci 123: 3061–3070. DOI jcs.066712 [pii] 10.1242/jcs.066712
47. Garrean S, Gao XP, Brovkovych V, Shimizu J, Zhao YY, Vogel SM, Malik AB (2006) Caveolin-1 regulates NF-kappaB activation and lung inflammatory response to sepsis induced by lipopolysaccharide. J Immunol 177: 4853–4860. DOI 177/7/4853 [pii]
48. Hunter I, Nixon GF (2006) Spatial compartmentalization of TNFR1-dependent signaling pathways in human airway smooth muscle cells: lipid rafts are essential for TNF-alpha -mediated activation of RhoA but dispensable for the activation of the NF-kappa B and MAPK pathways. J Biol Chem 281: 34705–34715
49. Le Saux CJ, Teeters K, Miyasato SK, Hoffmann PR, Bollt O, Douet V, Shohet RV, Broide DH, Tam EK (2008) Down-regulation of caveolin-1, an inhibitor of transforming growth factor-beta signaling, in acute allergen-induced airway remodeling. J Biol Chem 283: 5760–5768. DOI M701572200 [pii] 10.1074/jbc.M701572200

50. Sommer B, Montano LM, Carbajal V, Flores-Soto E, Ortega A, Ramirez-Oseguera R, Irles C, El-Yazbi AF, Cho WJ, Daniel EE (2009) Extraction of membrane cholesterol disrupts caveolae and impairs serotonergic (5-HT2A) and histaminergic (H1) responses in bovine airway smooth muscle: role of Rho-kinase. Can J Physiol Pharmacol 87: 180–195. DOI y08-114 [pii] 10.1139/y08-114
51. Hotta K, Emala CW, Hirshman CA (1999) TNF-alpha upregulates Gialpha and Gqalpha protein expression and function in human airway smooth muscle cells. American Journal of Physiology 276: L405–411
52. Oenema TA, Kolahian S, Nanninga JE, Rieks D, Hiemstra PS, Zuyderduyn S, Halayko AJ, Meurs H, Gosens R (2010) Pro-inflammatory mechanisms of muscarinic receptor stimulation in airway smooth muscle. Respir Res 11: 130. DOI 1465-9921-11-130 [pii] 10.1186/1465-9921-11-130
53. White TA, Xue A, Chini EN, Thompson M, Sieck GC, Wylam ME (2006) Role of transient receptor potential C3 in TNF-alpha-enhanced calcium influx in human airway myocytes. Am J Respir Cell Mol Biol 35: 243–251. DOI 2006-0003OC [pii]10.1165/rcmb.2006-0003OC
54. El-Yazbi AF, Cho WJ, Schulz R, Daniel EE (2008) Calcium extrusion by plasma membrane calcium pump is impaired in caveolin-1 knockout mouse small intestine. European journal of pharmacology 591: 80–87
55. Shakirova Y, Mori M, Ekman M, Erjefalt J, Uvelius B, Sward K (2010) Human urinary bladder smooth muscle is dependent on membrane cholesterol for cholinergic activation. European journal of pharmacology 634: 142–148
56. Polyak E, Boopathi E, Mohanan S, Deng M, Zderic SA, Wein AJ, Chacko S (2009) Alterations in caveolin expression and ultrastructure after bladder smooth muscle hypertrophy. J Urol 182: 2497–2503. DOI S0022-5347(09)01735-2 [pii] 10.1016/j.juro.2009.07.011
57. Chidlow JH, Jr., Sessa WC (2010) Caveolae, caveolins, and cavins: complex control of cellular signalling and inflammation. Cardiovasc Res 86: 219–225. DOI cvq075 [pii] 10.1093/cvr/cvq075

CD38-Cyclic ADP-Ribose-Mediated Calcium Signaling in Airway Myocytes

Deepak A. Deshpande, Alonso Guedes, Mythili Dileepan, Timothy F. Walseth, and Mathur S. Kannan

Contents

Abstract Nicotinamide adenine dinucleotide (NAD) metabolites, cyclic ADP-ribose (cADPR), and nicotinic acid adenine dinucleotide phosphate (NAADP) have been identified as calcium-releasing second messengers. In smooth muscle including that of airways, cADPR plays a vital role in the dynamic regulation of intracellular calcium and contraction. CD38, a 45 kDa bifunctional transmembrane protein, possesses enzymatic activities (ADP-ribosyl cyclase and cADPR hydrolase) necessary for the synthesis and degradation of cADPR. Together, CD38 and cADPR form a signaling cascade in agonist-induced calcium elevation

---

D.A. Deshpande, Ph.D.
Department of Medicine, Pulmonary and Critical Care Medicine, School of Medicine, University of Maryland, HSF-II, Rm-S143, 20 Penn Street, Baltimore, MD 21201, USA

A. Guedes, D.V.M., Ph.D.
Department of Anesthesiology, School of Veterinary Medicine, University of California, Davis, CA, USA

M. Dileepan • M.S. Kannan, B.V.Sc., M.Sc., Ph.D. (✉)
Department of Veterinary and Biomedical Sciences, College of Veterinary Medicine, University of Minnesota, 205 Veterinary Science, 1971 Commonwealth Ave., St. Paul, MN 55108, USA
e-mail: kanna001@umn.edu

T.F. Walseth, Ph.D.
Department of Pharmacology, University of Minnesota, St. Paul, MN 55108, USA

Y.-X. Wang (ed.), *Calcium Signaling In Airway Smooth Muscle Cells*,
DOI 10.1007/978-3-319-01312-1_13,

in airway smooth muscle (ASM) cells similar to well-established phospholipase C and inositol trisphosphate (PLC/IP3) pathway. CD38/cADPR is considered an endogenous activator of calcium release from the sarcoplasmic reticulum via ryanodine receptor channels. Most importantly, findings from ex vivo and in vivo studies have established the contribution of CD38/cADPR-mediated calcium release to the regulation of contractile responsiveness of airways and respiratory function. CD38 expression is regulated by inflammatory cytokines, microRNAs, and exogenous drugs such as corticosteroids. Changes in CD38 expression and cADPR production have significant consequences in ASM functions and also contribute to hyperresponsiveness seen during airway inflammatory conditions such as asthma. This chapter describes numerous studies that have established signaling, functional, and pathophysiological roles of CD38/cADPR in ASM.

Keywords CD38 • cADPR • Cyclase • Cytokine • miRNA

Cyclic ADP-ribose (cADPR), a nicotinamide adenine dinucleotide (NAD) metabolite, has been identified as a calcium-releasing second messenger in a variety of cell types including smooth muscle cells. In airway smooth muscle (ASM) cells, calcium release from intracellular stores such as the sarcoplasmic reticulum (SR) plays a prominent role during agonist responses. The principal pathways of SR calcium release involve activation of the inositol 1,4,5-trisphosphate receptors (IP_3Rs) and the ryanodine receptors (RyRs). Recent studies on the mechanisms underlying SR calcium release through RyRs have shown that both calcium and cADPR are involved in the activation of these channels. Furthermore, the capacity for $[Ca^{2+}]_i$ regulation through this mechanism in ASM cells is significantly altered during inflammation and appears to depend on the expression of a cell-surface protein CD38. CD38 expression and its function determine the magnitude of cADPR-mediated calcium responses in ASM cells to a variety of agonists that act via G-protein-coupled receptors (GPCRs). We will outline evidence from studies from our laboratory as well as from others that the CD38/cADPR signaling is relevant in calcium homeostasis in ASM cells and that this pathway is highly regulated by inflammatory cytokines that are relevant in asthma. The functional significance of CD38/cADPR-mediated calcium release in the regulation of respiratory functions under normal and airway diseases is also discussed.

ASM contraction and relaxation are regulated by a complex, coordinated action of receptors, ion channels, and effector molecules within a cell. When excited by neurotransmitters or agonists of inflammatory origin, smooth muscle contraction occurs via actin and myosin cross-bridge cycling. This excitation–contraction (EC) coupling is preceded by an elevation in $[Ca^{2+}]_i$. Relaxing agents, in contrast, in part mediate their effect by inhibiting calcium elevation in ASM. Elevation of $[Ca^{2+}]_i$ is a prerequisite for the subsequent steps in cross-bridge cycling and contraction. Under basal conditions, ASM cells maintain a low level of calcium (approx. 100–200 nM) resulting from a balance between calcium-elevating and reuptake mechanisms. Depending on the agonist and duration of stimulation, the magnitude of $[Ca^{2+}]_i$ rise varies. Multiple mechanisms mediate processes by which

ASM cells maintain $[Ca^{2+}]_i$ at a low baseline level, elevate $[Ca^{2+}]_i$ upon agonist stimulation, and then return $[Ca^{2+}]_i$ to resting levels upon withdrawal of the stimulus. The key elements dictating the rise and fall of global $[Ca^{2+}]_i$ include sources of calcium, calcium-release channels, second messengers, pumps, and exchangers. The resultant calcium level in the cell is a complex interplay of calcium source and second messengers that are released inside the ASM cell. Activation and inactivation kinetics of all these regulatory molecules render calcium homeostasis a very complex phenomenon. In this context, cADPR, a calcium-releasing second messenger, has been identified as a key player in ASM cells [13].

Calcium signaling in ASM can be initiated by electrical changes on the plasma membrane but is most commonly caused by stimulation of cells by select GPCR agonists [10]. Gq-coupled GPCR agonists are primary regulators of calcium signaling and contraction of ASM cells [23]. Each of these agonists (either released by nerve terminals or infiltrating inflammatory cells) activates its cognate receptors to induce calcium signaling and, subsequently, contraction of ASM. A known endogenous GPCR agonist is acetylcholine (ACh), which is released from parasympathetic nerve terminals and mediates its effect via Gq-coupled m3 muscarinic receptors (m3mAChRs). Other Gq-coupled receptors that ASM cells express and are physiologically and pathologically important include H1 histamine receptors (H1HRs), BK1 receptors (activated by bradykinin), CysLT1Rs (cysteinyl leukotrienes), PARs (thrombin), 5-HT2c receptors (serotonin), and D2 receptors (dopamine). Activation of all these receptors results in an elevation of free cytoplasmic calcium in ASM cells. Proton-sensing receptors such as OGR1 are also expressed on ASM, and activation by acidic pH results in an elevation of calcium and contraction of ASM cells [54]. Ligand-gated channels such as $GABA_A$, $GABA_B$, and glycine receptors are known to modulate ASM calcium homeostasis and contractility [43, 45, 64]. A recent study revealed the that expression of taste receptor type 2 (TAS2R) on human ASM cells and activation of ASM cells by a number of bitter tastants results in an elevation of $[Ca^{2+}]_i$ [12].

Activation of Gq-coupled receptors on ASM membrane by agonists results in the accumulation of calcium-releasing second messenger molecules. In the context of Gq-coupled GPCRs, agonists such as ACh, histamine, and LTD4 cause the accumulation of 1,4,5-trisphosphate (IP_3) in ASM cells [5]. Similar findings have been demonstrated for ASM strips obtained from trachea and bronchi [6, 7]. Studies from our laboratory and others revealed accumulation of cADPR in ASM upon stimulation with agonists such as ACh, bradykinin, endothelin-1, and histamine [33]. Similar observations were found in longitudinal smooth muscle of gastrointestinal (GI) tract and vascular smooth muscle [15, 35]. Agonist-induced increase in cADPR concentration in ASM cells suggests a potential second messenger role for cADPR in ASM, similar to IP_3.

cADPR is synthesized by the action of ADP-ribosyl cyclase and degraded by cADPR hydrolase [39]. Interestingly, both enzyme activities are associated with a single-membrane-bound protein, CD38 in ASM [62] (Fig. 1). Membrane-bound ADP-ribosyl cyclase and cADPR hydrolase activities are present in membrane preparations obtained from human, murine, and porcine ASM [11, 14, 62].

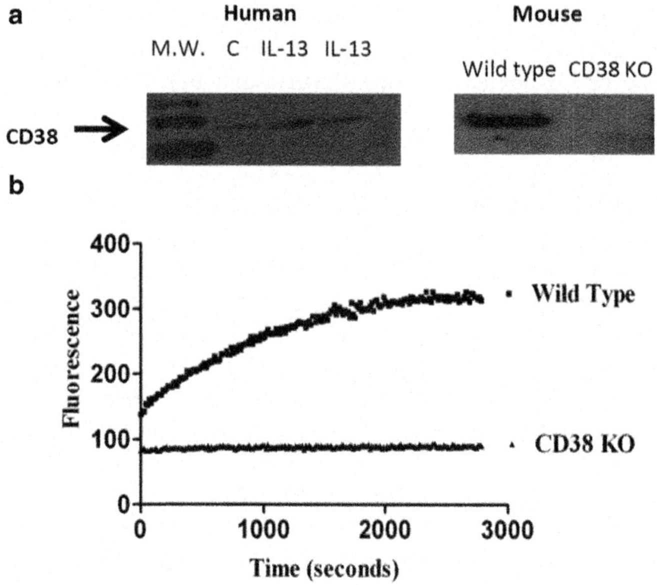

Fig. 1 *CD38 expression and ADP-ribosyl activity in airway smooth muscle (ASM).* Western blot analysis (**a**) using membrane lysates prepared from human and mouse ASM reveal expression of CD38. Treating ASM with cytokine IL-13 (*left panel*) results in increased expression of CD38. Incubating ASM membrane preparations isolated from wild-type, and not CD38 null, mice with NGD, a substrate for ADP-ribosyl cyclase, results in a time-dependent increase in the production of cGDPR (**b**). These data suggest that CD38 is the primary source of ADP-ribosyl cyclase activity in ASM (Figures reproduced with permission [9, 14])

CD38 is the primary source for cADPR production in ASM, although non-CD38 ADP-ribosyl activities have also been described in other cell types [42]. An increase in CD38 expression results in increased levels of cADPR in ASM. Furthermore, siRNA-mediated knockdown or genetic ablation of CD38 results in diminished levels of cADPR in ASM, suggesting a role of CD38 in mediating cADPR production [14, 55]. A low level of cADPR is present in ASM cells from CD38-deficient mice, suggesting a potential source of non-CD38 ADP-ribosyl cyclases in ASM [14]. Agonist stimulation results in increased ADP-ribosyl cyclase activity in ASM, corresponding to increased cADPR accumulation. The mechanisms that regulate the cADPR hydrolase activity of CD38 are not clearly understood. Future studies are needed to establish the time kinetics of cADPR synthesis and degradation in ASM cells. Most importantly, agonist stimulation seems to favor cADPR synthesis in ASM cells.

The role of G proteins in the activation of CD38 has been investigated extensively in smooth muscle cells. Incubation of small-intestine smooth muscle with GDP-γS resulted in inhibition of agonist-induced accumulation of cADPR [35, 36]. Similar findings were demonstrated in cardiac myocytes using GTP-γS or cholera toxin [22]. Agonist activation of CD38 activity and cADPR accumulation in ASM provide indirect evidence for the involvement of G proteins. However, no direct evidence for such an interaction has been established in ASM. Receptor subtype specificity for CD38/cADPR-mediated calcium release has also been investigated. cADPR accumulation is seen in ASM cells stimulated with both m2 and m3 mACh receptors [33]. Another study using 8-bromo-cADPR suggested sensitivity for cADPR antagonism for m2 mACh receptor activation [63]. Studies from our laboratory demonstrated inhibition of calcium response by ACh and ET-1 and not histamine by cADPR antagonist [63]. Although receptor subtype specificity for CD38/cADPR coupling has been shown in smooth muscles of arteries (m1 mAChR) [18] and seminiferous tubules (ETB) [3], the precise mechanism and significance in ASM need further investigation.

Agonist-stimulated increase in ADP-ribosyl cyclase (which potentiates synthesis of cADPR) has also been established in ASM cells [33]. Agonists such as ACh, bradykinin, endothelin, histamine, cholecystokinin, angiotensin, adenosine trisphosphate, nitric oxide, glucose, oxytocin, and isoproterenol modulate ADP-ribosyl activity. Phosphorylation, ADP-ribosylation, or S-nitrosylation of specific amino acid residues on ADP-ribosyl cyclase could result in an increase or decrease in enzyme activity. In bifunctional enzymes such as CD38, it is intriguing how specific enzyme activities are differentially and selectively modulated by agonist stimulation. More studies are needed to establish whether any or all of these mechanisms are involved in the activation of CD38 in ASM. Nevertheless, agonist stimulation appears to increase ADP-ribosyl activity and cADPR accumulation in ASM cells. Collectively, extensive investigations in ASM provide evidence for the CD38/cADPR second messenger pathway that is activated upon stimulation of ASM by a variety of calcium-elevating Gq-coupled GPCR agonists, similar to the well-established PLC/IP_3 pathway.

1 Calcium Regulation by cADPR

Three different modalities of calcium signals based on temporal and spatial features are described for smooth muscle cells, including ASM cells. These include global calcium increase, calcium oscillations, and calcium sparks [26]. A global calcium increase upon activation of ASM by an agonist such as ACh is characterized by a biphasic $[Ca^{2+}]_i$ response, an initial transient rise in calcium followed by a decline to a steady-state level that remains above the basal concentration. Using membrane-permeable cADPR antagonist and ASM cells obtained from CD38-deficient mice, studies have demonstrated the role of cADPR in mediating agonist-induced calcium responses in ASM cells [14]. Stimulation of human ASM cells with agonists such as ACh, histamine, bradykinin, or thrombin results in a biphasic increase in global calcium (Fig. 2). Preincubation of cells with 8-bromo-cADPR, a cell-permeable cADPR antagonist, significantly inhibits agonist-induced $[Ca^{2+}]_i$ responses in human ASM cells [11]. Similar findings were observed in freshly isolated porcine ASM cells [52, 63]. Although findings from one study [63] indicated agonist specificity for cADPR-mediated calcium release, this does not seem to be the

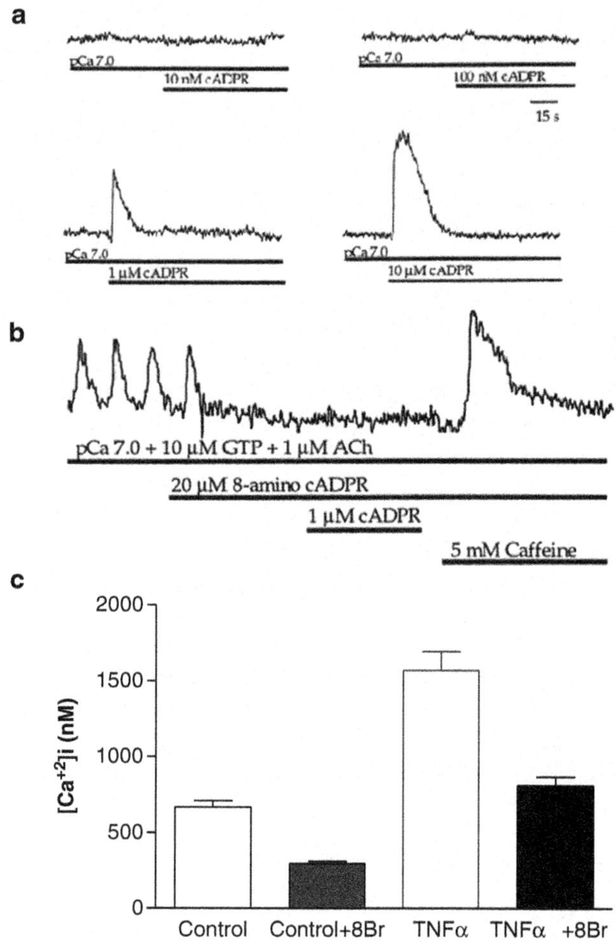

Fig. 2 *cADPR-mediated calcium responses in airway smooth muscle (ASM).* Addition of cADPR to permeabilized ASM cells results in a concentration-dependent increase in intracellular calcium concentration (**a**). Stimulating ASM cells with acetylcholine (*ACh*) generates an oscillatory calcium rise and fall in a localized region of the cell and the addition of 8-amino-cADPR, a cADPR antagonist, abrogates ACh-induced calcium oscillations (**b**). Preincubation of human ASM cells with 8-bromo-cADPR, a membrane-permeant cADPR antagonist, significantly inhibits ACh-induced calcium responses (**c**). cADPR antagonist is effective in inhibiting TNF-α-induced calcium hyperresponses in human ASM cells. These data provide pharmacological evidence for cADPR-mediated calcium release in ASM cells (Figures reproduced with permission [11, 52])

case in human ASM cells, where the CD38/cADPR pathway seems to be utilized by multiple agonists.

Another line of evidence in support of a role of cADPR in the regulation of $[Ca^{2+}]_i$ in ASM cells comes from studies using cADPR analogs (Fig. 2). Addition of cADPR to permeabilized porcine ASM cells causes a concentration-dependent increase in $[Ca^{2+}]_i$ [52]. Similar findings were observed in studies using bovine ASM cells [16]. Interestingly, cADPR-mediated calcium release in porcine ASM cells is not sensitive to heparin, an IP_3 receptor antagonist, whereas pretreatment with ryanodine or ruthenium red abolished cADPR-induced calcium release. These findings suggest that cADPR mediates calcium release via RyR channels and not IP_3R channels in ASM. The $[Ca^{2+}]_i$ responses induced by cADPR require the addition of micromolar concentrations of cADPR to permeabilized tracheal smooth muscle cells. A number of technical limitations and the purity of compounds used in the studies might account for the relatively low affinity of cADPR for calcium release. Interestingly, cADPR-mediated calcium responses were transient in nature. Nevertheless, these studies provide substantial direct evidence for the role of cADPR in regulating ASM calcium homeostasis. Overexpression of CD38 in ASM cells [55] or inflammatory cytokine-induced increase in CD38 expression [11, 55] results in enhanced calcium responses to GPCR agonists, suggesting the critical role of CD38 in the generation of cADPR and regulating dynamic calcium homeostasis in ASM cells. The augmented $[Ca^{2+}]_i$ responses to agonists following treatment with inflammatory cytokines in human ASM cells is inhibited by the membrane-permeant cADPR antagonist 8-bromo-cADPR (Fig. 2) [9, 11]. This provides further evidence for CD38-derived cADPR in the regulation of calcium homeostasis.

The contribution of cADPR to $[Ca^{2+}]_i$ regulation in ASM cells was also demonstrated in studies using airway myocytes from CD38-deficient mice [14]. Calcium responses to ACh and endothelin-1 were significantly diminished in ASM cells isolated from CD38-deficient mice trachea compared to wild-type counterparts (Fig. 3). Downregulation of CD38 expression by antisense CD38 expression [31] and siRNA transfection [55] in ASM cells significantly attenuates agonist-induced calcium responses in ASM cells (Fig. 3). Taken together these findings provide genetic evidence for the role of CD38/cADPR-mediated calcium release in ASM calcium homeostasis.

Our understanding of the mechanisms involved in dynamic $[Ca^{2+}]_i$ regulation in ASM cells has changed significantly with our ability to measure $[Ca^{2+}]_i$ in real time using confocal microscopy. The enhanced spatial and temporal resolution features of confocal microscopy has revealed that exposure of ASM cells to ACh induces regenerative and propagating intracellular calcium oscillations [51]. Recent studies using lung slices have also demonstrated oscillatory patterns of calcium release in ASM upon agonist stimulation [2, 4, 50]. These $[Ca^{2+}]_i$ oscillations originate at a location within an ASM cell and propagate to the ends of the cell. The peak-to-trough amplitude of a $[Ca^{2+}]_i$ oscillation within a given region of the ASM cell is constant but varies across different regions of the cell, regardless of the concentration of the agonist used to induce the event. However, as the concentration of the agonist increases, the basal $[Ca^{2+}]_i$ increases such that the apparent peak-to-trough amplitude of the oscillation decreases. Furthermore, the propagation velocity of $[Ca^{2+}]_i$ oscillations increases as the basal calcium increases with increasing agonist concentration. Raising the basal $[Ca^{2+}]_i$ by exposing the permeabilized ASM cells to increasing extracellular calcium concentration also increases the frequency and

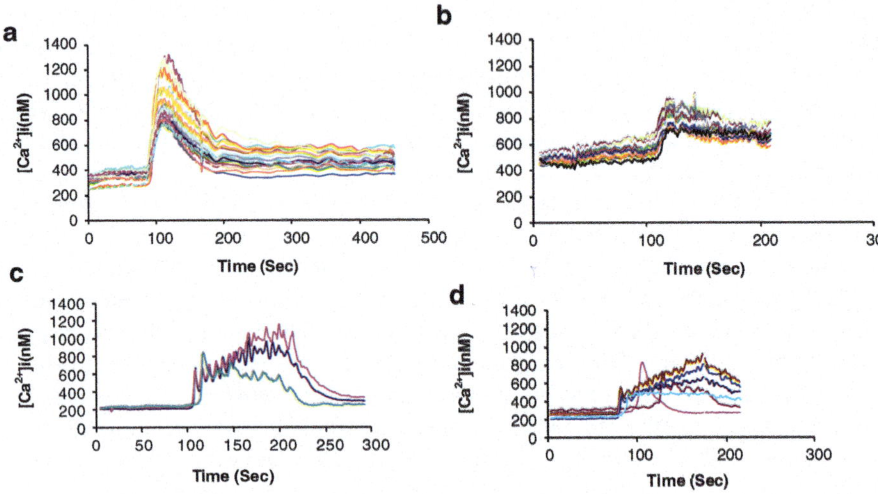

Fig. 3 *Role of CD38 in cADPR-mediated calcium release.* Downregulation of CD38 in human airway smooth muscle (ASM) cells using antisense CD38 transduced by viral infection method resulted in a significantly attenuated calcium responses to bradykinin in ASM cells (**b**) compared to controls (**a**). Similarly, stimulation of ASM cells obtained from CD38 null mice elicited a diminished calcium response to ACh stimulation (**d**) compared to the responses in ASM cells from wild-type mice (**c**). These data support the hypothesis that CD38 is the primary source of cADPR-mediated calcium release in ASM cells (Figures reproduced with permission [14, 31])

propagation velocity of $[Ca^{2+}]_i$ oscillations. The addition of cADPR to permeabilized ASM cells also increases the basal $[Ca^{2+}]_i$ and results in an increased frequency and propagation velocity of $[Ca^{2+}]_i$ oscillations [52]. In contrast, the frequency and amplitude of calcium oscillations are compromised when cells are pretreated with 8-bromo-cADPR, a cADPR antagonist. These observations provided some important mechanistic insights into dynamic $[Ca^{2+}]_i$ regulation by cADPR in ASM cells during agonist stimulation. The spatiotemporal integration of $[Ca^{2+}]_i$ oscillations in the entire cell is reflected as mean calcium, which increases with agonist concentration. The biphasic elevation of $[Ca^{2+}]_i$ measured in the whole cell and reported in earlier studies (described earlier) is in reality the spatiotemporal integration of calcium-release events in the entire cell.

Another modality of calcium signaling is called calcium sparks, which are defined as spontaneous, transient elevations of calcium in a very finite area of a cell. Calcium sparks in a cell are minimal and asynchronous in nature and reported in ASM cells [65, 66]. Calcium sparks are generated primarily by calcium release via RyR channels and not dependent on calcium influx from the extracellular space. However, the role of cADPR in mediating calcium spark activity needs further investigation. Calcium sparks are known to induce spontaneous transient inward currents (STICs) and spontaneous transient outward currents (STOCs) [41, 65]. The addition of cADPR to equine tracheal smooth muscle cells resulted in increased amplitude and frequency of STICs, suggesting spontaneous release of calcium by cADPR in ASM cells [61]. The calcium sparks in ASM cells appear to occur often in clusters within a given region of a cell. These sparks may coalesce into large $[Ca^{2+}]_i$ oscillations that propagate to other regions of the cell. The exposure of the ASM cells to ACh causes initiation of large $[Ca^{2+}]_i$ oscillations that often originate within regions of intense calcium spark activity. We believe that calcium sparks in ASM cells give rise to $[Ca^{2+}]_i$ oscillations, and regions of intense calcium spark activity may be the site of origin of these oscillations. Whether these regions are endowed with a high density of calcium-release channels in the SR is not currently known.

The studies described previously highlight the role of the CD38/cADPR pathway in mediating calcium release from intracellular stores, predominantly SR via RyR channels. Recent investigations have focused on identifying the role of cADPR in store-operated calcium entry (SOCE) in ASM cells. The findings of a study by Ay et al. [1] using porcine ASM cells suggest a role of both IP_3R- and RyR-mediated SR calcium release in store-operated calcium influx via plasma membrane store-operated calcium channels (SOCCs). Subsequent studies identified Transient Receptor Potential (TRP) channels as mediators of store-operated calcium influx [19]. SOCE in ASM is increased under inflammatory conditions (as demonstrated by treatment of ASM cells with TNF-α), and siRNA-mediated downregulation of CD38 expression attenuates cytokine-induced increase in SOCE [55]. These studies indirectly demonstrate the role of CD38/cADPR second messenger pathway in mediating SOCE in ASM.

2 cADPR Binding Sites

Several lines of evidence suggest that cADPR-mediated calcium release occurs via activation of RyR channels on the SR [19, 52] (Fig. 4). Unlike IP_3R-mediated calcium release that is dependent on the binding of IP_3 to its receptors on the SR, there is no well-understood endogenous RyR ligand. One proposed mechanism for RyR-mediated calcium release involves calcium-induced calcium release (CICR) wherein the initial calcium released via IP_3R acts as a ligand to open RyR channels [34]. Studies from different smooth muscle preparations clearly identified cADPR as an endogenous ligand for RyRs and a potential candidate to modulate the CICR mechanism [13]. This hypothesis is supported by the observation that cADPR added exogenously to ASM cells induces calcium release, and cADPR-induced calcium increase is blunted by RyR antagonists but not IP_3R antagonists [52]. High-affinity binding of ^{32}P-cADPR to permeabilized smooth muscle cells and tissue lysates provides the experimental evidence for cADPR–SR interaction [35]. Furthermore, cADPR binding to the SR membrane is inhibited by unlabeled cADPR, ryanodine, and ruthenium red, but not IP_3, suggesting an interaction of cADPR and RyRs [35]. Another line of evidence comes from studies carried out using reconstituted RyRs. Campell's group reconstituted RyRs in lipid bilayers, and when cADPR was added exogenously, the open probability of RyR channels was

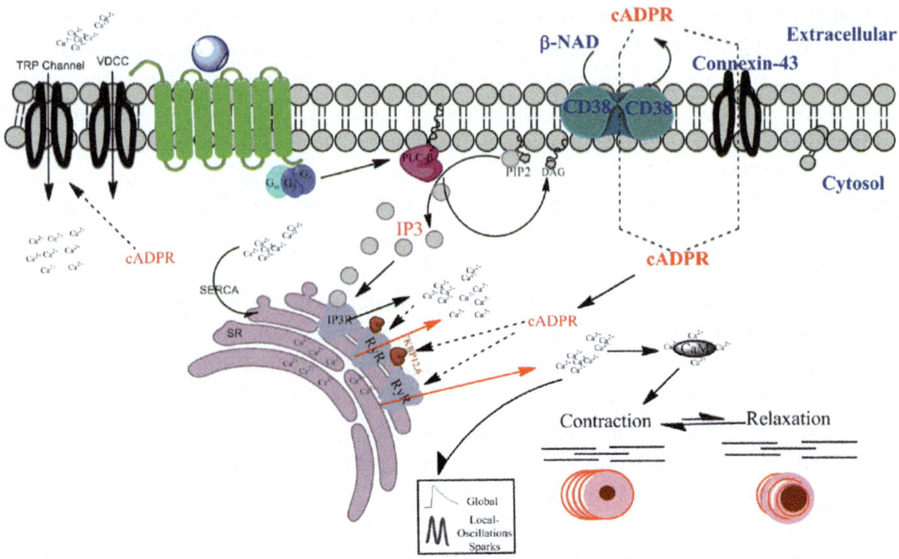

Fig. 4 *Model depicting CD38/cADPR-mediated calcium release and functional effect in airway smooth muscle (ASM):* Stimulation of ASM cells with GPCR agonists such as ACh results in the production of inositol 1,4,5-trisphosphate (IP_3), which in turn binds to its receptors on the sarcoplasmic reticulum (*SR*). cADPR produced by the catalytic action of ADP-ribosyl cyclase (part of membrane-bound protein CD38) acts as a calcium-releasing second messenger, presumably via ryanodine receptors (*RyRs*) on the SR membrane. Extracellularly produced cADPR is believed to enter the cell via membrane channel formed by dimer of CD38 or connexin-43. Accessory proteins such as FKBP12.6 and calmodulin (*CaM*) are known to be involved in cADPR-mediated calcium release. Recent studies have established the role of cADPR in store-operated calcium entry via TRP channels on the plasma membrane. cADPR-mediated calcium release is involved in the regulation of global as well as local/compartmentalized (e.g., oscillations) calcium homeostasis in ASM cells. In vivo studies using CD38 null mice confirmed the functional role of CD38/cADPR-mediated calcium release in the regulation of bronchomotor tone

increased [40]. Biochemical and pharmacological evidence suggests the interaction of cADPR and RyR on the SR membrane to mobilize calcium.

The involvement of accessory proteins, such as calmodulin (CaM) and FK506 binding protein (FKBP12.6), as cADPR binding partners on RyRs has also been investigated (Fig. 4). CaM was proposed based on studies using sea urchin egg microsomes [38], and pharmacological evidence was provided in pancreatic islet cells in which cADPR-mediated calcium release was inhibited by the incubation of islet cells with a CaM inhibitor, W-7 [46, 47, 57]. However, no evidence was obtained for the involvement of CaM in cADPR-mediated calcium release in ASM cells. Another intermediary protein of interest in the actions of cADPR is FKBP12.6. RyR channels are stabilized by the interaction with FKBP, and during activation of RyR, FKBP dissociates to increase the open probability of RyRs [37]. FKBP12.6 is the predominant isoform in ASM. Biochemical studies revealed dissociation of FKBP12.6 upon addition of cADPR to the arterial smooth muscle and pancreatic acinar cells [44, 56]. Furthermore, cADPR activation of RyR channels is significantly inhibited by pretreatment with FK506, suggesting a potential role of FKBP12.6 in cADPR–RyR interaction. Further evidence came when cADPR failed to release calcium in ASM isolated from FKBP12.6 knockout mice [61].

In summary, studies in ASM cells have demonstrated the following: (1) CD38/cADPR is activated upon agonist stimulation of ASM; (2) cADPR analogs increase whereas cADPR antagonists inhibit calcium release in ASM; (3) cADPR-mediated calcium release is involved in the regulation of store-operated calcium release, calcium oscillations, global calcium responses, and calcium sparks; and (4) cADPR-mediated calcium release can occur by direct or indirect (via accessory proteins) binding to RyRs.

3 Regulation of CD38 Expression and Effect on Calcium Homeostasis

In principle, any change in the expression of CD38 or enzyme activity associated with CD38 should result in the altered production of cADPR and calcium release from the SR. The altered expression and function of CD38 have been demonstrated in various pathological conditions. Diseases such as diabetes, hypoxia-induced vasoconstriction, hematopoietic malignancies, and thyroiditis are associated with altered CD38 expression or activity [13, 42]. We hypothesized that CD38/cADPR pathway in ASM played a critical role in aberrant calcium regulatory mechanisms seen during airway inflammation. For these studies, we used an in vitro cell culture model and well-established inflammatory and TH2 cytokines. Asthma is an inflammatory disease characterized by release of inflammatory and TH2 cytokines in the lungs, resulting in structural and functional changes in resident effector cells [53]. Pretreatment of ASM cells with different concentrations of inflammatory (TNF-α, IL-1β, IFN-γ) or TH2 (IL-13) cytokines resulted in increased CD38 expression and cADPR production [9, 11]. Calcium responses to different Gq-coupled GPCR agonists were significantly higher in cytokine-treated ASM cells compared to controls, suggesting ASM hyperresponsiveness. Interestingly, cell-permeable cADPR antagonist was able to reverse cytokine-induced ASM hyperresponsiveness. These studies for the first time demonstrated the potential pathological role of the CD38/cADPR pathway in airways (Table 1). The augmentation of CD38 expression, and therefore the capacity for CD38/cADPR signaling, is significantly greater following TNF-α exposure in ASM cells derived from asthmatics than in cells from nonasthmatics [29].

TNF-α-induced increase in CD38 expression in ASM cells has been demonstrated by other laboratories [55, 60]. Ex vivo treatment of tracheal rings with TNF-α resulted in increased contractility to ACh exposure in wild-type mice and not in CD38 null mice, suggesting a critical role of CD38 in cytokine-induced

Table 1 Regulation of CD38 expression and effect on calcium dynamics in ASM

Factors	CD38 expression	Calcium responses
Cytokines		
IL-1β	↑	↑↑
TNF-α	↑↑↑	↑↑↑↑
IFN-γ	↑	↑
IL-13	↑↑	↑↑↑
Glucocorticoids	↓↓	↓↓
Micro RNAs		
miR-140-3p	↓↓↓	ND
miR-708-5p	ND	ND
miR-499-5p	ND	ND
miR-155	ND	ND

Up arrow: increase, *down arrow*: decrease, *ND*: not determined

airway hyperresponsiveness (AHR) [25]. The *in vivo* relevance of cytokine-induced CD38 upregulation was also investigated using mouse models (described in detail in the next section). Similarly, inflammatory cytokine (TNF- α) treatment increased SOCE in ASM, and downregulation of TNF-α-induced CD38 expression by siRNA reversed a TNF-α-induced increase in SOCE [55]. These studies provided genetic evidence for CD38/cADPR pathway in regulating ASM hyperresponsiveness. Glucocorticoids (the drug of choice in asthma treatment) inhibit cytokine-induced CD38 expression [32] and cADPR-mediated calcium release (unpublished observation). This provides evidence for the clinical application of the CD38/cADPR pathway in airway diseases. Collectively, the studies described earlier demonstrate that an increase in CD38 expression in ASM by cytokines results in altered calcium homeostasis and may contribute to airway hyperresponsiveness.

4 Transcriptional and Posttranscriptional Regulation of CD38 Expression

Our laboratory undertook detailed investigations related to the regulation of CD38 expression in human ASM cells (Fig. 5 and Table 1). These investigations involved cells from both nonasthmatics and asthmatics. The latter group was comprised of cells from subjects with a history of clinical asthma as well as from fatal asthmatics. In initial studies, we examined the transcription factors and signaling mechanisms through MAP kinases in the regulation of CD38 expression. The promoter region of the CD38 gene has response elements for the transcription factors NF-κB and AP-1 [58]. In studies involving the transfection of ASM cells with a promoter-luciferase construct, we showed that NF-κB and AP-1 were critical for TNF-α-induced activation of the promoter [58]. The mutation of either of the transcription factor binding sites within the CD38 promoter resulted in a loss of TNF-α-induced activation of the promoter. Transcriptional regulation is critical in cytokine-induced

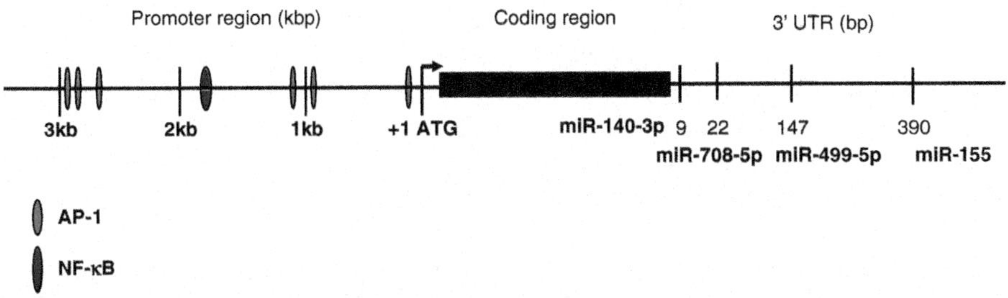

Fig. 5 *Organization of CD38 gene and potential regulatory sites.* Genomic analysis of CD38 gene reveals the presence of transcription factor binding sites upstream of the start codon, ATG. Full-length CD38 mRNA has binding sites for several microRNAs, suggesting potential posttranscriptional regulation of CD38 expression. An increase in the expression of CD38 results in altered calcium homeostasis, hypercontractility of ASM, and bronchoconstriction. Understanding transcriptional and posttrascriptional regulation of CD38 expression is critical to exploiting CD38 as a potential antiasthma therapeutic target

overexpression of the CD38 gene. Cytokine signaling in ASM involves multiple mechanisms. We investigated the role of various MAP kinases in cytokine-induced CD38 expression [28, 30]. The exposure of growth-arrested ASM cells to TNF-α caused a significant induction of CD38 expression in a concentration- and time-dependent fashion and activation of the MAP kinases. Pretreatment of the cells with inhibitors of various MAP kinases or transfection with dominant negative constructs for ERK and JNK MAP kinases resulted in a significant attenuation of CD38 expression induced by the inflammatory cytokine. Furthermore, in the presence of inhibitors of p38 or JNK MAP kinases, there was significant inhibition of transcription factor activation and CD38 expression. In the presence of inhibitors of p38 and ERK MAP kinases, there was significant attenuation of stability of CD38 transcript. These observations indicate that CD38 expression in human ASM cells is regulated by TNF-α through both transcriptional and posttranscriptional mechanisms, the latter involving specific MAP kinases.

There is considerable interest in the posttranscriptional regulation of gene expression in mammalian cells, especially by microRNAs as they pertain to physiological and pathological functions. We examined the specific roles of microRNAs in the posttranscriptional regulation of CD38 expression. Multiple Web-based target prediction algorithms were used to determine potential microRNA binding sites within the 3'UTR of CD38. Among the predicted microRNA binding sites, we initially evaluated the effect of miR-140-3p in the regulation of CD38 expression [27]. The rationale for pursuing this specific microRNA for the initial studies was that it is expressed at high levels in human ASM cells and its expression is regulated by TNF-α [11, 55, 60]. Exposure of ASM cells to TNF-α results in downregulation of miR-140-3p expression, and this downregulation is greater in cells from asthmatics. In human ASM cells transfected with a miR-140-3p mimic, there was a concentration-dependent inhibition of CD38 expression. The magnitude of inhibition of CD38 expression by the miR-140-3p mimic in cells from asthmatics and nonasthmatics was comparable. Thus, the differential induction of CD38 expression by TNF-α in asthmatic ASM cells that we reported earlier may not be due to the greater reduction of miR-140-3p expression in these cells but may arise from other mechanisms. We next undertook studies to examine the mechanisms by which miR-140-3p regulates CD38 expression. The binding of miR-140-3p mimic to 3'UTR of CD38 was studied by transfecting HEK290 or NIH3T3 cells with the 3'UTR-luciferase construct. Following transfection with the miR-140-3p mimic, there was slight reduction of luciferase activity that was reversed by mutation of the miR-140-3p binding site within the 3'UTR. Although the magnitude of reduction in luciferase activity by the miR-140-3p mimic was marginal, inhibition of CD38 expression in human ASM cells by the same concentration of the mimic was profound. Therefore, we examined whether activation of transcription factors and MAP kinases that regulate CD38 expression were involved in the effects of miR-140-3p in HASM cells. Our results indicated that in ASM cells transfected with the miR-140-3p mimic, there was marginal reduction of p38 MAP kinase activation as well as NF-κB activation [27]. Taken together, our results indicated that both direct binding of miR-140-3p to 3'UTR of CD38 and indirect mechanisms involving decreased activation of transcription factors and MAP kinases account for the microRNA regulation of CD38 expression in human ASM cells. Whether or not altered expression of CD38 by miR-140 results in changes in calcium homeostasis in ASM needs to be determined. Our results do not rule out the contribution of other transcription factors in the effects of miR-140-3p in regulation.

In summary, in ASM cells, (1) CD38 expression is regulated by inflammatory cytokines that are relevant in asthma; (2) cytokine-induced CD38 expression results in increased cADPR production and calcium release; (3) augmentation of CD38/cADPR by cytokines is significantly greater in ASM cells derived from asthmatics than in nonasthmatics; (4) transcriptional regulation of CD38 expression involves the activation of p38 and JNK MAP kinases; (5) posttranscriptional regulation via transcript stability involves the activation of p38 and ERK MAP kinases; (6) microRNAs play a critical role in the regulation of CD38 expression; and (7) the overexpression of miR-140-3p in HASM cells causes a significant reduction in CD38 expression induced by TNF-α, indicating a potential therapeutic role for this microRNA in reversing the asthmatic phenotype.

5 CD38 and Airway (Hyper) Responsiveness

The hallmarks of asthma include airway inflammation, remodeling, and AHR [24]. ASM plays an active role in the development and progression of asthmatic symptoms, and, most importantly, ASM assumes a hypercontractile, hyperproliferative, and hypersecretory phenotype [24, 67]. We and others have investigated the role of CD38/cADPR-mediated calcium release in asthma pathogenesis by determining the role of the CD38/cADPR signaling in cytokine-induced changes in ASM functions using cultured human ASM cells and animal models. First, human ASM cells exposed to TNF-α, IL-1β, or INF-γ showed higher CD38 expression and ADP-ribosyl cyclase activity compared with controls and developed exaggerated $[Ca^{2+}]_i$ responses to the contractile agonists acetylcholine, bradykinin, and thrombin (see earlier section) [11]. Subsequently, similar observations were made in cultured human ASM cells exposed to IL-13 with the additional finding that the cADPR antagonist 8-bromo-cADPR significantly attenuated the $[Ca^{2+}]_i$ responses to bradykinin, histamine, and thrombin [9]. Both IL-13 and TNF-α cause ASM hyperresponsiveness to carbachol and KCl and hyporesponsiveness to isoproterenol (IL-13 only) as demonstrated in murine tracheal ring experiments [20, 21, 59]. The change in responsiveness to carbachol, but not to isoproterenol, is significantly attenuated in tracheal rings from CD38-deficient (CD38 KO) mice compared to wild-type controls [20, 21]. It has also been demonstrated that the expression of proinflammatory genes (IL-6, RANTES) in ASM cells exposed to TNF-α is differentially regulated (IL-6 increased, RANTES decreased) by IFN-β-dependent activation of the CD38/cADPR signaling pathway [60]. Emerging evidence shows that TNF-α significantly increases CD38 expression, enzyme activity, and cADPR synthesis by human ASM cells obtained from asthmatic subjects as compared to ASM cells from nonasthmatics [29]. These studies clearly demonstrate that the CD38/cADPR signaling system is targeted by and contributes to the effects of key cytokines thought to mediate the asthmatic phenotype.

Taking advantage of the CD38 KO mice, studies have confirmed that CD38 is a critical source of cADPR in the lungs and contributes to bronchomotor tone (Fig. 6) [14]. The CD38/cADPR signaling pathway has an important role in cytokine-mediated AHR, as demonstrated in mice intranasally challenged with IL-13 or TNF-α [20, 21]. The absence of CD38 did not affect the magnitude or the nature of the inflammatory cell infiltrate in both the BAL fluid and in the lung parenchyma as well as cytokine/chemokine (IL-5, IL-6, IL-10, IL-12, IL-13, IFN-γ, eotaxin, and MCP-1) concentrations in BAL fluid. However, the length of exposure of the lungs to TNF-α did influence the inflammatory responses and airway responsiveness to methacholine in a CD38-dependent and independent manner. A single exposure to TNF-α caused no changes in inflammatory cells and cytokine/chemokine concentrations in the BAL fluid of both wild-type and CD38 KO mice but induced CD38-dependent AHR as occurred only in the wild-type animals. Extending the airway exposure to TNF-α to 1 week caused CD38-independent bronchoalveolar (predominantly neutrophilic) inflammation and AHR since both occurred similarly in wild-type and CD38 KO mice. A 4-week exposure caused bronchoalveolar inflammation (neutrophilic/eosinophilic) as well as parenchymal inflammation but with no AHR in either the wild-type or the CD38 KO mice. Taken together, these findings demonstrate that CD38 KO mice were able to mount robust inflammatory responses to key cytokines involved in the asthmatic phenotype, despite the fact that the CD38/cADPR signaling system plays important roles in the regulation of innate and adaptive immune responses in vivo [8, 48, 49]. Thus, CD38-mediated responses of the resident airway cells, likely the ASM, rather than its role in inflammatory cell functions, appear to be most relevant to AHR induced by directly exposing the airways to cytokines.

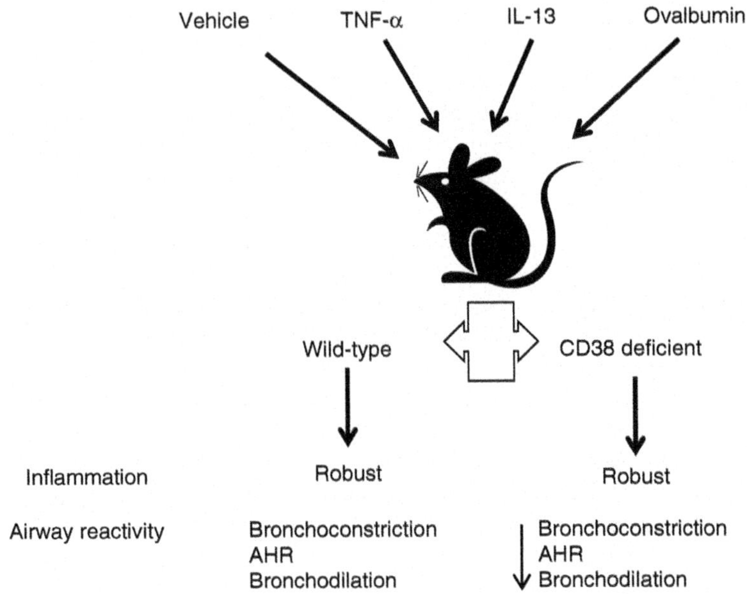

Fig. 6 *Functional role of CD38 in the regulation of airway tone during normal and asthmatic conditions*. Acetylcholine-induced bronchoconstriction is decreased in CD38-deficient mice compared to wild-type mice, suggesting a functional role of CD38 in the regulation of bronchomotor tone. The induction of asthma phenotype using allergen (*ovalbumin*) sensitization and challenge or direct administration of cytokines such as IL-13 and TNF-a resulted in a substantial increase in bronchoconstrictor response to cholinergic challenge in wild-type, and not CD38 null, mice. These studies suggest a pathological role of CD38 in airway inflammation

The function of CD38 in pulmonary inflammation and AHR associated with inhaled allergens has been examined in two distinct murine models of airway inflammation [17, 21]. First, airway inflammation and responsiveness to methacholine were determined in WT and CD38 KO mice intranasally challenged with IL-13 three times on alternate days. Following such a challenge, both wild-type and CD38 KO mice developed robust airway inflammation, characterized by increased numbers of cells, including eosinophils in the bronchoalveolar fluid. The concentrations of some chemokines in the lungs were similar following IL-13 challenge in both wild-type and CD38 KO mice. However, the methacholine responsiveness of the airways was significantly attenuated in the CD38 KO mice compared to wild-type mice, indicating a role of CD38 in the development of AHR.

The role of CD38 in AHR was also examined in a model of ovalbumin-induced airway inflammation [15]. Intranasal ovalbumin challenge following intraperitoneal sensitization caused increased total inflammatory cell numbers in the BAL fluid of both CD38 KO and wild-type mice [17]. However, here the inflammatory cell infiltrate was characterized predominantly by macrophages in CD38 KO and eosinophils and neutrophils in the wild-type mice. Differences were also found in the serum levels of OVA-specific IgE, in cytokine (IL-4, IL-5, and IL-13) concentrations in supernatants of cells obtained from peribronchial lymph nodes of sensitized and challenged mice and subsequently cultured for 4 days in the presence OVA, and in the parenchymal inflammatory response. Nonetheless, AHR to inhaled methacholine occurred only in the wild-type mice. Furthermore, reconstitution of CD38 KO mice with wild-type bone marrow did not restore AHR to the wild-type level. Together, these studies indicate that CD38 is not as critical for the inflammatory response as it is for the AHR induced by exposing the airways to allergens.

In summary, ASM can orchestrate and perpetuate pulmonary inflammation in addition to being an essential regulator of bronchomotor tone in healthy and asthmatic airways (Fig. 6). The CD38/cADPR signaling system has a prominent role in cytokine- and allergen-induced AHR that is independent of its function in inflammatory cell responses. The substantial evidence for a potential role of CD38 in the pathophysiology of human chronic inflammatory airway diseases such as asthma awaits direct confirmation.

References

1. Ay B, Prakash YS, Pabelick CM and Sieck GC. Store-operated Ca2+ entry in porcine airway smooth muscle. *Am J Physiol Lung Cell Mol Physiol* 286: L909–L917, 2004.
2. Bai Y, Zhang M and Sanderson MJ. Contractility and Ca2+ signaling of smooth muscle cells in different generations of mouse airways. *Am J Respir Cell Mol Biol* 36: 122–130, 2007.
3. Barone F, Genazzani AA, Conti A, Churchill GC, Palombi F, Ziparo E, Sorrentino V, Galione A and Filippini A. A pivotal role for cADPR-mediated Ca2+ signaling: regulation of endothelin-induced contraction in peritubular smooth muscle cells. *FASEB J* 16: 697–705, 2002.
4. Bergner A and Sanderson MJ. Acetylcholine-induced calcium signaling and contraction of airway smooth muscle cells in lung slices. *J Gen Physiol* 119: 187–198, 2002.
5. Billington CK and Penn RB. Signaling and regulation of G protein-coupled receptors in airway smooth muscle. *Respir Res* 4: 2, 2003.
6. Chilvers ER, Batty IH, Barnes PJ and Nahorski SR. Formation of inositol polyphosphates in airway smooth muscle after muscarinic receptor stimulation. *J Pharmacol Exp Ther* 252: 786–791, 1990.
7. Chilvers ER, Lynch BJ and Challiss RA. Phosphoinositide metabolism in airway smooth muscle. *Pharmacol Ther* 62: 221–245, 1994.
8. Cockayne DA, Muchamuel T, Grimaldi JC, Muller-Steffner H, Randall TD, Lund FE, Murray R, Schuber F and Howard MC. Mice deficient for the ecto-nicotinamide adenine dinucleotide glycohydrolase CD38 exhibit altered humoral immune responses. *Blood* 92: 1324–33, 1998.
9. Deshpande DA, Dogan S, Walseth TF, Miller SM, Amrani Y, Panettieri RA and Kannan MS. Modulation of calcium signaling by interleukin-13 in human airway smooth muscle: role of CD38/cyclic adenosine diphosphate ribose pathway. *Am J Respir Cell Mol Biol* 31: 36–42, 2004.
10. Deshpande DA and Penn RB. Targeting G protein-coupled receptor signaling in asthma. *Cell Signal* 18: 2105–2120, 2006.
11. Deshpande DA, Walseth TF, Panettieri RA and Kannan MS. CD38/cyclic ADP-ribose-mediated Ca2+ signaling contributes to airway smooth muscle hyper-responsiveness. *FASEB J* 17: 452–4, 2003.

12. Deshpande DA, Wang WC, McIlmoyle EL, Robinett KS, Schillinger RM, An SS, Sham JS and Liggett SB. Bitter taste receptors on airway smooth muscle bronchodilate by localized calcium signaling and reverse obstruction. *Nat Med* 16: 1299–1304, 2010.
13. Deshpande DA, White TA, Dogan S, Walseth TF, Panettieri RA and Kannan MS. CD38/cyclic ADP-ribose signaling: role in the regulation of calcium homeostasis in airway smooth muscle. *Am J Physiol Lung Cell Mol Physiol* 288: L773–L788, 2005.
14. Deshpande DA, White TA, Guedes AG, Milla C, Walseth TF, Lund FE and Kannan MS. Altered airway responsiveness in CD38-deficient mice. *Am J Respir Cell Mol Biol* 32: 149–156, 2005.
15. Dipp M and Evans AM. Cyclic ADP-ribose is the primary trigger for hypoxic pulmonary vasoconstriction in the rat lung in situ. *Circ Res* 89: 77–83, 2001.
16. Franco L, Bruzzone S, Song P, Guida L, Zocchi E, Walseth TF, Crimi E, Usai C, De Flora A and Brusasco V. Extracellular cyclic ADP-ribose potentiates ACh-induced contraction in bovine tracheal smooth muscle. *Am J Physiol Lung Cell Mol Physiol* 280: L98–L106, 2001.
17. Gally F, Hartney JM, Janssen WJ and Perraud AL. CD38 plays a dual role in allergen-induced airway hyperresponsiveness. *Am J Respir Cell Mol Biol* 40: 433–442, 2009.
18. Ge ZD, Zhang DX, Chen YF, Yi FX, Zou AP, Campbell WB and Li PL. Cyclic ADP-ribose contributes to contraction and Ca2+ release by M1 muscarinic receptor activation in coronary arterial smooth muscle. *J Vasc Res* 40: 28–36, 2003.
19. Gosling M, Poll C and Li S. TRP channels in airway smooth muscle as therapeutic targets. *Naunyn Schmiedebergs Arch Pharmacol* 371: 277–284, 2005.
20. Guedes AG, Jude JA, Paulin J, Kita H, Lund FE and Kannan MS. Role of CD38 in TNF-alpha-induced airway hyperresponsiveness. *Am J Physiol Lung Cell Mol Physiol* 294: L290–L299, 2008.
21. Guedes AG, Paulin J, Rivero-Nava L, Kita H, Lund FE and Kannan MS. CD38-deficient mice have reduced airway hyperresponsiveness following IL-13 challenge. *Am J Physiol Lung Cell Mol Physiol* 291: L1286–L1293, 2006.
22. Higashida H, Egorova A, Higashida C, Zhong ZG, Yokoyama S, Noda M and Zhang JS. Sympathetic potentiation of cyclic ADP-ribose formation in rat cardiac myocytes. *J Biol Chem* 274: 33348–54, 1999.
23. Hirota S, Helli P and Janssen LJ. Ionic mechanisms and Ca2+ handling in airway smooth muscle. *Eur Respir J* 30: 114–133, 2007.
24. Holtzman MJ. Asthma as a chronic disease of the innate and adaptive immune systems responding to viruses and allergens. *J Clin Invest* 122: 2741–2748, 2012.
25. Jain D, Keslacy S, Tliba O, Cao Y, Kierstein S, Amin K, Panettieri RA, Jr., Haczku A and Amrani Y. Essential role of IFNbeta and CD38 in TNFalpha-induced airway smooth muscle hyper-responsiveness. *Immunobiology* 213: 499–509, 2008.
26. Janssen LJ. Ionic mechanisms and Ca(2+) regulation in airway smooth muscle contraction: do the data contradict dogma? *Am J Physiol Lung Cell Mol Physiol* 282: 1161–78, 2002.
27. Jude JA, Dileepan M, Subramanian S, Solway J, Panettieri RA, Jr., Walseth TF and Kannan MS. miR-140-3p regulation of TNF-alpha-induced CD38 expression in human airway smooth muscle cells. *Am J Physiol Lung Cell Mol Physiol* 303: L460–L468, 2012.
28. Jude JA, Panettieri RA, Jr., Walseth TF and Kannan MS. TNF-alpha regulation of CD38 expression in human airway smooth muscle: role of MAP kinases and NF-kappaB. *Adv Exp Med Biol* 691: 449–459, 2011.
29. Jude JA, Solway J, Panettieri RA, Jr., Walseth TF and Kannan MS. Differential induction of CD38 expression by TNF-{alpha} in asthmatic airway smooth muscle cells. *Am J Physiol Lung Cell Mol Physiol* 299: L879–L890, 2010.
30. Jude JA, Tirumurugaan KG, Kang BN, Panettieri RA, Walseth TF and Kannan MS. Regulation of CD38 expression in human airway smooth muscle cells: role of class I phosphatidylinositol 3 kinases. *Am J Respir Cell Mol Biol* 47: 427–435, 2012.

31. Kang BN, Deshpande DA, Tirumurugaan KG, Panettieri RA, Walseth TF and Kannan MS. Adenoviral mediated anti-sense CD38 attenuates TNF-alpha-induced changes in calcium homeostasis of human airway smooth muscle cells. *Can J Physiol Pharmacol* 83: 799–804, 2005.
32. Kang BN, Tirumurugaan KG, Deshpande DA, Amrani Y, Panettieri RA, Walseth TF and Kannan MS. Transcriptional regulation of CD38 expression by tumor necrosis factor-alpha in human airway smooth muscle cells: role of NF-kappaB and sensitivity to glucocorticoids. *FASEB J* 20: 1000–1002, 2006.
33. Kip SN, Smelter M, Iyanoye A, Chini EN, Prakash YS, Pabelick CM and Sieck GC. Agonist-induced cyclic ADP ribose production in airway smooth muscle. *Arch Biochem Biophys* 452: 102–107, 2006.
34. Kotlikoff MI. Calcium-induced calcium release in smooth muscle: the case for loose coupling. *Prog Biophys Mol Biol* 83: 171–91, 2003.
35. Kuemmerle JF and Makhlouf GM. Agonist-stimulated cyclic ADP ribose. Endogenous modulator of Ca(2+)-induced Ca2+ release in intestinal longitudinal muscle. *Journal of Biological Chemistry* 270: 25488–94, 1995.
36. Kuemmerle JF, Murthy KS and Makhlouf GM. Longitudinal smooth muscle of the mammalian intestine. A model for Ca2+ signaling by cADPR. *Cell Biochem Biophys* 28: 31–44, 1998.
37. Lanner JT, Georgiou DK, Joshi AD and Hamilton SL. Ryanodine receptors: structure, expression, molecular details, and function in calcium release. *Cold Spring Harb Perspect Biol* 2: a003996, 2010.
38. Lee HC, Aarhus R and Graeff RM. Sensitization of calcium-induced calcium release by cyclic ADP-ribose and calmodulin. *Journal of Biological Chemistry* 270: 9060–6, 1995.
39. Lee HC, Galione A and Walseth TF. Cyclic ADP-ribose: metabolism and calcium mobilizing function. *Vitam Horm* 48: 199–257, 1994.
40. Li PL, Tang WX, Valdivia HH, Zou AP and Campbell WB. cADP-ribose activates reconstituted ryanodine receptors from coronary arterial smooth muscle. *Am J Physiol Heart Circ Physiol* 280: 208–15, 2001.
41. Lifshitz LM, Carmichael JD, Lai FA, Sorrentino V, Bellve K, Fogarty KE and ZhuGe R. Spatial organization of RYRs and BK channels underlying the activation of STOCs by Ca(2+) sparks in airway myocytes. *J Gen Physiol* 138: 195–209, 2011.
42. Malavasi F, Deaglio S, Funaro A, Ferrero E, Horenstein AL, Ortolan E, Vaisitti T and Aydin S. Evolution and function of the ADP ribosyl cyclase/CD38 gene family in physiology and pathology. *Physiol Rev* 88: 841–886, 2008.
43. Mizuta K, Xu D, Pan Y, Comas G, Sonett JR, Zhang Y, Panettieri RA, Jr., Yang J and Emala CW, Sr. GABAA receptors are expressed and facilitate relaxation in airway smooth muscle. *Am J Physiol Lung Cell Mol Physiol* 294: L1206–L1216, 2008.
44. Noguchi N, Takasawa S, Nata K, Tohgo A, Kato I, Ikehata F, Yonekura H and Okamoto H. Cyclic ADP-ribose binds to FK506-binding protein 12.6 to release Ca2+ from islet microsomes. *Journal of Biological Chemistry* 272: 3133–6, 1997.
45. Osawa Y, Xu D, Sternberg D, Sonett JR, D'Armiento J, Panettieri RA and Emala CW. Functional expression of the GABAB receptor in human airway smooth muscle. *Am J Physiol Lung Cell Mol Physiol* 291: L923–L931, 2006.
46. Ozawa T. Ryanodine-sensitive Ca(2+) release mechanism of rat pancreatic acinar cells is modulated by calmodulin. *Biochimica et Biophysica Acta* 1452: 254–62, 1999.
47. Ozawa T and Nishiyama A. Characterization of ryanodine-sensitive Ca2+ release from microsomal vesicles of rat parotid acinar cells: regulation by cyclic ADP-ribose. *Journal of Membrane Biology* 156: 231–9, 1997.
48. Partida-Sanchez S, Cockayne DA, Monard S, Jacobson EL, Oppenheimer N, Garvy B, Kusser K, Goodrich S, Howard M, Harmsen A, Randall TD and Lund FE. Cyclic ADP-ribose production by CD38 regulates intracellular calcium release, extracellular calcium

influx and chemotaxis in neutrophils and is required for bacterial clearance in vivo. *Nat Med* 7: 1209–16, 2001.
49. Partida-Sanchez S, Randall TD and Lund FE. Innate immunity is regulated by CD38, an ecto-enzyme with ADP-ribosyl cyclase activity. *Microbes Infect* 5: 49–58, 2003.
50. Perez JF and Sanderson MJ. The frequency of calcium oscillations induced by 5-HT, ACH, and KCl determine the contraction of smooth muscle cells of intrapulmonary bronchioles. *J Gen Physiol* 125: 535–553, 2005.
51. Prakash YS, Kannan MS and Sieck GC. Regulation of intracellular calcium oscillations in porcine tracheal smooth muscle cells. *Am J Physiol* 272: 966–75, 1997.
52. Prakash YS, Kannan MS, Walseth TF and Sieck GC. Role of cyclic ADP-ribose in the regulation of [Ca2+]i in porcine tracheal smooth muscle. *American Journal of Physiology* 274: 1653–60, 1998.
53. Riffo-Vasquez Y, Pitchford S and Spina D. Cytokines in airway inflammation. *Int J Biochem Cell Biol* 32: 833–53, 2000.
54. Saxena H, Deshpande DA, Tiegs BC, Yan H, Battafarano RJ, Burrows WM, Damera G, Panettieri RA, DuBose TD, Jr., An SS and Penn RB. The GPCR OGR1 (GPR68) mediates diverse signalling and contraction of airway smooth muscle in response to small reductions in extracellular pH. *Br J Pharmacol* 166: 981–990, 2012.
55. Sieck GC, White TA, Thompson MA, Pabelick CM, Wylam ME and Prakash YS. Regulation of store-operated Ca2+ entry by CD38 in human airway smooth muscle. *Am J Physiol Lung Cell Mol Physiol* 294: L378–L385, 2008.
56. Tang WX, Chen YF, Zou AP, Campbell WB and Li PL. Role of FKBP12.6 in cADPR-induced activation of reconstituted ryanodine receptors from arterial smooth muscle. *Am J Physiol Heart Circ Physiol* 282: 1304–10, 2002.
57. Thomas JM, Summerhill RJ, Fruen BR, Churchill GC and Galione A. Calmodulin dissociation mediates desensitization of the cADPR-induced Ca2+ release mechanism. *Curr Biol* 12: 2018–22, 2002.
58. Tirumurugaan KG, Kang BN, Panettieri RA, Foster DN, Walseth TF and Kannan MS. Regulation of the cd38 promoter in human airway smooth muscle cells by TNF-alpha and dexamethasone. *Respir Res* 9: 26, 2008.
59. Tliba O, Deshpande D, Chen H, Van Besien C, Kannan M, Panettieri RA and Amrani Y. IL-13 enhances agonist-evoked calcium signals and contractile responses in airway smooth muscle. *Br J Pharmacol* 140: 1159–62, 2003.
60. Tliba O, Panettieri RA, Jr., Tliba S, Walseth TF and Amrani Y. Tumor necrosis factor-alpha differentially regulates the expression of proinflammatory genes in human airway smooth muscle cells by activation of interferon-beta-dependent CD38 pathway. *Mol Pharmacol* 66: 322–329, 2004.
61. Wang YX, Zheng YM, Mei QB, Wang QS, Collier ML, Fleischer S, Xin HB and Kotlikoff MI. FKBP12.6 and cADPR regulation of Ca2+ release in smooth muscle cells. *Am J Physiol Cell Physiol* 286: C538–C546, 2004.
62. White TA, Johnson S, Walseth TF, Lee HC, Graeff RM, Munshi CB, Prakash YS, Sieck GC and Kannan MS. Subcellular localization of cyclic ADP-ribosyl cyclase and cyclic ADP-ribose hydrolase activities in porcine airway smooth muscle. *Biochim Biophys Acta* 1498: 64–71, 2000.
63. White TA, Kannan MS and Walseth TF. Intracellular calcium signaling through the cADPR pathway is agonist specific in porcine airway smooth muscle. *FASEB J* 17: 482–4, 2003.
64. Yim PD, Gallos G, Xu D, Zhang Y and Emala CW. Novel expression of a functional glycine receptor chloride channel that attenuates contraction in airway smooth muscle. *FASEB J* 25: 1706–1717, 2011.
65. ZhuGe R, Fogarty KE, Baker SP, McCarron JG, Tuft RA, Lifshitz LM and Walsh JV, Jr. Ca(2+) spark sites in smooth muscle cells are numerous and differ in number of ryanodine

receptors, large-conductance K(+) channels, and coupling ratio between them. *Am J Physiol Cell Physiol* 287: C1577–C1588, 2004.

66. ZhuGe R, Fogarty KE, Tuft RA, Lifshitz LM, Sayar K and Walsh JV, Jr. Dynamics of signaling between Ca(2+) sparks and Ca(2+)- activated K(+) channels studied with a novel image-based method for direct intracellular measurement of ryanodine receptor Ca(2+) current. *J Gen Physiol* 116: 845–864, 2000.
67. Zuyderduyn S, Sukkar MB, Fust A, Dhaliwal S and Burgess JK. Treating asthma means treating airway smooth muscle cells. *Eur Respir J* 32: 265–274, 2008.

Pathways and Signaling Crosstalk with Oxidant in Calcium Influx in Airway Smooth Muscle Cells

Lei Cai and Qinghua Hu

Contents

Abstract Influx of extracellular calcium through calcium channels in the plasma membrane contributes to the sustained phase of $[Ca^{2+}]_i$ elevation following agonist stimulation. The interactions between channel proteins like TRPC, STIM1, and Orai1, as well as protein and regulatory molecules including reactive oxygen species (ROSs), have been shown to be involved in the opening of calcium entry channels. The contribution of endogenous hydrogen peroxide (H_2O_2) to agonist-induced intracellular calcium oscillation was first reported as early as 2002 in vascular endothelial cells. While both the receptor-operated calcium channel and store-operated calcium channel proteins possess cysteine residues, they could also be potential targets of ROSs. Mitochondria-derived H_2O_2 was shown to work as a

L. Cai • Q. Hu (✉)
Department of Pathophysiology, Tongji Medical College, Huazhong University of Science and Technology, Wuhan 430030, People's Republic of China
e-mail: qinghuaa@mails.tjmu.edu.cn

Y.-X. Wang (ed.), *Calcium Signaling In Airway Smooth Muscle Cells*,
DOI 10.1007/978-3-319-01312-1_14,

cofactor together with STIM1 to mediate histamine-stimulated Ca^{2+} influx in airway smooth muscle cells. Furthermore, intracellular ROSs can interact with members of transient receptor potential (TRP) superfamilies and contribute to Ca^{2+} influx. ROS-induced TRP intracellular translocation has been suggested as a critical underlying mechanism of TRP activation, which could be a clue to explain how intracellular ROSs cooperate with STIM1 to mediate Ca^{2+} influx. The cooperation of ROSs and calcium signaling has magnified physiological or pathological effects.

Keywords Oxidant • Cooperation • Calcium influx

Abbreviations

$[Ca^{2+}]_i$	intracellular calcium concentration
ASMC	airway smooth muscle cell
SR	sarcoplasmic reticulum
IP_3	inositol 1,4,5-trisphosphate
ROS	reactive oxygen species
VDCC	voltage-dependent calcium channels
ROC	receptor-operated calcium channel
SOC	store-operated calcium channel
CCE	capacitative Ca^{2+} entry
PIP_2	phosphatidylinositol biphosphate
Tg	thapsigragin
I_{CRAC}	calcium-release-activated calcium current
TRP	transient receptor potential
STIM	stromal interaction molecule
TRPC	TRP canonical
Trx	thioredoxin
CRAC	Ca^{2+} release-activated channel
SCID	severe combined immune deficiency
NADPH	nicotinamide adenine dinucleotide phosphate
RNS	reactive nitrogen species
RyR	ryanodine receptor
mtDNA	mitochondrial DNA
EB	ethidium bromide
IGF-1	insulin-like growth factor 1
EGF	endothelial growth factor
AVP	arginine vasopressin
VCAM	vascular cell adhesion molecule

1 Introduction

In airway smooth muscle cells (ASMCs), a relatively low intracellular calcium concentration ($[Ca^{2+}]_i$) is maintained in resting conditions. The elevation of $[Ca^{2+}]_i$ is required for the contractile response of ASMCs, which is of fundamental importance to physiological respiratory function as well as airway hyperresponsiveness. The mechanisms underlying control of $[Ca^{2+}]_i$ homeostasis in ASMCs have been uncertain.

The elevation of $[Ca^{2+}]_i$ in ASMCs following contractile agonist stimulation is typically a biphasic process characterized by an initial rapid and transient rise in calcium followed by a sustained steady-state concentration above the basal level. Both the calcium influx through ion channels located in the plasma membrane and the calcium release from the sarcoplasmic reticulum (SR) are involved in this biphasic process. Earlier studies attributed the initial phase of $[Ca^{2+}]_i$ response to release from the SR via inositol 1,4,5-trisphosphate (IP_3) and ryanodine receptors (RyRs) [1], while the sustained phase of response was thought to be due to influx from the extracellular space via voltage- and receptor-gated channels in the plasma membrane, as well as by a store-operated calcium entry (SOCE) mechanism [2–4]. The opening of calcium entry channels involves the activation of diverse signaling pathways and the interactions between proteins and proteins as well as protein and regulatory molecules including reactive oxygen species (ROSs) [5, 6].

2 Channels Mediate Ca^{2+} Influx in the Plasma Membrane of Airway Smooth Muscle Cells

2.1 *Voltage-Dependent Calcium Channels*

Many studies have recorded inward Ca^{2+} currents using myocytes from both tracheal and bronchial smooth muscle from a variety of species. Among the various forms of voltage- dependent calcium channels (VDCCs), the L-type channel is probably the most important one in most smooth muscle types. The channels are formed by hetero-oligomeric complexes consisting of an $\alpha 1$ subunit, which forms the pore of the channel and provides the extracellular binding site for most agonists and antagonists together with a β, $\alpha 2$-δ, and possibly γ subunit. Croxton et al. [7] showed that peripheral airway smooth muscle was more resistant to dihydropyridine-sensitive (L-type) VDCC antagonists than tracheal smooth muscle, indicating that L-type VDCCs are the predominant mechanism for Ca^{2+} entry in tracheal smooth muscle, while Janssen [8] found T-type VDCCs as well as L-type

VDCCs in canine bronchial smooth muscle cells by using the whole-cell patch clamp technique. The presence of a T-type Ca^{2+} current is not ubiquitous in airway smooth muscle. The significance of T-type channels lies in two aspects of airways activity: the maintenance of smooth muscle tone and the physiological and pathological action of airway modulators. Spontaneous oscillations of membrane potential occur in bronchial smooth muscle cells in the range of the T-type window current [9], and this would support a route for Ca^{2+} influx and maintenance of tone. The role of T-type Ca^{2+} channels and their interaction with a Ca^{2+}-activated outward current in this activity, however, remains to be established.

Although VDCCs can be readily identified in freshly isolated airway myocytes and channel activity is increased by cholinergic agents such as methacholine, the overall contribution of these channels to agonist-induced contractile responses seems small [10]. The prolonged rise in $[Ca^{2+}]_i$ seen following stimulation with agonists such as histamine is insensitive to classical VDCC antagonists. In addition, these agents are poor bronchodilators. Unlike the importance of voltage-dependent channels in Ca^{2+} homeostasis of vascular smooth muscle, the sustained Ca^{2+} influx inspired by agonists in ASMCs is more dependent on non-voltage-dependent pathways, including receptor-operated calcium (ROC) channels and store-operated calcium (SOC) channels.

2.2 *Receptor-Operated Calcium (ROC) Channels*

The term receptor-operated channel was simultaneously proposed by Bolton [11] and Van Breemen et al. [12] to define any plasma channel that was opened as a result of agonist binding to its receptor and independent of a previous change in membrane potential. To date, in order to distinguish between ROC and SOC, it is generally admitted that a ROC corresponds to a Ca^{2+} permeable nonselective cation channel that is opened by an agonist but not by depleting intracellular Ca^{2+} stores. Inversely, a SOC can be activated without any agonist acting at membrane receptors by using Ca^{2+}-ATPase inhibitors (thapsigargin, cyclopiazonic acid) or caffeine that depletes intracellular Ca^{2+} stores [13]. Moreover, ROCs have a unitary conductance four to five times larger than that of SOC. ROC includes, on the one hand, the classic ligand-gated channel (P2X) in which the ligand binding site and the cationic pore are contained within the same protein and, on the other hand, the more diversified family of metabotropic receptors that are structurally separated from the cation channel (i.e., separate receptor-operated channels) and where the latter is activated by various transduction mechanisms. A main characteristic of ROCs is that, at the physiologic resting membrane potential, the channel current is a nonselective cation current (Icat) carrying both mono- and divalent cations with varying degrees of calcium selectivity.

2.3 Store-Operated Calcium (SOC) Channels

Store-operated calcium entry was originally proposed by Putney, who used the term capacitative Ca^{2+} entry (CCE) to describe the phenomenon of coupling between endoplasmic reticulum (ER) Ca^{2+} concentration and the activation of membrane Ca^{2+} entry channels in nonexcitable cells. In this respect, Ca^{2+} store depletion activates Ca^{2+} entry through plasma membrane channels to refill the store, as in a capacitor [14]. Hence, these channels have been called store-operated calcium (SOC) channels. This Ca^{2+} entry pathway was demonstrated to be ubiquitously present in all excitable and nonexcitable cells.

Under physiological conditions, SOCE is activated in response to stimulation of membrane receptors that lead to the hydrolysis of PIP_2, IP_3 generation, and IP_3-mediated Ca^{2+} release from the ER via activation of the IP_3 receptor. Use of the ER Ca^{2+} pump inhibitor thapsigargin (Tg) demonstrated that SOCE activation is regulated by depletion of the intracellular Ca^{2+} store rather than proximal events associated with receptor-dependent PIP_2 hydrolysis. Thus, the term store-operated calcium entry, or capacitative Ca^{2+} entry, was coined. Although the Ca^{2+} influx SOC channel replenishes the ER-Ca^{2+} store, it also regulates a number of critical physiological functions such as secretion, cell proliferation, endothelial cell migration, T-cell activation, and mast cell degranulation.

The first store-operated current to be measured and the one often studied in greatest detail is the calcium-release-activated calcium current (I_{CRAC}) identified in T lymphocytes and hematopoietic cells. Various SOC channels with diverse biophysical characteristics, ranging from nonselective to relatively Ca^{2+} selective, have been identified in many different cell types, suggesting the possibility that a variety of distinct channels may be involved in SOCE. Previous studies on vascular smooth muscle have shown a relation between depletion of internal Ca^{2+} stores and the rate of Ca^{2+} influx across the plasmalemmal membrane [15]. Subsequently, several functional studies in smooth muscle showed that there were Ca^{2+} permeable, non-voltage-dependent cation conductances that are activated following store depletion and carry a sufficient amount of Ca^{2+} ions to activate contraction. However, it is evident that our knowledge of SOC in smooth muscle cells is considerably less than in many other cell types due in part to the difficulty of recording specifically the ionic current activated by store depletion in this cell type.

Kume and colleagues [16] reported another Ca^{2+} influx pathway called stretch-activated cation channels in human bronchial smooth muscle cells, which is activated by mechanical stretch, and the inhibitory effect of ruthenium red on this stretch-induced Ca^{2+} response suggested the involvement of transient receptor potential vanilloid (TRPVs) in this pathway. However, intracellular second messengers, such as PLC and IP_3, have not been shown to be involved. Details of the Ca^{2+} channels are described in other chapters of this book.

3 Signal Cooperation in Calcium Influx in Airway Smooth Muscle Cells

3.1 Cooperation Among TRP, STIM1, and Orial

Agonist-stimulated Ca^{2+} influx in ASMCs is a complex process, and diverse channel proteins have been shown to be involved in and interact with each other for channel activation. The transient receptor potential (TRP), stromal interaction molecule (STIM), and Orai1 are the most key components of agonist-induced calcium entry channel reported in the last 10 years. They have been shown to work together cooperatively and contribute to Ca^{2+} influx in a variety of cell types including ASMCs.

The long search for the channel-mediating SOC first led to the identification of mammalian TRP channels. Members of the TRPC (TRP canonical) subfamily were proposed as candidate channels for SOC based on their activation by stimuli that lead to PIP_2 hydrolysis. The TRPC subfamily consists of seven members (TRPC1-7) that display diverse properties, modes of regulation, and physiological functions. It has also been suggested that they are assembled as homomeric or heteromeric channels, although there is little information regarding the status of endogenous TRPC channels. While it is generally accepted that TRPC channels are activated downstream of agonist-stimulated PIP_2 hydrolysis, there is considerable discrepancy regarding their activation mechanisms.

In ASMCs, the existence of the three biggest and most characterized subfamilies (TRPC, transient receptor potential melastatin, TRPV) of TRP has been proposed, and it has been suggested that they form agonist-activated, cation-permeable channels that contribute to Ca^{2+} influx to regulate contraction and proliferation [17]. The members of TRP families reported to be expressed in human ASMs are TRPC1, -3, -4, -5, -6, TRPM4, -7, and TRPV1, -2, -4 [18–21].

In other types of smooth muscle cells, TRPC1 is expressed in the vasculature and is linked to Ca^{2+} entry as a ROC and a SOC. Expression and knockdown data support the theory that TRPC1 forms heteromultimeric channels with TRPC3–5 and TRPP2 in the vasculature. Further, TRPC1 may be the key isoform responsible for SOC that causes pulmonary vasoconstriction and stimulates pulmonary artery smooth muscle cells (PASMCs) proliferation. Previous studies have demonstrated that TRPCs can function as SOCs as well as ROCs.

TRPC3 and 6 are expressed in vascular tissues and are believed to function primarily as ROCs, although TRPC6 may also form heterotetrameric channels with other TRP isoforms that can be activated by store depletion in pulmonary vascular smooth muscle and endothelial cells. TRPC6 is expressed in vascular smooth muscle cells (VSMCs) in the systemic and pulmonary vasculature. Knockout of TRPC6 abrogates the hypoxic pulmonary vasoconstriction and hypoxia-induced cation influx. TRPC4 is predominantly expressed in the endothelium and is important in regulating lung microvascular permeability, agonist-dependent vasorelaxation, and gene transcription. While TRPC4 is expressed at a lower level in VSMCs, it may play a role in regulating contraction and proliferation in both store- and receptor-mediated manners. TRPC5 expression in VSMCs is unclear. Some researchers describe TRPC5 protein and transcript expression in PASMCs and pulmonary artery endothelial cells, whereas other researchers show conflicting data. TRPC7 is thought to contribute to both ROC and SOC formation. The mechanism involved in transmitting the signal of store depletion to the channels located on the plasma membrane remained elusive until the stromal interaction molecule (STIM) was identified in independent groups in 2005 [22, 23]. STIM1 is a 685-amino-acid-long single-transmembrane protein that is expressed on the SR/ER membrane and plasma membrane. STIM1 is believed to be a Ca^{2+} sensor, which monitors the Ca^{2+} content of intracellular stores. Near the N-terminus of STIM1 is an EF-hand domain that senses $[Ca^{2+}]_i$. When Ca^{2+} is not bound to the EF-hand domain when, for example, the SR/ER Ca^{2+} store is depleted, STIM1 undergoes a conformational change, which allows it to multimerize and translocate to the SR/ER–plasma membrane junction (or puncta), binds with Orai1 tetramers on the plasma membrane, and activates SOC.

Orai1 is a four-transmembrane domain plasma membrane protein identified in 2006 through genome-wide RNAi screening in Drosophila S2 cell for Ca^{2+}-release-activated channel (CRAC)-associated proteins and by genetic linkage analysis in severe combined immune deficiency (SCID) patients [24, 25]. In Greek, Orai means the keepers of the gates of heaven. As a doorkeeper, knockdown of Orai1 decreased CRAC channel function in S2 cells, while a single point mutation in Orai1 (R91W) was associated with loss of I_{CRAC} and defect in the Ca^{2+} signaling of T lymphocytes obtained from SCID patients.

Strong evidence suggests that only when Orai1, STIM1, and TRP cooperate, they could form a functional calcium entry channel. Although reports on the interaction of these three molecules are very limited from investigations of ASMCs, some basic mechanisms of their cooperation have been revealed from studies of other tissues of smooth muscle cells.

Coexpression of STIM1 with Orai1 in HEK-293 cells resulted in large increases in SOCE and I_{CRAC} compared to vector control cells [26]. Coimmunoprecipitation data show that STIM1 and Orai1 interact, and store depletion significantly increases the amount for interaction. However, the localization of native STIM1 and Oria1 in resting and stimulated cells has not yet been conclusively described.

STIM1 also cooperates or interacts with and modulates the activity of TRPC1 and other members of the TRPC channel family [27–29]. Knockdown of STIM1 in several cell types reduces endogenous TRPC1 function stimulated by store depletion. Furthermore, coexpression of TRPC1 and STIM1 increases SOCE. STIM1 immunoprecipitates with TRPC1, and this association increases upon stimulation of cells either with an agonist or thapsigargin [30–32]. As in the case of Orai1, a mutant of STIM1 also induces the constitutive activation of TRPC1. Two regions of STIM1 have been suggested as interacting with TRPC1: the ERM (ezrin/radixin/moesin) domain [28], which is involved in interactions with TRPC1, and the STIM1 polybasic tail (684KK685) residues, which interact electrostatically with the negatively charged residues in the C-terminus of TRPC1 (639DD640), resulting in gating of the channel [32]. Thus, STIM1 has distinct gating mechanisms for TRPC1 and Orai1, and it appears to be a central molecule that mediates the dynamic assembly of a TRPC1-STIM1-Oria1complex in response to store depletion.

Together, these data highlight the critical contributions of STIM1, Orai1, and TRPC1 to SOCE. While the role of STIM1 in regulating TRPC1 has now been resolved, the exact functional cooperation or interaction between TRPC1 and Orai1 is not yet known. A study in human platelets cells suggested that Orai1 mediated the communication between STIM1 and TRPC1 since, in the absence of Orai1, TRPC1 functions as a store-independent channel [33]. Another study suggested that Orai1 acted as a regulatory subunit of TRPC channels based on findings that Orai1 physically interacts with the N- and C-termini of TRPC3 and TRPC6 to transform store-insensitive channels into store-operated channels [34]. Further studies will be required to elucidate the exact contribution of Orai1 to TRPC1–SOC function.

3.2 Oxidative Stress Cooperates with STIM1 in Calcium Influx in Airway Smooth Muscle Cells

Oxidative stress is an important pathogenic factor in airway diseases, particularly when inflammation is prominent. Inflammatory mediators such as histamine are contractile agonists playing critical roles in pathophysiological conditions with airway hyper-responsiveness.

A number of publications have suggested that, in a variety of cell types, intracellular ROSs may directly participate in intracellular calcium regulation. One of the earliest reports, from 2002, about the potential significance of endogenous hydrogen peroxide (H_2O_2) in agonist-induced intracellular calcium signaling was in vascular endothelial cells [35]. In that study, histamine-induced calcium oscillation in endothelial cells was inhibited by blocking of nicotinamide adenine dinucleotide phosphate (NADPH) oxidase-generated ROS production but restored by exposing the cell to 10 μM H_2O_2, which by itself did not alter intracellular calcium signaling. Thus, NADPH oxidase-derived ROSs are critical to the generation of calcium oscillations in human artery endothelial cells during histamine stimulation. However, the mechanism of how ROSs contribute to calcium oscillation has not been fully elucidated. ROSs may affect intracellular calcium signaling by directly causing extracellular Ca^{2+} influx or activating IP_3 or RyRs, leading to Ca^{2+} release from SR and a subsequent extracellular Ca^{2+} influx.

Concerning how H_2O_2 induced intracellular Ca^{2+} mobilization, the pathway by which Ca^{2+} was released from the internal store is better understood. Several studies reported stimulation of IP_3 receptor-mediated Ca^{2+} release from internal stores in response to ROSs [36]. The modification of cysteine residues in the IP_3 receptor by the thiol reagent, thimerosal, causes Ca^{2+} spikes through increasing IP_3

receptor sensitivity to the resting levels of IP_3 [37]. Subsequent reports showed that physiological levels of cellular ROSs produced through membrane-associated NADPH oxidase (NOXs) could increase the sensitivity of the IP_3 receptor to IP_3 [38]. Another major Ca^{2+}-release channel, RyRs, has been shown to be modified by compounds affecting sulfhydryl groups with consequences for channel function [36, 39, 40]. RyR isoforms can be activated by S-nitrosylation [41, 42]. Exogenous addition of redox compounds, as well as endogenous production of ROSs or reactive nitrogen species (RNSs), modifies RyRs on cysteine residues and causes their activation. Experimental evidence in skeletal muscle suggests that Cys3635 is involved in RyR1 redox sensing, and the modification of this residue causes RyR1 activation and provides protection from calmodulin-binding-mediated inhibition at high Ca^{2+} concentrations [43, 44]. Furthermore, S-nitrosylation of the Cys3635-containing calmodulin binding domain is responsible for RyR1 activation by NO at physiologically relevant oxygen tensions [41]. Other studies suggested that endogenous ROSs produced via NOX enzymes activate Ca^{2+} release through RyRs [45].

Meanwhile, to our knowledge, both the ROC and SOC channel proteins can be the potential targets of ROSs because STIM1 proteins possess several cysteine residues, and all three Orai isoforms possess predicted extracellular and intracellular cysteines that could be targets for modification. It is not hard to speculate that ROSs may also have direct effects on extracellular calcium entry.

Although investigations about the regulatory role of ROSs on ASMC Ca^{2+} influx is just starting, some promising signs have already appeared. Using the strategy of mitochondrial DNA (mtDNA) depletion, our investigation first reported the role of mitochondria-generated ROSs in ASMC Ca^{2+} influx. We established mtDNA-depleted ASMCs by exposing rat ASMCs to low levels (200 ng/ml) of ethidium bromide (EB) and supplemented with pyruvate and uridine. Mitochondria were shown to be the main source of intracellular ROS generation in histamine-stimulated ASMCs, and histamine-stimulated intracellular H_2O_2 generation was at a level equivalent to 0.1 μM extracellular H_2O_2 application. In mtDNA-depleted ASMCs, mitochondria-driven ROS production was diminished. Upon 2 μM histamine exposure, there was no sustained plateau following the peak $[Ca^{2+}]_i$, increasing as in most intact ASMCs but with an altered decrease phase of $[Ca^{2+}]_i$ peak response, and quickly returned to the basal level in mtDNA-depleted ASMCs. However, the external application of 0.1 μM H_2O_2, which does not affect $[Ca^{2+}]_i$ by itself, reversed almost all the alterations in $[Ca^{2+}]_i$ kinetics in mitochondria-depleted ASMs.

To the best of our knowledge, the H_2O_2 concentration used in our study was much lower than the lowest concentration reported to effectively activate PLC on cell membranes and subsequently initiate IP_3 cascades. It appears that IP_3 cascade turnover was not the mechanism underlying the role of 0.1 μM H_2O_2 in Ca^{2+} influx in our study. We suspect that a low level of H_2O_2-induced IP_3 sensitization or activation helped maintain the depletion status of the ER Ca^{2+} store and consequently facilitated STIM1-mediated extracellular Ca^{2+} influx. To explore the mechanism by which a low level of H_2O_2 contributes to the sustained phase of the histamine response, STIM1-sepecific shRNAi/GFP constructs were used in a mtDNA-depleted system. The interesting result was that the reversion of the abolished $[Ca^{2+}]_i$ plateau by extracellular application of H_2O_2 in mitochondria-depleted ASMCs could only happen in the presence of STIM1 but not in the absence of STIM1. These results indicate that H_2O_2 may be necessary but insufficient for a sustained histamine-stimulated $[Ca^{2+}]_i$ response. In other words, mitochondria-derived H_2O_2 may work as a cofactor together with STIM1 to mediate histamine-stimulated Ca^{2+} influx in ASMCs. But how mitochondria-derived H_2O_2 cooperates with STIM1 to maintain Ca^{2+} influx in ASMCs still needs to be elucidated by further investigations. Some recent studies have reported that intracellular ROSs can interact with members of TRP families and contribute to Ca^{2+} influx. They also suggested that ROS-induced TRP intracellular translocation was an important underlying mechanism of TRP activation, which may explain why intracellular H_2O_2 cooperates with STIM1 to mediate Ca^{2+} influx in ASMCs.

3.3 *Oxidative Stress Cooperation with TRPC in Calcium Influx*

The TRPC members expressed in ASMCs have been identified as TRPC1, 3, 4, and 6 [18]. TRPC6 has already been suggested as having a potentially important role in the control of contractile response for its involvement in receptor-stimulated calcium influx in a variety of VSMCs. Among the TRPCs that have been shown to express in human ASMCs, TRPC6 may also play a similar role in airway smooth muscle function.

In many cases, the amount of TRP channels at the plasma membrane is altered in response to physiological stimuli. Controlled translocation between intracellular compartments and the plasma membrane of TRP channels is one of the major mechanisms by which the number of active TRPs is regulated. TRP channel translocation between cell compartments was observed initially for murine TRPV2 upon stimulation with insulin-like growth factor-I (IGF-1) [46]. In recent years, regulated translocation from intracellular compartments to the plasma membrane for members of other TRP families has also been demonstrated, including most members of the TRPC family, such as TRPC1, TRPC3, TRPC4, TRPC5, and TRPC6. The stimulation of TRPC protein trafficking to the plasma membrane has been suggested as a downstream mechanism for physiological TRPC channel activation through a phospholipase C pathway following activation of G-protein-coupled receptors or receptor tyrosine kinases [47].

Thrombin has been shown to be one of the stimuli regulating TRPC translocation. In human pulmonary artery endothelial cells, application of thrombin activates IP_3 triggers calcium release from the ER and activates Rho, which in turn forms a complex with IP_3 receptors and TRPC1 that translocate to the plasma membrane in an actin polymerization-dependent manner and mediates sustained calcium influx [48]. In addition, endothelial growth factor (EGF) turns out to be a common stimulator that activates the translocation of TRPC3, 4, and 5, although using different signaling pathways. In COS-7 cells, TRPC4 appears to be activated and translocated to cell membranes depending on the Src family of tyrosine kinases [49]. Instead, the mechanism of TRPC5 translocation in HEK-293 cells was suggested by Bezzerides et al. to be attributable to EGF-facilitated exocytosis through PI_3 kinase-Rac1-PIP kinase pathway activation [50]. However, although EGF was also reported to induce translocation of TRPC3 to the cell membrane of HEK-293 cells, the translocation of TRPC3 appears not to be a major mechanism for regulating calcium influx via TRPC3 [51]. In addition, the translocation of TRPC6 to the plasma membrane was observed in HEK-293 cells following stimulation of carbachol through a G-protein cascade or stimulation by thapsigargin with ER-store depletion. In addition, TRPC6 translocation was supposed to be the main mechanism for controlling calcium influx since the surface accumulation of TRPC6 is rather fast and parallels the increase in $[Ca^{2+}]_i$ [52].

Recent studies have called attention to the modulation of TRP channels by cellular redox [53–56], suggesting a novel possible mechanism for agonist-initiated cellular response. A very recent investigation carried out by Graham et al. [57] showed that extracellular application of H_2O_2 from 0.01 to 100 μM did not affect calcium ER release, but it did evoke calcium entry response in a dose-dependent manner in TRPC6 over expressed HEK-293 cells. Their findings also suggest that both exogenous and physiologically generated H_2O_2 in response to arginine vasopressin (AVP) can activate TRPC6 channels through promotion of TRPC6 protein trafficking to the plasma membrane as well as through modification of protein thiol groups and enhancement of the TRPC6 sensitivity to diacylglycerol. Their results from two lines of experiments supported the physiological relevance of the redox regulation of TRPC6 channels. On one hand, ROS generated by activation of the AVP V1R pathway contributed to TRPC6 channel activation, including plasma membrane translocation and enhancement of sensitivity, in a heterologous expression system. On the other hand, endogenous TRPC6 in smooth muscle cells fully contributed to H_2O_2-stimulated membrane currents. An important clue from these findings is that full activation of TRPC6 by an agonist might require a certain level of intracellular H_2O_2. In other words, in agonist-stimulated physiological responses, TRPC6 channels should cooperate with endogenously low levels of H_2O_2 to carry out their full activation and mediate calcium influx.

The latest study in the same lab [58] demonstrated the essential role of the interplay of ROSs and TRPC6 in regulating agonist-induced calcium entry and cell contraction in VSMCs. They found that AVP-induced calcium entry was comparably suppressed by knockdown of TRPC6 or NADPH oxidase inhibitors in A7r5 cells. Moreover, H_2O_2 can mimic AVP response, and the ROS-dependent responses were significantly inhibited by either knockdown of TRPC6 or blocking of the TRPC6 channel with a specific antibody. They further proved the importance of the cooperative effect of these two signals in controlling vascular function by using primary aortic VSMCs from TRPC6 knockout (KO) or wild-type mice. Meanwhile, in another study, Weissmann and his colleagues [59], using NOX2 and TRPC6 gene KO mice, also proved the importance of TRPC6 activation by

NOX2-driven ROS production in ischemia-reperfusion-induced calcium entry in endothelial cells. Thus, the cooperation of ROS and TRPC6 might be a common mechanism of agonist-stimulated calcium entry. Besides the direct modification of specific groups of channel protein, stimulating the translocation of channel vesicles to the plasma membrane may also be a key pathway.

Several other members of the TRP subfamilies have also been demonstrated to be redox sensitive, including TRPA1, TRPM2, TRPC3, TRPC4, and TRPC5. Studies on TRPC3 and TRPC4 channels expressed in HEK-293 cells showed that they formed redox-sensitive cation channels upon H_2O_2 stimulation [60]. It has been suggested that H_2O_2 may indirectly activate TRPC3 and TRPC4 channels through activation of phospholipase C. However, it is unknown if the activation is dependent on the oxidation–reduction of cysteine thiol groups. In endothelial cells, TRPC3 and TRPC4 are believed to be capable of forming heteromeric channels that are redox sensitive. In these cells, tertbutyl hydroperoxide leads to membrane depolarization with currents that resemble TRP currents in terms of cation selectivity, La^{3+} sensitivity, and a lack of voltage dependence. Further, expression of the N-terminal fragment, but not the C-terminal fragment, of human TRPC3 abolished the oxidant-induced cation current. These data suggest that TRPC3 forms channels that are activated by ROSs in porcine aortic endothelial cells [61]. Later, in porcine aortic endothelial cells and HEK-293 cells, the same group showed that TRPC3 and TRPC4 associated to form heteromeric channels [62]. These TRPC3/4 heteromeric channels were activated by cholesterol oxidase, suggesting that they are regulated by redox state. However, no direct evidence links TRPC4 itself to redox regulation. TRPC5 homotetrameric channels and TRPC5/TRPC1 heterotetrameric channels have also been reported to be activated by extracellular Trx. The proposed mechanism was suggested to consist of reduced Trx's breaking a disulfide bond in the extracellular loop adjacent to the ion selectivity filter of TRPC5 and then increasing the ion conductivity of the channel. It is interesting to note that as EGF is a stimulus of TRPC3 and TRPC4 translocation, specific inhibition of H_2O_2 production nearly abolishes the response induced by EGF [63]. Thus, intracellular H_2O_2 may also contribute to the plasma translocation of TRPC3 and 4.

Together, evident from the previously cited studies on the regulatory effects of ROSs on TRPC activation, the translocation of TRPCs could be a key mechanism of intracellular H_2O_2 cooperating with STIM1 in ASMCs.

4 Biological Significance and Prospect of Signal Cooperation

ROSs and calcium are critical messenger molecules that mediate most of a cellular signal cascade, leading to diverse physiological or pathological cellular effects. It is generally accepted that various pathways activated by ROSs are convergent to influence intracellular calcium signaling, while changes in intracellular calcium might also impact ROS generation [64]. It is not hard to speculate that cooperation between two interacting signal molecules should be more notable than the linear summation of two separate signals. This was already partially proved by previous studies [65, 66]. It was found that histamine stimulation markedly enhanced the efficiency of calcium-oscillation-regulated gene expression of VCAM1 and NFκB versus calcium oscillations alone in the absence of histamine stimulation. In addition, blocking histamine-stimulated generation of intracellular ROSs abolished this effect. This result strongly suggests an augmented effect of signal cooperation on downstream events.

As two universal signal molecules, we can speculate that the crosstalk of ROSs and calcium could be involved in a wide range of intracellular signal transduction pathways. The augmented downstream event from signaling cooperation may have more profound significance for a wide range of diverse physiological or pathological process.

References

1. Sanders KM (2001) Invited review: mechanisms of calcium handling in smooth muscles. J Appl Physiol 91: 1438–1449.
2. Worley JF, 3rd, Kotlikoff MI (1990) Dihydropyridine-sensitive single calcium channels in airway smooth muscle cells. Am J Physiol 259: L468–480.
3. Murray RK, Kotlikoff MI (1991) Receptor-activated calcium influx in human airway smooth muscle cells. J Physiol 435: 123–144.
4. Parekh AB, Penner R (1997) Store depletion and calcium influx. Physiol Rev 77: 901–930.
5. Cheng KT, Ong HL, Liu X, Ambudkar IS (2011) Contribution of TRPC1 and Orai1 to Ca^{2+} entry activated by store depletion. Adv Exp Med Biol 704: 435–449.
6. Chen T, Zhu L, Wang T, Ye H, Huang K, et al. Mitochondria depletion abolishes agonist-induced Ca^{2+} plateau in airway smooth muscle cells: potential role of H_2O_2. Am J Physiol Lung Cell Mol Physiol 298: L178–188.
7. Croxton TL, Fleming C, Hirshman CA (1994) Expression of dihydropyridine resistance differs in porcine bronchial and tracheal smooth muscle. Am J Physiol 267: L106–112.
8. Janssen LJ (1997) T-type and L-type Ca^{2+} currents in canine bronchial smooth muscle: characterization and physiological roles. Am J Physiol 272: C1757–1765.
9. Janssen LJ, Daniel EE (1991) Depolarizing agents induce oscillations in canine bronchial smooth muscle membrane potential: possible mechanisms. J Pharmacol Exp Ther 259: 110–117.
10. Hall IP (2000) Second messengers, ion channels and pharmacology of airway smooth muscle. Eur Respir J 15: 1120–1127.
11. Bolton TB (1979) Mechanisms of action of transmitters and other substances on smooth muscle. Physiol Rev 59: 606–718.
12. Van Breemen C, Aaronson P, Loutzenhiser R (1978) Sodium-calcium interactions in mammalian smooth muscle. Pharmacol Rev 30: 167–208.
13. de la Fuente G, Palacios O, Villagra E, Villanueva ME (1995) Isolation of Coxsackieviruses B5 in a fatal case of meningoencephalitis. Rev Med Chil 123: 1510–1513.
14. Putney JW, Jr. (1986) A model for receptor-regulated calcium entry. Cell Calcium 7: 1–12.
15. Casteels R, Droogmans G (1981) Exchange characteristics of the noradrenaline-sensitive calcium store in vascular smooth muscle cells or rabbit ear artery. J Physiol 317: 263–279.

16. Ito S, Kume H, Naruse K, Kondo M, Takeda N, et al. (2008) A novel Ca^{2+} influx pathway activated by mechanical stretch in human airway smooth muscle cells. Am J Respir Cell Mol Biol 38: 407–413.
17. Gosling M, Poll C, Li S (2005) TRP channels in airway smooth muscle as therapeutic targets. Naunyn Schmiedebergs Arch Pharmacol 371: 277–284.
18. Corteling RL, Li S, Giddings J, Westwick J, Poll C, et al. (2004) Expression of transient receptor potential C6 and related transient receptor potential family members in human airway smooth muscle and lung tissue. Am J Respir Cell Mol Biol 30: 145–154.
19. Jia Y, Wang X, Varty L, Rizzo CA, Yang R, et al. (2004) Functional TRPV4 channels are expressed in human airway smooth muscle cells. Am J Physiol Lung Cell Mol Physiol 287: L272–278.
20. White TA, Xue A, Chini EN, Thompson M, Sieck GC, et al. (2006) Role of transient receptor potential C3 in TNF-alpha-enhanced calcium influx in human airway myocytes. Am J Respir Cell Mol Biol 35: 243–251.
21. Perez-Zoghbi JF, Karner C, Ito S, Shepherd M, Alrashdan Y, et al. (2009) Ion channel regulation of intracellular calcium and airway smooth muscle function. Pulm Pharmacol Ther 22: 388–397.
22. Zhang SL, Yu Y, Roos J, Kozak JA, Deerinck TJ, et al. (2005) STIM1 is a Ca^{2+} sensor that activates CRAC channels and migrates from the Ca^{2+} store to the plasma membrane. Nature 437: 902–905.
23. Roos J, DiGregorio PJ, Yeromin AV, Ohlsen K, Lioudyno M, et al. (2005) STIM1, an essential and conserved component of store-operated Ca^{2+} channel function. J Cell Biol 169: 435–445.
24. Feske S, Gwack Y, Prakriya M, Srikanth S, Puppel SH, et al. (2006) A mutation in Orai1 causes immune deficiency by abrogating CRAC channel function. Nature 441: 179–185.
25. Gwack Y, Sharma S, Nardone J, Tanasa B, Iuga A, et al. (2006) A genome-wide Drosophila RNAi screen identifies DYRK-family kinases as regulators of NFAT. Nature 441: 646–650.
26. Hogan PG, Lewis RS, Rao A Molecular basis of calcium signaling in lymphocytes: STIM and ORAI. Annu Rev Immunol 28: 491–533.
27. Cheng KT, Liu X, Ong HL, Ambudkar IS (2008) Functional requirement for Orai1 in store-operated TRPC1-STIM1 channels. J Biol Chem 283: 12935–12940.
28. Huang GN, Zeng W, Kim JY, Yuan JP, Han L, et al. (2006) STIM1 carboxyl-terminus activates native SOC, Icrac and TRPC1 channels. Nat Cell Biol 8: 1003–1010.
29. Ong HL, Cheng KT, Liu X, Bandyopadhyay BC, Paria BC, et al. (2007) Dynamic assembly of TRPC1-STIM1-Orai1 ternary complex is involved in store-operated calcium influx. Evidence for similarities in store-operated and calcium release-activated calcium channel components. J Biol Chem 282: 9105–9116.
30. Lopez JJ, Salido GM, Pariente JA, Rosado JA (2006) Interaction of STIM1 with endogenously expressed human canonical TRP1 upon depletion of intracellular Ca^{2+} stores. J Biol Chem 281: 28254–28264.
31. Yuan JP, Zeng W, Huang GN, Worley PF, Muallem S (2007) STIM1 heteromultimerizes TRPC channels to determine their function as store-operated channels. Nat Cell Biol 9: 636–645.
32. Zeng W, Yuan JP, Kim MS, Choi YJ, Huang GN, et al. (2008) STIM1 gates TRPC channels, but not Orai1, by electrostatic interaction. Mol Cell 32: 439–448.
33. Jardin I, Lopez JJ, Salido GM, Rosado JA (2008) Orai1 mediates the interaction between STIM1 and hTRPC1 and regulates the mode of activation of hTRPC1-forming Ca^{2+} channels. J Biol Chem 283: 25296–25304.
34. Liao Y, Erxleben C, Yildirim E, Abramowitz J, Armstrong DL, et al. (2007) Orai proteins interact with TRPC channels and confer responsiveness to store depletion. Proc Natl Acad Sci U S A 104: 4682–4687.
35. Hu Q, Yu ZX, Ferrans VJ, Takeda K, Irani K, et al. (2002) Critical role of NADPH oxidase-derived reactive oxygen species in generating Ca^{2+} oscillations in human aortic endothelial cells stimulated by histamine. J Biol Chem 277: 32546–32551.

36. Hidalgo C, Donoso P (2008) Crosstalk between calcium and redox signaling: from molecular mechanisms to health implications. Antioxid Redox Signal 10: 1275–1312.
37. Bootman MD, Taylor CW, Berridge MJ (1992) The thiol reagent, thimerosal, evokes Ca^{2+} spikes in HeLa cells by sensitizing the inositol 1,4,5-trisphosphate receptor. J Biol Chem 267: 25113–25119.
38. Hu Q, Zheng G, Zweier JL, Deshpande S, Irani K, et al. (2000) NADPH oxidase activation increases the sensitivity of intracellular Ca^{2+} stores to inositol 1,4,5-trisphosphate in human endothelial cells. J Biol Chem 275: 15749–15757.
39. Du W, Frazier M, McMahon TJ, Eu JP (2005) Redox activation of intracellular calcium release channels (ryanodine receptors) in the sustained phase of hypoxia-induced pulmonary vasoconstriction. Chest 128: 556S-558S.
40. Xi Q, Cheranov SY, Jaggar JH (2005) Mitochondria-derived reactive oxygen species dilate cerebral arteries by activating Ca^{2+} sparks. Circ Res 97: 354–362.
41. Sun J, Xin C, Eu JP, Stamler JS, Meissner G (2001) Cysteine-3635 is responsible for skeletal muscle ryanodine receptor modulation by NO. Proc Natl Acad Sci U S A 98: 11158–11162.
42. Xu L, Eu JP, Meissner G, Stamler JS (1998) Activation of the cardiac calcium release channel (ryanodine receptor) by poly-S-nitrosylation. Science 279: 234–237.
43. Porter Moore C, Zhang JZ, Hamilton SL (1999) A role for cysteine 3635 of RYR1 in redox modulation and calmodulin binding. J Biol Chem 274: 36831–36834.
44. Zhang JZ, Wu Y, Williams BY, Rodney G, Mandel F, et al. (1999) Oxidation of the skeletal muscle Ca^{2+} release channel alters calmodulin binding. Am J Physiol 276: C46–53.
45. Espinosa A, Leiva A, Pena M, Muller M, Debandi A, et al. (2006) Myotube depolarization generates reactive oxygen species through NAD(P)H oxidase; ROS-elicited Ca^{2+} stimulates ERK, CREB, early genes. J Cell Physiol 209: 379–388.
46. Kanzaki M, Zhang YQ, Mashima H, Li L, Shibata H, et al. (1999) Translocation of a calcium-permeable cation channel induced by insulin-like growth factor-I. Nat Cell Biol 1: 165–170.
47. Cayouette S, Boulay G (2007) Intracellular trafficking of TRP channels. Cell Calcium 42: 225–232.
48. Mehta D, Ahmmed GU, Paria BC, Holinstat M, Voyno-Yasenetskaya T, et al. (2003) RhoA interaction with inositol 1,4,5-trisphosphate receptor and transient receptor potential channel-1 regulates Ca^{2+} entry. Role in signaling increased endothelial permeability. J Biol Chem 278: 33492–33500.
49. Odell AF, Scott JL, Van Helden DF (2005) Epidermal growth factor induces tyrosine phosphorylation, membrane insertion, and activation of transient receptor potential channel 4. J Biol Chem 280: 37974–37987.
50. Bezzerides VJ, Ramsey IS, Kotecha S, Greka A, Clapham DE (2004) Rapid vesicular translocation and insertion of TRP channels. Nat Cell Biol 6: 709–720.
51. Smyth JT, Lemonnier L, Vazquez G, Bird GS, Putney JW, Jr. (2006) Dissociation of regulated trafficking of TRPC3 channels to the plasma membrane from their activation by phospholipase C. J Biol Chem 281: 11712–11720.
52. Cayouette S, Lussier MP, Mathieu EL, Bousquet SM, Boulay G (2004) Exocytotic insertion of TRPC6 channel into the plasma membrane upon Gq protein-coupled receptor activation. J Biol Chem 279: 7241–7246.
53. Song MY, Makino A, Yuan JX (2011) Role of reactive oxygen species and redox in regulating the function of transient receptor potential channels. Antioxid Redox Signal 15: 1549–1565.
54. Hecquet CM, Ahmmed GU, Vogel SM, Malik AB (2008) Role of TRPM2 channel in mediating H_2O_2-induced Ca^{2+} entry and endothelial hyperpermeability. Circ Res 102: 347–355.
55. Yoshida T, Inoue R, Morii T, Takahashi N, Yamamoto S, et al. (2006) Nitric oxide activates TRP channels by cysteine S-nitrosylation. Nat Chem Biol 2: 596–607.
56. Hara Y, Wakamori M, Ishii M, Maeno E, Nishida M, et al. (2002) LTRPC2 Ca^{2+}-permeable channel activated by changes in redox status confers susceptibility to cell death. Mol Cell 9: 163–173.

57. Graham S, Ding M, Ding Y, Sours-Brothers S, Luchowski R, et al. (2010) Canonical transient receptor potential 6 (TRPC6), a redox-regulated cation channel. J Biol Chem 285: 23466–23476.
58. Ding Y, Winters A, Ding M, Graham S, Akopova I, et al. (2011) Reactive oxygen species-mediated TRPC6 protein activation in vascular myocytes, a mechanism for vasoconstrictor-regulated vascular tone. J Biol Chem 286: 31799–31809.
59. Weissmann N, Sydykov A, Kalwa H, Storch U, Fuchs B, et al. (2012) Activation of TRPC6 channels is essential for lung ischaemia-reperfusion induced oedema in mice. Nat Commun 3: 649.
60. Groschner K, Rosker C, Lukas M (2004) Role of TRP channels in oxidative stress. Novartis Found Symp 258: 222–230; discussion 231–225, 263–226.
61. Groschner K, Hingel S, Lintschinger B, Balzer M, Romanin C, et al. (1998) Trp proteins form store-operated cation channels in human vascular endothelial cells. FEBS Lett 437: 101–106.
62. Poteser M, Graziani A, Rosker C, Eder P, Derler I, et al. (2006) TRPC3 and TRPC4 associate to form a redox-sensitive cation channel. Evidence for expression of native TRPC3-TRPC4 heteromeric channels in endothelial cells. J Biol Chem 281: 13588–13595.
63. Bae YS, Kang SW, Seo MS, Baines IC, Tekle E, et al. (1997) Epidermal growth factor (EGF)-induced generation of hydrogen peroxide. Role in EGF receptor-mediated tyrosine phosphorylation. J Biol Chem 272: 217–221.
64. Shimoda LA, Undem C (2010) Interactions between calcium and reactive oxygen species in pulmonary arterial smooth muscle responses to hypoxia. Respir Physiol Neurobiol 174: 221–229.
65. Zhu L, Luo Y, Chen T, Chen F, Wang T, et al. (2008) Ca^{2+} oscillation frequency regulates agonist-stimulated gene expression in vascular endothelial cells. J Cell Sci 121: 2511–2518.
66. Zhu L, Song S, Pi Y, Yu Y, She W, et al. Cumulated Ca^{2+} spike duration underlies Ca^{2+} oscillation frequency-regulated NFkappaB transcriptional activity. J Cell Sci 124: 2591–2601.

Role of RhoA/Rho-kinase and Calcium Sensitivity in Airway Smooth Muscle Functions

Satoru Ito

Contents

Abstract Changes in the contractility of airway smooth muscle (ASM) play important roles in bronchoconstriction and airway hyperresponsiveness associated with the pathophysiology of asthma and possibly, in part, chronic obstructive pulmonary disease (COPD). A common feature associated with ASM contraction is phosphorylation of myosin light chain (MLC), which is determined by the balance between MLC kinase (MLCK) and MLC phosphatase (MLCP) activities.

S. Ito, M.D., Ph.D. (✉)
Department of Respiratory Medicine, Nagoya University Graduate School of Medicine, 65 Tsurumai-cho, Showa-ku, Nagoya 466-8550, Japan
e-mail: itori@med.nagoya-u.ac.jp

Y.-X. Wang (ed.), *Calcium Signaling In Airway Smooth Muscle Cells*,
DOI 10.1007/978-3-319-01312-1_15,

MLCK is activated by increases in the intracellular Ca^{2+} concentration ($[Ca^{2+}]_i$), whereas MLCP is inactivated by Rho-kinase, a target protein of RhoA, independently of $[Ca^{2+}]_i$. The latter mechanism, contraction due to MLCP inhibition mediated by RhoA/Rho-kinase, is a major type of Ca^{2+} sensitization. Phosphorylation of the 17-kD myosin phosphatase inhibitor protein by protein kinase C is another mechanism of Ca^{2+} sensitization via MLCP suppression. This chapter focuses on recent evidence regarding regulation of ASM contraction by the RhoA/Rho-kinase pathway and the possible roles of this pathway in the pathogenesis of asthma and COPD.

Keywords Asthma • Chronic obstructive pulmonary disease • Ca^{2+} sensitization • RhoA • Rho-kinase • Myosin light chain phosphatase

Abbreviations

ASM	airway smooth muscle
$[Ca^{2+}]_i$	intracellular Ca^{2+} concentration
COPD	chronic obstructive pulmonary disease
CPI-17	17-kD myosin phosphatase inhibitor protein
GAP	GTPase-activating protein
GDI	guanine nucleotide dissociation inhibitor
GDP	guanosine diphosphate
GEF	guanine nucleotide exchange factor
G protein	GTP-binding protein
GTP	guanosine trisphosphate
IL	interleukin
IP_3	inositol 1,4,5-trisphosphate
MBS	myosin-binding subunit
MLC	myosin light chain
MLCK	myosin light chain kinase
MLCP	myosin light chain phosphatase
PDGF	platelet-derived growth factor
PKA	protein kinase A
PKC	protein kinase C
Ser	serine
S1P	sphingosine 1-phosphate
SR	sarcoplasmic reticulum
Thr	threonine
TNF-α	tumor necrosis factor-α
TRP	transient receptor potential

1 Introduction

Contraction of airway smooth muscle (ASM) plays a central role in airway narrowing during an acute asthma attack. An increase in intracellular Ca^{2+} concentrations ($[Ca^{2+}]_i$) followed by activation of myosin light chain kinase (MLCK) is the main initiation of ASM contraction in response to contractile stimuli, such as acetylcholine, inflammatory mediators, and cytokines. Phosphorylation of the 20-kDa regulatory myosin light chain (MLC) at serine (Ser)19 by MLCK generates the contraction of smooth muscle cells [50], and inactivation of MLCK abolishes ASM contractions mediated by $[Ca^{2+}]_i$ elevation [56, 62]. Therefore, Ca^{2+}/calmodulin-dependent MLCK plays a central role in Ca^{2+}-dependent smooth muscle contraction.

In contrast, dephosphorylation of MLC is mediated by myosin light chain phosphatase (MLCP). Inactivation of MLCP by RhoA and Rho- kinase, the downstream effector of RhoA, enhances smooth muscle contraction without affecting $[Ca^{2+}]_i$ [80, 149, 161]. This mechanism, referred to as "Ca^{2+} sensitization", has been intensively investigated for more than two decades. This chapter will summarize current information on the role of RhoA and Rho-kinase in ASM cell functions, specifically contraction, and describe possible involvement of Ca^{2+} sensitization in the pathophysiology of asthma and chronic obstructive pulmonary disease (COPD).

2 Mechanisms of Airway Smooth Muscle Contraction

2.1 *Role of Intracellular Ca^{2+} Concentrations in Smooth Muscle Contraction*

$[Ca^{2+}]_i$ regulates a variety of ASM cell functions, not only contraction but also proliferation, migration, and cytokine production [69, 70, 98, 99, 123, 153, 154, 174]. Agonists for G_q- or $G_{12/13}$-coupled receptor, cytokines, and growth factors, which are closely related to the pathogenesis of asthma, induce ASM contraction in conjunction with $[Ca^{2+}]_i$ elevation [36, 123, 128]. Ca^{2+} release from intracellular sarcoplasmic reticulum (SR) stores and Ca^{2+} influx through plasma membrane Ca^{2+} channels are the main elevators of $[Ca^{2+}]_i$. There are two different mechanisms for release from SR Ca^{2+} stores, inositol 1,4,5-trisphosphate (IP_3)-induced Ca^{2+} release via IP_3 receptors and Ca^{2+}-induced Ca^{2+} release via ryanodine receptors [76]. Activation of G_q-coupled receptors and subsequent activation of phospholipase C induces production of IP_3, which binds to IP_3 receptors on SR Ca^{2+} stores to promote Ca^{2+} release [55]. In addition, Ca^{2+} influx via voltage-dependent Ca^{2+} channels due to membrane depolarization causes Ca^{2+} release from ryanodine-sensitive SR Ca^{2+} stores [166].

There are multiple Ca^{2+} influx pathways through the plasma membrane, including voltage-dependent Ca^{2+} channels, receptor-operated Ca^{2+} entry, store-operated Ca^{2+} entry (originally introduced as capacitative Ca^{2+} entry [125]), and stretch-activated Ca^{2+} entry into ASM cells [63, 70, 106, 123]. The reverse mode of the Na^{+}/Ca^{2+} exchanger is another route for Ca^{2+} influx [93]. Stimulation of G_q-coupled receptors is closely linked to receptor-operated Ca^{2+} entry activation [18, 66, 123]. As in other cell types, Orai1 and STIM1 are the plasma membrane Ca^{2+} channel and SR Ca^{2+} sensor, respectively, for store-operated Ca^{2+} entry in ASM cells [120, 121, 153]. Transient receptor potential (TRP) superfamily channels are candidate molecules for receptor-operated Ca^{2+} entry and stretch-activated Ca^{2+} entry [18, 63, 68, 123, 167]. Please see *other chapters of this textbook* for details on Ca^{2+} regulation in ASM cells.

2.2 Ca^{2+} Sensitivity in Smooth Muscle Contraction

In addition to the Ca^{2+}-dependent pathway, Ca^{2+}-independent mechanisms are also involved in smooth muscle contraction [77, 149–151]. This process, referred to as Ca^{2+} sensitization, is mostly regulated by inhibition of MLCP. MLCP, a type-1 protein phosphatase, mediates dephosphorylation of MLC, which leads to smooth muscle relaxation [39, 124]. Thus, the degree of MLC phosphorylation that determines smooth muscle contractile force is regulated by the balance between MLCK and MLCP activities [148, 151] (Fig. 1).

The contraction of ASM occurs in a phasic manner, followed by a tonic (sustained) contractile response. The initial phasic response is mostly dependent on Ca^{2+} release from the SR, and the tonic response relies on Ca^{2+} influx as well as inactivation of MLCP [21, 61, 122, 127, 138]. MLCP activity is suppressed by Rho-kinase, the downstream effector of RhoA [28, 58, 80, 100]. Previous studies revealed that an increase in MLC phosphorylation levels via inhibition of MLCP by the RhoA/Rho-kinase pathway is the main mechanism of Ca^{2+} sensitization in smooth muscle contraction [19, 34, 80, 83, 161].

3 RhoA and Rho-kinase

3.1 Regulation of RhoA Activity

RhoA belongs to the Rho family of small GTPases that includes Rho, Rac1, and Cdc42 [142]. The Rho family proteins are ubiquitous key regulators of a variety of cell functions such as cytoskeletal dynamics, cell migration, adhesion, and contraction [75, 142]. Among them, RhoA is highly expressed in smooth muscle cells and plays the most important role in smooth muscle contraction. RhoA activity is

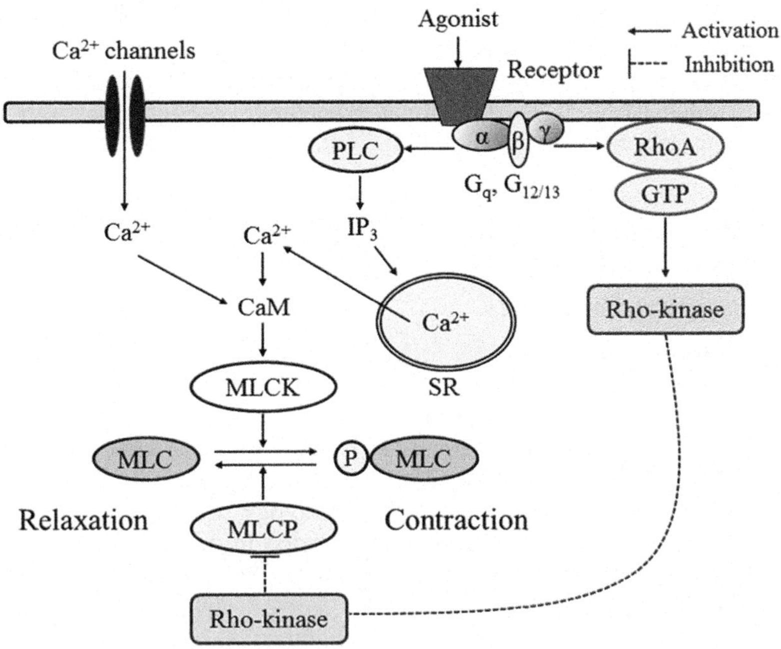

Fig. 1 Intracellular pathways for contraction of airway smooth muscle. The agonist for G_q- or $G_{12/13}$-coupled receptor induces intracellular Ca^{2+} mobilization, which leads to MLCK activation. On the other hand, activation of Rho-kinase by GTP-bound RhoA (active RhoA) inhibits MLCP. Both MLCK activation and MLCP inhibition result in an increase in the levels of MLC phosphorylation (P-MLC). The latter mechanism, inactivation of MLCP by RhoA/Rho-kinase independently of Ca^{2+}, is called Ca^{2+} sensitization

enhanced by the stimulation of receptors coupled to heterotrimeric GTP-binding (G) proteins, specifically $G_{12/13}$ and G_q proteins [28, 124, 142]. In ASM cells, stimulation of G_q- or $G_{12/13}$-coupled receptors, including M_3 muscarinic receptors, histamine H_1 receptors, cysteinyl leukotriene-1 receptors, sphingosine 1-phosphate (S1P) $S1P_2$ and $S1P_3$ receptors, prostanoid FP, TP, and EP_1 receptors, and endothelin-1 ET_A and ET_B receptors, induces contraction [8, 90, 129, 143].

Similar to other G proteins, RhoA has a high affinity for GTP and GDP. The GTP-bound RhoA (GTP-RhoA) located in the cell membrane is an active form, while the GDP-bound form (GDP-RhoA) located in the cytoplasm is inactive [28, 124]. Thus, RhoA activity is determined by translocation of GTP-RhoA to the cell membrane. Switching between the GDP- and GTP-bound forms is regulated by guanine nucleotide exchange factors (GEFs), GTPase-activating proteins (GAPs), and guanine nucleotide dissociation inhibitors (GDIs) [9, 22]. In resting (inactivated) cells, GDI binds to GDP-RhoA and transports GDP-RhoA from the plasma membrane to the cytosol. When a G_q- or $G_{12/13}$-coupled receptor on the plasma membrane is stimulated, GDP-RhoA is converted to GTP-RhoA by GEFs. On the other hand, GAPs are negative regulators of RhoA. GAPs accelerate the intrinsic GTPase activity of RhoA and convert it back to GDP-RhoA

[28, 124]. Moreover, geranylgeranylation of RhoA by geranylgeranyltransferases is essential for this membrane translocation and anchoring of RhoA [162]. Indeed, inhibitors of 3-hydroxy-3-methylglutaryl-coenzyme A (HMG-CoA) reductase (statins), which are used widely to treat hypercholesterolemia, inactivate RhoA by inhibiting its geranylgeranylation [92].

3.2 Rho-kinase and Myosin Light Chain Phosphatase Activity

Rho-kinase, also known as ROCα or ROCK, is a cytoplasmic Ser/threonine (Thr) protein kinase that was identified as an effector of RhoA [58, 91, 100]. Other effectors of RhoA include protein kinase N, p140 mDia, citron kinase, and rhotekin [28]. Rho-kinase has a kinase domain in its N-terminal, a coiled-coil domain in its middle portion, and a putative pleckstrin-homology domain in its C-terminal. The Rho-binding domain of Rho-kinase is located in the C-terminal portion of the coiled-coil domain. Binding of GTP-RhoA to the Rho-binding domain enhances Rho-kinase activity [100].

The main mechanism of smooth muscle contraction mediated by Rho-kinase is primarily based on the inhibition of MLCP (Fig. 1). As described earlier, MLCP is responsible for MLC dephosphorylation. MLCP consists of three subunits: a 37-kDa type-1 phosphatase catalytic subunit, a 130-kDa myosin-binding subunit (MBS), and a 20-kDa subunit (M20) [40]. MBS, also called myosin phosphatase target subunit (MYPT), was identified as a substrate of Rho-kinase [100]. When Rho-kinase is activated by the binding of GTP-RhoA to its Rho-binding domain, Rho-kinase phosphorylates MBS of MLCP, which results in MLCP inactivation [80] (Fig. 1). The major sites of MBS phosphorylation by Rho-kinase have been identified as Thr695/697, Ser849/854, and Thr850/855 [26, 79]. Moreover, Rho-kinase is able to directly phosphorylate MLC at Ser19 independently of Ca^{2+} and MLCK in vitro [1]. Other substrates of Rho-kinase are LIM kinase [115] and Na^{+}-H^{+} exchanger [159].

3.3 RhoA and Rho-kinase in Ca^{2+} Sensitization

Previous studies showed that contractions induced by G_q agonists were larger than those due to membrane depolarization by a high K^{+} solution at the same $[Ca^{2+}]_i$ in smooth muscle tissues [10, 44]. An extracellular solution containing high concentrations of K^{+} is widely used to induce membrane depolarization in studies assessing smooth muscle functions [65, 117]. Karaki and his coworkers measured contraction and $[Ca^{2+}]_i$ simultaneously using a Ca^{2+}-sensitive fluorescent dye, fura-2, in intact smooth muscle tissues [77, 118, 132, 133]. They showed that Ca^{2+}

sensitivity was increased by the muscarinic receptor agonist carbachol compared with a high K^+-induced contraction in canine ASM tissues [117]. Based on these findings, this phenomenon, Ca^{2+} sensitization, has been considered to be mediated by a G_q-coupled mechanism in smooth muscle contraction, including those in airways.

A series of studies by Somlyo and coworkers has clarified the mechanisms of Ca^{2+} sensitization using smooth muscle tissue preparations permeabilized by incubation with α-toxin or β-escin. Because the $[Ca^{2+}]_i$ becomes identical to the Ca^{2+} concentration of the extracellular solution in permeabilized (skinned) tissue, this technique is useful to compare Ca^{2+} sensitivity directly. G_q agonists and GTPγS, a nonhydrolyzable GTP analog, increase the contraction of permeabilized smooth muscle at a constant $[Ca^{2+}]_i$. These findings clearly reveal increased Ca^{2+} sensitivity in G_q-mediated smooth muscle contraction [11, 27, 29, 34, 53, 56, 82, 83, 170]. The G_q-mediated increase in Ca^{2+} sensitivity is blocked by inhibitory toxins specific for Rho family proteins, such as *Clostridium botulinum* exoenzyme C3, *Clostridium difficile* toxin B, and *Staphylococcal* toxin EDIN [34, 45, 96, 111, 116]. Furthermore, the introduction of GTPγS-Rho or the constitutively active form of RhoA into permeabilized smooth muscle increases contraction at a constant $[Ca^{2+}]_i$ [34, 45]. The findings from these earlier studies demonstrate the pivotal role of RhoA in the Ca^{2+} sensitization of smooth muscle.

Several compounds, such as Y-27632, fasudil (HA-1077), hydroxyfasudil (HA-1100), and H-1152P, have been developed as specific Rho-kinase inhibitors [23, 59, 131, 161]. In 1997, Y-27632 was introduced as the first Rho-kinase inhibitor that inhibits Ca^{2+} sensitization in vascular smooth muscle and reduces arterial blood pressure in spontaneously hypertensive rats [161]. In that study, fasudil and H-7, originally developed as an inhibitor of multiple Ser/Thr protein kinases in 1984 [43], were also demonstrated to be potent inhibitors of Rho-kinase. Y-27632 and fasudil selectively inhibit Rho-kinase in a manner competitive with adenosine triphosphate (ATP) [161]. Hydroxyfasudil, a metabolite of fasudil, is more specific for Rho-kinase than fasudil [60]. Use of these cell-permeable pharmacological inhibitors enables evaluation of Rho-kinase in smooth muscle cell functions. Recently, two aminofurazan-based compounds, GSK269962A and SB-7720770-B, were identified as a novel class of specific Rho-kinase inhibitor [23]. Although these new compounds are more potent in inhibiting Rho-kinase, it is still unknown how they affect ASM functions.

3.4 RhoA and Rho-kinase in Airway Smooth Muscle Contraction

As in other smooth muscle cells, activation of the RhoA/Rho-kinase pathway contributes to the contraction of ASM via Ca^{2+} sensitization. The RhoA inhibitor *Clostridium botulinum* C3 toxin inhibits contraction induced by acetylcholine, endothelin-1, or GTPγS in porcine tracheal smooth muscle tissues permeabilized with α-toxin or β-escin [19]. Pharmacological studies in various mammalian species, including humans, cows, pigs, guinea pigs, rabbits, and rats, have demonstrated that the Rho-kinase inhibitors Y-27632 and fasudil attenuate the contraction of ASM tissues mediated by these G_q agonists and GTPγS by reducing Ca^{2+} sensitivity [17, 56, 61, 66, 101, 168, 170]. It should be noted that the relative contribution of RhoA/Rho-kinase to contraction depends on species and agonists/ligands. Moreover, ATP or S1P enhances methacholine-induced contraction, which is inhibited by Y-27632, in guinea-pig tracheal smooth muscle [90, 113]. Therefore, Ca^{2+} sensitization mediated by Rho-kinase could be involved in the agonist synergism in ASM contraction [30, 90, 113].

RhoA/Rho-kinase is also activated by cytokines and growth factors, which couple receptors different from G-protein-coupled receptors. Hunter et al. demonstrated that tumor necrosis factor (TNF)-α increases Ca^{2+} sensitivity with RhoA activation in guinea-pig ASM [52]. Because the translocation of GTP-RhoA and phosphorylation of MYPT-1 by TNF-α are rapidly increased, stimulation of TNF-α receptors directly activates the RhoA/Rho-kinase. In human bronchial and guinea-pig tracheal smooth muscle, contraction induced by growth factors or insulin is dependent on Rho-kinase [36, 139, 140]. Similarly, activation of RhoA by receptor tyrosine kinase is found in vascular smooth muscle cells [112]. Regulation of Rho-GEFs by receptor tyrosine kinase activation is proposed for the mechanism underlying growth-factor-receptor-mediated RhoA activation [124]. Chiba et al. reported that treatment of mouse bronchi with IL-13 for 12 h induced RhoA protein expression with a transient STAT6 phosphorylation [16]. IL-17A, a cytokine derived from T helper 17 cells, enhances ASM contraction with upregulation of the RhoA/Rho-kinase pathway through NF-κB activation in mice [87]. Therefore, IL-13 and IL-17A, cytokines important for inflammatory responses in asthma pathogenesis, increase Ca^{2+} sensitivity by upregulating RhoA expression levels via activating transcription factors in mice.

Furthermore, several studies have proposed the involvement of the RhoA/Rho-kinase pathway in receptor-independent contraction of ASM. Inhibition of Rho-kinase reduces high K^+-induced contraction both in airway and vascular smooth muscle [36, 72, 95, 126]. Contraction mediated by oxidative stress due to hydrogen peroxide is also inhibited by Y-27632 in a concentration-dependent manner in guinea-pig tracheal smooth muscle [84]. Additionally, contractions induced by mechanical stretch, called resting tone, are blocked by Y-27632 or a potent inhibitor of RhoA, simvastatin, in guinea-pig ASM [64]. Consistent with these results, Schaafsma et al. [135, 138] found that Y-27632 decreased the basal resting tone in ASM of both allergen-sensitized and normal guinea pigs. One possible reason is that inhibition of constitutive Rho-kinase activity causes ASM relaxation. Another plausible reason is that the release of mediators or neurotransmitters such as prostanoids via autocrine or paracrine mechanisms may activate RhoA/Rho-kinase [124].

RhoA/Rho-kinase also regulates ion flux pathways such as Ca^{2+} homeostasis and ion channels. The inhibition of Rho-kinase by Y-27632 and fasudil partially but significantly reduces $[Ca^{2+}]_i$ due to agonists in guinea-pig tracheal smooth muscle [61, 66]. This is consistent with findings in vascular smooth muscle contraction mediated by α_1-adrenergic receptors [32, 165]. Regulation of TRPC5 by Rac1 and the Na^+-H^+ exchanger and volume-regulated anion channels by RhoA/Rho-kinase has been reported [6, 109, 159]. However, the mechanisms of Ca^{2+} channel regulation by RhoA/Rho-kinase in smooth muscle cells remain to be elucidated.

4 Other Mechanisms of Ca^{2+} Sensitization

4.1 Role of Protein Kinase C

A physiological role for protein kinase C (PKC) in agonist-induced Ca^{2+} sensitization in smooth muscle contraction has long been proposed [77, 150]. PKC, which was first characterized on the basis of its activation in vitro by Ca^{2+}, phospholipids, and diacylglycerol [110], represents a family of various lipid-dependent Ser/Thr kinases that can be divided into three categories: conventional, novel, and atypical PKCs [108]. Application of 4β-phorbol 12,13-dibutyrate, one of the phorbol ester PKC activators, directly increases Ca^{2+} sensitivity in permeabilized rabbit ASM tissues [24, 56]. Ca^{2+} sensitization induced by GTPγS and acetylcholine is partially blocked by PKC inhibitors in permeabilized ASM tissues of dogs, pigs, and rabbits [11, 56]. In intact murine lung slices, methacholine-induced ASM contraction is relaxed by GF-109203X, a PKC inhibitor, without altering $[Ca^{2+}]_i$ [4]. In contrast, methacholine-induced contraction or an increase in $[Ca^{2+}]_i$ was not affected by GF-109203X in nonpermeabilized guinea-pig tracheal smooth muscle tissues [61]. Therefore, the involvement of PKC in agonist-induced Ca^{2+} sensitization may depend on the species and agonists. The 17-kD myosin phosphatase–inhibitor protein CPI-17 was found to be both a target of PKC and negative regulator of MLCP [25]. PKC-dependent phosphorylation of CPI-17 inhibits MLCP activity, which leads to increased Ca^{2+} sensitivity in smooth muscle contraction [25, 81, 130]. Morin et al. demonstrated that a knockdown of CPI-17 with siRNA inhibited contraction of human bronchial smooth muscle by decreasing Ca^{2+} sensitivity [104]. Additionally, CPI-17 can be phosphorylated by Rho-kinase [86]. Taken together, PKC-CPI-17 activation is involved in the mechanisms of Ca^{2+} sensitization. However, its contribution appears to be less than that of RhoA/Rho-kinase in ASM contraction [56].

4.2 Ca^{2+} Desensitization by β-agonist and cAMP

Agonists activating β_2-adrenergic receptors, β_2-agonists, are the drugs of choice for inhibiting bronchoconstriction in patients with asthma or COPD. The β_2-adrenergic receptor, as well as prostanoid EP_2 and EP_4 receptors, couples to G_s protein, which links not to RhoA activation but to adenylate cyclase. Activation of adenylate cyclase increases intracellular cAMP generation and activates protein kinase A (PKA). PKA plays a central role in the regulation of ASM cell functions by β_2-adrenergic receptor [33, 85, 169]. In addition, Epac, a Rap1 GEF, another downstream effector of cAMP action [20], regulates ASM cell functions [78, 173]. Several studies have shown that β-agonists and other cAMP-increasing agents reduce not only $[Ca^{2+}]_i$ but also Ca^{2+} sensitivity in ASM contraction [7]. Janssen and colleagues showed that the β_1- and β_2-agonist isoproterenol reduces carbachol-mediated Ca^{2+} sensitization with enhanced MLCP activity in porcine and bovine ASM tissues [71]. In guinea-pig ASM, isoproterenol or an activator of adenylate cyclase forskolin inhibits methacholine-induced contraction with reduction of both $[Ca^{2+}]_i$ and Ca^{2+} sensitivity [114]. Moreover, formoterol, a long-acting β_2-agonist used for treatment of asthma and COPD, inhibits histamine-induced ASM contraction by decreasing Ca^{2+} sensitivity at low concentrations and by attenuating Ca^{2+} oscillations at higher concentrations in human lung slices [127]. Theophylline, which elevates cytosolic cAMP levels by inhibiting phosphodiesterase, significantly reduces carbachol-induced contraction without a significant decrease in $[Ca^{2+}]_i$ in porcine ASM [57]. In α-toxin permeabilized rabbit ASM tissues, cAMP analog, 8-bromo-cAMP, partially attenuates carbachol- or leukotriene D_4-induced contraction [24]. These findings further support a role for cAMP in Ca^{2+} desensitization in ASM contraction. Moreover, activation of Epac inhibits Ca^{2+} sensitization induced by carbachol, endothelin-1, and bradykinin in mouse bronchi [173]. Taken together, both PKA and Epac are likely involved in cAMP-mediated Ca^{2+} desensitization in ASM relaxation.

Liu et al. demonstrated that RhoA and Rho-kinase activities stimulated by carbachol were partially but significantly inhibited by isoproterenol and a long-acting β_2-agonist, salmeterol, in bovine tracheal smooth muscle tissues [94]. Furthermore, lysophosphatidylcholine decreases the relaxing effects of isoproterenol via Ca^{2+} sensitization mediated by Rho-kinase in guinea-pig tracheal smooth muscle [89]. Although the mechanism is still uncertain, reduced Ca^{2+} sensitivity by cAMP and a β_2-agonist and desensitization of β_2-adrenergic receptors may partially depend on RhoA/Rho-kinase inactivation.

Another possible mechanism of reduced sensitivity to Ca^{2+} by β-agonists is the inhibition of actin polymerization independently of the RhoA/Rho-kinase pathway. Indeed, activation of the cAMP/PKA pathway by isoproterenol and prostaglandin E_2 inhibits the actin cytoskeleton in ASM cells [3, 47, 48, 85]. Actin dynamics are essential for maintaining the contractile force of smooth muscle, and cytochalasin D, an inhibitor of F-actin polymerization, blocks ASM contraction without inhibiting $[Ca^{2+}]_i$ [66]. Komalavilas et al. showed that phosphorylation of

HSP20 by PKA decreased phosphorylation of cofilin, leading to actin disruption and relaxation of ASM [85]. Thus, actin depolymerization via HSP20 phosphorylation may be involved in the mechanism of cAMP/PKA-mediated Ca^{2+} desensitization in ASM cell contraction.

5 Role of RhoA/Rho-kinase in Airway Smooth Muscle Functions Other Than Contraction

5.1 Cytoskeleton and Mechanical Properties

The RhoA/Rho-kinase pathway participates in the regulation of the actin cytoskeleton [46, 75]. Contractile G-protein-coupled receptor agonists enhance actin polymerization by activating the RhoA/Rho-kinase pathway in ASM cells [46, 147]. The actin stress fiber formation mediated by RhoA/Rho-kinase activation is mostly regulated via MLC phosphorylation [28].

Cytoskeletal proteins, including actin, intermediate filaments, and microtubules, regulate the mechanical properties of cells [42, 68, 105, 152, 157]. Because ASM is dynamically stretched during tidal breathing, it is important to characterize the mechanical properties of ASM in a dynamic condition (i.e., stiffness) as well as a static condition (i.e., isometric force). An et al. demonstrated that cell stiffness is significantly reduced by the actin disruptor cytochalasin D, Y-27632, and C3 toxin in cultured rat ASM cells [2]. Similar to the findings in cultured cells, tissue stiffness is significantly reduced by fasudil and cytochalasin D in guinea pigs [67]. Thus, regulation of both cell and tissue stiffness by Rho-kinase is closely associated with actin cytoskeleton and contractile prestress.

5.2 Role of RhoA and Rho-kinase in Cell Proliferation and Migration

The increase in ASM mass due to enhanced proliferation of ASM cells has been implicated in the pathogenesis of airway hyperresponsiveness and remodeling in patients with asthma [35, 49, 73, 74, 155, 158]. Involvement of the RhoA/Rho-kinase pathway in ASM cell proliferation has been proposed. The proliferation stimulated by fetal bovine serum is attenuated by Y-27632 and a potent RhoA inhibitor, simvastatin, in cultured human ASM cells [155]. Similarly, Rho-kinase has been implicated in the proliferation of vascular smooth muscle cells [134]. In contrast, Y-27632 does not inhibit basal or platelet-derived growth factor (PDGF)-induced proliferation in bovine ASM cells [36]. Moreover, among the Rho family, Rac and Cdc42, rather than RhoA, regulate PDGF-induced proliferation in bovine ASM cells [5]. Therefore, the contribution of RhoA/Rho-kinase to ASM cell proliferation might depend on mitogen and species.

ASM cell migration toward the airway epithelium in response to inflammatory mediators might contribute to airway remodeling [31, 97]. The dynamics of the cytoskeleton and signal transduction pathways, including RhoA/Rho-kinase and $[Ca^{2+}]_i$, are involved in the processes of ASM cell migration [31, 153]. Rho-kinase inhibition by Y-27632 significantly inhibited cell migration induced by PDGF or leukotriene E_4 in cultured human ASM cells [119]. In contrast, interestingly, PDGF-induced migration of human ASM cells is not inhibited by MLCK inhibitors [12]. Thus, it is possible that RhoA/Rho-kinase regulates cell migration, possibly via the actin cytoskeleton, independently of MLC phosphorylation. Further studies are necessary to identify the downstream pathways of Ca^{2+} and RhoA/Rho-kinase in ASM cell migration.

6 Effects of Mechanical Stress on RhoA Activation

Mechanical stresses such as mechanical stretch and compression affect ASM cell functions by activating a signal transduction cascade and Ca^{2+} influx pathways [63, 105, 144, 145, 147]. Recent evidence that mechanical stress contributes to the pathogenesis of airway remodeling in asthma has been reported [38, 160]. Using a cell stretch apparatus, Smith et al. demonstrated that the RhoA/Rho-kinase pathway was activated in response to cyclic stretching in cultured canine ASM cells [147]. Stretch-mediated RhoA/Rho-kinase activation leads to actin polymerization, transforming growth factor-β1 expression, cell alignment, and contraction via Ca^{2+} sensitization of ASM cells [14, 103, 144, 146, 147]. Because mechanical stress within the airway is enhanced by bronchoconstriction, the RhoA/Rho-kinase activation of ASM cells may lead to further contraction and airway remodeling.

7 RhoA and Rho-kinase as Therapeutic Targets for Airway Diseases

Because RhoA/Rho-kinase plays a key role in ASM contraction, this pathway is expected to be a novel therapeutic target for asthma and COPD [15, 17, 88, 141]. In animal models of allergic asthma, RhoA protein expression is increased in bronchi of rats and guinea pigs [15, 138]. Administration of the Rho-kinase inhibitors Y-27632 and fasudil inhibits airway hyperresponsiveness in rodent models of ovalbumin-sensitized and challenged allergic asthma [41, 135, 138, 156]. Moreover, airway hyperresponsiveness and inflammation due to an ovalbumin challenge were significantly reduced in heterozygous Rho-kinase isoform (ROCK1, ROCK2) knockout mice ($ROCK1^{+/-}$, $ROCK2^{+/-}$) compared with wild-type mice [172]. These findings from animal studies suggest that RhoA/Rho-kinase is involved in the pathogenesis of airway hyperresponsiveness and inflammation in asthma. Nevertheless, the contribution of RhoA/Rho-kinase to the pathophysiology of asthma in humans is still unknown.

In animal models, the Rho-kinase inhibitors are effective as bronchodilators. Inhaled Y-27632 significantly reduced airway resistance induced by acetylcholine or ovalbumin in guinea pigs in vivo [54]. Moreover, inhalation of Y-27632 effectively reversed airway hyperresponsiveness to histamine and $PGF_{2\alpha}$ following an allergen challenge in guinea pigs [135]. Importantly, Y-27632 and the β_2-agonists salbutamol and terbutaline additively relaxed methacholine-induced contraction of bovine tracheal smooth muscle [107]. Thus, Rho-kinase inhibitors could be a novel bronchodilator, and the combination of a Rho-kinase inhibitor with a β_2-agonist could be more potent in inhibiting bronchoconstriction than β_2-agonist monotherapy during an acute asthma attack [13].

Statins such as simvastatin and atorvastatin, which inhibit RhoA activity, are also good candidate drugs for the treatment of asthma and COPD. Simvastatin inhibits cell proliferation and extracellular matrix synthesis in human ASM cells in vitro [88, 137, 155, 164]. However, two negative reports have been published regarding the administration of statins to patients with asthma [51, 102]. Hothersall et al. reported that the addition of atorvastatin to inhaled corticosteroids resulted in no short-term improvement in the control of atopic asthma [51]. In a study by Menzies et al., simvastatin did not have anti-inflammatory activity in patients with asthma [102]. In contrast, statin treatment could be beneficial for COPD by inhibiting inflammation and reducing cancer mortality [163, 171]. Thus, it is likely that the beneficial effects of statins on COPD are not caused by altering ASM functions.

8 Conclusion and Perspectives

Since Rho-kinase was identified as a target protein of RhoA, considerable progress has been made in characterizing the mechanisms of ASM contraction. It is now widely known that Ca^{2+} sensitization via the RhoA/Rho-kinase pathway in conjunction with $[Ca^{2+}]_i$ regulates ASM contraction. Therefore, RhoA/Rho-kinase could be a novel therapeutic target for the treatment of asthma and COPD. Nevertheless, it is still unclear whether RhoA/Rho-kinase is upregulated in the airway and lungs of patients with asthma and COPD. Moreover, the mechanisms of the Ca^{2+}-desensitizing effects of cAMP/PKA have not yet been elucidated. Future studies are required to establish the role of Rho/Rho-kinase and Ca^{2+} sensitization in the pathogenesis of asthma and COPD.

Acknowledgments This work was supported by grants-in-aid from the Ministry of Education, Culture, Sports, Science, and Technology of Japan (17790531, 19689017, and 22590837). We also thank Ms. Katherine Ono for providing language help.

References

1. Amano M, Ito M, Kimura K, Fukata Y, Chihara K, Nakano T, Matsuura Y, Kaibuchi K (1996) Phosphorylation and activation of myosin by Rho-associated kinase (Rho-kinase). J Biol Chem 271:20246–20249.
2. An SS, Laudadio RE, Lai J, Rogers RA, Fredberg JJ (2002) Stiffness changes in cultured airway smooth muscle cells. Am J Physiol Cell Physiol 283:C792–801.
3. Aso H, Ito S, Mori A, Suganuma N, Morioka M, Takahara N, Kondo M, Hasegawa Y (2013) Differential regulation of airway smooth muscle cell migration by prostanoid EP receptor subtypes. Am J Respir Cell Mol Biol 48:322–329.
4. Bai Y, Sanderson MJ (2006) Modulation of the Ca^{2+} sensitivity of airway smooth muscle cells in murine lung slices. Am J Physiol Lung Cell Mol Physiol 291:L208–221.
5. Bauerfeld CP, Hershenson MB, Page K (2001) Cdc42, but not RhoA, regulates cyclin D1 expression in bovine tracheal myocytes. Am J Physiol Lung Cell Mol Physiol 280:L974–982.
6. Bezzerides VJ, Ramsey IS, Kotecha S, Greka A, Clapham DE (2004) Rapid vesicular translocation and insertion of TRP channels. Nat Cell Biol 6:709–720.
7. Billington CK, Ojo OO, Penn RB, Ito S (2013) cAMP regulation of airway smooth muscle function. Pulm Pharmacol Ther 26:112–120.
8. Billington CK, Penn RB (2003) Signaling and regulation of G protein-coupled receptors in airway smooth muscle. Respir Res 4:2.
9. Bos JL, Rehmann H, Wittinghofer A (2007) GEFs and GAPs: critical elements in the control of small G proteins. Cell 129:865–877.
10. Bradley AB, Morgan KG (1987) Alterations in cytoplasmic calcium sensitivity during porcine coronary artery contractions as detected by aequorin. J Physiol 385:437–448.
11. Bremerich DH, Warner DO, Lorenz RR, Shumway R, Jones KA (1997) Role of protein kinase C in calcium sensitization during muscarinic stimulation in airway smooth muscle. Am J Physiol 273:L775–781.
12. Carlin SM, Roth M, Black JL (2003) Urokinase potentiates PDGF-induced chemotaxis of human airway smooth muscle cells. Am J Physiol Lung Cell Mol Physiol 284:L1020–1026.
13. Cazzola M, Page CP, Calzetta L, Matera MG (2012) Pharmacology and therapeutics of bronchodilators. Pharmacol Rev 64:450–504.
14. Chaudhuri S, Smith PG (2008) Cyclic strain-induced HSP27 phosphorylation modulates actin filaments in airway smooth muscle cells. Am J Respir Cell Mol Biol 39:270–278.
15. Chiba Y, Matsusue K, Misawa M (2010) RhoA, a possible target for treatment of airway hyperresponsiveness in bronchial asthma. J Pharmacol Sci 114:239–247.
16. Chiba Y, Nakazawa S, Todoroki M, Shinozaki K, Sakai H, Misawa M (2009) Interleukin-13 augments bronchial smooth muscle contractility with an up-regulation of RhoA protein. Am J Respir Cell Mol Biol 40:159–167.
17. Chiba Y, Takada Y, Miyamoto S, Mitsuisaito M, Karaki H, Misawa M (1999) Augmented acetylcholine-induced, Rho-mediated Ca^{2+} sensitization of bronchial smooth muscle contraction in antigen-induced airway hyperresponsive rats. Br J Pharmacol 127:597–600.
18. Corteling RL, Li S, Giddings J, Westwick J, Poll C, Hall IP (2004) Expression of transient receptor potential C6 and related transient receptor potential family members in human airway smooth muscle and lung tissue. Am J Respir Cell Mol Biol 30:145–154.
19. Croxton TL, Lande B, Hirshman CA (1998) Role of G proteins in agonist-induced Ca^{2+} sensitization of tracheal smooth muscle. Am J Physiol 275:L748–755.
20. De Rooij J, Zwartkruis FJ, Verheijen MH, Cool RH, Nijman SM, Wittinghofer A, Bos JL (1998) Epac is a Rap1 guanine-nucleotide-exchange factor directly activated by cyclic AMP. Nature 396:474–477.
21. Delmotte P, Sanderson MJ (2010) Effects of formoterol on contraction and Ca^{2+} signaling of mouse airway smooth muscle cells. Am J Respir Cell Mol Biol 42:373–381.
22. Dermardirossian C, Bokoch GM (2005) GDIs: central regulatory molecules in Rho GTPase activation. Trends Cell Biol 15:356–363.

23. Doe C, Bentley R, Behm DJ, Lafferty R, Stavenger R, Jung D, Bamford M, Panchal T, Grygielko E, Wright LL, Smith GK, Chen Z, Webb C, Khandekar S, Yi T, Kirkpatrick R, Dul E, Jolivette L, Marino JP, Jr., Willette R, Lee D, Hu E (2007) Novel Rho kinase inhibitors with anti-inflammatory and vasodilatory activities. J Pharmacol Exp Ther 320:89–98.
24. Endou K, Iizuka K, Yoshii A, Tsukagoshi H, Ishizuka T, Dobashi K, Nakazawa T, Mori M (2004) 8-Bromo-cAMP decreases the Ca^{2+} sensitivity of airway smooth muscle contraction through a mechanism distinct from inhibition of Rho-kinase. Am J Physiol Lung Cell Mol Physiol 287:L641–648.
25. Eto M, Senba S, Morita F, Yazawa M (1997) Molecular cloning of a novel phosphorylation-dependent inhibitory protein of protein phosphatase-1 (CPI17) in smooth muscle: its specific localization in smooth muscle. FEBS Lett 410:356–360.
26. Feng J, Ito M, Ichikawa K, Isaka N, Nishikawa M, Hartshorne DJ, Nakano T (1999) Inhibitory phosphorylation site for Rho-associated kinase on smooth muscle myosin phosphatase. J Biol Chem 274:37385–37390.
27. Fujihara H, Walker LA, Gong MC, Lemichez E, Boquet P, Somlyo AV, Somlyo AP (1997) Inhibition of RhoA translocation and calcium sensitization by in vivo ADP-ribosylation with the chimeric toxin DC3B. Mol Biol Cell 8:2437–2447.
28. Fukata Y, Amano M, Kaibuchi K (2001) Rho-Rho-kinase pathway in smooth muscle contraction and cytoskeletal reorganization of non-muscle cells. Trends Pharmacol Sci 22:32–39.
29. Gailly P, Gong MC, Somlyo AV, Somlyo AP (1997) Possible role of atypical protein kinase C activated by arachidonic acid in Ca^{2+} sensitization of rabbit smooth muscle. J Physiol 500:95–109.
30. Gerthoffer WT (1996) Agonist synergism in airway smooth muscle contraction. J Pharmacol Exp Ther 278:800–807.
31. Gerthoffer WT (2008) Migration of airway smooth muscle cells. Proc Am Thorac Soc 5:97–105.
32. Ghisdal P, Vandenberg G, Morel N (2003) Rho-dependent kinase is involved in agonist-activated calcium entry in rat arteries. J Physiol 551:855–867.
33. Goncharova EA, Goncharov DA, Zhao H, Penn RB, Krymskaya VP, Panettieri RA, Jr. (2012) β_2-adrenergic receptor agonists modulate human airway smooth muscle cell migration via vasodilator-stimulated phosphoprotein. Am J Respir Cell Mol Biol 46:48–54.
34. Gong MC, Iizuka K, Nixon G, Browne JP, Hall A, Eccleston JF, Sugai M, Kobayashi S, Somlyo AV, Somlyo AP (1996) Role of guanine nucleotide-binding proteins–ras-family or trimeric proteins or both–in Ca^{2+} sensitization of smooth muscle. Proc Natl Acad Sci U S A 93:1340–1345.
35. Gosens R, Roscioni SS, Dekkers BG, Pera T, Schmidt M, Schaafsma D, Zaagsma J, Meurs H (2008) Pharmacology of airway smooth muscle proliferation. Eur J Pharmacol 585:385–397.
36. Gosens R, Schaafsma D, Grootte Bromhaar MM, Vrugt B, Zaagsma J, Meurs H, Nelemans SA (2004) Growth factor-induced contraction of human bronchial smooth muscle is Rho-kinase-dependent. Eur J Pharmacol 494:73–76.
37. Gosens R, Schaafsma D, Meurs H, Zaagsma J, Nelemans SA (2004) Role of Rho-kinase in maintaining airway smooth muscle contractile phenotype. Eur J Pharmacol 483:71–78.
38. Grainge CL, Lau LC, Ward JA, Dulay V, Lahiff G, Wilson S, Holgate S, Davies DE, Howarth PH (2011) Effect of bronchoconstriction on airway remodeling in asthma. N Engl J Med 364:2006–2015.
39. Hartshorne DJ, Ito M, Erdodi F (1998) Myosin light chain phosphatase: subunit composition, interactions and regulation. J Muscle Res Cell Motil 19:325–341.
40. Hartshorne DJ, Ito M, Erdodi F (2004) Role of protein phosphatase type 1 in contractile functions: myosin phosphatase. J Biol Chem 279:37211–37214.
41. Hashimoto K, Peebles RS, Jr., Sheller JR, Jarzecka K, Furlong J, Mitchell DB, Hartert TV, Graham BS (2002) Suppression of airway hyperresponsiveness induced by ovalbumin sensitisation and RSV infection with Y-27632, a Rho kinase inhibitor. Thorax 57:524–527.

42. Hayakawa K, Tatsumi H, Sokabe M (2008) Actin stress fibers transmit and focus force to activate mechanosensitive channels. J Cell Sci 121:496–503.
43. Hidaka H, Inagaki M, Kawamoto S, Sasaki Y (1984) Isoquinolinesulfonamides, novel and potent inhibitors of cyclic nucleotide dependent protein kinase and protein kinase C. Biochemistry 23:5036–5041.
44. Himpens B, Casteels R (1990) Different effects of depolarization and muscarinic stimulation on the Ca^{2+}/force relationship during the contraction-relaxation cycle in the guinea pig ileum. Pflugers Arch : Eur J Physiol 416:28–35.
45. Hirata K, Kikuchi A, Sasaki T, Kuroda S, Kaibuchi K, Matsuura Y, Seki H, Saida K, Takai Y (1992) Involvement of rho p21 in the GTP-enhanced calcium ion sensitivity of smooth muscle contraction. J Biol Chem 267:8719–8722.
46. Hirshman CA, Emala CW (1999) Actin reorganization in airway smooth muscle cells involves Gq and Gαi-2 activation of Rho. Am J Physiol 277:L653–661.
47. Hirshman CA, Zhu D, Panettieri RA, Emala CW (2001) Actin depolymerization via the beta-adrenoceptor in airway smooth muscle cells: a novel PKA-independent pathway. Am J Physiol Cell Physiol 281:C1468–1476.
48. Hirshman CA, Zhu D, Pertel T, Panettieri RA, Emala CW (2005) Isoproterenol induces actin depolymerization in human airway smooth muscle cells via activation of an Src kinase and GS. Am J Physiol Lung Cell Mol Physiol 288:L924–931.
49. Hirst SJ, Martin JG, Bonacci JV, Chan V, Fixman ED, Hamid QA, Herszberg B, Lavoie JP, Mcvicker CG, Moir LM, Nguyen TT, Peng Q, Ramos-Barbon D, Stewart AG (2004) Proliferative aspects of airway smooth muscle. J Allergy Clin Immunol 114:S2–17.
50. Horowitz A, Menice CB, Laporte R, Morgan KG (1996) Mechanisms of smooth muscle contraction. Physiol Rev 76:967–1003.
51. Hothersall EJ, Chaudhuri R, Mcsharry C, Donnelly I, Lafferty J, Mcmahon AD, Weir CJ, Meiklejohn J, Sattar N, Mcinnes I, Wood S, Thomson NC (2008) Effects of atorvastatin added to inhaled corticosteroids on lung function and sputum cell counts in atopic asthma. Thorax 63:1070–1075.
52. Hunter I, Cobban HJ, Vandenabeele P, Macewan DJ, Nixon GF (2003) Tumor necrosis factor-alpha-induced activation of RhoA in airway smooth muscle cells: role in the Ca^{2+} sensitization of myosin light chain20 phosphorylation. Mol Pharmacol 63:714–721.
53. Iizuka K, Dobashi K, Yoshii A, Horie T, Suzuki H, Nakazawa T, Mori M (1997) Receptor-dependent G protein-mediated Ca^{2+} sensitization in canine airway smooth muscle. Cell Calcium 22:21–30.
54. Iizuka K, Shimizu Y, Tsukagoshi H, Yoshii A, Harada T, Dobashi K, Murozono T, Nakazawa T, Mori M (2000) Evaluation of Y-27632, a rho-kinase inhibitor, as a bronchodilator in guinea pigs. Eur J Pharmacol 406:273–279.
55. Iizuka K, Yoshii A, Dobashi K, Horie T, Mori M, Nakazawa T (1998) InsP3, but not novel Ca^{2+} releasers, contributes to agonist-initiated contraction in rabbit airway smooth muscle. J Physiol 511:915–933.
56. Iizuka K, Yoshii A, Samizo K, Tsukagoshi H, Ishizuka T, Dobashi K, Nakazawa T, Mori M (1999) A major role for the rho-associated coiled coil forming protein kinase in G-protein-mediated Ca^{2+} sensitization through inhibition of myosin phosphatase in rabbit trachea. Br J Pharmacol 128:925–933.
57. Ise S, Nishimura J, Hirano K, Hara N, Kanaide H (2003) Theophylline attenuates Ca^{2+} sensitivity and modulates BK channels in porcine tracheal smooth muscle. Br J Pharmacol 140:939–947.
58. Ishizaki T, Maekawa M, Fujisawa K, Okawa K, Iwamatsu A, Fujita A, Watanabe N, Saito Y, Kakizuka A, Morii N, Narumiya S (1996) The small GTP-binding protein Rho binds to and activates a 160 kDa Ser/Thr protein kinase homologous to myotonic dystrophy kinase. EMBO J 15:1885–1893.

59. Ishizaki T, Uehata M, Tamechika I, Keel J, Nonomura K, Maekawa M, Narumiya S (2000) Pharmacological properties of Y-27632, a specific inhibitor of rho-associated kinases. Mol Pharmacol 57:976–983.
60. Ito K, Shimomura E, Iwanaga T, Shiraishi M, Shindo K, Nakamura J, Nagumo H, Seto M, Sasaki Y, Takuwa Y (2003) Essential role of rho kinase in the Ca^{2+} sensitization of prostaglandin F(2alpha)-induced contraction of rabbit aortae. J Physiol 546:823–836.
61. Ito S, Kume H, Honjo H, Katoh H, Kodama I, Yamaki K, Hayashi H (2001) Possible involvement of Rho kinase in Ca^{2+} sensitization and mobilization by MCh in tracheal smooth muscle. Am J Physiol Lung Cell Mol Physiol 280:L1218-1224.
62. Ito S, Kume H, Honjo H, Kodama I, Katoh H, Hayashi H, Shimokata K (2004) ML-9, a myosin light chain kinase inhibitor, reduces intracellular Ca^{2+} concentration in guinea pig trachealis. Eur J Pharmacol 486:325–333.
63. Ito S, Kume H, Naruse K, Kondo M, Takeda N, Iwata S, Hasegawa Y, Sokabe M (2008) A novel Ca^{2+} influx pathway activated by mechanical stretch in human airway smooth muscle cells. Am J Respir Cell Mol Biol 38:407–413.
64. Ito S, Kume H, Oguma T, Ito Y, Kondo M, Shimokata K, Suki B, Naruse K (2006) Roles of stretch-activated cation channel and Rho-kinase in the spontaneous contraction of airway smooth muscle. Eur J Pharmacol 552:135–142.
65. Ito S, Kume H, Shiraki A, Kondo M, Makino Y, Kamiya K, Hasegawa Y (2008) Inhibition by the cold receptor agonists menthol and icilin of airway smooth muscle contraction. Pulm Pharmacol Ther 21:812–817.
66. Ito S, Kume H, Yamaki K, Katoh H, Honjo H, Kodama I, Hayashi H (2002) Regulation of capacitative and noncapacitative receptor-operated Ca^{2+} entry by Rho-kinase in tracheal smooth muscle. Am J Respir Cell Mol Biol 26:491–498.
67. Ito S, Majumdar A, Kume H, Shimokata K, Naruse K, Lutchen KR, Stamenovic D, Suki B (2006) Viscoelastic and dynamic nonlinear properties of airway smooth muscle tissue: roles of mechanical force and the cytoskeleton. Am J Physiol Lung Cell Mol Physiol 290:L1227-1237.
68. Ito S, Suki B, Kume H, Numaguchi Y, Ishii M, Iwaki M, Kondo M, Naruse K, Hasegawa Y, Sokabe M (2010) Actin cytoskeleton regulates stretch-activated Ca^{2+} influx in human pulmonary microvascular endothelial cells. Am J Respir Cell Mol Biol 43:26–34.
69. Iwata S, Ito S, Iwaki M, Kondo M, Sashio T, Takeda N, Sokabe M, Hasegawa Y, Kume H (2009) Regulation of endothelin-1-induced interleukin-6 production by Ca^{2+} influx in human airway smooth muscle cells. Eur J Pharmacol 605:15–22.
70. Janssen LJ (2002) Ionic mechanisms and Ca^{2+} regulation in airway smooth muscle contraction: do the data contradict dogma? Am J Physiol Lung Cell Mol Physiol 282:L1161–1178.
71. Janssen LJ, Tazzeo T, Zuo J (2004) Enhanced myosin phosphatase and Ca(2+)-uptake mediate adrenergic relaxation of airway smooth muscle. Am J Respir Cell Mol Biol 30:548–554.
72. Janssen LJ, Tazzeo T, Zuo J, Pertens E, Keshavjee S (2004) KCl evokes contraction of airway smooth muscle via activation of RhoA and Rho-kinase. Am J Physiol Lung Cell Mol Physiol 287:L852–858.
73. Johnson PR, Roth M, Tamm M, Hughes M, Ge Q, King G, Burgess JK, Black JL (2001) Airway smooth muscle cell proliferation is increased in asthma. Am J Respir Crit Care Med 164:474–477.
74. Joubert P, Hamid Q (2005) Role of airway smooth muscle in airway remodeling. J Allergy Clin Immunol 116:713–716.
75. Kaibuchi K, Kuroda S, Amano M (1999) Regulation of the cytoskeleton and cell adhesion by the Rho family GTPases in mammalian cells. Annu Rev Biochem 68:459–486.
76. Kannan MS, Prakash YS, Brenner T, Mickelson JR, Sieck GC (1997) Role of ryanodine receptor channels in Ca2+ oscillations of porcine tracheal smooth muscle. Am J Physiol 272: L659–664.

77. Karaki H, Ozaki H, Hori M, Mitsui-Saito M, Amano K, Harada K, Miyamoto S, Nakazawa H, Won KJ, Sato K (1997) Calcium movements, distribution, and functions in smooth muscle. Pharmacol Rev 49:157–230.
78. Kassel KM, Wyatt TA, Panettieri RA, Jr., Toews ML (2008) Inhibition of human airway smooth muscle cell proliferation by beta 2-adrenergic receptors and cAMP is PKA independent: evidence for EPAC involvement. Am J Physiol Lung Cell Mol Physiol 294:L131–138.
79. Kawano Y, Fukata Y, Oshiro N, Amano M, Nakamura T, Ito M, Matsumura F, Inagaki M, Kaibuchi K (1999) Phosphorylation of myosin-binding subunit (MBS) of myosin phosphatase by Rho-kinase in vivo. J Cell Biol 147:1023–1038.
80. Kimura K, Ito M, Amano M, Chihara K, Fukata Y, Nakafuku M, Yamamori B, Feng J, Nakano T, Okawa K, Iwamatsu A, Kaibuchi K (1996) Regulation of myosin phosphatase by Rho and Rho-associated kinase (Rho-kinase). Science 273:245–248.
81. Kitazawa T, Eto M, Woodsome TP, Brautigan DL (2000) Agonists trigger G protein-mediated activation of the CPI-17 inhibitor phosphoprotein of myosin light chain phosphatase to enhance vascular smooth muscle contractility. J Biol Chem 275:9897–9900.
82. Kitazawa T, Gaylinn BD, Denney GH, Somlyo AP (1991) G-protein-mediated Ca^{2+} sensitization of smooth muscle contraction through myosin light chain phosphorylation. J Biol Chem 266:1708–1715.
83. Kitazawa T, Masuo M, Somlyo AP (1991) G protein-mediated inhibition of myosin light-chain phosphatase in vascular smooth muscle. Proc Natl Acad Sci U S A 88:9307–9310.
84. Kojima K, Kume H, Ito S, Oguma T, Shiraki A, Kondo M, Ito Y, Shimokata K (2007) Direct effects of hydrogen peroxide on airway smooth muscle tone: roles of Ca^{2+} influx and Rho-kinase. Eur J Pharmacol 556:151–156.
85. Komalavilas P, Penn RB, Flynn CR, Thresher J, Lopes LB, Furnish EJ, Guo M, Pallero MA, Murphy-Ullrich JE, Brophy CM (2008) The small heat shock-related protein, HSP20, is a cAMP-dependent protein kinase substrate that is involved in airway smooth muscle relaxation. Am J Physiol Lung Cell Mol Physiol 294:L69–78.
86. Koyama M, Ito M, Feng J, Seko T, Shiraki K, Takase K, Hartshorne DJ, Nakano T (2000) Phosphorylation of CPI-17, an inhibitory phosphoprotein of smooth muscle myosin phosphatase, by Rho-kinase. FEBS Lett 475:197–200.
87. Kudo M, Melton AC, Chen C, Engler MB, Huang KE, Ren X, Wang Y, Bernstein X, Li JT, Atabai K, Huang X, Sheppard D (2012) IL-17A produced by alphabeta T cells drives airway hyper-responsiveness in mice and enhances mouse and human airway smooth muscle contraction. Nat Med 18:547–554.
88. Kume H (2008) RhoA/Rho-kinase as a therapeutic target in asthma. Curr Med Chem 15:2876–2885.
89. Kume H, Ito S, Ito Y, Yamaki K (2001) Role of lysophosphatidylcholine in the desensitization of beta-adrenergic receptors by Ca^{2+} sensitization in tracheal smooth muscle. Am J Respir Cell Mol Biol 25:291–298.
90. Kume H, Takeda N, Oguma T, Ito S, Kondo M, Ito Y, Shimokata K (2007) Sphingosine 1-phosphate causes airway hyper-reactivity by rho-mediated myosin phosphatase inactivation. J Pharmacol Exp Ther 320:766–773.
91. Leung T, Manser E, Tan L, Lim L (1995) A novel serine/threonine kinase binding the Ras-related RhoA GTPase which translocates the kinase to peripheral membranes. J Biol Chem 270:29051–29054.
92. Liao JK, Laufs U (2005) Pleiotropic effects of statins. Annu Rev Pharmacol Toxicol 45:89–118.
93. Liu B, Peel SE, Fox J, Hall IP (2010) Reverse mode Na+/Ca2+ exchange mediated by STIM1 contributes to Ca2+ influx in airway smooth muscle following agonist stimulation. Respir Res 11:168.
94. Liu C, Zuo J, Janssen LJ (2006) Regulation of airway smooth muscle RhoA/ROCK activities by cholinergic and bronchodilator stimuli. Eur Respir J 28:703–711.

95. Liu C, Zuo J, Pertens E, Helli PB, Janssen LJ (2005) Regulation of Rho/ROCK signaling in airway smooth muscle by membrane potential and [Ca2+]i. Am J Physiol Lung Cell Mol Physiol 289:L574–582.
96. Lucius C, Arner A, Steusloff A, Troschka M, Hofmann F, Aktories K, Pfitzer G (1998) Clostridium difficile toxin B inhibits carbachol-induced force and myosin light chain phosphorylation in guinea-pig smooth muscle: role of Rho proteins. J Physiol 506:83–93.
97. Madison JM (2003) Migration of airway smooth muscle cells. Am J Respir Cell Mol Biol 29:8–11.
98. Mahn K, Hirst SJ, Ying S, Holt MR, Lavender P, Ojo OO, Siew L, Simcock DE, Mcvicker CG, Kanabar V, Snetkov VA, O'connor BJ, Karner C, Cousins DJ, Macedo P, Chung KF, Corrigan CJ, Ward JP, Lee TH (2009) Diminished sarco/endoplasmic reticulum Ca2+ ATPase (SERCA) expression contributes to airway remodelling in bronchial asthma. Proc Natl Acad Sci U S A 106:10775–10780.
99. Mahn K, Ojo OO, Chadwick G, Aaronson PI, Ward JP, Lee TH (2010) Ca^{2+} homeostasis and structural and functional remodelling of airway smooth muscle in asthma. Thorax 65:547–552.
100. Matsui T, Amano M, Yamamoto T, Chihara K, Nakafuku M, Ito M, Nakano T, Okawa K, Iwamatsu A, Kaibuchi K (1996) Rho-associated kinase, a novel serine/threonine kinase, as a putative target for small GTP binding protein Rho. EMBO J 15:2208–2216.
101. Matsumoto H, Moir LM, Oliver BG, Burgess JK, Roth M, Black JL, Mcparland BE (2007) Comparison of gel contraction mediated by airway smooth muscle cells from patients with and without asthma. Thorax 62:848–854.
102. Menzies D, Nair A, Meldrum KT, Fleming D, Barnes M, Lipworth BJ (2007) Simvastatin does not exhibit therapeutic anti-inflammatory effects in asthma. J Allergy Clin Immunol 119:328–335.
103. Mohamed JS, Boriek AM (2010) Stretch augments TGF-beta1 expression through RhoA/ROCK1/2, PTK, and PI3K in airway smooth muscle cells. Am J Physiol Lung Cell Mol Physiol 299:L413–424.
104. Morin C, Sirois M, Echave V, Rousseau E (2008) CPI-17 silencing-reduced responsiveness in control and TNF-alpha-treated human bronchi. Am J Respir Cell Mol Biol 39:638–643.
105. Morioka M, Parameswaran H, Naruse K, Kondo M, Sokabe M, Hasegawa Y, Suki B, Ito S (2011) Microtubule dynamics regulate cyclic stretch-induced cell alignment in human airway smooth muscle cells. PLoS One 6:e26384.
106. Murray RK, Kotlikoff MI (1991) Receptor-activated calcium influx in human airway smooth muscle cells. J Physiol 435:123–144.
107. Nakahara T, Moriuchi H, Yunoki M, Sakamato K, Ishii K (2000) Y-27632 potentiates relaxant effects of beta 2-adrenoceptor agonists in bovine tracheal smooth muscle. Eur J Pharmacol 389:103–106.
108. Newton AC (1995) Protein kinase C: structure, function, and regulation. J Biol Chem 270:28495–28498.
109. Nilius B, Voets T, Prenen J, Barth H, Aktories K, Kaibuchi K, Droogmans G, Eggermont J (1999) Role of Rho and Rho kinase in the activation of volume-regulated anion channels in bovine endothelial cells. J Physiol 516:67–74.
110. Nishizuka Y (1992) Intracellular signaling by hydrolysis of phospholipids and activation of protein kinase C. Science 258:607–614.
111. Noda M, Yasuda-Fukazawa C, Moriishi K, Kato T, Okuda T, Kurokawa K, Takuwa Y (1995) Involvement of rho in GTP gamma S-induced enhancement of phosphorylation of 20 kDa myosin light chain in vascular smooth muscle cells: inhibition of phosphatase activity. FEBS Lett 367:246–250.
112. Ogita H, Kunimoto S, Kamioka Y, Sawa H, Masuda M, Mochizuki N (2003) EphA4-mediated Rho activation via Vsm-RhoGEF expressed specifically in vascular smooth muscle cells. Circ Res 93:23–31.

113. Oguma T, Ito S, Kondo M, Makino Y, Shimokata K, Honjo H, Kamiya K, Kume H (2007) Roles of P2X receptors and Ca^{2+} sensitization in extracellular adenosine triphosphate-induced hyperresponsiveness in airway smooth muscle. Clin Exp Allergy 37:893–900.
114. Oguma T, Kume H, Ito S, Takeda N, Honjo H, Kodama I, Shimokata K, Kamiya K (2006) Involvement of reduced sensitivity to Ca^{2+} in β-adrenergic action on airway smooth muscle. Clin Exp Allergy 36:183–191.
115. Ohashi K, Nagata K, Maekawa M, Ishizaki T, Narumiya S, Mizuno K (2000) Rho-associated kinase ROCK activates LIM-kinase 1 by phosphorylation at threonine 508 within the activation loop. J Biol Chem 275:3577–3582.
116. Otto B, Steusloff A, Just I, Aktories K, Pfitzer G (1996) Role of Rho proteins in carbachol-induced contractions in intact and permeabilized guinea-pig intestinal smooth muscle. J Physiol 496:317–329.
117. Ozaki H, Kwon SC, Tajimi M, Karaki H (1990) Changes in cytosolic Ca2+ and contraction induced by various stimulants and relaxants in canine tracheal smooth muscle. Pflugers Arch : Eur J Physiol 416:351–359.
118. Ozaki H, Sato K, Satoh T, Karaki H (1987) Simultaneous recordings of calcium signals and mechanical activity using fluorescent dye fura 2 in isolated strips of vascular smooth muscle. Jpn J Pharmacol 45:429–433.
119. Parameswaran K, Cox G, Radford K, Janssen LJ, Sehmi R, O'Byrne PM (2002) Cysteinyl leukotrienes promote human airway smooth muscle migration. Am J Respir Crit Care Med 166:738–742.
120. Peel SE, Liu B, Hall IP (2006) A key role for STIM1 in store operated calcium channel activation in airway smooth muscle. Respir Res 7:119.
121. Peel SE, Liu B, Hall IP (2008) ORAI and store-operated calcium influx in human airway smooth muscle cells. Am J Respir Cell Mol Biol 38:744–749.
122. Penn RB, Benovic JL (2008) Regulation of heterotrimeric G protein signaling in airway smooth muscle. Proc Am Thorac Soc 5:47–57.
123. Perez-Zoghbi JF, Karner C, Ito S, Shepherd M, Alrashdan Y, Sanderson MJ (2009) Ion channel regulation of intracellular calcium and airway smooth muscle function. Pulm Pharmacol Ther 22:388–397.
124. Puetz S, Lubomirov LT, Pfitzer G (2009) Regulation of smooth muscle contraction by small GTPases. Physiology 24:342–356.
125. Putney JW, Jr. (1986) A model for receptor-regulated calcium entry. Cell Calcium 7:1–12.
126. Ratz PH, Berg KM, Urban NH, Miner AS (2005) Regulation of smooth muscle calcium sensitivity: KCl as a calcium-sensitizing stimulus. Am J Physiol Cell Physiol 288:C769–783.
127. Ressmeyer AR, Bai Y, Delmotte P, Uy KF, Thistlethwaite P, Fraire A, Sato O, Ikebe M, Sanderson MJ (2010) Human airway contraction and formoterol-induced relaxation is determined by Ca2+ oscillations and Ca2+ sensitivity. Am J Respir Cell Mol Biol 43:179–191.
128. Risse PA, Jo T, Suarez F, Hirota N, Tolloczko B, Ferraro P, Grutter P, Martin JG (2011) Interleukin-13 inhibits proliferation and enhances contractility of human airway smooth muscle cells without change in contractile phenotype. Am J Physiol Lung Cell Mol Physiol 300:L958–966.
129. Rosenfeldt HM, Amrani Y, Watterson KR, Murthy KS, Panettieri RA, Jr., Spiegel S (2003) Sphingosine-1-phosphate stimulates contraction of human airway smooth muscle cells. FASEB J 17:1789–1799.
130. Sakai H, Chiba Y, Hirano T, Misawa M (2005) Possible involvement of CPI-17 in augmented bronchial smooth muscle contraction in antigen-induced airway hyper-responsive rats. Mol Pharmacol 68:145–151.
131. Sasaki Y, Suzuki M, Hidaka H (2002) The novel and specific Rho-kinase inhibitor (S)-(+)-2-methyl-1-[(4-methyl-5-isoquinoline)sulfonyl]-homopiperazine as a probing molecule for Rho-kinase-involved pathway. Pharmacol Ther 93:225–232.

132. Sato K, Ozaki H, Karaki H (1988) Changes in cytosolic calcium level in vascular smooth muscle strip measured simultaneously with contraction using fluorescent calcium indicator fura 2. J Pharmacol Exp Ther 246:294–300.
133. Sato K, Ozaki H, Karaki H (1990) Differential effects of carbachol on cytosolic calcium levels in vascular endothelium and smooth muscle. J Pharmacol Exp Ther 255:114–119.
134. Satoh K, Fukumoto Y, Shimokawa H (2011) Rho-kinase: important new therapeutic target in cardiovascular diseases. Am J Physiol Heart Circ Physiol 301:H287–296.
135. Schaafsma D, Bos IS, Zuidhof AB, Zaagsma J, Meurs H (2006) Inhalation of the Rho-kinase inhibitor Y-27632 reverses allergen-induced airway hyperresponsiveness after the early and late asthmatic reaction. Respir Res 7:121.
136. Schaafsma D, Bos IS, Zuidhof AB, Zaagsma J, Meurs H (2008) The inhaled Rho kinase inhibitor Y-27632 protects against allergen-induced acute bronchoconstriction, airway hyperresponsiveness, and inflammation. Am J Physiol Lung Cell Mol Physiol 295:L214–219.
137. Schaafsma D, Dueck G, Ghavami S, Kroeker A, Mutawe MM, Hauff K, Xu FY, Mcneill KD, Unruh H, Hatch GM, Halayko AJ (2011) The mevalonate cascade as a target to suppress extracellular matrix synthesis by human airway smooth muscle. Am J Respir Cell Mol Biol 44:394–403.
138. Schaafsma D, Gosens R, Bos IS, Meurs H, Zaagsma J, Nelemans SA (2004) Allergic sensitization enhances the contribution of Rho-kinase to airway smooth muscle contraction. Br J Pharmacol 143:477–484.
139. Schaafsma D, Gosens R, Bos IS, Meurs H, Zaagsma J, Nelemans SA (2005) Role of contractile prostaglandins and Rho-kinase in growth factor-induced airway smooth muscle contraction. Respir Res 6:85.
140. Schaafsma D, Gosens R, Ris JM, Zaagsma J, Meurs H, Nelemans SA (2007) Insulin induces airway smooth muscle contraction. Br J Pharmacol 150:136–142.
141. Schaafsma D, Gosens R, Zaagsma J, Halayko AJ, Meurs H (2008) Rho kinase inhibitors: a novel therapeutical intervention in asthma? Eur J Pharmacol 585:398–406.
142. Seasholtz TM, Majumdar M, Brown JH (1999) Rho as a mediator of G protein-coupled receptor signaling. Mol Pharmacol 55:949–956.
143. Shiraki A, Kume H, Oguma T, Makino Y, Ito S, Shimokata K, Honjo H, Kamiya K (2009) Role of Ca^{2+} mobilization and Ca^{2+} sensitization in 8-iso-$PGF_{2\alpha}$-induced contraction in airway smooth muscle. Clin Exp Allergy 39:236–245.
144. Smith PG, Deng L, Fredberg JJ, Maksym GN (2003) Mechanical strain increases cell stiffness through cytoskeletal filament reorganization. Am J Physiol Lung Cell Mol Physiol 285:L456–463.
145. Smith PG, Moreno R, Ikebe M (1997) Strain increases airway smooth muscle contractile and cytoskeletal proteins in vitro. Am J Physiol 272:L20–27.
146. Smith PG, Roy C, Fisher S, Huang QQ, Brozovich F (2000) Selected contribution: mechanical strain increases force production and calcium sensitivity in cultured airway smooth muscle cells. J Appl Physiol 89:2092–2098.
147. Smith PG, Roy C, Zhang YN, Chauduri S (2003) Mechanical stress increases RhoA activation in airway smooth muscle cells. Am J Respir Cell Mol Biol 28:436–442.
148. Somlyo AP (1997) Signal transduction. Rhomantic interludes raise blood pressure. Nature 389:908–909, 911.
149. Somlyo AP, Somlyo AV (1994) Signal transduction and regulation in smooth muscle. Nature 372:231–236.
150. Somlyo AP, Somlyo AV (2000) Signal transduction by G-proteins, rho-kinase and protein phosphatase to smooth muscle and non-muscle myosin II. J Physiol 522:177–185.
151. Somlyo AP, Somlyo AV (2003) Ca2+ sensitivity of smooth muscle and nonmuscle myosin II: modulated by G proteins, kinases, and myosin phosphatase. Physiol Rev 83:1325–1358.
152. Stamenovic D, Mijailovich SM, Tolic-Norrelykke IM, Chen J, Wang N (2002) Cell prestress. II. Contribution of microtubules. Am J Physiol Cell Physiol 282:C617–624.

153. Suganuma N, Ito S, Aso H, Kondo M, Sato M, Sokabe M, Hasegawa Y (2012) STIM1 regulates platelet-derived growth factor-induced migration and Ca^{2+} influx in human airway smooth muscle cells. PLoS One 7:e45056.
154. Sweeney M, Mcdaniel SS, Platoshyn O, Zhang S, Yu Y, Lapp BR, Zhao Y, Thistlethwaite PA, Yuan JX (2002) Role of capacitative Ca2+ entry in bronchial contraction and remodeling. J Appl Physiol 92:1594–1602.
155. Takeda N, Kondo M, Ito S, Ito Y, Shimokata K, Kume H (2006) Role of RhoA inactivation in reduced cell proliferation of human airway smooth muscle by simvastatin. Am J Respir Cell Mol Biol 35:722–729.
156. Taki F, Kume H, Kobayashi T, Ohta H, Aratake H, Shimokata K (2007) Effects of Rho-kinase inactivation on eosinophilia and hyper-reactivity in murine airways by allergen challenges. Clin Exp Allergy 37:599–607.
157. Tang DD, Bai Y, Gunst SJ (2005) Silencing of p21-activated kinase attenuates vimentin phosphorylation on Ser-56 and reorientation of the vimentin network during stimulation of smooth muscle cells by 5-hydroxytryptamine. Biochem J 388:773–783.
158. Tliba O, Panettieri RA, Jr. (2009) Noncontractile functions of airway smooth muscle cells in asthma. Annu Rev Physiol 71:509–535.
159. Tominaga T, Ishizaki T, Narumiya S, Barber DL (1998) p160ROCK mediates RhoA activation of Na-H exchange. EMBO J 17:4712–4722.
160. Tschumperlin DJ, Drazen JM (2006) Chronic effects of mechanical force on airways. Annu Rev Physiol 68:563–583.
161. Uehata M, Ishizaki T, Satoh H, Ono T, Kawahara T, Morishita T, Tamakawa H, Yamagami K, Inui J, Maekawa M, Narumiya S (1997) Calcium sensitization of smooth muscle mediated by a Rho-associated protein kinase in hypertension. Nature 389:990–994.
162. Van Aelst L, D'souza-Schorey C (1997) Rho GTPases and signaling networks. Genes Dev 11:2295–2322.
163. Van Gestel YR, Hoeks SE, Sin DD, Huzeir V, Stam H, Mertens FW, Van Domburg RT, Bax JJ, Poldermans D (2009) COPD and cancer mortality: the influence of statins. Thorax 64:963–967.
164. Vigano T, Hernandez A, Corsini A, Granata A, Belloni P, Fumagalli R, Paoletti R, Folco G (1995) Mevalonate pathway and isoprenoids regulate human bronchial myocyte proliferation. Eur J Pharmacol 291:201–203.
165. Villalba N, Stankevicius E, Simonsen U, Prieto D (2008) Rho kinase is involved in Ca2+ entry of rat penile small arteries. Am J Physiol Heart Circ Physiol 294:H1923–1932.
166. Wang YX, Zheng YM, Mei QB, Wang QS, Collier ML, Fleischer S, Xin HB, Kotlikoff MI (2004) FKBP12.6 and cADPR regulation of Ca2+ release in smooth muscle cells. Am J Physiol Cell Physiol 286:C538–546.
167. Xiao JH, Zheng YM, Liao B, Wang YX (2010) Functional role of canonical transient receptor potential 1 and canonical transient receptor potential 3 in normal and asthmatic airway smooth muscle cells. Am J Respir Cell Mol Biol 43:17–25.
168. Yamagata S, Ichinose M, Sugiura H, Koarai A, Koike K, Shirato K (2000) Effect of a calcium sensitization modulator, Y-27632, on isolated human bronchus and pulmonary artery. Pulm Pharmacol Ther 13:25–29.
169. Yan H, Deshpande DA, Misior AM, Miles MC, Saxena H, Riemer EC, Pascual RM, Panettieri RA, Penn RB (2011) Anti-mitogenic effects of beta-agonists and PGE2 on airway smooth muscle are PKA dependent. FASEB J 25:389–397.
170. Yoshii A, Iizuka K, Dobashi K, Horie T, Harada T, Nakazawa T, Mori M (1999) Relaxation of contracted rabbit tracheal and human bronchial smooth muscle by Y-27632 through inhibition of Ca2+ sensitization. Am J Respir Cell Mol Biol 20:1190–1200.
171. Young RP, Hopkins R, Eaton TE (2009) Pharmacological actions of statins: potential utility in COPD. Eur Respir Rev 18:222–232.

172. Zhu M, Liu PY, Kasahara DI, Williams AS, Verbout NG, Halayko AJ, Fedulov A, Shoji T, Williams ES, Noma K, Shore SA, Liao JK (2011) Role of Rho kinase isoforms in murine allergic airway responses. Eur Respir J 38:841–850.
173. Zieba BJ, Artamonov MV, Jin L, Momotani K, Ho R, Franke AS, Neppl RL, Stevenson AS, Khromov AS, Chrzanowska-Wodnicka M, Somlyo AV (2011) The cAMP-responsive Rap1 guanine nucleotide exchange factor, Epac, induces smooth muscle relaxation by down-regulation of RhoA activity. J Biol Chem 286:16681–16692.
174. Zou JJ, Gao YD, Geng S, Yang J (2011) Role of STIM1/Orai1-mediated store-operated Ca^{2+} entry in airway smooth muscle cell proliferation. J Appl Physiol 110:1256–1263.

Role of Integrins in the Regulation of Calcium Signaling

Thai Tran and Chun Ming Teoh

Contents

Abstract Integrins are a large family of transmembrane proteins that constitute the main receptors for extracellular components and are important in mediating intracellular signaling events that govern cell adhesion, shape, polarity, growth, differentiation, migration, and cell survival. Less well known is their involvement in regulating calcium signaling that may influence cell contraction. This chapter provides an overview of the current state of knowledge of the signaling mechanisms by which integrins may regulate calcium signaling with a focus on airway smooth muscle cells.

Keywords Integrins • Calcium • Airway smooth muscle • GPCR • Crosstalk • Asthma

---

T. Tran, Ph.D. (✉) • C.M. Teoh
Department of Physiology, National University of Singapore, Block MD9, 2 Medical Drive, Singapore 117597, Singapore
e-mail: tran_thai@nuhs.edu.sg

Y.-X. Wang (ed.), *Calcium Signaling In Airway Smooth Muscle Cells*,
DOI 10.1007/978-3-319-01312-1_16,

1 Integrins

Integrins are a family of heterodimeric transmembrane receptors consisting of noncovalently linked α and β subunits. Currently there are 19 α- and 8 β-subunits that combine to form 25 different integrin receptors in mammals [26]. The most common extracellular matrix (ECM) ligands for integrin binding are shown in Fig. 1. The domain structure of the α- and β-subunits contain three parts: an extracellular domain that contains a divalent cation binding site required for ligand-receptor interactions, a hydrophobic transmembrane domain required for membrane anchoring, and a cytoplasmic domain responsible for mediating downstream signaling [3, 30].

Integrins bind to ECM components in a manner that is dependent upon both affinity and avidity and is influenced by the composition of the integrin heterodimer and ECM conformation. ECM-integrin binding affinity is determined by several factors. First, there are integrin recognition sequences in the ECM. The most recognizable sequence found in the majority of ECM proteins, including collagen, fibronectin, and vitronectin, is identified as an Arg-Gly-Asp (RGD) sequence. Other sites in the ECM that are responsible for binding with certain integrins include, but are not limited to, YIGSR (laminin), DGEA (native collagen, laminin), KQAGV (fibrinogen), and EILDV (fibronectin, VCAM). The presence of these unique

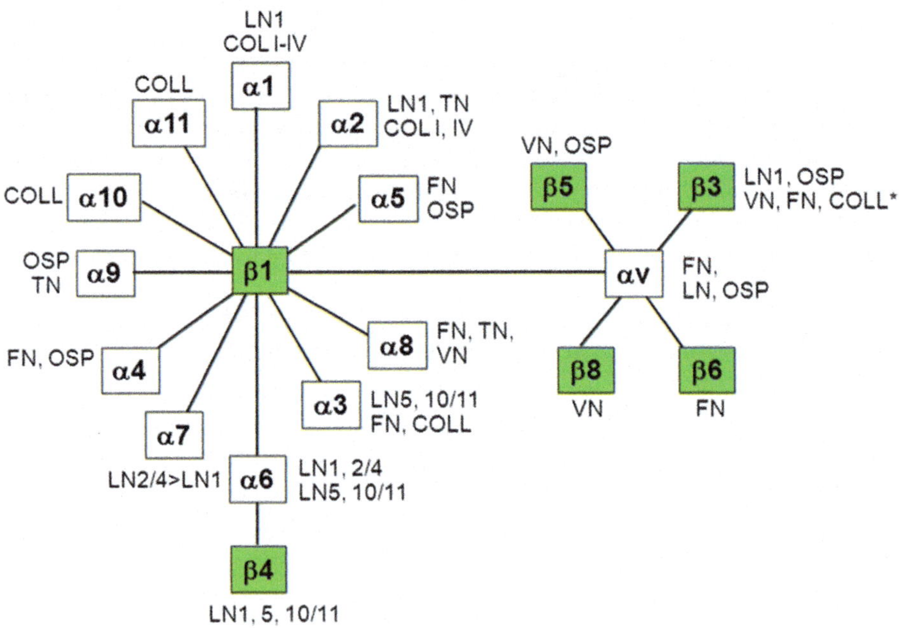

Fig. 1 Schema showing ECM binding partners for integrins. For clarity, those integrins that bind other adhesion proteins (e.g., VCAM) are omitted. *COLL* collagen, *COLL** denatured collagen, *FN* fibronectin, *LN* laminin, *OSP* osteopontin, *TN* tenascin, *VN* vitronectin

sequence-binding sites in the ECM provides additional binding selectivity for ECM families. Second, the presence of a region that binds divalent cations such as Mg^{2+} and Mn^{2+} allows for physiological ligand binding at a broad range of concentrations. Third, the activation state of the integrin can also affect the ECM-integrin binding affinity. Integrins must be in the activated state to undergo adhesion and binding to the ECM at appropriate times to induce "outside-in" signals to the cell. This is achieved through the switching of integrins from a low to a high affinity state. The switching of affinity state is in turn determined by "inside-out" signals acting to release intracellular constraints exerted on the integrin cytoplasmic tails [25, 43].

Integrin knockout studies have provided some insights into the role of integrin in the regulation of calcium signaling in muscle cells. In cardiomyocytes, integrin β1 knockout mice showed calcium deposition in the ventricular wall [45]. In support, integrin β1 and dystrophin double knockout mice showed more calcified lesions and precipitation on cardiomyocytes compared with dystrophin knockout mice alone [18]. In skeletal muscle cells, Guo and colleagues demonstrated that integrin α7 and dystrophin double knockout mice had massive calcification in the diaphragm, whereas muscle calcification was only occasionally seen in dystrophin knockout mice. Collectively, the authors suggest that the absence of integrin may affect calcium homeostasis in muscle cells [22].

2 Integrin Expression in Airway Smooth Muscle

Integrins play an essential role in lung development. For example, integrin α3β1 is the main laminin-binding integrin that connects alveolar and bronchial epithelial cells to the basement membrane at all stages of human fetal and adult lung development. Under basal, unstimulated conditions, airway smooth muscle (ASM) cells in culture express a plethora of integrins. However, the proportions of integrin subunits that are expressed on ASM appear to vary significantly between studies. This is likely due to variations in cell culture conditions. Nonetheless, the range in expression of individual integrin that has been reported for ASM cells, relative to the ubiquitous expression of the β1-subunit (considered as 100 %), is 30–60 % for α1, 60–80 % for α2, 30–60 % for α3, approx. 30 % for α4, 100 % for α5, approx. 40 % for α6, approx. 45 % for α7, 60–100 % for αv, 100 % for β1, and less than 1 % for β2-4 [19, 20, 35, 48, 49]. It is worth noting that changes in receptor expression have been reported under stimulated conditions (e.g., in response to the growth factor PDGF [49]). Moreover, we have shown that the laminin-binding α7 integrin subunit is exclusively needed to promote contractile phenotype acquisition in differentiated human ASM cells [48]. It remains to be explored whether changes in integrin receptor abundance and expression profile for ECM components in ASM cells are associated with disease progression.

3 Integrin Signaling and ASM Function

Integrin activation via ECM binding promotes the assembly of actin filaments and reorganization of actin filaments into large stress fibers (Fig. 2). The intracellular actin cytoskeleton, together with integrins and signaling molecules, forms aggregates known as focal adhesion [10, 21]. Focal adhesion consists of various signaling molecules in which integrin transmits signals through protein tyrosine kinases, such as focal adhesion kinase (FAK) and integrin-linked kinase (ILK), that regulate various downstream signaling pathways. This includes ASM phenotype and functional plasticity, transduction of mechanical force, proliferation, hypertrophy, apoptosis, immune modulation, and cell adhesion and migration [48].

Integrins not only function independently but can associate with growth factor receptors to produce a response. Integrin $\alpha v\beta 3$ has been shown to associate with insulin, PDGF, and VEGF receptors, whereas integrin $\alpha 5\beta 1$ associates with EGF receptors [41, 47, 53]. There is also a link between integrins with various types of ion channels (e.g., Na/H antiporter and potassium channels). But the physiological importance of this connection remains to be determined [2, 39, 42]. Collectively, these findings suggest that crosstalk of integrins with other receptors exists and may regulate various functions of ASM cells.

4 Calcium Signaling in ASM Contraction

In the lung, smooth muscle is the major contractile tissue of the airways. Contraction of ASM cells is regulated through active force generation and cell shortening. Most importantly, the shortening ability of unloaded bronchial smooth muscle cells from asthmatics is increased and can therefore contribute to airway hyperresponsiveness in asthma [31]. In support, passive sensitization of human bronchi with atopic serum increases maximal contractility and sensitivity to agonists with a concomitant increase in myosin light chain kinase (MLCK) expression [1]. In addition, ASM retains a capacity for other functions, including growth, migration, and synthesis of a vast array of regulatory molecules as part of the normal contractile and repair programming of ASM [24]. Under disease conditions such as asthma, these functions become dysregulated.

Smooth muscle contraction is regulated by phosphorylation and dephosphorylation of myosin and smooth muscle α-actin in response to cytosolic calcium concentration. Levels of cytosolic calcium concentration are determined by calcium influx from the extracellular space and calcium release from the intracellular calcium stores on the sarcoplasmic reticulum via the IP_3-activated calcium channel and ryanodine receptor (RyR) activation [13, 27]. Cytosolic calcium then binds to calmodulin, and the complex formed then activates MLCK. Activated MLCK phosphorylates myosin light chain (MLC_{20}), which is a prerequisite for ASM cross-bridge cycling and contraction [13, 27]. The extent of MLC_{20}

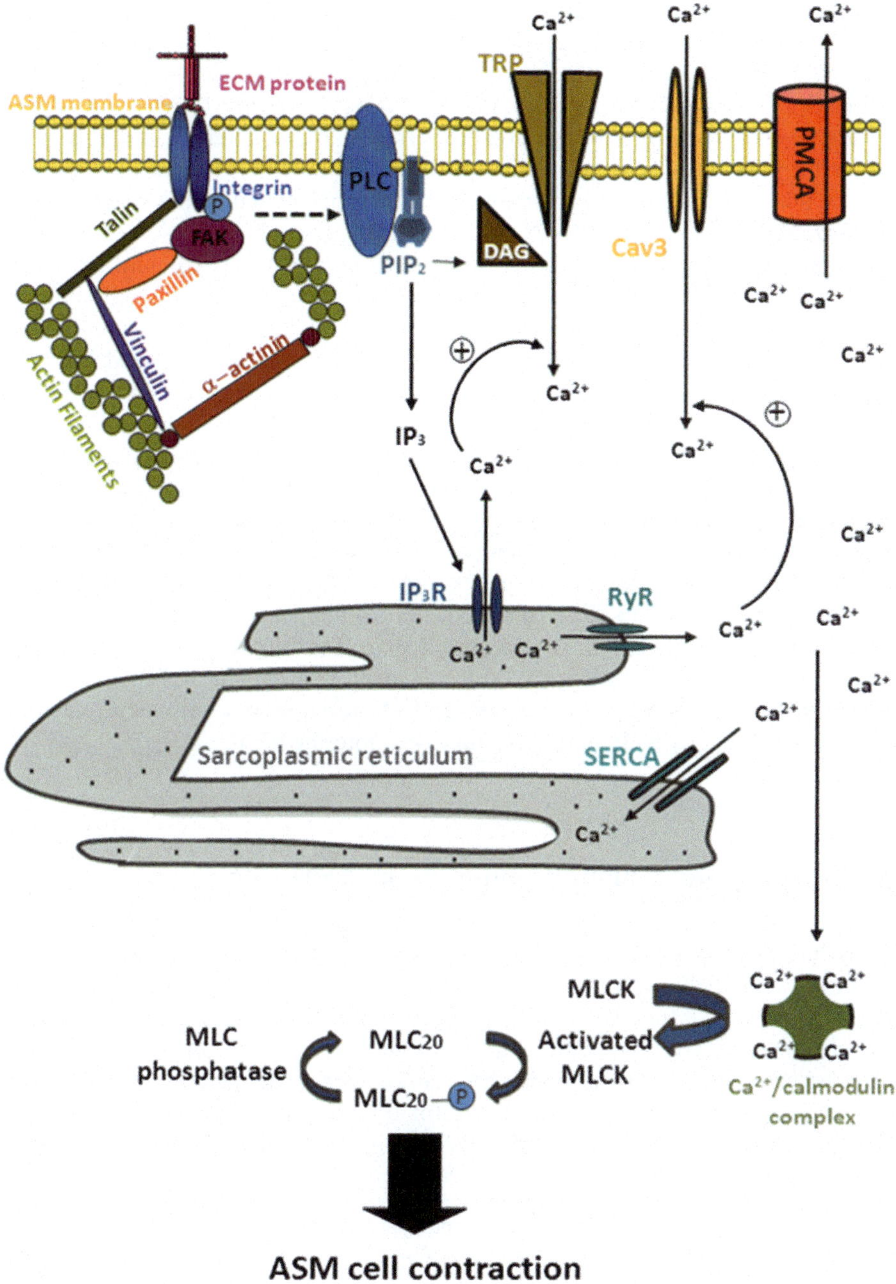

Fig. 2 Schema showing possible pathways for which integrins may regulate calcium signaling to influence contraction signaling. Integrin activation causes the phosphorylation of FAK and association with paxillin, which is essential for actin polymerization and the recruitment of linker proteins for tension development. Integrin activation also increases intracellular calcium

phosphorylation is determined by the balance of MLCK and MLC phosphatase (MLCP) activities, which causes MLC dephosphorylation [13]. Two main pathways regulate MLCP: (1) activation of RhoA and Rho-kinase by G-protein-coupled receptor (GPCR) agonists and (2) by diacylglycerol (DAG) and protein kinase C, via CPI-17 phosphorylation [13]. These pathways inhibit MLCP, and the net result is an enhancement of MLC phosphorylation and an augmented level of contraction at a fixed concentration of calcium, a phenomenon known as calcium sensitization.

Rhythmic contraction of ASM cells is driven by a cytosolic calcium oscillator. The cytosolic calcium oscillator is responsible for the periodic release of calcium from the sarcoplasmic reticulum. The initiation of calcium oscillator is modulated by neurotransmitters (e.g., acetylcholine) and hormones (e.g., endothelin-1) [6]. Neurotransmitters and hormones enhance the formation of IP_3 and DAG that are necessary for oscillatory activity [6]. Oscillatory activity is largely dependent on calcium entry into the cells. Calcium influx varies between cell types. For example, in ASM cells, members of the Cav3 family of T-type channels are responsible for the calcium entry [37]. Calcium entry sensitizes the IP_3 and RyR receptors on the sarcoplasmic reticulum to initiate calcium release. This is then amplified by regenerative calcium release either by IP_3 or RyR receptor, resulting in global calcium signal [15, 40]. This global calcium signal then activates signaling events that culminate in ASM cell contraction. The recovery phase of calcium signaling depends on two pumps: (1) plasma membrane calcium-ATPase (PMCA), which pumps calcium out of the cell, and (2) sarcoplasmic reticulum calcium-ATPase (SERCA), which pumps calcium back into the SR [6].

5 Integrin and Calcium Signaling in ASM cells

Integrin-induced calcium signaling is observed in numerous different cell types (neurons, endothelial cells, T cells, vascular smooth muscle). Table 1 summaries some of the known functional consequences of integrin-induced calcium signaling in various cell types. What remains to be further investigated is the role of integrin-induced calcium signaling in ASM cells. However, the fact that the same integrins are expressed on ASM cells is impetus for further study. Whether integrins can induce calcium signaling in ASM cells remains to be investigated. However, a

Fig. 2 (continued) concentration to cause phosphorylation of MLCK and activation of myosin ATPase activity and cross-bridge cycling. *Cav3* Cav3 family of T-type channels, *DAG* diacylglycerol, *ECM* extracellular matrix, *FAK* focal adhesion kinase, *IP_3* inositol 1,4,5-triphosphate, *MLC20* 20-kDa myosin light chain, *MLCK* myosin light chain kinase, *PIP2* phosphatidylinositol 4,5-bisphosphate, *PLC* phospholipase C, *PMCA* plasma membrane calcium-ATPase, *RyR* ryanodine receptor, *SERCA* sarco-endoplasmic reticulum calcium-ATPase, *TRP* transient receptor potential

Table 1 Functional consequences of integrin-induced calcium signaling in different cell types

Cell type	Integrin	Mechanism of calcium signaling	Functional consequences	References
MDCK	αvβ3	IP_3 dependent and IP_3 independent	Adhesion of MDCK cells	[46]
Neurons	α5β1	Voltage-dependent NMDA receptor and voltage-sensitive calcium channel	Synaptic plasticity in forebrain neurons	[29]
Jurkat	α6β1	IP_3 and cADPR	T cell binding to endothelial cells, migration	[44]
Endothelial	αvβ3	IP_3	Increased vascular permeability, cytokine release, leukocyte adhesion	[7]
Skeletal muscle	α7β1	IP_3 and L-type calcium channels	Acetylcholine clustering required for neuromuscular junction formation	[8, 9, 28]
Vascular smooth muscle	αvβ3	RGD dependent	Vasodilation	[33]

number of signaling mechanisms may be inferred from studies taken from vascular smooth muscle cells. Studies from vascular smooth muscle cells showed that integrin-induced calcium signaling required focal adhesion proteins (Fig. 2) [14, 50, 54]. The focal adhesion proteins consist of paxillin, talin, Src, and FAK [14, 54]. This is supported by a study done by Wu and colleagues [54], who showed that FAK blocking antibody was able to block the regulation of calcium current by the integrin α5β1. This suggests that regulation of calcium current by integrins involves the activation of an integrin and recruitment of focal adhesion proteins. Similarly, cytoskeleton linkage is required for integrin-induced calcium signaling in vascular smooth muscle cells [4, 32, 54]. The mechanisms by which integrins induce calcium signaling in vascular smooth muscle cells (which may also be relevant to ASM cells) include calcium release from ryanodine-sensitive calcium store and calcium influx through an L-type calcium channel [4, 11, 50, 54, 55]. Calcium influx is dependent on the type of integrins expressed on arteriolar smooth muscle cells. A study done on cremaster muscle arterioles showed that activation of the integrin α5β1 increases the calcium influx, whereas activation of the integrin αvβ3 decreases the calcium current [55]. Whether differences in the expression of integrins on ASM cells differentially regulate calcium influx remains to be determined.

Dekkers and colleagues showed that exogenous laminin sustains a contractile phenotype of ASM cells and prevents proliferation of bovine tracheal smooth muscle cells [17]. Using a guinea-pig model of chronic asthma to further explore the role of laminins in ASM remodeling in vivo, the authors treated the animals with the specific soluble laminin competing peptide YIGSR and showed that YIGSR did inhibit allergen-induced ASM accumulation [16]. However, in contrast to their previous in vitro studies, soluble YIGSR promoted a hypercontractile phenotype. Whether this process involved calcium regulation was not investigated. The authors reasoned that the microenvironment of the peptide is an important factor; thus, even though the effects of the peptide are consistent with disrupting laminin-ASM interactions in vitro and ex vivo conditions, the response to artificial laminin competing mimetic in tissue (in vivo) may not be as straightforward because there are multiple cell types involved and the in vivo system is subject to immune regulation. Thus, as with any pharmacological intervention, the development of peptide compounds that target ASM-laminin interactions will require careful assessment of the impact on all aspects of airway hyperresponsiveness, airway inflammation, and remodeling.

5.1 Integrin and GPCR Crosstalk in the Regulation of ASM Calcium Signaling

There is emerging interest in crosstalk between integrins and GPCRs in calcium signaling. These studies have been carried out in endothelial cells and myocytes, but similar findings might be found in ASM cells. In endothelial cells, the regulation of MAPK activity by integrins and the P2Y class of $G_{q/11}$-coupled receptors involved the activation of PKC and calcium [34]. As for atrial myocytes, β2-adrenergic receptor stimulation of calcium current was shown to be enhanced by the β1 integrin via inhibition of cAMP/PKA and activation of G_i/ERK/cPLA$_2$/AA signaling [36]. Cheng and coworkers also elegantly demonstrated the relationship between β1 integrin and β-adrenergic receptor regulation of L-type calcium current in neonatal rat ventricular myocytes. Moreover, overexpression of the β1 integrin impedes β-adrenergic-receptor-induced calcium current via inhibition of AC/cAMP activity [12]. A similar observation was also obtained in adult cat atrial myocytes [51]. It has been shown that a diseased heart is linked with increased integrin expression, suggesting an important role for integrins and β-adrenergic receptor crosstalk in this diseased state. GPCRs and integrins are both expressed on ASM cells. Either one alone is linked to ASM contraction signaling. Given that crosstalk between integrins and GPCRs in calcium signaling is found in endothelial cells and mycocytes, it is likely that crosstalk between integrins and GPCRs can exist to regulate ASM contraction. However, studies have not been carried out in ASM cells as yet. Moreover, contractile human ASM cells exhibit omega-shaped plasma invaginations known as caveolae (developed from lipid rafts that bind caveolin-1 protein) [23]. Caveolin-1 in caveolae is able to bind to integrin α-subunits and has been shown to regulate GPCR-mediated signaling such as cAMP signaling [38, 52]. Moreover, caveolae are found in close proximity to peripheral sarcoplasmic reticulum and mitochondria [5]. This suggests that crosstalk between integrin and GPCR may exist in caveolae and play a role in the spatial coordination of calcium-handling channels and organelles. Collectively, these suggest that caveolae may mediate ASM contractile response by aiding integrin and GPCR crosstalk in the regulation of calcium signaling.

In summary, integrin-induced calcium signaling is observed in many different cell types. What remains to be further investigated in detail is the role of integrin-induced calcium signaling in ASM cells and the functional relevance of this signaling mechanism to diseases such as asthma. Further understanding of the role of integrins in the regulation of calcium signaling will enhance the development of more tailored therapies for diseases where ASM plays a role.

References

1. Ammit, A., Armour, C., and Black, J. (2000). Smooth-muscle myosin light-chain kinase content is increased in human sensitized airways. Am J Respir Crit Care Med *161*, 257–263.
2. Arcangeli, A., Becchetti, A., Mannini, A., Mugnai, G., De Filippi, P., Tarone, G., Del Bene, M. R., Barletta, E., Wanke, E., and Olivotto, M. (1993). Integrin-mediated neurite outgrowth in neuroblastoma cells depends on the activation of potassium channels. J Cell Biol *122*, 1131–1143.
3. Argraves, W.S., Suzuki, S., Arai, H., Thompson, K., Pierschbacher, M.D., and Ruoslahti, E. (1987). Amino acid sequence of the human fibronectin receptor. J Cell Biol *105*, 1183–1190.
4. Balasubramanian, L., Ahmed, A., Lo, C.M., Sham, J.S., and Yip, K.P. (2007). Integrin-mediated mechanotransduction in renal vascular smooth muscle cells: activation of calcium sparks. Am J Physiol Regul Integr Comp Physiol *293*, R1586–1594.
5. Bergdahl, A., and Sward, K. (2004). Caveolae-associated signalling in smooth muscle. Can J Physiol Pharmacol *82*, 289–299.
6. Berridge, M.J. (2008). Smooth muscle cell calcium activation mechanisms. J Physiol *586*, 5047–5061.
7. Bhattacharya, S., Ying, X., Fu, C., Patel, R., Kuebler, W., Greenberg, S., and Bhattacharya, J. (2000). alpha(v)beta(3) integrin induces tyrosine phosphorylation-dependent Ca(2+) influx in pulmonary endothelial cells. Circ Res *86*, 456–462.
8. Burkin, D.J., Gu, M., Hodges, B.L., Campanelli, J.T., and Kaufman, S.J. (1998). A functional role for specific spliced variants of the alpha7beta1 integrin in acetylcholine receptor clustering. J Cell Biol *143*, 1067–1075.
9. Burkin, D.J., Kim, J.E., Gu, M., and Kaufman, S.J. (2000). Laminin and alpha7beta1 integrin regulate agrin-induced clustering of acetylcholine receptors. J Cell Sci *113 (Pt 16)*, 2877–2886.
10. Burridge, K., and Chrzanowska-Wodnicka, M. (1996). Focal adhesions, contractility, and signaling. Annu Rev Cell Dev Biol *12*, 463–518.
11. Chan, W.L., Holstein-Rathlou, N.H., and Yip, K.P. (2001). Integrin mobilizes intracellular Ca (2+) in renal vascular smooth muscle cells. Am J Physiol Cell Physiol *280*, C593–603.
12. Cheng, Q., Ross, R.S., and Walsh, K.B. (2004). Overexpression of the integrin beta (1A) subunit and the beta(1A) cytoplasmic domain modifies the beta-adrenergic regulation of the cardiac L-type Ca(2+)current. J Mol Cell Cardiol *36*, 809–819.
13. Chiba, Y., and Misawa, M. (2004). The role of RhoA-mediated Ca2+ sensitization of bronchial smooth muscle contraction in airway hyperresponsiveness. J Smooth Muscle Res *40*, 155–167.
14. Clark, E.A., and Brugge, J.S. (1995). Integrins and signal transduction pathways: the road taken. Science *268*, 233–239.

15. Dai, J.M., Kuo, K.H., Leo, J.M., van Breemen, C., and Lee, C.H. (2006). Mechanism of ACh-induced asynchronous calcium waves and tonic contraction in porcine tracheal muscle bundle. Am J Physiol Lung Cell Mol Physiol *290*, L459–469.
16. Dekkers, B.G., Bos, I.S., Halayko, A.J., Zaagsma, J., and Meurs, H. (2010). The laminin beta1-competing peptide YIGSR induces a hypercontractile, hypoproliferative airway smooth muscle phenotype in an animal model of allergic asthma. Respir Res *11*, 170.
17. Dekkers, B.G., Schaafsma, D., Nelemans, S.A., Zaagsma, J., and Meurs, H. (2007). Extracellular matrix proteins differentially regulate airway smooth muscle phenotype and function. Am J Physiol Lung Cell Mol Physiol *292*, L1405–1413.
18. Elsherif, L., Huang, M.S., Shai, S.Y., Yang, Y., Li, R.Y., Chun, J., Mekany, M.A., Chu, A.L., Kaufman, S.J., and Ross, R.S. (2008). Combined deficiency of dystrophin and beta1 integrin in the cardiac myocyte causes myocardial dysfunction, fibrosis and calcification. Circ Res *102*, 1109–1117.
19. Fernandes, D.J., Bonacci, J.V., and Stewart, A.G. (2006). Extracellular matrix, integrins, and mesenchymal cell function in the airways. Curr Drug Targets *7*, 567–577.
20. Freyer, A.M., Johnson, S.R., and Hall, I.P. (2001). Effects of growth factors and extracellular matrix on survival of human airway smooth muscle cells. Am J Respir Cell Mol Biol *25*, 569–576.
21. Giancotti, F., and Ruoslahti, E. (1999). Integrin signaling. Science *285*, 1028–1032.
22. Guo, C., Willem, M., Werner, A., Raivich, G., Emerson, M., Neyses, L., and Mayer, U. (2006). Absence of alpha 7 integrin in dystrophin-deficient mice causes a myopathy similar to Duchenne muscular dystrophy. Hum Mol Genet *15*, 989–998.
23. Halayko, A., Tran, T., and Gosens, R. (2008). Phenotype and functional plasticity of airway smooth muscle: role of caveolae and caveolins. Proc Am Thorac Soc *5*, 80–88.
24. Hirst, S.J., Walker, T.R., and Chilvers, E.R. (2000). Phenotypic diversity and molecular mechanisms of airway smooth muscle proliferation in asthma. Eur Respir J *16*, 159–177.
25. Hughes, P.E., Diaz-Gonzalez, F., Leong, L., Wu, C., McDonald, J.A., Shattil, S.J., and Ginsberg, M.H. (1996). Breaking the integrin hinge. A defined structural constraint regulates integrin signaling. J Biol Chem *271*, 6571–6574.
26. Humphries, M. (2000). Integrin structure. Biochem Soc Trans *28*, 311–339.
27. Janssen, L.J., and Killian, K. (2006). Airway smooth muscle as a target of asthma therapy: history and new directions. Respir Res *7*, 123.
28. Kwon, M.S., Park, C.S., Choi, K., Ahnn, J., Kim, J.I., Eom, S.H., Kaufman, S.J., and Song, W.K. (2000). Calreticulin couples calcium release and calcium influx in integrin-mediated calcium signaling. Mol Biol Cell *11*, 1433–1443.
29. Lin, C.Y., Hilgenberg, L.G., Smith, M.A., Lynch, G., and Gall, C.M. (2008). Integrin regulation of cytoplasmic calcium in excitatory neurons depends upon glutamate receptors and release from intracellular stores. Mol Cell Neurosci *37*, 770–780.
30. Löster, K., Vossmeyer, D., Hofmann, W., Reutter, W., and Danker, K. (2001). alpha1 Integrin cytoplasmic domain is involved in focal adhesion formation via association with intracellular proteins. Biochem J *356*, 233–240.
31. Ma, X., Cheng, Z., Kong, H., Wang, Y., Unnah, H., Stephens, N.L., and Laviolette, M. (2002). Changes in biophysical and biochemical properties of single branchial smooth muscle cells from asthmatic subjects. Am J Physiol Lung Cell Mol Physiol *283*, L1181–1189.
32. Maniotis, A.J., Chen, C.S., and Ingber, D.E. (1997). Demonstration of mechanical connections between integrins, cytoskeletal filaments, and nucleoplasm that stabilize nuclear structure. Proc Natl Acad Sci U S A *94*, 849–854.
33. Mogford, J.E., Davis, G.E., Platts, S.H., and Meininger, G.A. (1996). Vascular smooth muscle alpha v beta 3 integrin mediates arteriolar vasodilation in response to RGD peptides. Circ Res *79*, 821–826.
34. Montiel, M., de la Blanca, E.P., and Jimenez, E. (2005). Angiotensin II induces focal adhesion kinase/paxillin phosphorylation and cell migration in human umbilical vein endothelial cells. Biochem Biophys Res Commun *327*, 971–978.

35. Nguyen, T.T., Ward, J.P., and Hirst, S.J. (2005). beta1-Integrins mediate enhancement of airway smooth muscle proliferation by collagen and fibronectin. Am J Respir Crit Care Med *171*, 217–223.
36. Pabbidi, M.R., Ji, X., Samarel, A.M., and Lipsius, S.L. (2009). Laminin enhances beta(2)-adrenergic receptor stimulation of L-type Ca(2+) current via cytosolic phospholipase A(2) signalling in cat atrial myocytes. J Physiol *587*, 4785–4797.
37. Perez, J.F., and Sanderson, M.J. (2005). The frequency of calcium oscillations induced by 5-HT, ACH, and KCl determine the contraction of smooth muscle cells of intrapulmonary bronchioles. J Gen Physiol *125*, 535–553.
38. Rybin, V.O., Xu, X., Lisanti, M.P., and Steinberg, S.F. (2000). Differential targeting of beta -adrenergic receptor subtypes and adenylyl cyclase to cardiomyocyte caveolae. A mechanism to functionally regulate the cAMP signaling pathway. J Biol Chem *275*, 41447–41457.
39. Sadoshima, J., Takahashi, T., Jahn, L., and Izumo, S. (1992). Roles of mechano-sensitive ion channels, cytoskeleton, and contractile activity in stretch-induced immediate-early gene expression and hypertrophy of cardiac myocytes. Proc Natl Acad Sci U S A *89*, 9905–9909.
40. Sanderson, M.J., Delmotte, P., Bai, Y., and Perez-Zogbhi, J.F. (2008). Regulation of airway smooth muscle cell contractility by Ca2+ signaling and sensitivity. Proc Am Thorac Soc *5*, 23–31.
41. Schneller, M., Vuori, K., and Ruoslahti, E. (1997). Alphavbeta3 integrin associates with activated insulin and PDGFbeta receptors and potentiates the biological activity of PDGF. EMBO J *16*, 5600–5607.
42. Schwartz, M.A., Lechene, C., and Ingber, D.E. (1991). Insoluble fibronectin activates the Na/H antiporter by clustering and immobilizing integrin alpha 5 beta 1, independent of cell shape. Proc Natl Acad Sci U S A *88*, 7849–7853.
43. Schwartz, M.A., Schaller, M.D., and Ginsberg, M.H. (1995). Integrins: emerging paradigms of signal transduction. Annu Rev Cell Dev Biol *11*, 549–599.
44. Schöttelndreier, H., Potter, B.V., Mayr, G.W., and Guse, A.H. (2001). Mechanisms involved in alpha6beta1-integrin-mediated Ca(2+) signalling. Cell Signal *13*, 895–899.
45. Shai, S.Y., Harpf, A.E., Babbitt, C.J., Jordan, M.C., Fishbein, M.C., Chen, J., Omura, M., Leil, T.A., Becker, K.D., Jiang, M., *et al.* (2002). Cardiac myocyte-specific excision of the beta1 integrin gene results in myocardial fibrosis and cardiac failure. Circ Res *90*, 458–464.
46. Sjaastad, M.D., Lewis, R.S., and Nelson, W.J. (1996). Mechanisms of integrin-mediated calcium signaling in MDCK cells: regulation of adhesion by IP3- and store-independent calcium influx. Mol Biol Cell *7*, 1025–1041.
47. Soldi, R., Mitola, S., Strasly, M., Defilippi, P., Tarone, G., and Bussolino, F. (1999). Role of alphavbeta3 integrin in the activation of vascular endothelial growth factor receptor-2. EMBO J *18*, 882–892.
48. Tran, T., Ens-Blackie, K., Rector, E., Stelmack, G., McNeill, K., Tarone, G., Gerthoffer, W., Unruh, H., and Halayko, A. (2007). Laminin-binding integrin alpha7 is required for contractile phenotype expression by human airway myocytes. Am J Respir Cell Mol Biol *37*, 668–680.
49. Tran, T., McNeill, K.D., Gerthoffer, W.T., Unruh, H., and Halayko, A.J. (2006). Endogenous laminin is required for human airway smooth muscle cell maturation. Respir Res *7*, 117.
50. Umesh, A., Thompson, M.A., Chini, E.N., Yip, K.P., and Sham, J.S. (2006). Integrin ligands mobilize Ca2+ from ryanodine receptor-gated stores and lysosome-related acidic organelles in pulmonary arterial smooth muscle cells. J Biol Chem *281*, 34312–34323.
51. Wang, Y.G., Samarel, A.M., and Lipsius, S.L. (2000). Laminin acts via beta 1 integrin signalling to alter cholinergic regulation of L-type Ca(2+) current in cat atrial myocytes. J Physiol *526 Pt 1*, 57–68.
52. Wary, K.K., Mariotti, A., Zurzolo, C., and Giancotti, F.G. (1998). A requirement for caveolin-1 and associated kinase Fyn in integrin signaling and anchorage-dependent cell growth. Cell *94*, 625–634.

53. Woodard, A.S., García-Cardeña, G., Leong, M., Madri, J.A., Sessa, W.C., and Languino, L.R. (1998). The synergistic activity of alphavbeta3 integrin and PDGF receptor increases cell migration. J Cell Sci *111 (Pt 4)*, 469–478.
54. Wu, X., Davis, G.E., Meininger, G.A., Wilson, E., and Davis, M.J. (2001). Regulation of the L-type calcium channel by alpha 5beta 1 integrin requires signaling between focal adhesion proteins. J Biol Chem *276*, 30285–30292.
55. Wu, X., Mogford, J.E., Platts, S.H., Davis, G.E., Meininger, G.A., and Davis, M.J. (1998). Modulation of calcium current in arteriolar smooth muscle by alphav beta3 and alpha5 beta1 integrin ligands. J Cell Biol *143*, 241–252.

Sex Steroid Signaling in the Airway

Y.S. Prakash, Venkatachalem Sathish, and Elizabeth A. Townsend

Contents

Abstract The influence of sex hormones (estrogen, progesterone, testosterone) in the biology and pathophysiology of various organ systems is being increasingly recognized. While sex steroid effects, especially estrogens, on the cardiovascular system have been well studied, clinical and emerging bench research evidence exists for the role of sex hormones in airway and lung physiology and in diseases such as asthma, chronic obstructive pulmonary disease, and pulmonary fibrosis. In this chapter, we briefly define what sex differences exist in airway or lung structure and function under normal conditions, the potential role of sex steroids at different time points in life and in specific disease states, setting the stage to

Y.S. Prakash, M.D., Ph.D. (✉) • V. Sathish, Ph.D.
Departments of Physiology and Biomedical Engineering, Mayo Clinic, 4-184 W Jos SMH
200 First St SW, Rochester, MN 55905, USA

Anesthesiology College of Medicine, Mayo Clinic, Rochester, MN, USA
e-mail: prakash.ys@mayo.edu

E.A. Townsend, Ph.D.
Department of Anesthesiology, Columbia University, New York, NY, USA

Y.-X. Wang (ed.), *Calcium Signaling In Airway Smooth Muscle Cells*,
DOI 10.1007/978-3-319-01312-1_17, © Springer International Publishing Switzerland 2014

explore whether and how sex hormone signaling mechanisms may be contributory. We focus on the mechanisms by which estrogens, progresterone, and testosterone influence airway smooth muscle and epithelium in regulating airway tone and the potential contribution of these hormones in airway remodeling.

Keywords Lung disease • Asthma • Female • Gender • Estrogen

1 Introduction

The Institute of Medicine has highlighted the idea that sex is or should be a biological variable in bench research, clinical research, and clinical practice norms ("Exploring Biological Contributions to Human Health: Does Sex Matter?"; http://www.nap.edu/openbook.php?isbn=0309072816). The importance of sex or gender differences was also highlighted in a recent article entitled "Putting Gender on the Agenda" [1]. In this regard, sex differences in cardiovascular structure and function under normal conditions and in disease states are well recognized [2], and the physiological effects of sex steroids (especially estrogens) have been examined extensively. However, surprisingly less attention is paid to the respiratory system and to airway and lung diseases despite the wide range of diseases, such as asthma and chronic obstructive pulmonary disease (COPD), that show sex differences in incidence and severity. Here, the role of sex steroids on airway structure and function in mediating these differences is also still under investigation. The importance of sex differences or steroid effects in the airway lies in known inherent differences in airway structure and function from early in life throughout the human lifespan and the impact of hormonal changes (especially in women). Furthermore, epidemiological evidence suggests a role for sex in the incidence, susceptibility, and severity of a variety of airway diseases and highlights the importance of recognizing and understanding the potential contribution of sex steroids. In this chapter, we introduce the concept of sex differences in the airway under normal conditions at different life points and in asthma and explore the mechanisms by which sex steroids (estrogen, progesterone, and testosterone) may influence airway structure and function under these conditions. The summary of current knowledge on sex steroid influences on the airway is illustrated in Fig. 1.

2 Sex Differences in the Airway

Sex differences in the airway occur as early as 16–24 weeks' gestation [3], with the number of bronchi being fewer in female fetuses compared to males. In addition to inherent sex differences in utero, maternal or fetal sex steroids can also have effects

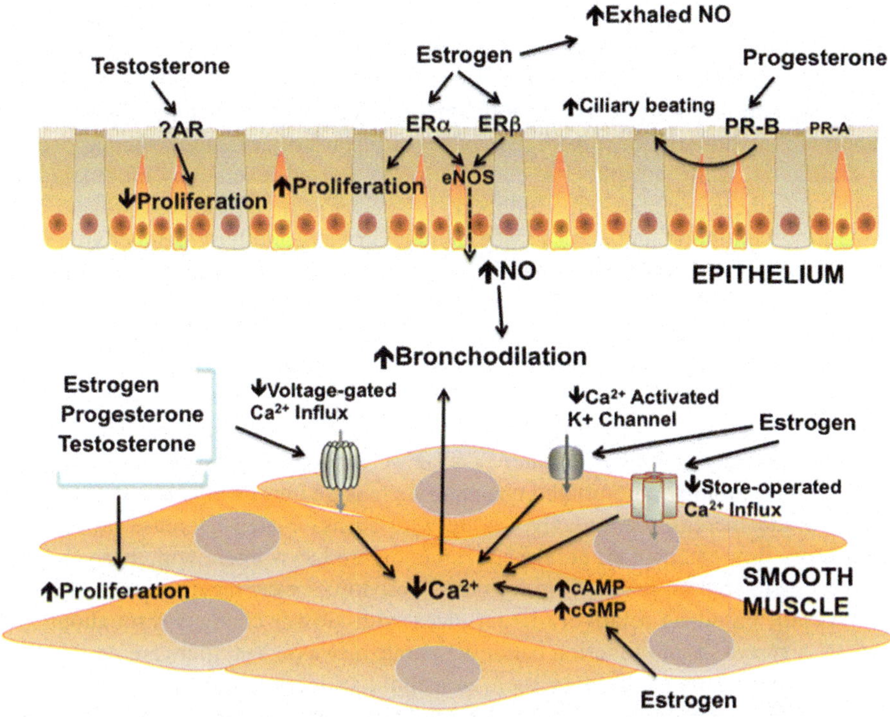

Fig. 1 Sex steroid effects on airway epithelium and smooth muscle. Estrogens may have detrimental effects on airway structure/function by enhancing epithelial and airway smooth muscle proliferation, but they may alleviate bronchoconstriction by reducing $[Ca^{2+}]_i$ in smooth muscle via multiple mechanisms, as well as enhancing epithelial NO. However, these effects may be complicated by the sometimes additive, sometimes opposing effects of progesterone, which also enhance cell proliferation but can reduce exhaled NO. Overall, testosterone has effects opposite to those of female sex steroids, at least in the epithelium, but does enhance ASM proliferation. *AR* androgen receptor, *ER* estrogen receptor, *PR* progesterone receptor

(see [4–6] for an in-depth review of this fascinating topic briefly presented here). The fetal lung is exposed to circulating steroid precursors produced by maternal and fetal adrenal glands [e.g., androstenedione, dehydroepiandrosterone (DHEA)], placental estradiol, and fetal testosterone [7]. Circulating testicular androgen levels peak prior to lung surfactant production, with male lungs being exposed to higher levels compared to females [4]. Later, both males and females are exposed to increased levels of maternal estradiol. Branching lung morphogenesis involves androgens [8], but androgens can delay lung maturation. In contrast, estrogens have stimulatory effects on lung maturation. Thus, the airways of female fetuses mature faster than those of male fetuses, and small airways in females maintain patency better than males due to the earlier production of surfactants. As a result, neonatal females have higher expiratory flow rates at birth, which are maintained throughout their lifespan. The reader is referred to other in-depth reviews on this fascinating topic [4, 5]. The relevance of such intrinsic and sex-steroid-induced differences in airway/lung development lies in the greater susceptibility of premature male infants to respiratory distress syndrome [9] and bronchopulmonary dysplasia. Furthermore, early establishment of differences in lung structure and function may underlie or impact the course of respiratory diseases, such as asthma in childhood, while postpubertal changes in sex steroid levels may modulate lung function and disease states in adulthood.

During childhood and adolescence, female airways and lung parenchyma grow proportionally, whereas males exhibit dysanaptic growth wherein airway growth lags behind growth of the lung (disproportionately fewer alveoli per number of airways). Accordingly, dysanaptic growth results in narrower airways in males, while, conversely, proportional airway growth in females contributes to lower specific airway resistance and larger forced expiratory flow rates. Following puberty, the greater expiratory flow rate is less obvious in young women, but still women generally maintain a greater FEV1/FVC ratio. As adults age, independently of disease pathology, age-related changes, such as decreased elastic recoil of large airways and increased connective tissue with an associated decrease in maximal expiratory flow rate, are all greater in men compared to women, with these changes also occurring but more slowly in women. These topics were recently reviewed [10].

Overall, based on relatively larger sizes with slower changes throughout life, female airways should function better than male airways. Yet, there are paradoxical differences in airway diseases such as asthma in boys versus girls and women versus men, underlining a potential modulating role for sex steroids in airway structure and function. There is currently limited information regarding such effects, which are discussed here.

3 Sex Differences in Airway Diseases

Prepubertal boys have a greater incidence of asthma compared to girls, while following the onset of puberty and during the reproductive years in females, women have more asthma [10, 11]. Even with COPD, women may be more susceptible to cigarette smoke and may respond with greater airway disease as opposed to the parenchymal changes observed in males. It is of course difficult to attribute these data to intrinsic versus sex-steroid-induced differences in airway diseases of adults, and even here sex steroid effects on different cell types in the airway may be contributory (see [10] for a recent review). Nonetheless, some key features potentially relevant to sex steroid effects on the airway are noteworthy, especially in women.

Girls with early menarche are at higher risk for developing asthma [12]. However, girls with Turner's syndrome (characterized by low serum levels of estrogen) display greater airway responsiveness to carbachol compared to normal age-matched girls. Furthermore, treatment with exogenous estrogen resolves airway hyperresponsiveness in girls with Turner's syndrome. Thus, even in the peripubertal phase, the relationship between sex steroids and airway reactivity is not all that clear. In adult women, cyclical variations in estrogens and progesterone during the menstrual cycle also influence asthma symptoms (premenstrual asthma occurring in approx. 40 % of asthmatic women), although the temporal relationship between fluctuations in progesterone and estrogen and worsening of symptoms is not clear. Regardless, airway responsiveness to bronchodilator is lower [11], while premenstrual asthma exacerbations can be reduced by suppressing hormone fluctuations with oral contraceptives. During pregnancy, estrogen and progesterone concentrations rise steadily and peak during the third trimester [11], but only approx. 10 % of asthmatic women have acute exacerbations, approx. 30 % show improvement, another 30 % demonstrate worsening, and the remainder show no changes [13]. At menopause, asthma in women decreases [14], while hormone replacement generally increases asthma symptoms. Compared to women, the severity of asthma in men is relatively stable from puberty until later in life, when decreasing serum testosterone at age >50 years can increase asthma. These important differences between males and females at different time points in life were recently reviewed by our group [10].

Overall, the previously cited data suggest that in addition to inherent sex differences, sex steroids may play a modulatory role in airway diseases such as asthma, this topic has been examined in animal models [15], but these data are not entirely consistent with clinical human data, as highlighted in our recent review [10]. In animal models, androgens have a protective role, with estrogens being proinflammatory. Ovalbumin-sensitized/challenged female mice show greater airway reactivity, higher IgE levels, greater inflammation, and relative resistance to glucocorticoid. Conversely, castration in males exacerbates disease, while ovariectomy or estrogen receptor blockade in females alleviates it. However, exogenous progesterone worsens allergic airway disease. Furthermore, estrogens appear to have opposing effects on airway inflammation (increase) versus reactivity (reduction) [16, 17]. Overall, these animal data verify a role for sex steroids in asthma as observed in humans, but the discrepant inflammation versus responsiveness data suggest that differential effects of sex steroids on specific airway cell types (e.g., epithelium vs. smooth muscle) should be considered.

4 Sex Steroid Signaling

The pathways involved in the synthesis of sex steroids are well known [18, 19]. Local tissue concentrations are determined by both circulating (gonadally derived) and locally produced sex steroids [20]. In tissues other than the airways, both genomic and nongenomic effects of sex steroids have been studied [21, 22]. Classically, estrogens activate two estrogen receptors (ERα and ERβ), progesterone targets two receptors as well (PR-A and PR-B), and testosterone targets the androgen receptor (AR), all of which are members of the superfamily of nuclear receptors that also includes the classical glucocorticoid receptor. Ligand binding results in cytoplasmic to nuclear receptor translocation, with homo- or heterodimerization. Typically, ERα is a transcriptional activator, whereas ERβ may antagonize ERα, although this may not be the case in the airways (see below). Nuclear heterodimers can occur between ERα and ERβ, and even ERα with AR, making sex steroid signaling fairly complex. With progesterone, PR-B is the main transcriptional activator, and PR-A is a repressor. Both testosterone and the more active metabolite 5α-dihydrotestosterone (DHT) can bind the AR.

Rapid, nongenomic sex steroid signaling involves membrane-localized or cytoplasmic steroid receptors or the more recently recognized G-protein-coupled receptors (GPCRs), which can bind sex steroids (e.g., the estrogen-sensitive GPCR30). The topic of nongenomic sex steroid signaling has been reviewed extensively [21, 22]. A major aspect of nongenomic signaling is the regulation of intracellular calcium ($[Ca^{2+}]_i$) [21, 23], which is highly relevant to the airway as well. $[Ca^{2+}]_i$ regulation by sex steroids can involve increasing Ca^{2+} via the PLC-DAG-IP_3 signaling cascade or, more commonly, reducing Ca^{2+} by inhibiting influx mechanisms such as L-type Ca^{2+} channels [24], K^+ channels [25], or chloride currents [26].

Sex steroids can mediate their cellular effects via activation of cyclic nucleotides, PKC, PKA, and PKG as well as a myriad of signaling pathways such as mitogen-activated protein kinases (MAPKs), tyrosine kinases, and lipid kinases. These pathways can provide feedback regulation subsequent to steroid receptor activation via ligand-independent activation or via direct phosphorylation of receptors. Estradiol activates ERK 1/2, p38, and JNK pathways, resulting in both c-Jun and c-Fos transcription. Thus, rapid, nongenomic actions of estrogens can potentially exhibit genomic downstream effects as well. PR-B (usually an activator) can crosstalk with ERs by priming them to activate the Src-Ras-ERK pathways. Additionally, PRs can activate p42, MAPK, and PI3K in some cases. ARs activate the c-Src, Raf-1, and ERK-2 pathways, and thus MAPKs. In addition to these signaling intermediates, genomic effects can be fine-tuned by a range of coregulators such as heat shock proteins and other chaperones and recently identified coactivators and corepressors – a topic of growing importance, but beyond the scope of this chapter. Nonetheless, it is clear that the different sex steroid receptors can activate a number of pathways relevant to the regulation of airway structure and function under both normal conditions and disease states, especially those involving inflammation.

5 Sex Steroid Effects in the Airway

5.1 Bronchial Epithelial Cells

The bronchial epithelium serves as a barrier and early responder to external stimuli (allergens, pollutants, and pathogens) and as a modulator of airway tone. Furthermore, epithelial inflammation is involved in the pathogenesis of airway diseases, such as asthma and bronchitis. Accordingly, the sex steroid effects on bronchial epithelial cells (BECs) are important to understand but has hardly been examined.

Airway epithelial cells express both ERα and ERβ [27]. In vascular endothelium, estrogens facilitate dissociation of endothelial nitric oxide synthase (eNOS) from caveolae and activate the NO pathway, potentiating vasodilation. Epithelial NO can be a bronchodilator [28, 29], and exhaled NO is an indicator of airway inflammation typically via inducible iNOS from BECs. Using receptor-specific ligands (R,R)-5,11-diethyl-5,6,11,12-tetrahydro-2,8-chrysenediol ((R,R)-THC) and diaryl-propionitrile (DPN) for ERα and ERβ, respectively, we recently reported that ER activation acutely (within minutes) increases NO production in human BECs [27] via eNOS phosphorylation. These effects appear to involve increases in $[Ca^{2+}]_i$ via enhanced sarcoplasmic reticulum Ca^{2+} release through IP_3 receptor channels and involve caveolin [27]. In an epithelial cell line, estradiol increases conversion of L-arginine to L-citrulline through eNOS, an effect inhibited by the ER antagonist ICI 182,780 [30]. Furthermore, we found that in bronchial rings derived from female patients, physiologically relevant concentrations of estrogens (<10 nM) produce substantial bronchodilation, an effect blunted but not eliminated by epithelial denudation [27]. These limited data suggest that estrogens are capable of inducing NO in bronchial epithelium via nongenomic mechanisms and producing bronchodilation, akin to vasodilation.

Whether PRs or ARs are expressed in airway epithelium has barely been examined. A single recent study [31] reported that both male and female human airway epithelial cells expressed comparable levels of PR-B (with much less expression of PR-A). Interestingly, PR expression was found to be restricted to the proximal region of cilia in these cells, and progesterone decreased ciliary beat frequency. Perhaps more interesting is the fact that the inclusion of 17β-estradiol prevented progesterone-induced reduction of ciliary motility in these studies. These very limited data suggest a need to further explore sex steroid effects on airway epithelium relevant to airway function and the interactions between sex steroids.

A few studies that have explored testosterone have found epithelium-independent effects on airway tone (see below), but that does not necessarily rule out expression of ARs or other actions of male sex hormones on the epithelium. Whether or not airway epithelium expresses ARs is unknown.

In other organs, estrogens, progesterone, and testosterone are all known to modulate cell proliferation [22]. There is currently little information on airway epithelium. A single study found that estrogen increases the proliferation of some types of immortalized epithelial cells [32]. There are no data on the effects of androgens or progesterone on epithelial proliferation or migration.

5.2 *Airway Smooth Muscle*

Given clinical data on sex differences in asthma and airway-predominant COPD, an increasing, albeit limited, number of studies has been conducted on the role of sex steroids in the modulation of airway smooth muscle (ASM) structure and function. Among the steroids, estrogens have probably been examined the most in depth.

The general consensus from in vitro work, including our own, is that estrogens are bronchodilatory, but these data should be interpreted carefully based on the concentrations of agonist used as well as whether acute or chronic effects are being reported. This issue was recently reviewed [10]. For example, it has been reported that estradiol, testosterone, and progesterone can acutely potentiate isoprenaline-induced relaxation in isolated pig bronchus, suggesting an enhancement of β_2-agonist-induced bronchodilation. However, if anything, supraphysiologic concentrations of estradiol appear to be effective. On the other hand, in rats, chronic exposure to low levels of estradiol decreased acetylcholine (ACh)-induced enhancement of airway resistance, but high-dose estradiol (100 μg/kg) increased responsiveness to ACh. None of these earlier studies specifically examined the mechanisms by which estradiol imposes its effects, and an epithelium-independent versus epithelium-dependent mechanism needs to be distinguished. Estradiol-induced relaxation of rabbit trachea parallels findings in the rat and probably involves direct effects on ASM. For example, relaxation of ACh-precontracted rabbit trachea by 100 μM estradiol (a supraphysiologic concentration) is not prevented by NOS inhibition or epithelial denudation, suggesting a direct effect on ASM. The underlying mechanisms appear to involve prostaglandin synthesis and cGMP modulation of ASM tone [33]. This is significant because both cGMP and estrogens can influence Ca^{2+} influx channels and, thus, synergistically enhance bronchodilation. Similarly, prostaglandins modulate cAMP levels, which can also be influenced by estrogens, thus potentiating bronchodilation. Indeed, in a recent study, we reported that physiologically relevant concentrations of estradiol (1–10 nM) increase cAMP levels in human ASM cells and potentiate the effects of isoproterenol on intracellular Ca^{2+}, suggesting an enhancement of bronchodilation [10]. Accordingly, these limited in vitro data highlight a potentially important role for the direct effects of estradiol on ASM in producing bronchodilation.

The mechanisms underlying the direct effects of estrogens on ASM tone have been examined only to a limited extent. In mouse ASM, estrogens enhance the activity of Ca^{2+}-activated K^+ channels via the NO-cGMP-PKG pathway producing membrane hyperpolarization and thus indirectly reducing $[Ca^{2+}]_i$ and blunting carbachol-induced bronchoconstriction [16]. Conversely, female mice lacking ERα displayed enhanced airway responsiveness to methacholine [15], but the effect was considered to be indirect. However, this is in contrast to effects in human ASM. We have reported that in human ASM cells, concentrations as low as 100 pM can substantially decrease Ca^{2+} responses to ACh and other bronchoconstrictor agonists [24], working predominantly via ERα, although human ASM expresses comparable levels of ERβ as well. At least in human ASM, estrogens appear to inhibit L-type channels as well as store-operated Ca^{2+} entry (mediated by mechanisms such as STIM1 and Orai1) [24]. Interestingly, estrogens do not appear to significantly affect intracellular Ca^{2+} release or reuptake in human ASM.

These limited data suggest that estrogens can nongenomically produce bronchodilation by reducing ASM $[Ca^{2+}]_i$. Furthermore, given the clinical findings of sex differences in airway reactivity between males and females and the findings reported earlier in ASM, one potential explanation may be the differences in ASM

ER expression. In ongoing studies, we have found that both male and female human ASM expresses ERα and ERβ to comparable extents, as in the lungs of male and female mice [34]. Unfortunately, in spite of in vitro data suggesting estrogen-induced bronchodilation, in vivo studies in mice are not entirely consistent, especially given the confounding effects of steroids on other airway cell types in models of airway inflammation. Male C57BL/6 mice show more hyperresponsiveness than females, but a direct protective effect of estrogens cannot be proven. Indeed, female mice conversely exhibit more airway inflammation. Overall, the limited in vitro and in vivo data suggest that estrogen effects on ASM may be complex and context-dependent and remain to be further explored in the context of airway inflammation.

Compared to estrogen, the effects of progesterone on ASM have been studied even less. In isolated, constricted pig bronchial rings, high levels of progesterone (likely nonphysiological) potentiates isoprenaline-induced bronchodilation [35], albeit not to the same extent as estradiol. In guinea-pig trachea, both progesterone and, to a lesser extent, 5β-pregnenalone reverse agonist-induced contraction [36] by direct inhibition of Ca^{2+} influx. Thus, like estrogens, progesterone can be bronchodilatory at least in vitro. However, as with estrogen, in vivo data are not consistent. For example, in ovalbumin-sensitized male mice, progesterone exacerbates airway hyperreactivity [37], but direct effects on ASM per se are not clear. Furthermore, it is currently unknown whether ASM of mice express PR.

Recent studies have examined the nongenomic effects of androgens on ASM [38, 39] and have found that physiologic concentrations of testosterone can also produce bronchodilation. However, epithelium denudation can attenuate the response to testosterone, while inhibition of AR with flutamide has no effect. On the other hand, in male guinea-pig and bovine trachea, DHT-induced relaxation is epithelium independent [39]. These effects of testosterone are thought to involve the inhibition of voltage-gated Ca^{2+} channels [39]. As with female sex steroids, in vivo animal data are confounding. Male C57BL/6 mice are more hyperresponsive than female mice [40], while castrated male mice have lower airway responsiveness to methacholine. These data suggest that airway responsiveness is controlled through different physiologic mechanisms in male and female mice.

Increased ASM mass and hyperplasia are also important aspects of airway remodeling in diseases such as asthma. What would be interesting, but has been examined only to a limited extent, is whether sex steroids also had genomic effects on ASM in terms of airway remodeling. Although some cues may be taken from steroid effects on vascular smooth muscle, effects in the airway are not entirely the same. Pretreatment of ASM with physiologic concentrations of testosterone, estradiol, and progesterone (1 nM–1 μM) has been reported to have no effect on thrombin-induced ASM cell proliferation [41]. Furthermore, in rat tracheal smooth muscle, testosterone and the estrogen precursor dehydroepiandrosterone (DHEA) inhibit platelet-derived growth-factor-induced proliferation. On the other hand, a recent study reported that physiologic concentrations of both testosterone and 17β-estradiol substantially enhanced proliferation of rabbit ASM cells [42]. In ongoing studies using human ASM cells, we have also found that chronic exposure (days) to 1–10 nM estradiol can enhance cell proliferation. The contrast between the acute bronchodilatory effects of female sex steroids versus chronic worsening of airway remodeling by cell proliferation should be noted, but it remains to be determined whether sex steroids are actually beneficial or detrimental to airway structure and function.

6 Implications, Future Directions, and Conclusions

Although data are still emerging, it is clear that sex differences exist in the airways of males versus females, and more importantly, sex steroids have substantial effects on airways, particularly epithelium and ASM. Based on animal data, androgens may have a protective role while estrogens may be proinflammatory yet bronchodilatory. These dual effects may be particularly important in women undergoing large variations in sex steroid levels such as during the menstrual cycle or pregnancy. Furthermore, estrogens and progesterone seem to sometimes produce similar effects on airway epithelium and ASM and could therefore synergize in their overall effect, a topic that remains to be explored. However, in terms of cell proliferation, the two sex steroids may have opposing effects. Accordingly, depending on their relative concentrations, receptor expression, and the context of exposure, interactions between estrogen and progesterone may contribute to asthma exacerbations or alleviation. Furthermore, sex steroid effects on airway cells may be modulated by inflammation. These topics have not yet been systematically examined.

Acknowledgments This research was supported by the Flight Attendants Medical Research Institute (FAMRI; VS), a Small Grants Program by the Clinical Translational Science Activities (CTSA) of the Mayo Clinic (YSP), and an NIGMS T32 Institutional Training grant to the Columbia University Department of Anesthesiology (EAT).

References

1. No authors (2010) Putting gender on the agenda. Nature 465:665. doi: 10.1038/465665a
2. Miller VM (2010) Sex-based differences in vascular function. Womens Health (Lond Engl) 6:737–52. doi: 10.2217/whe.10.53
3. Hepper PG, Shannon EA and Dornan JC (1997) Sex differences in fetal mouth movements. Lancet 350:1820.
4. Seaborn T, Simard M, Provost PR, Piedboeuf B and Tremblay Y (2010) Sex hormone metabolism in lung development and maturation. Trends Endocrinol Metab 21:729–38.
5. Becklake MR and Kauffmann F (1999) Gender differences in airway behaviour over the human life span. Thorax 54:1119–38.
6. Carey MA, Card JW, Voltz JW, Arbes SJ, Jr., Germolec DR, Korach KS and Zeldin DC (2007) It's all about sex: gender, lung development and lung disease. Trends Endocrinol Metab 18:308–13.

7. Abramovich DR (1974) Human sexual differentiation–in utero influences. J Obstet Gynaecol Br Commonw 81:448–53.
8. Kimura Y, Suzuki T, Kaneko C, Darnel AD, Akahira J, Ebina M, Nukiwa T and Sasano H (2003) Expression of androgen receptor and 5alpha-reductase types 1 and 2 in early gestation fetal lung: a possible correlation with branching morphogenesis. Clin Sci (Lond) 105:709–13.
9. Martin JA, Hamilton BE, Sutton PD, Ventura SJ, Menacker F and Kirmeyer S (2006) Births: final data for 2004. Natl Vital Stat Rep 55:1–101.
10. Townsend EA, Miller VM and Prakash YS (2012) Sex differences and sex steroids in lung health and disease. Endocr Rev 33:1–47. doi: 10.1210/er.2010-0031
11. Melgert BN, Ray A, Hylkema MN, Timens W and Postma DS (2007) Are there reasons why adult asthma is more common in females? Curr Allergy Asthma Rep 7:143–50.
12. Postma DS (2007) Gender differences in asthma development and progression. Gend Med 4 Suppl B:S133–46.
13. Schatz M, Dombrowski MP, Wise R, Thom EA, Landon M, Mabie W, Newman RB, Hauth JC, Lindheimer M, Caritis SN, Leveno KJ, Meis P, Miodovnik M, Wapner RJ, Paul RH, Varner MW, O'Sullivan M J, Thurnau GR, Conway D and McNellis D (2003) Asthma morbidity during pregnancy can be predicted by severity classification. J Allergy Clin Immunol 112:283–8.
14. Skobeloff EM, Spivey WH, St Clair SS and Schoffstall JM (1992) The influence of age and sex on asthma admissions. JAMA 268:3437–40.
15. Carey MA, Card JW, Voltz JW, Germolec DR, Korach KS and Zeldin DC (2007) The impact of sex and sex hormones on lung physiology and disease: lessons from animal studies. Am J Physiol Lung Cell Mol Physiol %R 10.1152/ajplung.00174.2007 293:L272–278.
16. Dimitropoulou C, White RE, Ownby DR and Catravas JD (2005) Estrogen reduces carbachol-induced constriction of asthmatic airways by stimulating large-conductance voltage and calcium-dependent potassium channels. Am J Respir Cell Mol Biol 32:239–47.
17. Matsubara S, Swasey CH, Loader JE, Dakhama A, Joetham A, Ohnishi H, Balhorn A, Miyahara N, Takeda K and Gelfand EW (2008) Estrogen determines sex differences in airway responsiveness after allergen exposure. Am J Respir Cell Mol Biol 38:501–8.
18. Ghayee HK and Auchus RJ (2007) Basic concepts and recent developments in human steroid hormone biosynthesis. Reviews in endocrine & metabolic disorders 8:289–300. doi: 10.1007/s11154-007-9052-2
19. Payne AH and Hales DB (2004) Overview of steroidogenic enzymes in the pathway from cholesterol to active steroid hormones. Endocr Rev 25:947–70.
20. Luu-The V and Labrie F (2010) The intracrine sex steroid biosynthesis pathways. Prog Brain Res 181:177–92. doi: S0079-6123(08)81010-2 [pii] 10.1016/S0079-6123(08)81010-2
21. Miller VM and Duckles SP (2008) Vascular actions of estrogens: functional implications. Pharmacol Rev 60:210–41.
22. Simoncini T and Genazzani AR (2003) Non-genomic actions of sex steroid hormones. Eur J Endocrinol 148:281–92.
23. Tofovic SP (2010) Estrogens and Development of Pulmonary Hypertension – Interaction of Estradiol Metabolism and Pulmonary Vascular Disease. J Cardiovasc Pharmacol 56:696–708.
24. Townsend EA, Thompson MA, Pabelick CM and Prakash YS (2010) Rapid effects of estrogen on intracellular Ca^{2+} regulation in human airway smooth muscle. Am J Physiol Lung Cell Mol Physiol 298:L521 – 30.
25. Valverde MA, Rojas P, Amigo J, Cosmelli D, Orio P, Bahamonde MI, Mann GE, Vergara C and Latorre R (1999) Acute activation of Maxi-K channels (hSlo) by estradiol binding to the beta subunit. Science 285:1929–31.
26. Lambert JJ, Belelli D, Peden DR, Vardy AW and Peters JA (2003) Neurosteroid modulation of GABAA receptors. Prog Neurobiol 71:67–80.
27. Townsend E, Meuchel L, Thompson M, Pabelick C and Prakash Y (2011) Estrogen increased nitric oxide production in human bronchial epithelium. J Pharmacol Exp Therap 339:815–824.

28. Hamad AM, Clayton A, Islam B and Knox AJ (2003) Guanylyl cyclases, nitric oxide, natriuretic peptides, and airway smooth muscle function. Am J Physiol Lung Cell Mol Physiol 285:L973 – 83.
29. Folkerts G and Nijkamp FP (2006) Nitric oxide in asthma therapy. Curr Pharm Des 12:3221–32.
30. Kirsch EA, Yuhanna IS, Chen Z, German Z, Sherman TS and Shaul PW (1999) Estrogen acutely stimulates endothelial nitric oxide synthase in H441 human airway epithelial cells. Am J Respir Cell Mol Biol 20:658–66.
31. Jain R, Ray JM, Pan JH and Brody SL (2012) Sex hormone-dependent regulation of cilia beat frequency in airway epithelium. Am J Respir Cell Mol Biol 46:446–53. doi: 10.1165/rcmb.2011-0107OC
32. Ivanova MM, Mazhawidza W, Dougherty SM and Klinge CM (2010) Sex Differences in Estrogen Receptor Subcellular Location and Activity in Lung Adenocarcinoma Cells. Am. J. Respir. Cell Mol. Biol. 42:320–330.
33. Pang JJ, Xu XB, Li HF, Zhang XY, Zheng TZ and Qu SY (2002) Inhibition of beta-estradiol on trachea smooth muscle contraction in vitro and in vivo. Acta Pharmacol Sin 23:273–7.
34. Jia S, Zhang X, He DZ, Segal M, Berro A, Gerson TG, Wang Z and Casale TB (2011) Expression and Function of a Novel Variant of Estrogen Receptor-ER-{alpha}36 in Mouse Airway. Am J Respir Cell Mol Biol. doi: 2010-0268OC [pii] 10.1165/rcmb.2010-0268OC
35. Foster PS, Goldie RG and Paterson JW (1983) Effect of steroids on beta-adrenoceptor-mediated relaxation of pig bronchus. Br J Pharmacol 78:441–5.
36. Perusquia M, Hernandez R, Montano LM, Villalon CM and Campos MG (1997) Inhibitory effect of sex steroids on guinea-pig airway smooth muscle contractions. Comp Biochem Physiol C Pharmacol Toxicol Endocrinol 118:5–10.
37. Hellings PW, Vandekerckhove P, Claeys R, Billen J, Kasran A and Ceuppens JL (2003) Progesterone increases airway eosinophilia and hyper-responsiveness in a murine model of allergic asthma. Clin Exp Allergy 33:1457–63.
38. Kouloumenta V, Hatziefthimiou A, Paraskeva E, Gourgoulianis K and Molyvdas PA (2006) Non-genomic effect of testosterone on airway smooth muscle. Br J Pharmacol 149:1083–91.
39. Bordallo J, de Boto MJ, Meana C, Velasco L, Bordallo C, Suarez L, Cantabrana B and Sanchez M (2008) Modulatory role of endogenous androgens on airway smooth muscle tone in isolated guinea-pig and bovine trachea; involvement of beta2-adrenoceptors, the polyamine system and external calcium. Eur J Pharmacol 601:154–62.
40. Card JW, Carey MA, Bradbury JA, DeGraff LM, Morgan DL, Moorman MP, Flake GP and Zeldin DC (2006) Gender differences in murine airway responsiveness and lipopolysaccharide-induced inflammation. J Immunol 177:621–30.
41. Stewart AG, Fernandes D and Tomlinson PR (1995) The effect of glucocorticoids on proliferation of human cultured airway smooth muscle. Br J Pharmacol 116:3219–26.
42. Stamatiou R, Paraskeva E, Papagianni M, Molyvdas PA and Hatziefthimiou A (2011) The mitogenic effect of testosterone and 17beta-estradiol on airway smooth muscle cells. Steroids 76:400–8. doi: S0039-128X(10)00299-0 [pii] 10.1016/j.steroids.2010.12.010

Regulation of Contractility in Immature Airway Smooth Muscle

Y.S. Prakash, Christina M. Pabelick, and Richard J. Martin

Contents

Abstract Airway smooth muscle (ASM) plays an important role in the regulation of airway tone at all stages of life. While there has been much examination of the mechanisms by which adult ASM contracts and relaxes, especially given its importance in diseases such as asthma, there has been substantially less exploration of immature ASM. Here, given the fact that embryonic/fetal and postnatal lung development is associated with substantial and rapid changes in airway structure and size, the immature ASM must maintain the capability to contract/relax as well as proliferate. Beyond this obvious physiological relevance, the clinical importance of understanding the developing ASM lies in its potential role in diseases of the newborn that predispose infants to adult problems such as asthma. In this chapter we briefly describe the current state of knowledge regarding immature ASM and mechanisms that regulate its contractility.

Y.S. Prakash, M.D., Ph.D. (✉) • C.M. Pabelick, M.D.
Mayo Clinic, 4-184 W Jos SMH 200 First St SW, Rochester, MN 55905, USA

Departments of Anesthesiology and Physiology and Biomedical Engineering,
College of Medicine, Mayo Clinic, Rochester, MN, USA
e-mail: prakash.ys@mayo.edu

R.J. Martin, M.D.
Department of Pediatrics, Division of Neonatology, Rainbow Babies Children's Hospital,
Case Western Reserve University, Cleveland, OH, USA

Y.-X. Wang (ed.), *Calcium Signaling In Airway Smooth Muscle Cells*,
DOI 10.1007/978-3-319-01312-1_18,

Keywords Lung • Embryology • Premature birth • Asthma • Bronchopulmonary dysplasia

1 Airway Smooth Muscle in Developing Lung

In the developing lung, airway smooth muscle (ASM) appears early on epithelial tubules as the lung buds from the foregut in the pseudoglandular phase. Interestingly, even at this early stage of differentiation, the developing ASM has intrinsic contractile function, although neural control develops more slowly as nerves follow the growth of the branching airways. Unlike the tonic contractility of postnatal ASM, fetal ASM exhibits intrinsic rhythmic contractions [1] that may contribute to the further formation and remodeling of the developing airway and growth of the lung. Beginning as thin sheets encircling the developing airways, ASM cell proliferation and migration along the extending airways results in compact, crisscrossed bundles that eventually resemble postnatal airway bundles [2]. Interestingly, the maturation of ASM begins at the terminal tubules and progresses toward the trachea. Overall, the ASM layer forms a continuous, syncytial network of interlocking ASM bundles that continues to grow postnatally.

Given species differences in sizes of developing airways at different maturational stages, most studies on functional changes in ASM during development have used maximum contractility measurements (see [3] for a review). In general, there appears to be significant differences between species in the extent and pattern of age-related changes in ASM contractility, some of which may be attributable to intrinsic changes in ASM G-protein-coupled receptor maturations (which may themselves be agonist specific) or to differences in the relative contribution of contractility per se versus the opposing elastic forces within the airway (a more complex relationship). Nonetheless, in most species, there is an initial increase in contractility at birth, with subsequent increases or decreases with maturation. In the pig trachea, unit force production increases early during development but then decreases. Interestingly, airways of 4-week postnatal piglets show higher sensitivity to cholinergic and histaminergic stimulation compared to fetal pigs or even adult swine [4]. Force responses to ACh are also higher in early postnatal piglets, increasing in the first few weeks of life but then becoming smaller in young swine [3]. This effect of increased contractility (even normalized for ASM size) appears to be intrinsic to the neonatal ASM, mediated partly by enhanced acetylcholine (ACh) release by parasympathetic nerves. Downstream muscarinic receptor coupling may in fact be impaired at birth and develops postnatally, while reductions in ACh responsiveness with age appears to involve increased acetylcholinesterase activity. These in vitro data are matched by in vivo airway resistance measurements in response to histamine [3]. A similar pattern of early increased sensitivity to bronchoconstrictor agonist followed by later reductions in sensitivity has been reported in guinea-pig airway both in vitro and in vivo. For example, LTD4 is a more potent bronchoconstrictor in immature guinea-pig airway compared to mature animals [5], an effect inhibited by indomethacin. Importantly, in both porcine and guinea-pig airways, responsiveness to KCl is not altered with age, highlighting the maturational changes in bronchoconstrictor agonist expression and receptor signaling per se. Similar observations of initially higher contractility have been observed in dog, sheep, and rabbit (see [3] for a review). However, response to ACh continues to increase in sheep, while responses to histamine increase in dogs.

Regardless of the species differences in age-related changes to ASM contractility, what is clear is that prior to and at birth, ASM is capable of contraction. What is less clear is whether the mechanisms that control contractility, particularly $[Ca^{2+}]_i$, are just quantitatively different in the developing ASM or whether the relative contribution of specific mechanisms differ. In this chapter, we briefly discuss what is known regarding these issues.

2 $[Ca^{2+}]_i$ Regulation in ASM

Many, if not all, of the chapters in this book extensively report on the mechanisms that regulate $[Ca^{2+}]_i$ in adult ASM. Briefly, both plasma membrane Ca^{2+} influx mechanisms, as well as intracellular Ca^{2+} release and reuptake, are involved in the $[Ca^{2+}]_i$ responses of ASM cells to agonist. Although the relative contribution of specific mechanisms may differ between ASMs of different species, in general Ca^{2+} influx is mediated by L-type Ca^{2+} channels, receptor-operated channels, nonspecific cation channels, Na^+/Ca^{2+} exchange, and store-operated Ca^{2+} entry [6, 7]. The last influx mechanism may involve a variety of transient receptor potential (TRP) channels, especially TRPC1, 3, and 6 [8]. Intracellular Ca^{2+} release can occur via both IP_3 receptor (IP_3R) and ryanodine receptor (RyR) channels on the sarcoplasmic reticulum (SR), the former being activated by the PLC-IP_3 pathway, and the latter by the CD38/cADPR mechanism. Reduction in $[Ca^{2+}]_i$ can occur through SR reuptake via the Sarco-endoplasmic reticulum calcium ATPase (SERCA) pump, or extrusion across the plasma membrane via Na^+/Ca^{2+} exchange and plasma membrane Ca^{2+} ATPase. Other mechanisms, such as mitochondria and lysosomes, may also play a role in modulating and buffering $[Ca^{2+}]_i$. Overall, adult ASM has a myriad of $[Ca^{2+}]_i$ regulatory mechanisms by which bronchoconstrictor agonists, for example ACh, histamine, and endothelin-1, can mediate their effects.

3 $[Ca^{2+}]_i$ Regulation in Developing ASM

While a large number of studies, including our own, have examined the roles of different mechanisms in adult ASM of different species under normal conditions and under conditions of inflammation, oxidant stress, and other insults relevant to airway diseases, (e.g., [6, 7, 9, 10]) there is currently limited information on which of the $[Ca^{2+}]_i$ regulatory mechanisms are involved in mediating bronchoconstriction in the developing airway. Here, the interesting findings that airway contractility is enhanced in early postnatal development would suggest an increased contribution of $[Ca^{2+}]_I$, but this has not been systematically examined. An alternative mechanism may be enhanced Ca^{2+} sensitivity of the contractile apparatus (the Ca^{2+}-calmodulin-myosin light chain kinase/RhoA-MLC pathways); however, this has also not been systematically examined.

Based on the observation that early fetal airways demonstrate spontaneous contractions (airway peristalsis, presumably to facilitate fluid movement within the developing passages), a few studies have explored such contractions [11–13]. For example, McCray [11] used explants of developing tubules from postmortem fetuses during the first trimester and examined such contractions under light microscopy. Interestingly, contractions were observed to originate at the same region of a tubule and spread in the same direction at regular intervals. Furthermore, inhibition of neural input using tetrodotoxin had no effect on spontaneous contractions, suggesting an intrinsic mechanism. However, inhibition of L-type Ca^{2+} channels with nifedipine or nonspecific inhibition of influx with CdCl2 stopped spontaneous contractility, suggesting a large role for influx. On the other hand, muscarinic stimulation produced tetany, suggesting a functional cholinergic system early in airway development. In a subsequent, elegant study using embryonic rat lungs loaded with fluorescent Ca^{2+} indicator fluo-4, Featherstone et al. [12] reported that spontaneous airway contractions that spread along the tracheobronchial tree involves travelling $[Ca^{2+}]_i$ waves. Intercellular communication and propagation of these waves along the airway were thought to be mediated by action potentials and gap junctions. Inhibition of L-type Ca^{2+} channels as well TEA-sensitive K + channels (i.e., voltage-gated or large-conductance BK channels) or removal of extracellular Ca^{2+} prevented spontaneous $[Ca^{2+}]_i$ waves, suggesting a large role for plasma membrane Ca^{2+} influx. However, inhibition of IP_3R channels with 2-APB or RyR channels with ryanodine were also effective in blocking spontaneous waves, suggesting the additional involvement of SR Ca^{2+} mechanisms. These findings highlight the fact that even in the first trimester, the early developing ASM contains many of the $[Ca^{2+}]_i$ regulatory mechanisms that are present in adult ASM, which can contribute to spontaneous and ACh-induced $[Ca^{2+}]_i$ oscillations, as we and others have reported in adult ASM [10, 14].

Changes in postnatal ASM have also been largely examined in terms of contractility as described earlier. However, a few studies have examined the underlying mechanisms. For example, in 2-week-old rabbit ASM, where maximal contractility and sensitivity to carbachol is higher than in adult, accumulation of IP_3 is also higher, largely due to the lesser breakdown of this second messenger [15]. While binding of IP_3 to IP_3R is comparable between these ages, the effect of Ca^{2+} is different than that in adult rabbit ASM; higher levels of $[Ca^{2+}]_i$ will inhibit IP_3R but enhance IP_3R activity in immature ASM [16]. Thus the combination of increased IP_3 and IP_3R activity likely contribute to greater contractility in the immature ASM. Furthermore, the potent bronchoconstrictor endothelin-1 also appears to have greater effects in immature rabbit ASM, working via enhanced IP_3 production, greater Ca^{2+} influx via L-type Ca^{2+} channels, and an increased role of PKC [17].

There is currently no information on maturational changes in RyR channels or their regulation in ASM.

In addition to intracellular regulatory mechanisms, studies in ASM of various animals suggest a significant role for plasma membrane, especially those determined by membrane potential. For example, 2-week-old bovine tracheal ASM has a resting membrane potential similar to that of adult animals, but the contribution of the electrogenic Na+/K + pump is greater in the immature airway [18]. Separately, in guinea pig, both fetal and newborn extrathoracic tracheal ASMs are more sensitive to the L-type Ca^{2+} channel agonist BAYK8644, suggesting a greater role for this influx mechanism [19]. Overall, these data suggest that the early postnatal airway may be more sensitive to bronchoconstrictor agonists. The teleological relevance of such increased airway sensitivity to stimuli could be to prevent aspiration, especially of gastric contents.

While spontaneous $[Ca^{2+}]_i$ oscillations and contractions have been noted early in airway development, ASM for later gestational stages have been less examined, particularly in humans, due to a lack of tissue availability and our increasing ability to rescue premature infants at even earlier gestations such as 23 weeks. In a recent study, we characterized $[Ca^{2+}]_i$ responses in human fetal ASM cells from 18- to 22-week aborted fetuses (canalicular stage) [20]. This stage of airway development is particularly important given the fact that premature babies are particularly susceptible to neonatal respiratory distress syndrome and are at risk of developing bronchopulmonary dysplasia due to exposure to high inspired oxygen levels (hyperoxia) or mechanical ventilation [21]. Here, detrimental effects on airway structure and function can have immediate and long-term effects on airway reactivity with predisposition to asthma. While many studies use adult human ASM cells or tissues with interventions such as hyperoxia, this may not be a proper surrogate. In our study, we examined the regulation of $[Ca^{2+}]_i$ and cellular proliferation in response to hyperoxia and found that fetal ASM cells expressed substantial smooth muscle actin and myosin as well as several Ca^{2+} regulatory proteins, including M3 muscarinic and H1 histaminergic receptors, IP_3R, RyR, and SERCA protein, and some of the store-operated Ca^{2+} influx mechanisms. However, compared to adult human ASM, these proteins were relatively less abundant, especially SR Ca^{2+} release proteins, suggesting a greater dependence on Ca^{2+} influx mechanisms. Correspondingly, $[Ca^{2+}]_i$ responses to ACh and histamine were robust but substantially smaller and slower than in adult ASM. On the other hand, the proliferative potential of such fetal ASM cells was substantially greater than for adult ASM, consistent with their role in the growing airway. Interestingly, moderate levels of hyperoxia, less than 60 %, enhanced proliferation (also increasing $[Ca^{2+}]_i$ responses to agonist), while higher levels of oxygen induced apoptosis. These data suggest that oxygen can have profound effects on ASM $[Ca^{2+}]_i$ and airway structure during a critical period of airway development. Such effects of hyperoxia have also been noted in the early postnatal airway in animal models. For example, exposure of 21-day-old rats to 95 % hyperoxia for 8 days (a substantial exposure) resulted in increased in vivo airway responsiveness to methacholine, with changes in airway architecture akin to asthma: increased epithelial and ASM thickness [22].

4 Relaxation in Developing Airways

Compared to the substantial examination of bronchodilatory mechanisms in adults (given their alleviating role in asthma), much less is known regarding mechanisms that contribute to relaxation in immature airways. In many species, a parasympathetic nonadrenergic, noncholinergic (NANC) mechanism has been demonstrated, while in some species such as guinea pig an additional adrenergic response might also occur. In humans, NANC-based relaxation is particularly important.

In contrast to the relatively higher bronchoconstrictive capability of immature airways, relaxation appears to be less, with gradual maturation in bronchodilatory mechanisms. For example, in immature rabbits, relaxation of precontracted trachea by electrical field stimulation is almost absent at 1 week postnatal, present at 2–4 weeks, but is much greater in adults [23]. Similarly, in rats, substance P-induced relaxation in precontracted airways is greater in adults than in neonates [24], an effect that appears to involve NANC-derived NO. Here, there is strong evidence from multiple species that SP, working via NK1 receptors, induces NO release, especially by tracheal epithelium. Furthermore, even if the role of NO as a bronchodilator in adults is less obvious (unlike its well-known role in the vasculature), there is a known role for NO in bronchodilation in early postnatal ages. In addition to this induced relaxation following bronchoconstrictor agonist, in guinea pig, which shows baseline airway tone, spontaneous relaxation of ASM during sustained electrical field stimulation is less in 1–3 week-old animals than in adults, a difference that is abolished by inhibiting cyclooxygenase [25]. This could also involve NO due to prostaglandin activation of NO synthase. Overall, the limited data suggest that bronchodilatory mechanisms such as NO may be active, albeit to a lesser extent, in immature airway, with some of these mechanisms being enhanced with adulthood. Thus, the caliber and tone of immature airways are perhaps driven to a greater extent by constrictor mechanisms as relaxant pathways develop.

5 Conclusions

Clearly, even from the early stages of development, immature ASM contains many of the mechanisms required for the regulation of contraction. What appears to be interesting is the relatively larger role of Ca^{2+} influx mechanisms in the immature airway, perhaps reflecting a lesser or slower development of the SR. In this regard, it is possible that the plasma membrane is more susceptible to insults, such as inflammation, hyperoxia, and reactive oxygen species, that can contribute to airway diseases of the newborn. Further examination is required to more thoroughly understand these mechanisms and the potential interactions with other cells within the airway, especially the epithelium.

Acknowledgments This work was supported by National Institutes of Health Grant R01 HL056470 (YSP, RM) and the Flight Attendants Medical Research Institute (FAMRI; CMP).

References

1. Jesudason EC (2009) Airway smooth muscle: an architect of the lung? Thorax 64:541–5. doi: 10.1136/thx.2008.107094
2. Sparrow MP, Weichselbaum M and McCray PB (1999) Development of the innervation and airway smooth muscle in human fetal lung. Am J Respir Cell Mol Biol 20:550–60.
3. Chitano P and Murphy TM (2003) Maturational changes in airway smooth muscle shortening and relaxation. Implications for asthma. Respir Physiol Neurobiol 137:347–59.
4. Sparrow MP and Mitchell HW (1990) Contraction of smooth muscle of pig airway tissues from before birth to maturity. J Appl Physiol 68:468–77.
5. Duncan PG and Douglas JS (1985) Influences of gender and maturation on responses of guinea-pig airway tissues to LTD4. Eur J Pharmacol 112:423–7.
6. Jude JA, Wylam ME, Walseth TF and Kannan MS (2008) Calcium signaling in airway smooth muscle. Proc Am Thorac Soc 5:15–22. doi: 5/1/15 [pii] 10.1513/pats.200704-047VS
7. Pabelick CM, Sieck GC and Prakash YS (2001) Invited review: significance of spatial and temporal heterogeneity of calcium transients in smooth muscle. J Appl Physiol 91:488–96.
8. White TA, Xue A, Chini EN, Thompson M, Sieck GC and Wylam ME (2006) Role of transient receptor potential C3 in TNF-alpha-enhanced calcium influx in human airway myocytes. Am J Respir Cell Mol Biol 35:243–51. doi: 2006-0003OC [pii] 10.1165/rcmb.2006-0003OC
9. Mahn K, Ojo OO, Chadwick G, Aaronson PI, Ward JP and Lee TH (2010) Ca^{2+} homeostasis and structural and functional remodelling of airway smooth muscle in asthma. Thorax 65:547–52. doi: 65/6/547 [pii] 10.1136/thx.2009.129296
10. Sanderson MJ, Delmotte P, Bai Y and Perez-Zogbhi JF (2008) Regulation of airway smooth muscle cell contractility by Ca^{2+} signaling and sensitivity. Proc Am Thorac Soc 5:23–31. doi: 5/1/23 [pii] 10.1513/pats.200704-050VS
11. McCray PB, Jr. (1993) Spontaneous contractility of human fetal airway smooth muscle. Am J Respir Cell Mol Biol 8:573–80.
12. Featherstone NC, Jesudason EC, Connell MG, Fernig DG, Wray S, Losty PD and Burdyga TV (2005) Spontaneous propagating calcium waves underpin airway peristalsis in embryonic rat lung. Am J Respir Cell Mol Biol 33:153–60. doi: 10.1165/rcmb.2005-0137OC
13. Jesudason EC, Smith NP, Connell MG, Spiller DG, White MR, Fernig DG and Losty PD (2006) Peristalsis of airway smooth muscle is developmentally regulated and uncoupled from hypoplastic lung growth. Am J Physiol Lung Cell Mol Physiol 291:L559-65. doi: 10.1152/ajplung.00498.2005
14. Sieck GC, Han YS, Pabelick CM and Prakash YS (2001) Temporal aspects of excitation-contraction coupling in airway smooth muscle. J Appl Physiol 91:2266–74.
15. Rosenberg SM, Berry GT, Yandrasitz JR and Grunstein MM (1991) Maturational regulation of inositol 1,4,5-trisphosphate metabolism in rabbit airway smooth muscle. J Clin Invest 88:2032–8. doi: 10.1172/JCI115531
16. Schramm CM, Chuang ST and Grunstein MM (1992) Maturation of inositol 1,4,5-trisphosphate receptor binding in rabbit tracheal smooth muscle. Am J Physiol 263:L501–5.
17. Grunstein MM, Rosenberg SM, Schramm CM and Pawlowski NA (1991) Mechanisms of action of endothelin 1 in maturing rabbit airway smooth muscle. Am J Physiol 260:L434–43.
18. Souhrada M, Rothberg KG and Douglas JS (1988) Membrane properties of bovine airway smooth muscle cells: effects of maturation. Pulm Pharmacol 1:47–52.

19. Pichoff BE, Uyehara CF and Nakamura KT (1993) Effect of calcium agonists, BAY K 8644 and CGP 28392, on guinea pig airway smooth muscle function during development. J Pharmacol Exp Ther 265:524–8.
20. Hartman WR, Smelter DF, Sathish V, Karass M, Kim S, Aravamudan B, Thompson MA, Amrani Y, Pandya HC, Martin RJ, Prakash YS and Pabelick CM (2012) Oxygen-Dose Responsiveness of Human Fetal Airway Smooth Muscle Cells. Am J Physiol Lung Cell Mol Physiol. doi: 10.1152/ajplung.00037.2012
21. Jobe AH (2010) The new bronchopulmonary dysplasia. Current opinion in pediatrics. doi: 10.1097/MOP.0b013e3283423e6b
22. Hershenson MB, Aghili S, Punjabi N, Hernandez C, Ray DW, Garland A, Glagov S and Solway J (1992) Hyperoxia-induced airway hyperresponsiveness and remodeling in immature rats. The American journal of physiology 262:L263–9.
23. Colasurdo GN, Loader JE, Graves JP and Larsen GL (1994) Maturation of nonadrenergic noncholinergic inhibitory system in normal and allergen-sensitized rabbits. Am J Physiol 267:L739–44.
24. Mhanna MJ, Dreshaj IA, Haxhiu MA and Martin RJ (1999) Mechanism for substance P-induced relaxation of precontracted airway smooth muscle during development. Am J Physiol 276:L51–6.
25. Chitano P, Cox CM and Murphy TM (2002) Relaxation of guinea pig trachealis during electrical field stimulation increases with age. J Appl Physiol 92:1835–42. doi: 10.1152/japplphysiol.00688.2001

Mathematical Modeling of Calcium Dynamics in Airway Smooth Muscle Cells

James Sneyd, Pengxing Cao, Xiahui Tan, and Michael J. Sanderson

Contents

Abstract Oscillations in the concentration of free cytoplasmic calcium ($[Ca^{2+}]_i$) play a vital role in the generation and maintenance of force by airway smooth muscle (ASM) cells. Mathematical models have an important role to play in the study of such complex dynamic phenomena, and can be used to construct and test hypotheses for how such oscillations might occur, and how properties such as the oscillation period might be controlled. We briefly discuss the underlying principles of the construction of mathematical models of calcium dynamics, and show how our current model can be used to understand how oscillations of $[Ca^{2+}]_i$ in ASM are

J. Sneyd (✉) • P. Cao
Department of Mathematics, University of Auckland, Auckland, New Zealand
e-mail: sneyd@math.auckland.ac.nz; pcao002@aucklanduni.ac.nz

X. Tan • M.J. Sanderson
Department of Microbiology and Physiological Systems, University of Massachusetts Medical School, Worcester, MA, USA
e-mail: xiahui.tan@umassmed.edu; michael.sanderson@umassmed.edu

Y.-X. Wang (ed.), *Calcium Signaling In Airway Smooth Muscle Cells*,
DOI 10.1007/978-3-319-01312-1_19,

the result of a complex interplay between inositol trisphosphate receptors (IP_3R) and ryanodine receptors (RyRs). Agonist-stimulated production of inositol trisphosphate (IP_3) opens IP_3R, resulting in the release of Ca^{2+} from the endoplasmic reticulum (ER). This released Ca^{2+} stimulates the release of additional Ca^{2+} from both IP_3R and RyR, leading to cycles of Ca^{2+} release and reuptake from the ER. In the absence of IP_3 (no agonist), when the ER is overloaded with Ca^{2+} these cycles of release and reuptake are mediated primarily by the RyR. Conversely, in the presence of IP_3 (with agonist), when the ER is partially depleted of Ca^{2+}, these cycles are mediated primarily by the IP_3R. Thus, an understanding of both IP_3R and RyR is required for an understanding of how $[Ca^{2+}]_i$ oscillations are controlled in ASM.

Keywords Calcium oscillations • Calcium waves • Inositol trisphosphate receptor • Ryanodine receptor

1 Introduction

Airway smooth muscle (ASM) is one of the most important cell types in the study of airway hyperresponsiveness, as major breathing difficulties occur when the contraction of ASM is either too strong or too sensitive.

The Ca^{2+} dynamics of ASM are particularly important for the understanding of contraction in ASM because it is a rise in Ca^{2+} that causes contraction of ASM. Thus, it is vital to understand the mechanisms involved in the control of Ca^{2+} in ASM.

Fortunately for mathematical modelers, the concentration of free cytosolic Ca^{2+} ($[Ca^{2+}]_i$) in ASM exhibits a number of complex properties that can only be understood with a combination of theoretical and experimental approaches. In response to agonist stimulation, $[Ca^{2+}]_i$ displays a range of complex spatiotemporal behaviors, including periodic waves of varying period [2, 19, 23, 25]. It is simply not possible to understand such complex behavior without recourse to a quantitative model, although any such model must, of course, be based on experimental results. In addition to helping us understand experimental results, models can also guide further experiments.

Although there is already a large body of literature on how to model generic Ca^{2+} dynamics in a variety of cell types [8, 13], there is a considerably smaller body of literature on the modeling of Ca^{2+} dynamics in ASM. Previous work in this area is summarized in [15, 16, 24] and focused on the electrical properties of ASM. We, however, take a rather different point of view because our experiments indicate only a minor role for membrane electrical potential in contraction. Instead, we focus on the interactions between the two major internal Ca^{2+} release channels – the inositol trisphosphate receptor (IP_3R) and the ryanodine receptor (RyR).

Our primary goal is to show how the experimental data implicating IP_3R and RyR can be explained by a relatively simple model in which Ca^{2+} oscillations are mediated primarily by IP_3R but significantly modulated by RyR.

In early studies, agonist-induced Ca^{2+} oscillations were observed in isolated smooth muscle cells, but these oscillations were thought to be mediated via the RyR with the IP_3R having only a minor modulating or initiating role [18, 21, 22]. To correlate smooth muscle Ca^{2+} dynamics with airway contraction, we examined ASM cells in lung slices and found that increased frequencies of Ca^{2+} oscillations and waves correlated with airway contraction. However, the mechanisms driving these Ca^{2+} oscillations were highly dependent on IP_3 and IP3R, with RyRs having little impact on ongoing Ca^{2+} oscillations [2, 3, 20, 23, 26, 30]. The experimental evidence for these conclusions is also reviewed in another chapter in this volume (Sanderson). Perhaps the differences lie in the cell preparation (isolated or in lung slices) or more simply are due to the experimental technique. We have learned to appreciate, as will be addressed later, that the interaction of these two receptor types can thwart experiments designed to elucidate the responses of just a single receptor. In this respect, mathematical modeling has provided the necessary tool to disentangle such interactions.

Before we address ASM in particular, we first present a short description of the methods for constructing generic Ca^{2+} models, including both spatially homogeneous and spatially distributed models.

2 Generic Modeling of Calcium Dynamics

There are two major types of model of Ca^{2+} dynamics: a spatially homogeneous model, which assumes a well-mixed cell, and a spatially inhomogeneous model, which allows for spatial variation of $[Ca^{2+}]_i$. Within each of these divisions, models can be deterministic or stochastic and can contain a large number of parameters and dependent variables.

It is important to note that the type of model one constructs is not necessarily dependent on what is believed to be the "real" situation. For example, it is well known that cells are not well mixed and that $[Ca^{2+}]_i$ is not homogeneous. Nevertheless, a well-mixed model can still be a useful tool, guiding new experimental results and making testable predictions. Similarly, we also know that, at a higher level of detail, the release of Ca^{2+} through either IP_3R or RyR is inherently a stochastic, not a deterministic, process. In some situations this matters, and stochastic models must be used. In other cases, stochastic aspects are less important.

In other words, we construct models, not to be the most detailed and accurate representation of what we believe is the true situation, but to be useful tools to guide our understanding.

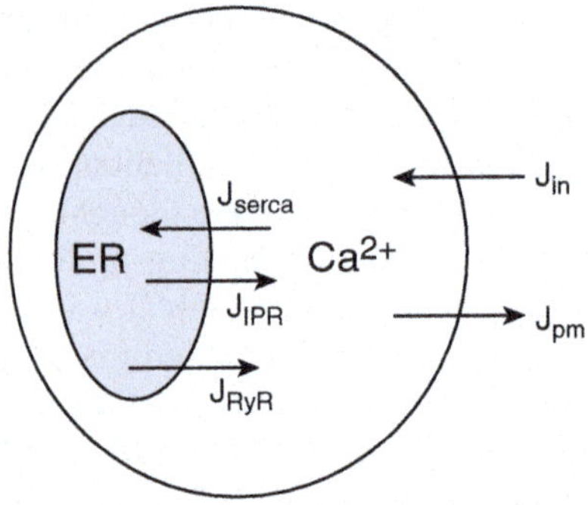

Fig. 1 Schematic diagram of a simple spatially homogeneous Ca^{2+} dynamics model, with five fluxes. In this model, Ca^{2+} is pumped into the endoplasmic reticulum (*ER*) from the cytoplasm by SERCA ATPase pumps (J_{serca}), is pumped out of the cell by plasma membrane ATPase pumps (J_{pm}), enters the cytoplasm from the outside through some unspecified entry pathway (J_{in}), and enters the cytoplasm from the ER through two channels, the IP_3R and the RyR

2.1 *Spatially Homogeneous Models*

If the cell is assumed to be well mixed, a typical equation for $[Ca^{2+}]_i$ expresses simply the conservation of Ca^{2+}. Thus, if we let c denote the concentration of Ca^{2+} in the cytoplasm, $[Ca^{2+}]_i$, then we have

$$\frac{dc}{dt} = J_{\text{into}} - J_{\text{out}}, \tag{1}$$

where J_{into} and J_{out} denote, respectively, all the fluxes of Ca^{2+} into and out of the cytoplasm.

It remains simply to specify which are the fluxes into the cytoplasm and which are the fluxes out of the cytoplasm. Each flux comes with its own model description, which can be substituted into (1) to obtain the complete equation for c.

A simple example is shown in Fig. 1, in which there are five fluxes into or out of the cytoplasm. Two of those fluxes, J_{in} (a generic influx of Ca^{2+}, possibly through store-operated channels, agonist-operated channels, or voltage-dependent Ca^{2+} channels) and J_{pm} [the flux through the plasma membrane (PM) ATPase pumps], are across the PM, while the other three, J_{RyR} (the flux through RyR), J_{PR} (the flux through IP_3R) , and J_{serca} (the flux through the sarcoplasmic/endoplasmic reticulum calcium ATPase (SERCA) pumps) are across the ER membrane.

In this case, (1) becomes

$$\frac{dc}{dt} = J_{\text{in}} - J_{\text{pm}} + \gamma(J_{\text{IPR}} + J_{\text{RyR}} - J_{\text{serca}}), \tag{2}$$

where the factor γ is required because all the fluxes have units of concentration per unit time and the volume of the ER is different from the volume of the cytoplasm. To this equation it is usually necessary to add a second equation describing the rate of change of the Ca^{2+} concentration in the ER (c_e);

$$\frac{dc_e}{dt} = -J_{\mathrm{IPR}} - J_{\mathrm{RyR}} + J_{\mathrm{serca}}. \tag{3}$$

Now one selects whichever model one wishes for each of the individual fluxes to complete the model construction. In general, each of these fluxes will involve other dynamic variables, which increases the total number of differential equations. Simpler models will have only two equations, while more complex models typically have as many as eight, or even more.

We emphasize that, although this simple model omits a vast amount of the known complexity in Ca^{2+} signalling (such as microdomains, the influence of the mitochondria, and direct effects of the membrane potential), it is still (as we shall see) a useful tool for the study of the mechanisms underlying Ca^{2+} oscillations, in some conditions.

One important hidden feature of the foregoing simple model is the inclusion of Ca^{2+} buffering. Although it appears that Ca^{2+} buffers are not included in (1), this is in fact not necessarily true. If the buffers are fast, linear, and unsaturated, then they contribute only the same constant factor to each flux. Hence, the fluxes should be interpreted as *effective* Ca^{2+} fluxes, i.e., that part of the actual flux that contributes to a change in the concentration of free Ca^{2+}.

2.2 *Inclusion of Diffusion*

To turn this simple spatially homogeneous model into a model that allows for a spatially varying $[Ca^{2+}]_i$ (to reflect reality), the model equations must be adapted to include the diffusion of Ca^{2+}. Although this can be complicated by the presence of Ca^{2+} buffers, if we assume again that the buffers are fast, linear, and unsaturated, then the diffusion equation for c remains linear, using only an effective diffusion coefficient (which is typically around 10 $\mu m^2/s$, instead of the approximately 300 $\mu m^2/s$ that is the diffusion coefficient for unbuffered Ca^{2+} in an aqueous solution).

Making this simplifying assumption gives

$$\frac{\partial c}{\partial t} = D_{\mathrm{eff}} \nabla^2 c + J_{\mathrm{in}} - J_{\mathrm{pm}} + \gamma(J_{\mathrm{IPR}} + J_{\mathrm{RyR}} - J_{\mathrm{serca}}), \tag{4}$$

where it is assumed that the cell is thin enough so that Ca^{2+} entering across the membrane is effectively seen by the entire width of the cell immediately. If one wishes to model a fully three-dimensional cell, with fluxes across the PM that are effectively separated from the fluxes into and out of the ER, then more complicated boundary conditions are required. However, this is not relevant for the discussion here, and so these mathematical details are omitted.

3 Modeling of Calcium Dynamics in Airway Smooth Muscle

To turn this simple generic model into a model specific for ASM involves the choice of which fluxes to include and which models to use for these fluxes. Ideally, all these choices should be governed by specific experimental data, but in this case, as in most cases, there are significant gaps in the data that need to filled by convenient assumptions.

For our initial spatially homogeneous model we make the following assumptions:

- Calcium is removed from the cytoplasm in two ways. It is pumped out of the cells by ATPase pumps on the PM, and it is pumped into the ER by SERCA ATPases. This assumption means that we ignore the influence of mitochondria on Ca^{2+} removal. Although it has been shown that mitochondria probably play an important role in Ca^{2+} removal [15], we have yet to determine how inclusion of these effects will affect our results. Our focus throughout is more on the interplay between the Ca^{2+} entry pathways, and, for an initial approximation, we believe it to be sufficient to model Ca^{2+} removal in a simpler way. The PM pumps and the SERCA pumps are both modeled by Hill functions, with Hill coefficients of four and two, respectively.
- Calcium entry into the cell from outside is via two pathways, one voltage-dependent, the other not. The voltage-dependent pathway is a voltage-gated Ca^{2+} channel, with the reversal potential given by the Goldman–Hodgkin–Katz equation and a single activation gating variable that is a time-independent increasing function of the membrane potential.
- The voltage-independent pathway is via Ca^{2+} channels that are opened by agonist stimulation (most likely using the arachadonic acid pathway, although this is not necessary), ignoring any effects of store-operated channels. This is an important assumption of the model and deserves slightly more discussion. Although it is known that store-operated channels are activated when the Ca^{2+} concentration in the ER gets very low, this is often not the case for the types of agonist stimulation that result in periodic waves mediated by the IP_3R [17, 27]. At lower agonist concentrations there is indeed increased Ca^{2+} entry, but this is controlled by chemical pathways that go through arachadonic acid. Thus, for simplicity, we model this influx as a simple increasing function of agonist concentration, ignoring the complexities of the intermediate biochemistry.

  However, although Ca^{2+} influx is an important model component, it remains true that the Ca^{2+} oscillations and waves are caused by the release and reuptake of Ca^{2+} from the ER, not by Ca^{2+} transport across the PM. This is clear from the fact that the oscillations persist in the absence of extracellular Ca^{2+} [19].

- Calcium moves from the ER to the cytoplasm through RyRs and IP_3Rs. The Ca^{2+} flux through the IP_3R is the most difficult and complex part of the model. There is still controversy over which model of the IP_3R should be used. Recent models of the IP_3R are based on single-channel data [10, 28], but these data might well miss important nonstationary effects, leading to model inaccuracies. However, older models, such as those of DeYoung and Keizer [5], Atri et al. [1], LeBeau et al. [14], and Sneyd and Dufour [29], suffer from their own problems, such as inactivation time constants that differ significantly from those measured experimentally. We do not yet have specific data about the IP_3R in ASM to be able to resolve this controversy with any authority.

  Even if the details of the IP_3R model turn out to be incorrect, if the model makes predictions that can be tested and confirmed, it is likely that the model has abstracted some important properties of ASM, incorrect details notwithstanding.

  Thus, for simplicity, much of our work on Ca^{2+} dynamics in ASM has used the model of DeYoung and Keizer [5] to describe the IP_3R.
- There are two principal components of the RyR model. First, it exhibits Ca^{2+}-induced Ca^{2+} release (CICR), as the flux through the receptor is an increasing function of c. This CICR can be modeled in a number of ways; we have chosen to use the model of Friel [9], which is an excellent quantitative description of CICR through RyR, although in a different cell type. The second important component of the RyR model is the fact that the RyR Ca^{2+} flux is dependent on the concentration of Ca^{2+} in the ER, c_e. As c_e increases, so does the RyR flux, whereas, conversely, as the ER gets depleted, the RyR flux decreases rapidly.

  This feature of the RyR is borrowed from similar models in cardiac cells, where it is well known that overloading of the SR leads to a much greater spontaneous release of Ca^{2+} through the RyR [4].
- We do not include a detailed model of the ionic fluxes across the cell membrane, or of the membrane potential. This is because our focus here is on the Ca^{2+} dynamics rather than the electrical properties of the cell. Although the membrane potential affects the Ca^{2+} dynamics via the flux through voltage-gated channels, this effect can be incorporated in a relatively simple way, without requiring detailed models of all the ionic currents.

Given the foregoing assumptions, one can substitute the various submodels into (2)–(4) to obtain the overall model of the Ca^{2+} dynamics. We do not give the full details here; they can all be found in [34].

3.1 Interplay Between IP_3R and RyR

Typical model solutions showing oscillations and waves of increased $[Ca^{2+}]_i$ are shown in Fig. 2. In response to an agonist, there is an initial release of Ca^{2+} through the IP_3R. This initial release causes a larger release of Ca^{2+} through the RyR, via a

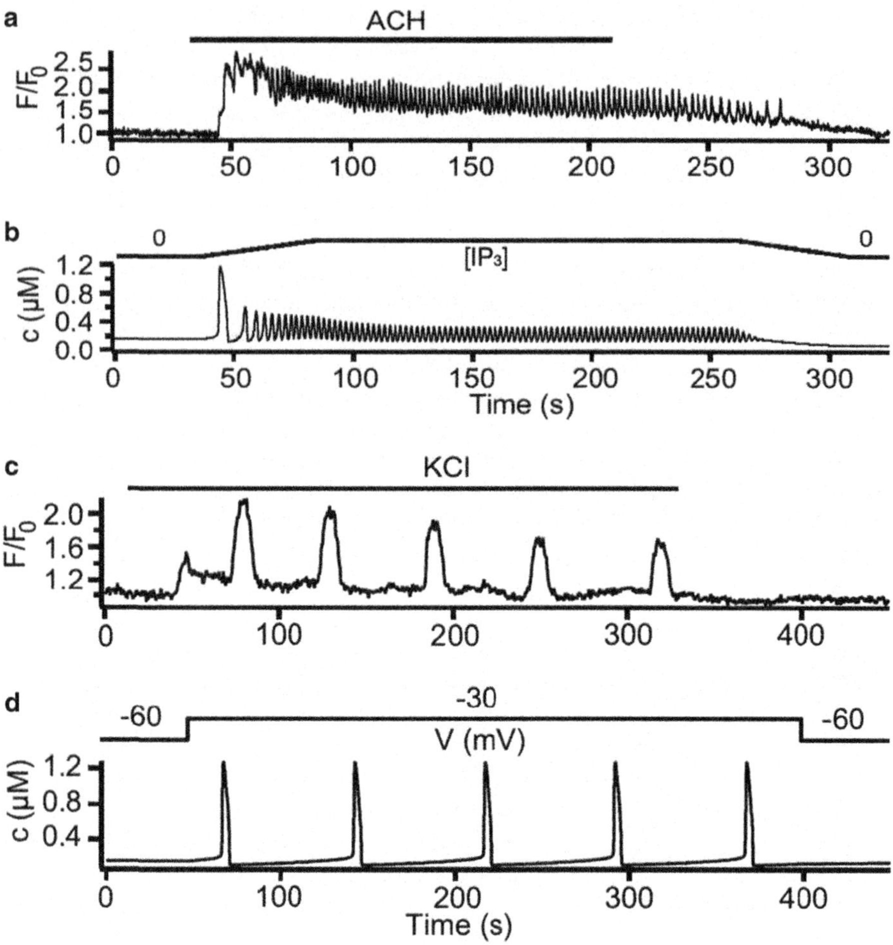

Fig. 2 Experimental results and model simulations. Basic oscillations under control conditions. (**a**) Application of acetylcholine (ACh) causes sustained oscillations in $[Ca^{2+}]_i$ in mouse ASM. These oscillations occur on a raised baseline. (**b**) Model simulation of ACh-induced $[Ca^{2+}]_i$ oscillations. ACh is simulated by an increase in $[IP_3]$. Although the model shows good qualitative agreement with the experimental results in panel **a**, the quantitative agreement is poor, with a large initial peak, and with the oscillations not occurring on a raised baseline. (**c**) Application of KCl to the ASM cell causes depolarization of cell membrane, entry of Ca^{2+} through voltage-gated channels, and cycles of Ca^{2+} release and reuptake. These Ca^{2+} oscillations are quite different from ACh-stimulated oscillations (panel **a**), being much wider, with a larger period. (**d**) Model simulation of response to depolarization. Ca^{2+} oscillations occur with a much larger period

process of CICR, which depletes the ER Ca^{2+} concentration, at least partially. This decrease in ER Ca^{2+} concentration inactivates the RyR but has much less effect on the IP_3R, which initiates a cycle of Ca^{2+} release and reuptake to and from the ER, leading to Ca^{2+} oscillations and periodic waves. The Ca^{2+} flux through the RyR continues to oscillate, but only with very small amplitude.

The cycles of Ca^{2+} release through the IP_3R occur by the same mechanism as in all previous models of Ca^{2+} oscillations based on the De Young–Keizer model [5]. An increase in $[Ca^{2+}]_i$ increases the open probability of the IP_3R, leading to a positive feedback cycle that operates on a subsecond time scale. However, the IP_3R also inactivates in response to increased $[Ca^{2+}]_i$, but on a slower time scale. The interaction between the fast activation and the slower inactivation leads to cycles of IP_3R opening and closing. The small-amplitude oscillations in the RyR flux are purely a response to the IP_3R-based oscillations and play little role in the control of the oscillation period. The function of these oscillations and waves shall be considered later.

This model also reproduces the response to KCl (Fig. 2), which depolarizes the cell, leading to an increased influx of Ca^{2+}. In this case, the ER overloads with Ca^{2+}, leading to spontaneous release of Ca^{2+} through the RyR, subsequent depletion of the ER, inhibition of the RyR, and termination of the Ca^{2+} transient. Once the Ca^{2+} transient ends, the ER can recover and the cycle can repeat, leading to long-period oscillations, quite different in nature from those caused by agonists that activate the IP_3R. In this case, because the concentration of IP_3 is zero (or at least very low), the IP_3R are not able to sustain high-frequency Ca^{2+} oscillations.

There is thus a close interplay between the RyR and the IP_3R. In some cases oscillations are governed by one receptor type, in other cases the other receptor type is more important, and during the initial response to agonist both RyR and IP_3R play a role. The dominant receptor type is essentially determined by the concentration of Ca^{2+} in the ER. When the ER is overfilled, the RyR predominate; when the ER Ca^{2+} concentration is lower, the IP_3R predominate.

By applying agonist, KCl, and ryanodine in various orders, we can explore these interactions further. If KCl is applied first, followed by methacholine (MCh), then the model predicts that long-period oscillations (mediated by the RyR) will be replaced by short-period oscillations (mediated by IP_3R). Conversely, if MCh is applied first, followed by KCl, the KCl will have no significant effect on the oscillatory pattern. Model simulations and the experimental verification are shown in Fig. 3. The intuitive explanation of these results is as follows.

- Upon application of KCl, Ca^{2+} entry increases, which overloads the SR, leading to spontaneous activation of the RyR and long-period spiking. Although there is some temporary depletion of the SR at the peak of the Ca^{2+} spike, the SR, on average, remains overloaded.
- When MCh is added, the IP_3R also opens, leading to increased Ca^{2+} flux from the SR into the cytoplasm. Because of this, the ER is, on average, depleted, which inactivates the RyR, leading to oscillations mediated almost soley by the IP_3R.
- Conversely, if MCh is added first, then the SR is depleted. Subsequent addition of KCl, despite increasing the Ca^{2+} influx, is unable to cause an overload of the SR, and thus the RyR remains inactivated.

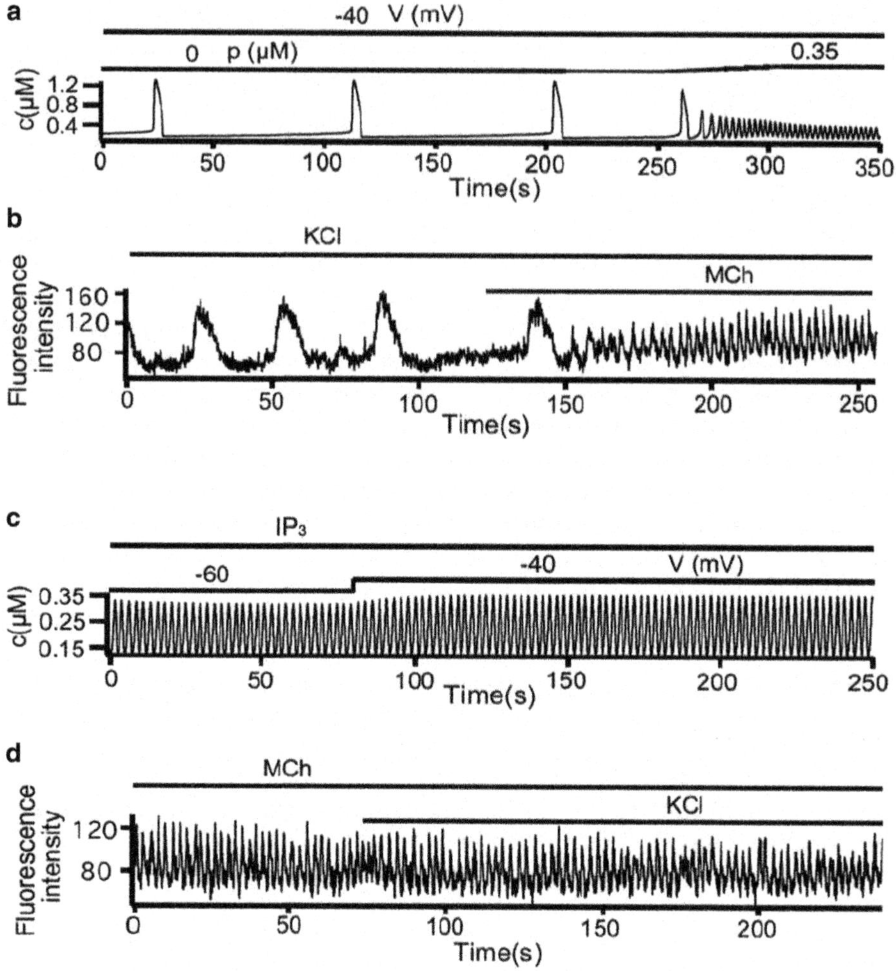

Fig. 3 (**a**) Model simulations, predicting response when KCl is applied first, followed by MCh (simulated by an increase in $[IP_3]$). The model predicts that the Ca^{2+} oscillations will speed up and decrease in magnitude. (**b**) Experimental confirmation of model prediction. Note that the model is predicting $[Ca^{2+}]_i$, while the experiments measure the fluorescence ratio. This difference is the cause of some of the quantitative discrepancy. (**c**) Model simulations predicting the response of the application of MCh first, followed by KCl. The model predicts that the oscillations will remain practically unchanged. (**d**) Experimental confirmation of model prediction

We can also use the model to predict the effects of adding ryanodine (Fig. 4a). Ryanodine acts in a slightly unusual manner in that it locks the RyR open, but only once it has already been opened – ryanodine has no effect on a closed RyR. If we include that effect in the model, then we see that pretreatment with ryanodine will eliminate agonist-induced Ca^{2+} oscillations. This is because the initial agonist-induced release of Ca^{2+} through the IP_3R causes the RyR to open (via Ca^{2+}-induced

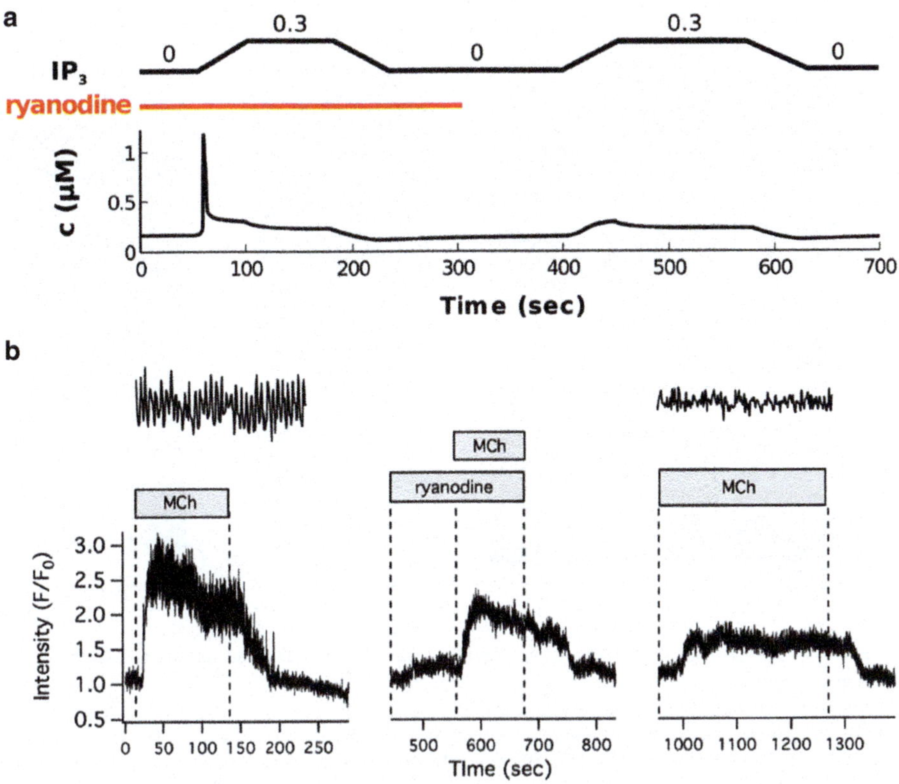

Fig. 4 (**a**) Model simulations predicting response when ryanodine is applied first, followed by MCh. After an initial peak, the Ca^{2+} response shows no oscillations in response to MCh. (**b**) Experimental confirmation of model predictions

Ca^{2+} release), but these RyRs are then locked in the open state by ryanodine, resulting in a large leak of Ca^{2+} out of the ER, which can then not support oscillations through the IP_3Rs. Again, this model prediction is confirmed experimentally (Fig. 4b).

Conversely, the addition of ryanodine has surprisingly little effect on MCh-induced oscillations (see the chapter by Sanderson in this volume). This is due to the fact that the addition of MCh results in a depletion of the ER and subsequent closure of most of the RyRs. Thus, when ryanodine is added to a cell exhibiting MCh-induced oscillations, there are few open RyRs upon which to act, and so the effect of the ryanodine is minimal. When ryanodine is added *before* MCh, this is not the case; the initial Ca^{2+} spike opens up a substantial fraction of the RyRs, which are then locked open by the ryanodine, leading to ER depletion and cessation of the oscillations. Hence the order of application of MCh and RyR is of crucial importance.

There is thus a close interplay between the IP_3R and the RyR in the control of Ca^{2+} oscillations, and this interplay occurs since both channel types are releasing

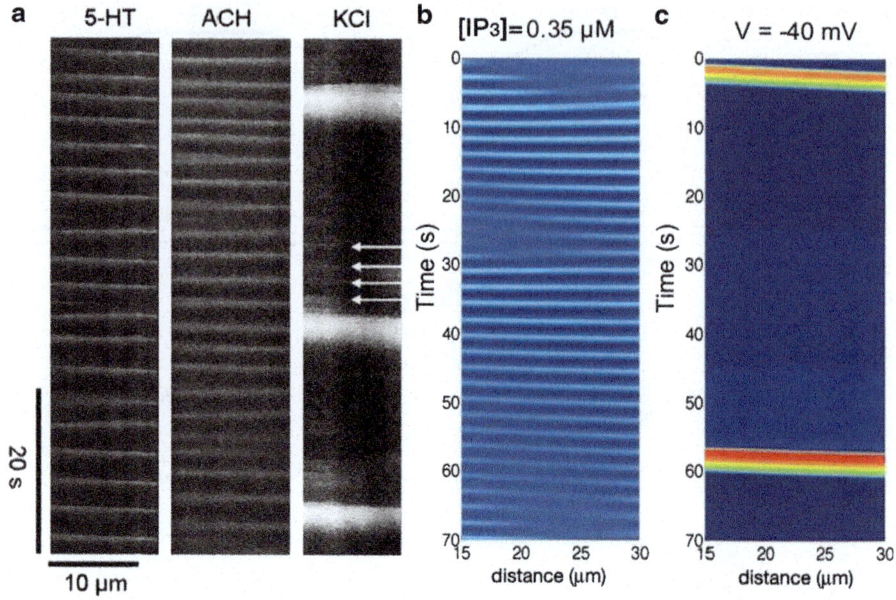

Fig. 5 (**a**) Experimental periodic waves of Ca^{2+} in response to three agonists. Lighter colors denote higher $[Ca^{2+}]_i$. In the response to 5-HT one can see that the wave reverses direction. Often, the waves in different directions are separated by abortive waves that do not travel the full length of the cell. Such abortive waves can also be seen in the response to KCl, where they are marked by *arrows*. (**b**) Model simulation of MCh-induced periodic Ca^{2+} waves. The model waves also reverse direction periodically. Lighter colors denote higher $[Ca^{2+}]_i$. (**c**) Model simulations of KCl-induced periodic Ca^{2+} waves. Lighter colors (i.e., *red* and *orange*) denote higher $[Ca^{2+}]_i$ (Reproduced (with permission) from [34])

Ca^{2+} from the same internal store, the ER. The amount of Ca^{2+} in the ER is controlled by both IP_3R and RyR, and, in turn, the ER Ca^{2+} concentration modulates release through both these Ca^{2+} channels. In this way, although MCh-induced oscillations are mediated primarily by the IP_3R, they are significantly affected by the RyR, via the control of how much Ca^{2+} is in the ER, and vice versa.

3.2 *Periodic Waves in Model*

When Ca^{2+} diffusion is included in the model, assuming that the Ca^{2+} buffers are fast and unsaturated, these oscillations turn into periodic waves (Fig. 5). In this case, each part of the cytoplasm is an autonomous oscillator, weakly linked to its neighbors by diffusion of Ca^{2+}. The appearance of periodic waves is a standard mathematical feature of such models.

However, there is one feature here of particular interest: in the cell, the periodic waves are often observed changing direction, as can be seen in the leftmost panel of Fig. 5a. Waves in different directions are separated by abortive waves that do not travel the full length of the cell. It is not possible (without additional intervention) to reproduce this qualitative feature in the model if the density of IP_3R is the same throughout the cell. Instead, the model predicts that the density of IP_3R will be higher at one end of the cell than the other. With such an IP_3R density gradient, one end of the cell is trying to oscillate at a slightly higher frequency than the other end. Although the higher-frequency oscillation can drive the lower-frequency oscillation for a short time, eventually it gets so far ahead that the wave traveling from the high-frequency end encounters a refractory region propagating from the low-frequency end, which aborts the wave. Then, before the high-frequency end can initiate the next wave, the low-frequency end initiates the next wave, resulting in a wave that moves in the opposite direction. However, this wave direction cannot be maintained for long because of the intrinsic frequency differences between the ends.

Given the observed symmetry of the ASM cells, this prediction is somewhat surprising because there is no obvious reason to suppose that the IP_3R density is asymmetrical in this way. Only additional experimental work will be able to answer this question.

3.3 Remaining Questions

Although the model appears to have considerable predictive power, and although it does seem to describe well the interactions between IP_3Rs and RyRs, many questions remain.

3.3.1 Model Accuracy

The model oscillations, although they seem to capture a number of qualitative features, suffer from poor quantitative agreement with experiment. This is not necessarily a significant disadvantage for the present study, but when we wish to use the Ca^{2+} model to drive a cross-bridge model to generate force (see below), the shape of the oscillations becomes important. Thus, more work is needed to construct a Ca^{2+} model that is more quantitatively accurate. One likely explanation for at least some of the quantitative discrepancy is the fact that the model output is the Ca^{2+} concentration, while experiments measure instead the fluorescence ratio. Since dye fluorescence is not linearly related to $[Ca^{2+}]_i$ but will instead show such effects as saturation and nonlinear binding, we expect a priori for there to be significant quantitative differences between the model predictions and the experimental results. However, this is almost certainly not the only cause of these differences.

3.3.2 Model Choice

The most recent models of the IP_3R [28] are based on single-channel data from single receptors in vivo and thus have a much more solid biophysical basis than any of the earlier models, such as the one used here. However, these recent models have shown that the slow time scale that is so important in the earlier models, i.e., the slow inactivation of the IP_3R by Ca^{2+}, does not appear to exist. All the observed time scales in these recent models are much faster, and thus use of these models does not lead to Ca^{2+} oscillations of the correct period. There are a number of possible resolutions to this conundrum. First, it may be that the single-channel data, having been collected over a relatively short time, cannot measure the longer time constants, even though they are an important part of the IP_3R response. Second, the steady-state nature of the single-channel data precludes measurement of nonstationary data, such as the transient response to a step increase in $[Ca^{2+}]_i$. Third, it may be that other mechanisms, in addition to simple modulation of the IP_3R by Ca^{2+}, play a major role in the control of oscillations in ASM. Which, if any, of these possibilities will turn out to be the most useful one is left for future work.

3.3.3 Stochastic or Deterministic

IP_3Rs and RyRs, by their intrinsic nature, are stochastic. It is known that stochastic processes can organize themselves in such a way as to appear to be deterministic oscillations [7, 31, 32], and it might be the case here that the Ca^{2+} oscillations are in fact the result of underlying stochastic opening of the IP_3Rs and RyRs. We are currently doing simulations to test this hypothesis, and preliminary results suggest this is not the case for ASM, but as yet we have no clear answer.

3.3.4 Species Differences

Human ASM, the cell type we are ultimately interested in, behave differently from mouse or rat ASM. In addition, oscillation periods at room temperature are longer than at body temperature. Thus, to construct a quantitatively accurate model for use in a multiscale model of ASM contraction, we need to modify the current model to take both species and temperature into account. It is not yet clear whether this can be accomplished simply by changes in parameter values or whether more structural changes to the model will be required.

4 Downstream of Calcium

The principal reason we are interested in the dynamics of Ca^{2+} in ASM is that an increase in Ca^{2+} activates the myosin cross bridges [via the activation of myosin light chain kinase (MLCK)], which bind to actin and, thus, generate a contractile force.

It is outside the scope of this chapter to discuss cross-bridge models in detail. Suffice it to say that myosin cross bridges are usually modeled by a variant of the Hai–Murphy model [11], which is itself an adaptation of the Huxley cross-bridge model originally designed for skeletal muscle [12].

The most detailed model of the activation of MLCK by Ca^{2+} is that of [6]. In this model, four Ca^{2+} ions bind to calmodulin (CaM) to activate it, and then Ca_4CaM activates MLCK, which then phosphorylates the myosin light chain to allow binding to actin. Their model derives rate constants in a detailed binding model of CaM.

The strength of ASM contraction is governed by a number of different factors, in which both Ca^{2+} oscillation period and average amplitude are important [33]. As the agonist concentration increases, so does both the Ca^{2+} oscillation frequency and the average Ca^{2+} amplitude (averaged over one oscillation cycle), and these increases translate to a stronger contraction. Increasing the oscillation frequency while keeping the same average amplitude also results in increased force, and thus force generation is frequency dependent, at least to some extent.

Most interestingly, periodic waves of Ca^{2+} generate a different force than do spatially homogeneous oscillations of the same period [33]. This is because the exact shape of the wave – for example, its rate of rise and rate of decay – have significant effects on force generation because the Ca^{2+} signal is translated to force through a highly nonlinear process.

5 Conclusions

We have constructed a mathematical model of Ca^{2+} oscillations in ASM cells and used the model to show how IP_3Rs and RyRs interact to control the oscillations. When Ca^{2+} entry is high and the IP_3Rs are closed (as is the case upon the application of KCl), the ER becomes overloaded with Ca^{2+}, resulting in cycles of Ca^{2+} release and reuptake from the ER via the RyRs. Such oscillations are large and wide, with large period. However, when the IP_3Rs open (upon, say, the application of MCh or similar agonist), the ER empties, the RyRs are mostly shut, and the Ca^{2+} oscillations are mediated by the opening and closing of the IP_3R. Such MCh-induced Ca^{2+} oscillations have a much smaller period than those induced by KCl and occur on a raised baseline.

The IP_3Rs and RyRs thus interact principally via their common access to the ER. Depletion of the ER by Ca^{2+} flux through one receptor type will affect the Ca^{2+} flux through the other receptor type, leading to interdependence and complex dynamic behavior. Thus, for example, even though application of ryanodine prior to application of MCh will eliminate Ca^{2+} oscillations, it is not the case that MCh-induced oscillations are mediated by RyRs. Instead, elimination of the oscillations occurs via depletion of the ER.

Although the model, in some aspects, lacks quantitative accuracy, it is nevertheless a useful predictive tool. A series of model predictions has been confirmed experimentally and led to a greatly increased understanding of how Ca^{2+} oscillations in ASM are controlled. Other model predictions, such as the expected spatial distribution of IP_3R and RyR densities, remain untested as of yet but provide a clear guide to experiments that will play an important role in advancing our understanding of the Ca^{2+} dynamics of ASM.

Acknowledgements This work was supported by National Institutes of Health Grant R01 HL103405 and by the University of Auckland, New Zealand.

References

1. A. Atri, J. Amundson, D. Clapham and J. Sneyd (1993) A single-pool model for intracellular calcium oscillations and waves in the Xenopus laevis oocyte, Biophys J, **65**, 1727–39
2. Y. Bai, M. Edelmann and M. J. Sanderson (2009) The contribution of inositol 1,4,5-trisphosphate and ryanodine receptors to agonist-induced Ca^{2+} signaling of airway smooth muscle cells, Am J Physiol Lung Cell Mol Physiol, **297**, L347–61
3. Y. Bai and M. J. Sanderson (2006) Airway smooth muscle relaxation results from a reduction in the frequency of Ca^{2+} oscillations induced by a cAMP-mediated inhibition of the IP_i3 receptor, Respir Res, **7**, 34
4. H. Cheng, W. J. Lederer and M. B. Cannell (1993) Calcium sparks: elementary events underlying excitation-contraction coupling in heart muscle, Science, **262**, 740–4
5. G. W. De Young and J. Keizer (1992) A single pool IP_3-receptor based model for agonist stimulated Ca^{2+} oscillations, Proc Natl Acad Sci U S A, **89**, 9895–9899
6. A. Fajmut, M. Brumen and S. Schuster (2005) Theoretical model of the interactions between Ca^{2+}, calmodulin and myosin light chain kinase, FEBS Lett, **579**, 4361–6
7. M. Falcke (2003) On the role of stochastic channel behavior in intracellular Ca^{2+} dynamics, Biophys J, **84**, 42–56
8. M. Falcke (2004) Reading the patterns in living cells - the physics of Ca^{2+} signaling, Advances in Physics, **53**, 255–440
9. D. Friel (1995) $[Ca^{2+}]_i$ oscillations in sympathetic neurons: an experimental test of a theoretical model, Biophys J, **68**, 1752–1766
10. E. Gin, M. Falcke, L. E. Wagner, D. I. Yule and J. Sneyd (2009) Markov chain Monte Carlo fitting of single-channel data from inositol trisphosphate receptors, J Theor Biol, **257**, 460–74
11. C. M. Hai and R. A. Murphy (1988) Cross-bridge phosphorylation and regulation of latch state in smooth muscle, Am J Physiol, **254**, C99–106
12. A. F. Huxley (1957) Muscle structure and theories of contraction, Progress in Biophysics, **7**, 255–318
13. J. P. Keener and J. Sneyd (2009) Mathematical Physiology, Second Edition, Springer-Verlag, New York
14. A. P. LeBeau, D. I. Yule, G. E. Groblewski and J. Sneyd (1999) Agonist-dependent phosphorylation of the inositol 1,4,5-trisphosphate receptor: A possible mechanism for agonist-specific calcium oscillations in pancreatic acinar cells, J Gen Physiol, **113**, 851–72
15. M. Marhl, D. Noble and E. Roux (2006) Modeling of molecular and cellular mechanisms involved in Ca^{2+} signal encoding in airway myocytes, Cell Biochem Biophys, **46**, 285–302
16. P. Mbikou, A. Fajmut, M. Brumen and E. Roux (2006) Theoretical and experimental investigation of calcium-contraction coupling in airway smooth muscle, Cell Biochem Biophys, **46**, 233–52
17. O. Mignen, J. L. Thompson and T. J. Shuttleworth (2001) Reciprocal regulation of capacitative and arachidonate-regulated noncapacitative Ca^{2+} entry pathways, J Biol Chem, **276**, 35676–83

18. C. M. Pabelick, Y. S. Prakash, M. S. Kannan and G. C. Sieck (1999) Spatial and temporal aspects of calcium sparks in porcine tracheal smooth muscle cells, Am J Physiol, **277**, L1018–25
19. J. F. Perez and M. J. Sanderson (2005) The Frequency of Calcium Oscillations Induced by 5-HT, ACH, and KCl Determine the Contraction of Smooth Muscle Cells of Intrapulmonary Bronchioles, J Gen Physiol, **125**, 535–53
20. J. F. Perez-Zoghbi, Y. Bai and M. J. Sanderson (2010) Nitric oxide induces airway smooth muscle cell relaxation by decreasing the frequency of agonist-induced Ca^{2+} oscillations, J Gen Physiol, **135**, 247–59
21. Y. S. Prakash, M. S. Kannan and G. C. Sieck (1997) Regulation of intracellular calcium oscillations in porcine tracheal smooth muscle cells, Am J Physiol, **272**, C966–75
22. Y. S. Prakash, C. M. Pabelick, M. S. Kannan and G. C. Sieck (2000) Spatial and temporal aspects of ACh-induced $[Ca^{2+}]_i$ oscillations in porcine tracheal smooth muscle, Cell Calcium, **27**, 153–62
23. A. R. Ressmeyer, Y. Bai, P. Delmotte, K. F. Uy, P. Thistlethwaite, A. Fraire, O. Sato, M. Ikebe and M. J. Sanderson (2010) Human airway contraction and formoterol-induced relaxation is determined by Ca^{2+} oscillations and Ca^{2+} sensitivity, Am J Respir Cell Mol Biol, **43**, 179–91
24. E. Roux, P. J. Noble, D. Noble and M. Marhl (2006) Modelling of calcium handling in airway myocytes, Prog Biophys Mol Biol, **90**, 64–87
25. M. J. Sanderson, Y. Bai and J. Perez-Zoghbi (2010) Ca^{2+} oscillations regulate contraction of intrapulmonary smooth muscle cells, Adv Exp Med Biol, **661**, 77–96
26. M. J. Sanderson, P. Delmotte, Y. Bai and J. F. Perez-Zogbhi (2008) Regulation of airway smooth muscle cell contractility by Ca^{2+} signaling and sensitivity, Proc Am Thorac Soc, **5**, 23–31
27. T. J. Shuttleworth, J. L. Thompson and O. Mignen (2007) STIM1 and the noncapacitative ARC channels, Cell Calcium, **42**, 183–91
28. I. Siekmann, L. E. Wagner, 2nd, D. Yule, C. Fox, D. Bryant, E. J. Crampin and J. Sneyd (2011) MCMC estimation of Markov models for ion channels, Biophys J, **100**, 1919–29
29. J. Sneyd and J. F. Dufour (2002) A dynamic model of the type-2 inositol trisphosphate receptor, Proc Natl Acad Sci U S A, **99**, 2398–403
30. J. Sneyd, K. Tsaneva-Atanasova, V. Reznikov, Y. Bai, M. J. Sanderson and D. I. Yule (2006) A method for determining the dependence of calcium oscillations on inositol trisphosphate oscillations, Proc Natl Acad Sci U S A, **103**, 1675–80
31. K. Thurley, A. Skupin, R. Thul and M. Falcke (2012) Fundamental properties of Ca^{2+} signals, Biochim Biophys Acta, **1820**, 1185–94
32. K. Thurley, I. F. Smith, S. C. Tovey, C. W. Taylor, I. Parker and M. Falcke (2011) Timescales of IP_3-evoked Ca^{2+} spikes emerge from Ca^{2+} puffs only at the cellular level, Biophys J, **101**, 2638–44
33. I. Wang, A. Z. Politi, N. Tania, Y. Bai, M. J. Sanderson and J. Sneyd (2008) A mathematical model of airway and pulmonary arteriole smooth muscle, Biophys J, **94**, 2053–64
34. I. Y. Wang, Y. Bai, M. J. Sanderson and J. Sneyd (2010) A mathematical analysis of agonist- and KCl-induced Ca^{2+} oscillations in mouse airway smooth muscle cells, Biophys J, **98**, 1170–81

Effects of Inflammatory Cytokines on Ca^{2+} Homeostasis in Airway Smooth Muscle Cells

Hisako Matsumoto

Contents

Abstract Crosstalk between airway inflammation and airway smooth muscle cells (ASMCs) contributes to airway hyperresponsiveness, a cardinal feature of asthma. The main putative mechanism underlying the agonist-induced intracellular Ca^{2+}

H. Matsumoto, M.D., Ph.D. (✉)
Department of Respiratory Medicine, Kyoto University, 54, Kawahara-cho, Shogoin, Sakyo-ku, Kyoto 606-8507, Japan
e-mail: hmatsumo@kuhp.kyoto-u.ac.jp

Y.-X. Wang (ed.), *Calcium Signaling In Airway Smooth Muscle Cells*,
DOI 10.1007/978-3-319-01312-1_20,

($[Ca^{2+}]_i$) transients in ASMCs is Ca^{2+} release from the sarcoplasmic reticulum (SR) via the inositol 1,4,5-trisphosphate (IP_3) receptor and ryanodine receptor (RyR). Ca^{2+} depletion in SR then triggers store-operated Ca^{2+} entry (SOCE), Ca^{2+} influx from extracellular space. These mechanisms are modulated by inflammatory cytokines, such as tumor necrosis factor (TNF)-α and interleukin (IL)-13, which have pivotal roles in asthma and chronic obstructive pulmonary disease (COPD).

TNF-α upregulates Gq and Gi protein expression, and interleukin (IL)-13 enhances histamine H1 receptor and cysteinyl leukotriene receptor 1, which enhances agonist-induced IP_3/IP_3 receptor signaling. Expression of CD38, which affects Ca^{2+} release from the SR via RyR, is upregulated with TNF-α, IL-13, and, to a lesser extent, interferon-γ and IL-1β pretreatment. TNF-α and IL-13 also augment SOCE and expression of caveolin-1, a scaffolding protein in caveolae, flask-shaped plasma membrane invaginations, which play a key role in Ca^{2+} signaling. Furthermore, both TNF-α and IL-13 decrease the expression of sarcoendoplasmic reticulum Ca^{2+} ATPase SERCA2, which transfers Ca^{2+} from the cytosol of the cell to the lumen of the SR to replenish Ca^{2+} in the SR. The downregulation of SERCA2 mimics altered Ca^{2+} homeostasis observed in asthma. This chapter describes the mechanisms that underlie the inflammatory cytokine-mediated modulation of $[Ca^{2+}]_i$ in ASMCs.

Keywords Cytokines • Interleukin-13 • Tumor necrosis factor-α • Thymic stromal lymphopoietin • CD38

Abbreviations

ACh	Acetylcholine
AHR	Airway hyperresponsiveness
ASMC	Airway smooth muscle cell
β_2AR	β_2 adrenergic receptor
BK_{Ca}	High-conductance Ca^{2+}-activated potassium
$[Ca^{2+}]_i$	Intracellular Ca^{2+}
cADPR	Cyclic ADP-ribose
CICR	Ca^{2+}-induced Ca^{2+} release
COPD	Chronic obstructive pulmonary disease
CysLT1R	Cysteinyl leukotriene receptor
DAG	1,2-diacylglycerol
ERK	Extracellular signal-regulated kinase
HASMC	Human ASMC
IFN	Interferon
IL	Interleukin
IP_3	Inositol 1,4,5-trisphosphate

JAK	Janus kinase
JNK	c-Jun N-terminal kinase
LTD4	Leukotriene D4
MLCK	Myosin light chain kinase
MLCP	Myosin light chain phosphatase
PKA	Protein kinase A
PKC	Protein kinase C
ROCC	Receptor-operated Ca^{2+} channels
ROCE	Receptor-operated Ca^{2+} entry
RyR	Ryanodine receptor
SERCA	Sarcoendoplasmic reticulum Ca^{2+} ATPase
STAT	Signal transducer and activator of transcription
STIM	Stromal-interacting molecule
SOCC	Store-operated Ca^{2+} channel
SOCE	Store-operated Ca^{2+} entry
SR	Sarcoplasmic reticulum
TGF	Transforming growth factor
TNF	Tumor necrosis factor
TNFR	Tumor necrosis factor receptor
TRPC	Transient receptor potential channels
TSLP	Thymic stromal lymphopoietin
VOCC	Voltage-operated Ca^{2+} channel

1 Introduction

Airway hyperresponsiveness (AHR) is a cardinal feature of asthma. The airway smooth muscle cells (ASMCs) from patients with asthma may be intrinsically different, which potentially contributes to airway hypersensitivity in asthma. The ASMCs derived from patients with asthma are more contractile than those from healthy subjects, when assessed using freshly isolated single cells [52] or in gel contraction assays where cultured ASMCs are embedded [58]. Ca^{2+} homeostasis of cultured ASMCs derived from patients with asthma differs from those derived from control subjects [92, 54]. More importantly, however, AHR is enhanced by crosstalk between airway inflammation and ASMCs [84]. Recent advances in this research area have uncovered several mechanisms that underlie the causal links between inflammatory cytokines and increased intracellular Ca^{2+} ($[Ca^{2+}]_i$) or Ca^{2+} sensitivity in ASMCs that exhibit ASM contraction [5, 81]. This chapter describes the mechanisms that underlie the inflammatory cytokine-mediated modulation of $[Ca^{2+}]_i$ and Ca^{2+} sensitivity in ASMCs, in particular via tumor necrosis factor (TNF)-α and interleukin (IL)-13, which have pivotal roles in asthma and chronic obstructive pulmonary disease (COPD), where glucocorticoids are largely ineffective.

Unlike gastrointestinal smooth muscle cells, bladder smooth muscle cells, or cardiac myocytes, ASMCs are relatively electrically quiescent and do not generate an action potential, although voltage-operated Ca^{2+} channels (VOCCs) are present in human ASMCs (HASMCs), and excitation of ASMCs is accompanied by membrane depolarization. However, agonist-induced sustained contraction and Ca^{2+} influx into ASMCs are relatively insensitive to VOCC inhibitors, in particular those directed at L-type VOCCs [56, 36]. Thus, the contribution of excitation-contraction coupling via VOCCs is limited in ASMCs. Instead, Ca^{2+} is released from the intracellular Ca^{2+} store in the sarcoplasmic reticulum (SR) via store-operated Ca^{2+} entry (SOCE), which is often caused by physiologically activated substances. Therefore, pharmacomechanical coupling is the major cause of elevated $[Ca^{2+}]_i$ and ASM contraction.

1.1 Ca^{2+} Release from Intracellular Ca^{2+} Store

The main putative mechanism underlying increased $[Ca^{2+}]_i$ in ASMCs is Ca^{2+} release from the SR [38, 74]. Major agonists, including acetylcholine (Ach), histamine, leukotriene D_4/E_4, and neurokinin, are relevant to the pathophysiology of asthma by binding Gq or Gi-protein-coupled receptors [15]. After agonists bind to these receptors, phospholipase C is activated to generate 1,2-diacylglycerol (DAG) and inositol 1,4,5-trisphosphate (IP_3). IP_3 then acts on IP_3 receptors to release Ca^{2+} from the SR, which leads to an initial peak in $[Ca^{2+}]_i$. A subsequent steady-state level of $[Ca^{2+}]_i$ or an intracellular Ca^{2+} oscillation is maintained by intracellular Ca^{2+} release from the SR via the IP_3 receptor [12] or ryanodine receptor (RyR) as well as extracellular Ca^{2+} influx [9, 42, 68]. Details of the Ca^{2+} oscillation are described in another chapter.

Several lines of evidence confirm an important role for CD38/cyclic ADP-ribose (cADPR) signaling during Ca^{2+} release from the SR via RyR in HASMCs [21]. In addition to IP_3, cADPR is a calcium-mobilizing second messenger molecule [68]. CD38, a 45-kDa transmembrane glycoprotein, possesses an ADP-ribosyl cyclase activity that converts β-nicotinamide adenine dinucleotide to cADPR and is involved in the synthesis of cADPR. CD38 also has a cADPR hydrolase activity that converts cADPR to ADPR. In CD38-deficient mice where the ADP-ribosyl cyclase activity is absent [21], the AHR to methacholine and airway eosinophilic inflammation are attenuated [25], which suggests that CD38 and cADPR have important roles in AHR and airway inflammation in asthma. In ASMCs, Ca^{2+} responses to ACh and endothelin-1 are also dependent on CD38 [21].

In cultured HASMCs, TNF-α [82], IL-13 [20], and, to a lesser extent, interferon (IFN)-γ and IL-1β [19] increase CD38 expression and ADP-ribosyl cyclase activity. Exposure to these cytokines also increases the Ca^{2+} response to contractile agonists, which is significantly attenuated by the cADPR antagonist 8Br-cADPR [19]. Details are provided later.

1.2 Store-Operated Ca^{2+} Entry

Ca^{2+} influx from extracellular space can be mediated by VOCCs, receptor-operated Ca^{2+} channels (ROCCs), store-operated Ca^{2+} channels (SOCCs), Ca^{2+}-dependent Cl^-, nonselective cation channels, and Na^+/Ca^{2+} exchangers. Details of the Ca^{2+} channels are described in another chapter. Of these, SOCC is thought to be essential for the regulation of Ca^{2+} homeostasis in ASMCs. SOCE is triggered by Ca^{2+} depletion in the SR. The mechanisms underlying SOCE are not fully understood because of the lack of highly selective SOCE blockers [56]. Instead, inhibitors of sarcoendoplasmic reticulum Ca^{2+} ATPase (SERCA) have been used to determine the mechanisms associated with SOCE, i.e., thapsigargin or cyclopiazonic acid, which inhibit the active uptake of Ca^{2+} into the SR, thereby leading to the passive depletion of the SR.

Stromal-interacting molecule (STIM)1 and Orai1 were recently identified SOCE-related molecules [51]. The SR membrane protein STIM1 acts as a sensor of Ca^{2+} release from the SR, whereas Orai1 is a pore-forming subunit in Ca^{2+}-release-activated Ca^{2+} channels. After sensing $[Ca^{2+}]_i$ and $[Ca^{2+}]_{SR}$, STIM1 is thought to translocate to SOCC, where it activates STIM1-Orai1 (Ca^{2+}-selective channels) and STIM1-transient receptor potential channels (TRPC) (nonselective cation channels). Both TNF-α [82, 95] and IL-13 [26] augment SOCE, and the contribution of SOCE-related molecules to enhanced SOCE has been elucidated in TNF-α signaling in ASMCs [78].

1.3 Ca^{2+} Sensitivity and Caveolae

Ca^{2+} sensitivity is an important factor in AHR and force generation in ASMCs, which is independent of $[Ca^{2+}]_i$ [94]. Ca^{2+} sensitivity in ASMCs is determined mostly by the balance between myosin light chain kinase (MLCK)-dependent phosphorylation of myosin light chain (MLC) and its dephosphorylation by MLC phosphatase (MLCP). The details of its regulatory mechanisms in ASMCs remain unclear, but the Rho-kinase or protein kinase C (PKC)/CPI-17 pathways play a key role in Ca^{2+} sensitivity in ASMCs. Activation of Gq, Gi, or $G_{12/13}$-coupled receptors stimulates the RhoA/Rho-kinase pathway to phosphorylate MYPT1, a subunit of MLCP, to inhibit MLCP activity [15, 79]. The PKC pathway activated by DAG phosphorylates CPI-17, a 17-kDa inhibitory protein, which also inhibits MLCP activity and increases Ca^{2+} sensitivity [74, 94].

Caveolae are 50–100-nm flask-shaped plasma membrane invaginations that are rich in cholesterol, sphingolipids, and integral membrane proteins known as caveolins. Human and canine ASMCs express caveolin-1 and 2 but not caveolin-3. Caveolin-2 associates with caveolin-1 and is thought to modulate the function of caveolin-1, a major scaffolding protein in caveolae. Caveolae contain several proteins, including Orai1 [78], TRPC [69], and RhoA [77], which are important in

Ca^{2+} signaling [35]. Depletion of cellular cholesterol using cyclodextrins leads to the disruption of caveolae, which significantly decreases isometric force generation in response to ACh in porcine ASM strips [77].

Both TNF-α [34] and IL-13 [17] enhance Ca^{2+} sensitivity via the RhoA/Rho-kinase pathway and caveolin-1 expression via the NF-κB and mitogen-activated protein kinase (MAPK) pathways [76]. RhoA activation depends on caveolae, so there is expected to be a close relationship between Ca^{2+} sensitivity and caveolae. However, the link between Ca^{2+} sensitivity, RhoA, and caveolae has only been observed for TNF-α, as described below.

1.4 SERCA

SERCA, a Ca^{2+} ATPase that resides in the SR, transfers Ca^{2+} from the cytosol of the cell to the lumen of the SR via ATP hydrolysis against a 10,000-fold gradient difference in $[Ca^{2+}]$ across the membrane. In HASMCs, the replenishment of Ca^{2+} in the SR depends solely on the activity of SERCA. Of the three SERCA isoforms, the main isoform in ASMCs is SERCA2. In cultured ASMCs derived from patients with asthma, the expression of SERCA2 is lower compared with healthy subjects, and the extent of its reduction is associated with the severity of the asthma. In ASMCs from patients with asthma, the return to baseline $[Ca^{2+}]_i$ after stimulation with bradykinin is delayed compared with ASMCs from healthy subjects. Moreover, the knockdown of SERCA2 using siRNA in ASMCs from healthy subjects increases cell spreading, eotaxin-1 release, and cell proliferation, which shows that the downregulation of SERCA2 in ASMCs is involved in the pathophysiology of asthma [54].

Exposure to TNF-α or IL-13 (20 ng/mL, overnight for each cytokine) decreases the protein expression of SERCA2 and slows Ca^{2+} reuptake in HASMCs (with zero extracellular Ca^{2+} and blocking plasma membrane Ca^{2+} fluxes) [75] (Figs. 1 and 2), which also shows that SERCA2 downregulation is involved in altered Ca^{2+} homeostasis observed in asthma [55, 70].

1.5 Relaxation

ASMC relaxation is one of the ultimate goals of asthma treatment. The activation of adenylyl cyclase via β_2 adrenergic receptor (β_2AR), the only G stimulatory-protein-coupled receptor, increases cAMP, which activates protein kinase A (PKA). The activation of G stimulatory (Gs) stimulates high-conductance Ca^{2+}-activated potassium (BK_{Ca}) channels in cAMP-dependent and independent manners, which allows the efflux of K^+ from the cell [49]. Furthermore, cAMP attenuates SOCE in a PKA-dependent manner in HASMCs [10]. An elegant study of Ca^{2+} oscillations by Bai and Sanderson demonstrated the interaction between cAMP and Ca^{2+} homeostasis during airway relaxation. cAMP-elevating agents decreased the frequency of

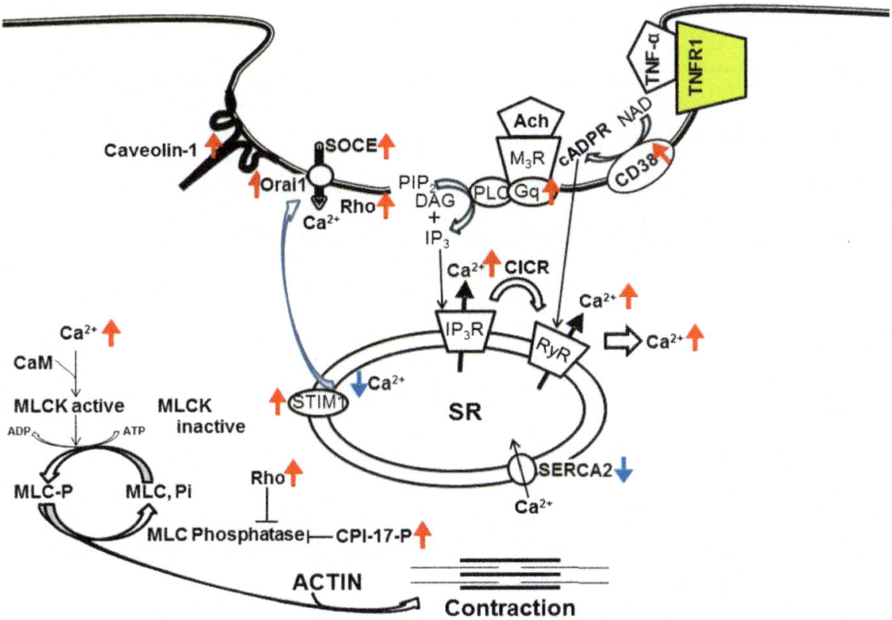

Fig. 1 Putative mechanisms underlying the potentiating effect of TNF-α on agonist-induced $[Ca^{2+}]_i$ in airway smooth muscle cells. *ACh* acetylcholine, *cADPR* cyclic ADP-ribose, *CaM* calmodulin, *CICR* Ca^{2+}-induced Ca^{2+} release, *DAG* 1,2-diacylglycerol, *IP_3* inositol 1,4,5-trisphosphate, *IP_3R* IP_3 receptor, *M_3R* muscarinic acetylcholine receptor M3, *MLCK* myosin light chain kinase, *NAD* nicotinamide adenine dinucleotide, *PIP_2* phosphatidylinositol 4,5-bisphosphate, *PLC* phospholipase C, *RyR* ryanodine receptor, *SERCA* sarcoendoplasmic reticulum Ca^{2+} ATPase, *SOCE* store-operated Ca^{2+} entry, *SR* sarcoplasmic reticulum, *STIM* stromal-interacting molecule, *TNFR* tumor necrosis factor receptor

Ca^{2+} oscillation and relaxed the airways of mice lung slices potentially via a reduction in Ca^{2+} release from the Ca^{2+} store via the IP_3 receptor [11].

Asthma is characterized by airway obstructions, which are variable and reversible after the inhalation of β_2AR agonists. However, an impaired response to β_2AR agonists is observed during virus-associated exacerbations. If HASMCs are incubated with media conditioned by rhinovirus-infected primary human airway epithelial cells, the β_2AR agonist-induced cAMP formation in HASMCs is attenuated without changing the number of β_2ARs [93]. The direct inoculation of rabbit ASMCs with rhinovirus also led to an impaired response to β_2AR, which was mediated via the upregulation of Gi protein [31].

IL-1β [80], IL-13 [50], TGF-β [65], and TNF-α [33, 61] are known to induce β_2AR desensitization in HASMCs (Fig. 3). These cytokines attenuate the isoproterenol-induced accumulation of cAMP and the isoproterenol-induced relaxation of ASMCs via the PKA-dependent phosphorylation of β_2AR, leading to its uncoupling from Gs-protein-mediated cAMP production or via the upregulation of Gi protein [30, 33]. IL-4 does not affect the isoproterenol-induced formation of cAMP.

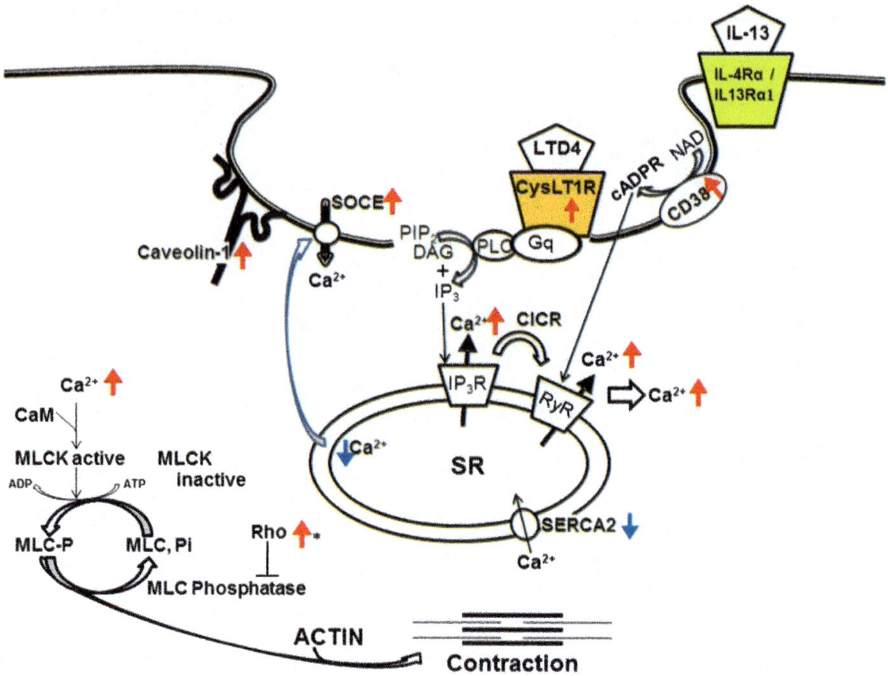

Fig. 2 Putative mechanisms underlying the potentiating effect of IL-13 on agonist-induced $[Ca^{2+}]_i$ in airway smooth muscle cells (ASMCs). *CysLT1R* cysteinyl leukotriene receptor 1, *LTD4* leukotriene D4. Observed in *murine ASMCs

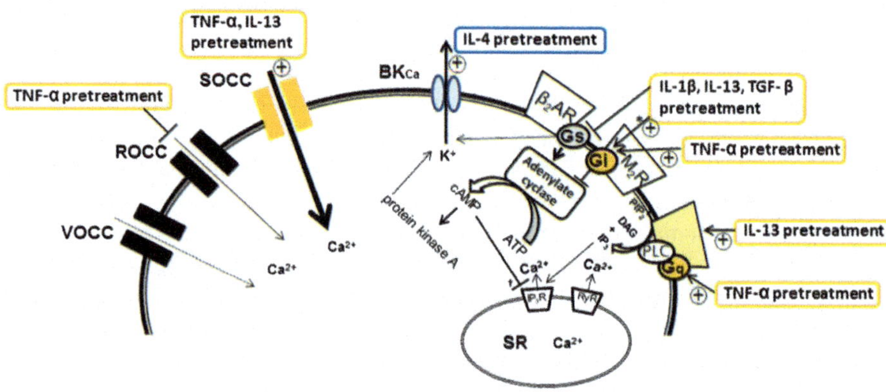

Fig. 3 Effects of inflammatory cytokines on membrane channels and G-protein-coupled receptors in airway smooth muscle cells (ASMCs). *β_2AR* β_2 adrenergic receptor, *BK_{Ca}* high-conductance Ca^{2+}-activated potassium channel, *M_2R* muscarinic acetylcholine receptor M2, *ROCC* receptor-operated Ca^{2+} channel, *SOCC* store-operated Ca^{2+} channel, *VOCC* voltage-operated Ca^{2+} channel. Observed in *rabbit ASMCs and †murine ASMCs

2 TNF-α

TNF-α is a representative proinflammatory cytokine that is elevated in the airways of patients with severe asthma and COPD. It is also secreted by many cells including macrophages, T-cells, and mast cells. Mast cells infiltrate the airway smooth muscle layer in patients with asthma [16], so interactions between TNF-α and ASMCs are relevant to asthma. TNF-α is a pleiotropic cytokine that is also involved in the induction of AHR. In normal subjects [86] or patients with mild asthma [87], AHR to inhaled methacholine and airway inflammation was enhanced 24 h after inhalation of recombinant human TNF-α compared with the control. Most evidence shows that TNF-α potentiates the agonist-induced increase in contractile responses [6, 85] and $[Ca^{2+}]_i$ [6], while it also increases the contractile response to electrical field stimulation [8]. Exposure to TNF-α (25 ng/mL for 16 h) potentiates the contractile response to ACh in isolated human bronchial segments (HASMs) by approximately 27 % compared with the vehicle [85].

Of the many inflammatory cytokines, TNF-α may be the most potent modulator of Ca^{2+} homeostasis in ASMCs. Several mechanisms that underlie the potentiating effect of TNF-α on agonist-induced $[Ca^{2+}]_i$ are described below (Fig. 1).

2.1 *Ca^{2+} Release from Intracellular Ca^{2+} Store*

Since the early 1990s, Amrani et al. have shown that TNF-α has direct effects on HASMC contractility in a nonspecific and TNF-α-receptor-dependent manner. TNF-α (10 ng/mL for 24 h) enhances carbachol- [2], bradykinin- [1, 4], and thrombin-induced [4] increases in $[Ca^{2+}]_i$, mainly via its cell-surface receptor TNFR1 (55 kDa) [3]. The nonspecific potentiation of agonist-induced $[Ca^{2+}]_i$ by TNF-α may be explained partly by its direct activation of G-protein-mediated signal transduction in TNF-α-treated HASMCs. Ca^{2+} mobilization in response to NaF, an agent that directly activates G proteins, is enhanced by TNF-α. Furthermore, the incubation of HASMCs with TNF-α (10 ng/mL for 72 h) significantly upregulates Gqα protein expression by 39 % and $G\alpha i_2$ protein expression by 107 % [33].

Kannan et al. highlighted the role of CD38 in the potentiation of agonist-induced $[Ca^{2+}]_i$ by TNF-α. As described earlier, CD38 expression is upregulated in TNF-α-treated HASMCs. One of the studies of the mechanisms underlying CD38 induction by TNF-α [39, 40, 88, 89] showed that the extent of CD38 induction by TNF-α in ASMCs from asthma patients was different from that in ASMCs from healthy subjects [37]. In basal conditions, there was no detectable CD38 mRNA expression or ADP-ribosyl cyclase activity in the ASMCs of subjects with or without asthma. After stimulation with 10 and 20 ng/mL TNF-α, however, CD38 expression and ADP-ribosyl cyclase activity in ASMCs from patients with asthma were significantly higher than those in ASMCs from healthy subjects. Thus, the upregulation of CD38 by TNF-α could be a target for asthma treatments.

Glucocorticoid, a key treatment used for asthma, attenuates TNF-α-induced CD38 expression and ADP-ribosyl cyclase activity, potentially via the inhibition of NF-κB and the induction of MKP-1, which dephosphorylates and inactivates activated MAPKs [41]. It is notable that the steroid sensitivity of CD38 induced by TNF-α is abrogated when HASMCs are treated concurrently with TNF-α and IFN-γ, i.e., cytokines that are known to be elevated in COPD. The concurrent treatment with TNF-α and IFN-γ induces the decoy glucocorticoid receptor β. As a consequence, fluticasone fails to induce the transcription of the glucocorticoid-responsive element-dependent gene and suppresses CD38 expression [91].

2.2 Store-Operated Ca^{2+} Entry

Exposure to TNF-α enhances SOCE. The enhanced SOCE in TNF-α-treated HASMCs (20 ng/mL, 24 h) is attenuated by knockdown of the expression of TRPC3 [95] and CD38 [82], but not TRPC1, 4, 5, or 6. Thus, TRPC3 and CD38 are responsible for SOCE in TNF-α-treated HASMCs. In contrast to SOCE, the receptor-operated Ca^{2+} entry (ROCE) activated by DAG that elicits nonselective cation entry (Na^+ and Ca^{2+} entry) is reduced in TNF-α-treated HASMCs [95]. TRPC is colocalized with caveolin-1 [69] and interacts with Orai1 [45], so the association between caveolae and SOCE has been explored further, as described below.

Exposure to TNF-α (20 ng/mL, 48 h) significantly increases SOCE after SR Ca^{2+} depletion, which is accompanied by an increase in the expression of two major SOCE-related molecules, Orai1 and STIM1, without changing their distributions [78]. Orai1 is localized mainly in the plasma membrane, while STIM1 is localized in the SR. The TNF-α-mediated enhancement of SOCE is thought to be mediated via Orai1 because Orai1 blocking with siRNA substantially attenuates SOCE. The expression of Orai1 and SOCE is decreased significantly when caveolin-1 expression is inhibited using siRNA, whereas caveolin-1 overexpression increases the expression of Orai1 and SOCE [78]. These findings support the potential interactions between TNF-α, SOCE-related proteins, and caveolae.

2.3 Ca^{2+} Sensitivity and Caveolae

Exposure to TNF-α also enhances Ca^{2+} sensitivity. Long-term exposure and brief exposure (30–45 min) to TNF-α enhances the agonist-induced contraction of bovine [64] or guinea-pig [66] tracheal smooth muscle. In a study of rats, AHR was increased by as early as 90 min after exposure to aerosolized recombinant human TNF-α [46]. These observations indicate that the induction of Ca^{2+} sensitivity by TNF-α is not necessarily mediated by de novo protein synthesis or by the Ca^{2+} mobilization induced by longer pretreatment. In HASMCs, TNF-α increases the Rho-kinase-mediated phosphorylation of MYPT-1, a subunit of MLCP, which is followed by an increase in Ca^{2+}-activated MLC phosphorylation [34]. MYPT-1 phosphorylation, which inhibits MLCP activity and increases Ca^{2+} sensitivity, reaches its maximal 4.7-fold activation within 5 min, while there is a slight decrease by 15 min, although it remains elevated even after 60 min. This TNF-α-induced Ca^{2+} sensitization is inhibited by the RhoA/Rho-kinase inhibitor Y27632. CPI-17, another pathway of Ca^{2+} sensitivity, is also involved in the induction of Ca^{2+} sensitivity by TNF-α [62].

A topological study of TNFR1 showed that the lipid rafts of caveolae were essential for the TNF-α-mediated activation of RhoA [35]. In HASMCs, TNFR1 is localized to the lipid raft and nonraft regions of the plasma membrane, while the TNFR1 from lipid rafts coprecipitated with caveolin-1. After TNF-α binding to TNFR1, the specific adaptor proteins TRADD, TRAF-2, and RIP are recruited into the lipid rafts, which then activate the RhoA/Rho-kinase pathway because RhoA is constitutively localized in lipid rafts. Depletion of cholesterol from the rafts causes a redistribution of TNFR1 to the nonraft plasma membrane and prevents TNF-α-mediated activation of RhoA. The NF-κB and MAPK pathways are not affected by the depletion of cholesterol, suggesting that these pathways are independent of the lipid rafts, although they are similarly activated by TNF-α.

Caveolin-1 is involved in TNF-α-induced Ca^{2+} sensitization, which is mediated via RhoA. Pretreatment of porcine ASM strips with TNF-α (50 ng/mL, overnight) enhances Ca^{2+} sensitization and force generation in response to Ach, which is decreased by RhoA inhibition. Disruption of caveolae further attenuates Ca^{2+} sensitization and force generation induced by TNF-α pretreatment [77]. In HASMCs, exposure to TNF-α (20 ng/mL for 24 h) increases the expression of caveolin-1 and 2, while disruption of caveolae leads to a significant decrease in agonist-induced Ca^{2+} responses [76]. Overall, lipid rafts and caveolin-1 play a significant role in TNF-α-induced Ca^{2+} sensitization.

3 IL-13

IL-13, a representative Th2 cytokine that is crucial in asthma, is involved in airway inflammation, remodeling, and steroid insensitivity [73]. The clinical administration of anti-IL-13 antibody improves airflow limitations in patients with asthma who are inadequately controlled by inhaled corticosteroid treatments [18]. It has been established that IL-13 plays a critical role in allergen-induced AHR and Ca^{2+} mobilization in HASMCs (Fig. 2).

3.1 Ca^{2+} Release from Intracellular Ca^{2+} Store

With the exception of one report [26], it has not been shown that IL-13 alone can change the baseline levels of $[Ca^{2+}]_i$ [63] or the resting stiffness of HASMCs [72], whereas exposure to IL-13 increases agonist-induced Ca^{2+} transients and force generation. Exposure to IL-13 (100 ng/mL, 24 h) increases the carbachol- and KCl-induced maximal force generation in murine tracheal rings. In cultured HASMCs, exposure to IL-13 (50 ng/mL, 24 h) augments bradykinin-, histamine-, and carbachol-induced increases in $[Ca^{2+}]_i$ by 60 %, 35 %, and 26 %, respectively [90]. Pretreatment with a lower concentration of IL-13 (10 ng/mL, 24 h) also increases histamine-induced $[Ca^{2+}]_i$ transients and increases cell stiffness, which is a proxy for force generation in cells, in cultured HASMCs [72]. Taken together, IL-13 and TNF-α may prime ASMCs for nonspecific increases in $[Ca^{2+}]_i$.

TNF-α does not increase the cell-surface expression of muscarinic receptor [2], whereas exposure to IL-13 upregulates the expression of the histamine H1 receptor [72] and cysteinyl leukotriene receptor (CysLT1R) [22, 59], which contributes to increases in agonist-induced $[Ca^{2+}]_i$ by IL-13.

As with TNF-α, CD38 plays a crucial role in potentiating the effect of IL-13 on agonist-induced $[Ca^{2+}]_i$ [20, 29, 59]. Exposure to IL-13 (50 ng/mL, 22 h) increases the expression of CD38 protein and the ADP-ribosyl cyclase activity in HASMCs, so the bradykinin-, thrombin-, and histamine-induced increase in $[Ca^{2+}]_i$ is enhanced in IL-13-pretreated HASMCs. This potentiating effect of IL-13 on agonist-induced $[Ca^{2+}]_i$ is abolished by the cADPR antagonist 8Br-cADPR [20]. In CD38-deficient mice, intranasal exposure to IL-13 reduces airway reactivity to inhaled methacholine and the rate of force generation with carbachol compared with wild-type mice[29].

The effect of IL-13 on $[Ca^{2+}]_i$ transients is bell-shaped, with its maximum at 50 ng/mL, followed by a decline when HASMCs are preincubated with high doses (100 ng/mL) of IL-13 [43]. HASMCs constitutively express the receptor complex IL-4Rα/IL-13Rα1, which mediates IL-4/IL-13 signaling. HASMCs also express IL-13Rα2, a decoy receptor with high affinity for IL-13. If HASMCs are incubated with IL-13 (0, 50, and 100 ng/mL), the expression of IL-13Rα2 increases in a concentration-dependent manner, whereas the expression of IL-13Rα1 does not differ with 50 or 100 ng/mL of IL-13. An imbalance between the expression of IL-13Rα1 and IL-13Rα2 may reduce the effect of high doses of IL-13 on $[Ca^{2+}]_i$ transients [43].

The Janus kinase (JAK)/signal transducer and activator of transcription (STAT) 6 pathway has been proposed as the main pathway in IL-13 signaling. However, the JAK2 inhibitor AG490, STAT6 inhibitor leflunomide, and antisense oligonucleotide for STAT6 fail to inhibit the potentiation of histamine-induced increase in $[Ca^{2+}]_i$ by IL-13. Instead, concurrent inhibition of c-Jun N-terminal kinases (JNK) and extracellular-signal-regulated kinases (ERKs) prevents the IL-13-induced potentiation of $[Ca^{2+}]_i$ transients. These findings suggest little involvement of the JAK/STAT6 pathway in the potentiation of histamine-induced increases in $[Ca^{2+}]_i$ by IL-13 compared with the secretion of eotaxin by IL-13 [67].

3.2 Store-Operated Ca^{2+} Entry

SOCE is also upregulated in T-bet knockout mice that mimic an IL-13-enriched milieu compared with wild-type mice [14]. In HASMCs, exposure to IL-13 (10 ng/mL, 24 h) augments the basal $[Ca^{2+}]_i$, Ca^{2+} release from the SR, and the Ca^{2+} influx via SOCE. Unlike TNF-α, IL-13 does not increase Orai1, STIM1, or TRPC1 at the mRNA level. The enhanced Ca^{2+} release from SR and SOCE with IL-13 is attenuated by dexamethasone and salbutamol, but not salmeterol. The inhibitory effect of salbutamol on SOCE, which is augmented by IL-13, is mediated by the cAMP/PKA pathway [26].

Matsumoto and Hirata et al. demonstrated that exposure to IL-13 (10 ng/mL, 24 h) enhanced the frequency of leukotriene D4 (LTD4)-induced Ca^{2+} oscillations in cultured HASMCs and increased the number of oscillating cells with upregulated expression of CysLT1R and CD38. Both xestospongin C, a specific inhibitor of IP_3 receptors, and ryanodine or ruthenium red, inhibitors of RyR, partially blocked LTD4-induced Ca^{2+} oscillations, while LTD4-induced Ca^{2+} oscillations were inhibited completely by 50 μM of 2-APB, which dominantly blocks SOCE at this concentration. These findings suggest that IL-13 cooperatively modulates the IP_3 receptor and RyR systems, and possibly SOCE, which leads to enhanced frequency of LTD4-induced Ca^{2+} oscillations [59].

3.3 Ca^{2+} Sensitivity and Caveolae

Ca^{2+} sensitization is also involved in the potentiation of the agonist-induced increases in $[Ca^{2+}]_i$ by IL-13, although the evidence is limited. Exposure to IL-13 increases RhoA protein expression by approximately threefold in cultured HASMCs. The IL-13-induced increase in RhoA expression is accompanied by the augmentation of ACh-induced contraction in murine bronchial tissues and the enhancement of AHR in naïve BALB/c mice. Increased RhoA expression is inhibited by the STAT6 inhibitor leflunomide [17].

Similar to TNF-α pretreatment, pretreatment with IL-13 (50 ng/mL for 24 h) enhances caveolin-1 and 2 expression, while suppression of caveolin-1 decreases the agonist (ACh, histamine, and bradykinin)-induced increase of $[Ca^{2+}]_i$ in IL-13-treated HASMCs [76]. However, the extent of the modulation of caveolin-1 expression by IL-13 is lower than that with TNF-α pretreatment (20 ng/mL, 24 h). The difference between TNF-α and IL-13 may reflect differences in their distribution or the regulation of receptors for each cytokine. TNFR1 is localized to the lipid raft/caveolae and upregulated after 24 h exposure to TNF-α, whereas the receptor complex L-4Rα/IL-13Rα1, a receptor of IL-13, is absent from caveolae and is not upregulated by exposure to TNF-α or IL-13.

4 IL-4

IL-4 is another representative Th2 cytokine that shares a receptor complex IL-4Rα/IL-13Rα1 with IL-13. In contrast to IL-13, however, the effects of IL-4 on Ca^{2+} homeostasis are largely inhibitory.

In bovine ASMCs, brief exposure to IL-4 (50 ng/mL for 20 min), but not IL-13, inhibits Ca^{2+} transients in response to carbachol by depleting the SR Ca^{2+} via RyR [23, 24, 53]. The inhibitory effect of IL-4 may be partly explained by its stimulatory effect on rapid large increases in BK_{Ca} channels. BK_{Ca} channels are plasma membrane ion channels that are highly sensitive to an increase in $[Ca^{2+}]_i$. When they are activated, the BK_{Ca} channels allow the efflux of K^+ from the cell, which leads to cell hyperpolarization and a decrease in cell excitability (Fig. 3). It is notable that IL-13 has little effect on BK_{Ca} channel activity and that it inhibits the effect of IL-4 [57].

5 Transforming Growth Factor (TGF)-β

TGF-β, a pleiotropic cytokine, induces a contractile phenotype in HASMCs. TGF-β increases the cell size, total protein synthesis, expression of α-smooth muscle actin and smooth muscle myosin heavy chain, formation of actomyosin filaments, and cell shortening in response to ACh in HASMCs [28]. However, there is limited evidence of its effects on Ca^{2+} homeostasis.

Exposure to TGF-β (12.5 ng/mL, 18 h) significantly enhances bradykinin-induced Ca^{2+} signals in cultured HASMCs [44]. This potentiating effect of TGF-β on agonist-induced $[Ca^{2+}]_i$ potentially occurs via the upregulation of bradykinin B2 receptor protein expression.

6 IL-1β

IL-1β is another proinflammatory cytokine that is particularly important in rhinovirus infections [27], which are a common trigger of asthma and COPD exacerbations.

The effect of IL-1β on HASMC contractility is inconsistent and depends on the agonists applied. Exposure to IL-1β increases the contractile response to substance P [13] but has no effect with ACh [85], while it decreases the contractile response to histamine [71]. IL-1β has potent inhibitory effects on the isoproterenol-induced accumulation of cAMP and relaxation in HASMCs via the desensitization of β_2AR [80, 81] or via the upregulation of Gi protein [30]. However, the effects of IL-1β on Ca^{2+} homeostasis remain elusive.

7 IFN-γ

IFN-γ is a representative Th1 cytokine that plays an important role in COPD, which is also critical for innate and adaptive immunity against viral and intracellular bacterial infections.

IFN-γ increases the neurokinin A/neurokinin 2 receptor expression and increases methacholine-induced AHR in mice [47]. Exposure to IFN-γ also increases LTD4-induced, but not bradykinin-induced, cell stiffness by increasing the cell-surface expression of CysLT1R in HASMCs [7]. However, the effects of IFN-γ on Ca^{2+} homeostasis remain unknown.

8 Thymic Stromal Lymphopoietin (TSLP)

TSLP, a recently identified IL-7-like cytokine, is genetically associated with asthma [32]. TSLP leads to dendritic cells favoring and maintaining Th2 immunity, but TSLP is induced via smoking in airway epithelial cells and is also upregulated in the airways of patients with COPD [96].

Both overnight exposure to cigarette smoking extract and exposure to TSLP potentiated a histamine-induced increase in $[Ca^{2+}]_i$. Each response was attenuated by TSLP and TSLP-receptor antibodies, which suggests there is a functional linkage between the cigarette-smoke-induced increase in ASM contractility and TSLP [83].

9 Conclusions

Recently, our understanding of the mechanisms underlying the altered Ca^{2+} homeostasis in TNF-α- and IL-13-enriched conditions has deepened. However, the alterations of Ca^{2+} homeostasis modulated by other relevant cytokines in severe asthma and COPD such as IL-17A, which enhances HASMCs contraction [48] and mediates steroid-resistant airway inflammation and AHR [60], remain unknown. Further pharmacological studies are warranted that target the altered Ca^{2+} homeostasis induced by crosstalk between inflammatory cytokines and HASMCs, particularly in severe asthma and COPD.

References

1. Amrani Y, Bronner C (1993) Tumor necrosis factor α potentiates the increase in cytosolic free calcium induced by bradykinin in guinea-pig tracheal smooth muscle cells. C R Acad Sci III 316: 1489–1494.
2. Amrani Y, Martinet N, Bronner C (1995) Potentiation by tumour necrosis factor-α of calcium signals induced by bradykinin and carbachol in human tracheal smooth muscle cells. Br J Pharmacol 114: 4–5.

3. Amrani Y, Panettieri RA, Jr., Frossard N, Bronner C (1996) Activation of the TNF α-p55 receptor induces myocyte proliferation and modulates agonist-evoked calcium transients in cultured human tracheal smooth muscle cells. Am J Respir Cell Mol Biol 15: 55–63.
4. Amrani Y, Krymskaya V, Maki C, Panettieri RA, Jr. (1997) Mechanisms underlying TNF-α effects on agonist-mediated calcium homeostasis in human airway smooth muscle cells. Am J Physiol 273: L1020–1028.
5. Amrani Y, Panettieri RA, Jr. (1998) Cytokines induce airway smooth muscle cell hyperresponsiveness to contractile agonists. Thorax 53: 713–716.
6. Amrani Y, Chen H, Panettieri RA, Jr. (2000) Activation of tumor necrosis factor receptor 1 in airway smooth muscle: a potential pathway that modulates bronchial hyper-responsiveness in asthma? Respir Res 1: 49–53.
7. Amrani Y, Moore PE, Hoffman R, Shore SA, Panettieri RA, Jr. (2001) Interferon-γ modulates cysteinyl leukotriene receptor-1 expression and function in human airway myocytes. Am J Respir Crit Care Med 164: 2098–2101.
8. Anticevich SZ, Hughes JM, Black JL, Armour CL (1995) Induction of human airway hyperresponsiveness by tumour necrosis factor-α. Eur J Pharmacol 284: 221–225.
9. Ay B, Prakash YS, Pabelick CM, Sieck GC (2004) Store-operated Ca^{2+} entry in porcine airway smooth muscle. Am J Physiol Lung Cell Mol Physiol 286: L909–917.
10. Ay B, Iyanoye A, Sieck GC, Prakash YS, Pabelick CM (2006) Cyclic nucleotide regulation of store-operated Ca^{2+} influx in airway smooth muscle. Am J Physiol Lung Cell Mol Physiol 290: L278–283.
11. Bai Y, Sanderson MJ (2006) Airway smooth muscle relaxation results from a reduction in the frequency of Ca^{2+} oscillations induced by a cAMP-mediated inhibition of the IP_3 receptor. Respir Res 7: 34.
12. Bai Y, Edelmann M, Sanderson MJ (2009) The contribution of inositol 1,4,5-trisphosphate and ryanodine receptors to agonist-induced Ca^{2+} signaling of airway smooth muscle cells. Am J Physiol Lung Cell Mol Physiol 297: L347–361.
13. Barchasz E, Naline E, Molimard M, Moreau J, Georges O, Emonds-Alt X, Advenier C (1999) Interleukin-1β-induced hyperresponsiveness to [Sar^9,$Met(O_2)^{11}$]substance P in isolated human bronchi. Eur J Pharmacol 379: 87–95.
14. Bergner A, Kellner J, Silva AK, Gamarra F, Huber RM (2006) Ca^{2+}-signaling in airway smooth muscle cells is altered in T-bet knock-out mice. Respir Res 7: 33.
15. Billington CK, Penn RB (2003) Signaling and regulation of G protein-coupled receptors in airway smooth muscle. Respir Res 4: 2.
16. Brightling CE, Bradding P, Symon FA, Holgate ST, Wardlaw AJ, Pavord ID (2002) Mast-cell infiltration of airway smooth muscle in asthma. N Engl J Med 346: 1699–1705.
17. Chiba Y, Nakazawa S, Todoroki M, Shinozaki K, Sakai H, Misawa M (2009) Interleukin-13 augments bronchial smooth muscle contractility with an up-regulation of RhoA protein. Am J Respir Cell Mol Biol 40: 159–167.
18. Corren J, Lemanske RF, Hanania NA, Korenblat PE, Parsey MV, Arron JR, Harris JM, Scheerens H, Wu LC, Su Z, Mosesova S, Eisner MD, Bohen SP, Matthews JG (2011) Lebrikizumab treatment in adults with asthma. N Engl J Med 365: 1088–1098.
19. Deshpande DA, Walseth TF, Panettieri RA, Kannan MS (2003) CD38/cyclic ADP-ribose-mediated Ca^{2+} signaling contributes to airway smooth muscle hyper-responsiveness. FASEB J 17: 452–454.
20. Deshpande DA, Dogan S, Walseth TF, Miller SM, Amrani Y, Panettieri RA, Kannan MS (2004) Modulation of calcium signaling by interleukin-13 in human airway smooth muscle: role of CD38/cyclic adenosine diphosphate ribose pathway. Am J Respir Cell Mol Biol 31: 36–42.
21. Deshpande DA, White TA, Dogan S, Walseth TF, Panettieri RA, Kannan MS (2005) CD38/cyclic ADP-ribose signaling: role in the regulation of calcium homeostasis in airway smooth muscle. Am J Physiol Lung Cell Mol Physiol 288: L773–788.

22. Espinosa K, Bosse Y, Stankova J, Rola-Pleszczynski M (2003) CysLT1 receptor upregulation by TGF-β and IL-13 is associated with bronchial smooth muscle cell proliferation in response to LTD4. J Allergy Clin Immunol 111: 1032–1040.
23. Ethier MF, Cappelluti E, Madison JM (2005) Mechanisms of interleukin-4 effects on calcium signaling in airway smooth muscle cells. J Pharmacol Exp Ther 313: 127–133.
24. Ethier MF, Madison JM (2006) IL-4 inhibits calcium transients in bovine trachealis cells by a ryanodine receptor-dependent mechanism. Faseb J 20: 154–156.
25. Gally F, Hartney JM, Janssen WJ, Perraud AL (2009) CD38 plays a dual role in allergen-induced airway hyperresponsiveness. Am J Respir Cell Mol Biol 40: 433–442.
26. Gao YD, Zou JJ, Zheng JW, Shang M, Chen X, Geng S, Yang J (2010) Promoting effects of IL-13 on Ca^{2+} release and store-operated Ca^{2+} entry in airway smooth muscle cells. Pulm Pharmacol Ther 23: 182–189.
27. Gern JE, Busse WW (1999) Association of rhinovirus infections with asthma. Clin Microbiol Rev 12: 9–18.
28. Goldsmith AM, Bentley JK, Zhou L, Jia Y, Bitar KN, Fingar DC, and Hershenson MB (2006) Transforming growth factor-β induces airway smooth muscle hypertrophy. Am J Respir Cell Mol Biol 34: 247–254.
29. Guedes AG, Paulin J, Rivero-Nava L, Kita H, Lund FE, Kannan MS (2006) CD38-deficient mice have reduced airway hyperresponsiveness following IL-13 challenge. Am J Physiol Lung Cell Mol Physiol 291: L1286–1293.
30. Hakonarson H, Herrick DJ, Serrano PG, Grunstein MM (1996) Mechanism of cytokine-induced modulation of β-adrenoceptor responsiveness in airway smooth muscle. J Clin Invest 97: 2593–2600.
31. Hakonarson H, Maskeri N, Carter C, Hodinka RL, Campbell D, Grunstein MM (1998) Mechanism of rhinovirus-induced changes in airway smooth muscle responsiveness. J Clin Invest 102: 1732–1741.
32. Hirota T, Takahashi A, Kubo M, Tsunoda T, Tomita K, Doi S, Fujita K, Miyatake A, Enomoto T, Miyagawa T, Adachi M, Tanaka H, Niimi A, Matsumoto H, Ito I, Masuko H, Sakamoto T, Hizawa N, Taniguchi M, Lima JJ, Irvin CG, Peters SP, Himes BE, Litonjua AA, Tantisira KG, Weiss ST, Kamatani N, Nakamura Y, Tamari M (2011) Genome-wide association study identifies three new susceptibility loci for adult asthma in the Japanese population. Nat Genet 43: 893–896.
33. Hotta K, Emala CW, Hirshman CA (1999) TNF-α upregulates Giα and Gqα protein expression and function in human airway smooth muscle cells. Am J Physiol 276: L405–411.
34. Hunter I, Cobban HJ, Vandenabeele P, MacEwan DJ, Nixon GF (2003) Tumor necrosis factor-α-induced activation of RhoA in airway smooth muscle cells: role in the Ca^{2+} sensitization of myosin light chain20 phosphorylation. Mol Pharmacol 63: 714–721.
35. Hunter I, Nixon GF (2006) Spatial compartmentalization of tumor necrosis factor (TNF) receptor 1-dependent signaling pathways in human airway smooth muscle cells. Lipid rafts are essential for TNF-α-mediated activation of RhoA but dispensable for the activation of the NF-κB and MAPK pathways. J Biol Chem 281: 34705–34715.
36. Janssen LJ, Killian K (2006) Airway smooth muscle as a target of asthma therapy: history and new directions. Respir Res 7: 123.
37. Jude JA, Solway J, Panettieri RA, Jr., Walseth TF, Kannan MS Differential induction of CD38 expression by TNF-α in asthmatic airway smooth muscle cells. Am J Physiol Lung Cell Mol Physiol 299: L879–890.
38. Jude JA, Wylam ME, Walseth TF, Kannan MS (2008) Calcium signaling in airway smooth muscle. Proc Am Thorac Soc 5: 15–22.
39. Kang BN, Deshpande DA, Tirumurugaan KG, Panettieri RA, Walseth TF, Kannan MS (2005) Adenoviral mediated anti-sense CD38 attenuates TNF-α-induced changes in calcium homeostasis of human airway smooth muscle cells. Can J Physiol Pharmacol 83: 799–804.
40. Kang BN, Tirumurugaan KG, Deshpande DA, Amrani Y, Panettieri RA, Walseth TF, Kannan MS (2006) Transcriptional regulation of CD38 expression by tumor necrosis factor-α in human

airway smooth muscle cells: role of NF-κB and sensitivity to glucocorticoids. FASEB J 20: 1000–1002.
41. Kang BN, Jude JA, Panettieri RA, Jr., Walseth TF, Kannan MS (2008) Glucocorticoid regulation of CD38 expression in human airway smooth muscle cells: role of dual specificity phosphatase 1. Am J Physiol Lung Cell Mol Physiol 295: L186–193.
42. Kannan MS, Prakash YS, Brenner T, Mickelson JR, Sieck GC (1997) Role of ryanodine receptor channels in Ca^{2+} oscillations of porcine tracheal smooth muscle. Am J Physiol 272: L659–664.
43. Kellner J, Gamarra F, Welsch U, Jorres RA, Huber RM, Bergner A (2007) IL-13Rα2 reverses the effects of IL-13 and IL-4 on bronchial reactivity and acetylcholine-induced Ca^{2+} signaling. Int Arch Allergy Immunol 142: 199–210.
44. Kim JH, Jain D, Tliba O, Yang B, Jester WF, Jr., Panettieri RA, Jr., Amrani Y, Pure E (2005) TGF-β potentiates airway smooth muscle responsiveness to bradykinin. Am J Physiol Lung Cell Mol Physiol 289: L511–520.
45. Kim MS, Zeng W, Yuan JP, Shin DM, Worley PF, Muallem S (2009) Native Store-operated Ca^{2+} Influx Requires the Channel Function of Orai1 and TRPC1. J Biol Chem 284: 9733–9741.
46. Kips JC, Tavernier J, Pauwels RA (1992) Tumor necrosis factor causes bronchial hyperresponsiveness in rats. Am Rev Respir Dis 145: 332–336.
47. Kobayashi M, Ashino S, Shiohama Y, Wakita D, Kitamura H, Nishimura T (2012) IFN-γ elevates airway hyper-responsiveness via up-regulation of neurokinin A/neurokinin-2 receptor signaling in a severe asthma model. Eur J Immunol 42: 393–402.
48. Kudo M, Melton AC, Chen C, Engler MB, Huang KE, Ren X, Wang Y, Bernstein X, Li JT, Atabai K, Huang X, Sheppard D (2012) IL-17A produced by αβ T cells drives airway hyper-responsiveness in mice and enhances mouse and human airway smooth muscle contraction. Nat Med 18: 547–554.
49. Kume H, Hall IP, Washabau RJ, Takagi K, Kotlikoff MI (1994) β-adrenergic agonists regulate K_{Ca} channels in airway smooth muscle by cAMP-dependent and -independent mechanisms. J Clin Invest 93: 371–379.
50. Laporte JC, Moore PE, Baraldo S, Jouvin MH, Church TL, Schwartzman IN, Panettieri RA, Jr., Kinet JP, Shore SA (2001) Direct effects of interleukin-13 on signaling pathways for physiological responses in cultured human airway smooth muscle cells. Am J Respir Crit Care Med 164: 141–148.
51. Lewis RS (2007) The molecular choreography of a store-operated calcium channel. Nature 446: 284–287.
52. Ma X, Cheng Z, Kong H, Wang Y, Unruh H, Stephens NL, Laviolette M (2002) Changes in biophysical and biochemical properties of single bronchial smooth muscle cells from asthmatic subjects. Am J Physiol Lung Cell Mol Physiol 283: L1181–1189.
53. Madison JM, Ethier MF (2001) Interleukin-4 rapidly inhibits calcium transients in response to carbachol in bovine airway smooth muscle cells. Am J Respir Cell Mol Biol 25: 239–244.
54. Mahn K, Hirst SJ, Ying S, Holt MR, Lavender P, Ojo OO, Siew L, Simcock DE, McVicker CG, Kanabar V, Snetkov VA, O'Connor BJ, Karner C, Cousins DJ, Macedo P, Chung KF, Corrigan CJ, Ward JP, Lee TH (2009) Diminished sarco/endoplasmic reticulum Ca^{2+} ATPase (SERCA) expression contributes to airway remodelling in bronchial asthma. Proc Natl Acad Sci U S A 106: 10775–10780.
55. Mahn K, Ojo OO, Chadwick G, Aaronson PI, Ward JP, Lee TH (2010) Ca^{2+} homeostasis and structural and functional remodelling of airway smooth muscle in asthma. Thorax 65: 547–552.
56. Marthan R (2004) Store-operated calcium entry and intracellular calcium release channels in airway smooth muscle. Am J Physiol Lung Cell Mol Physiol 286: L907–908.
57. Martin G, O'Connell RJ, Pietrzykowski AZ, Treistman SN, Ethier MF, Madison JM (2008) Interleukin-4 activates large-conductance, calcium-activated potassium (BK_{Ca}) channels in human airway smooth muscle cells. Exp Physiol 93: 908–918.

58. Matsumoto H, Moir LM, Oliver BG, Burgess JK, Roth M, Black JL, McParland BE (2007) Comparison of gel contraction mediated by airway smooth muscle cells from patients with and without asthma. Thorax 62: 848–854.
59. Matsumoto H, Hirata Y, Otsuka K, Iwata T, Inazumi A, Niimi A, Ito I, Ogawa E, Muro S, Sakai H, Chin K, Oku Y, Mishima M (2012) Interleukin-13 enhanced Ca^{2+} oscillations in airway smooth muscle cells. Cytokine 57: 19–24.
60. McKinley L, Alcorn JF, Peterson A, Dupont RB, Kapadia S, Logar A, Henry A, Irvin CG, Piganelli JD, Ray A, Kolls JK (2008) TH17 cells mediate steroid-resistant airway inflammation and airway hyperresponsiveness in mice. J Immunol 181: 4089–4097.
61. Moore PE, Lahiri T, Laporte JD, Church T, Panettieri RA, Jr., Shore SA (2001) Selected contribution: synergism between TNF-α and IL-1 β in airway smooth muscle cells: implications for β-adrenergic responsiveness. J Appl Physiol 91: 1467–1474.
62. Morin C, Sirois M, Echave V, Rousseau E (2008) CPI-17 silencing-reduced responsiveness in control and TNF-α-treated human bronchi. Am J Respir Cell Mol Biol 39: 638–643.
63. Moynihan B, Tolloczko B, Michoud MC, Tamaoka M, Ferraro P, Martin JG (2008) MAP kinases mediate interleukin-13 effects on calcium signaling in human airway smooth muscle cells. Am J Physiol Lung Cell Mol Physiol 295: L171–177.
64. Nakatani Y, Nishimura Y, Nishiuma T, Maeda H, Yokoyama M (2000) Tumor necrosis factor-α augments contraction and cytosolic Ca^{2+} sensitivity through phospholipase A_2 in bovine tracheal smooth muscle. Eur J Pharmacol 392: 175–182.
65. Nogami M, Romberger DJ, Rennard SI, Toews ML (1994) TGF-β 1 modulates β-adrenergic receptor number and function in cultured human tracheal smooth muscle cells. Am J Physiol 266: L187–191.
66. Parris JR, Cobban HJ, Littlejohn AF, MacEwan DJ, Nixon GF (1999) Tumour necrosis factor-α activates a calcium sensitization pathway in guinea-pig bronchial smooth muscle. J Physiol 518 (Pt 2): 561–569.
67. Peng Q, Matsuda T, Hirst SJ (2004) Signaling pathways regulating interleukin-13-stimulated chemokine release from airway smooth muscle. Am J Respir Crit Care Med 169: 596–603.
68. Prakash YS, Kannan MS, Walseth TF, Sieck GC (1998) Role of cyclic ADP-ribose in the regulation of $[Ca^{2+}]i$ in porcine tracheal smooth muscle. Am J Physiol 274: C1653–1660.
69. Prakash YS, Thompson MA, Vaa B, Matabdin I, Peterson TE, He T, Pabelick CM (2007) Caveolins and intracellular calcium regulation in human airway smooth muscle. Am J Physiol Lung Cell Mol Physiol 293: L1118–1126.
70. Prakash YS, Sathish V, Thompson MA, Pabelick CM, Sieck GC (2009) Asthma and sarcoplasmic reticulum Ca^{2+} reuptake in airway smooth muscle. Am J Physiol Lung Cell Mol Physiol 297: L794.
71. Pype JL, Xu H, Schuermans M, Dupont LJ, Wuyts W, Mak JC, Barnes PJ, Demedts MG, Verleden GM (2001) Mechanisms of interleukin 1β-induced human airway smooth muscle hyporesponsiveness to histamine. Involvement of p38 MAPK NF-κB. Am J Respir Crit Care Med 163: 1010–1017.
72. Risse PA, Jo T, Suarez F, Hirota N, Tolloczko B, Ferraro P, Grutter P, Martin JG (2011) Interleukin-13 inhibits proliferation and enhances contractility of human airway smooth muscle cells without change in contractile phenotype. Am J Physiol Lung Cell Mol Physiol 300: L958–966.
73. Saha SK, Berry MA, Parker D, Siddiqui S, Morgan A, May R, Monk P, Bradding P, Wardlaw AJ, Pavord ID, Brightling CE (2008) Increased sputum and bronchial biopsy IL-13 expression in severe asthma. J Allergy Clin Immunol 121: 685–691.
74. Sanderson MJ, Delmotte P, Bai Y, Perez-Zogbhi JF (2008) Regulation of airway smooth muscle cell contractility by Ca^{2+} signaling and sensitivity. Proc Am Thorac Soc 5: 23–31.
75. Sathish V, Thompson MA, Bailey JP, Pabelick CM, Prakash YS, Sieck GC (2009) Effect of proinflammatory cytokines on regulation of sarcoplasmic reticulum Ca^{2+} reuptake in human airway smooth muscle. Am J Physiol Lung Cell Mol Physiol 297: L26–34.

76. Sathish V, Abcejo AJ, VanOosten SK, Thompson MA, Prakash YS, Pabelick CM (2011) Caveolin-1 in cytokine-induced enhancement of intracellular Ca^{2+} in human airway smooth muscle. Am J Physiol Lung Cell Mol Physiol 301: L607–614.
77. Sathish V, Yang B, Meuchel LW, VanOosten SK, Ryu AJ, Thompson MA, Prakash YS, Pabelick CM (2011) Caveolin-1 and force regulation in porcine airway smooth muscle. Am J Physiol Lung Cell Mol Physiol 300: L920–929.
78. Sathish V, Abcejo AJ, Thompson MA, Sieck GC, Prakash YS, Pabelick CM (2012) Caveolin-1 regulation of store-operated Ca^{2+} influx in human airway smooth muscle. Eur Respir J 40: 470–478.
79. Seasholtz TM, Majumdar M, Brown JH (1999) Rho as a mediator of G protein-coupled receptor signaling. Mol Pharmacol 55: 949–956.
80. Shore SA, Laporte J, Hall IP, Hardy E, Panettieri RA, Jr. (1997) Effect of IL-1 β on responses of cultured human airway smooth muscle cells to bronchodilator agonists. Am J Respir Cell Mol Biol 16: 702–712.
81. Shore SA, Moore PE (2002) Effects of cytokines on contractile and dilator responses of airway smooth muscle. Clin Exp Pharmacol Physiol 29: 859–866.
82. Sieck GC, White TA, Thompson MA, Pabelick CM, Wylam ME, Prakash YS (2008) Regulation of store-operated Ca^{2+} entry by CD38 in human airway smooth muscle. Am J Physiol Lung Cell Mol Physiol 294: L378–385.
83. Smelter DF, Sathish V, Thompson MA, Pabelick CM, Vassallo R, Prakash YS (2010) Thymic stromal lymphopoietin in cigarette smoke-exposed human airway smooth muscle. J Immunol 185: 3035–3040.
84. Sterk PJ, Bel EH (1989) Bronchial hyperresponsiveness: the need for a distinction between hypersensitivity and excessive airway narrowing. Eur Respir J 2: 267–274.
85. Sukkar MB, Hughes JM, Armour CL, Johnson PR (2001) Tumour necrosis factor-α potentiates contraction of human bronchus in vitro. Respirology 6: 199–203.
86. Thomas PS, Yates DH, Barnes PJ (1995) Tumor necrosis factor-α increases airway responsiveness and sputum neutrophilia in normal human subjects. Am J Respir Crit Care Med 152: 76–80.
87. Thomas PS, Heywood G (2002) Effects of inhaled tumour necrosis factor α in subjects with mild asthma. Thorax 57: 774–778.
88. Tirumurugaan KG, Jude JA, Kang BN, Panettieri RA, Walseth TF, Kannan MS (2007) TNF-α induced CD38 expression in human airway smooth muscle cells: role of MAP kinases and transcription factors NF-κB and AP-1. Am J Physiol Lung Cell Mol Physiol 292: L1385–1395.
89. Tirumurugaan KG, Kang BN, Panettieri RA, Foster DN, Walseth TF, Kannan MS (2008) Regulation of the CD38 promoter in human airway smooth muscle cells by TNF-α and dexamethasone. Respir Res 9: 26.
90. Tliba O, Deshpande D, Chen H, Van Besien C, Kannan M, Panettieri RA, Jr., Amrani Y (2003) IL-13 enhances agonist-evoked calcium signals and contractile responses in airway smooth muscle. Br J Pharmacol 140: 1159–1162.
91. Tliba O, Cidlowski JA, Amrani Y (2006) CD38 expression is insensitive to steroid action in cells treated with tumor necrosis factor-α and interferon-γ by a mechanism involving the up-regulation of the glucocorticoid receptor β isoform. Mol Pharmacol 69: 588–596.
92. Trian T, Benard G, Begueret H, Rossignol R, Girodet PO, Ghosh D, Ousova O, Vernejoux JM, Marthan R, Tunon-de-Lara JM, Berger P (2007) Bronchial smooth muscle remodeling involves calcium-dependent enhanced mitochondrial biogenesis in asthma. J Exp Med 204: 3173–3181.
93. Trian T, Moir LM, Ge Q, Burgess JK, Kuo C, King NJ, Reddel HK, Black JL, Oliver BG, McParland BE (2010) Rhinovirus-induced exacerbations of asthma: How is the β_2-adrenoceptor implicated? Am J Respir Cell Mol Biol 43: 227–233.
94. Ward JP (2006) On Ca^{2+} sensitivity and the airways: Not just any smooth muscle. Eur Respir J 28: 680–682.

95. White TA, Xue A, Chini EN, Thompson M, Sieck GC, Wylam ME (2006) Role of transient receptor potential C3 in TNF-α-enhanced calcium influx in human airway myocytes. Am J Respir Cell Mol Biol 35: 243–251.
96. Ying S, O'Connor B, Ratoff J, Meng Q, Fang C, Cousins D, Zhang G, Gu S, Gao Z, Shamji B, Edwards MJ, Lee TH, Corrigan CJ (2008) Expression and cellular provenance of thymic stromal lymphopoietin and chemokines in patients with severe asthma and chronic obstructive pulmonary disease. J Immunol 181: 2790–2798.

Ca^{2+} Signaling and P2 Receptors in Airway Smooth Muscle

Luis M. Montaño, Edgar Flores-Soto, and Carlos Barajas-López

Contents

Abstract Adenosine 5′-triphosphate (ATP) acts as an extracellular mediator with direct biological effects mediated by purinergic 2 (P2) receptors in different tissues, including the respiratory system. ATP is hydrolyzed by ectonucleotidases, enzymes bound to the plasma membrane. There are three major families of ATP-hydrolyzing ectonucleotidases: the ectonucleotide triphosphate diphosphohydrolases, the ectonucleotide pyrophosphatase/phosphodiesterases, and the alkaline phosphatases. ATP induces a biphasic response (a contraction followed by a relaxation) in airway smooth muscle of different species (guinea pig and rat), including human bronchial rings. In the majority of the species, ATP-induced contraction is associated to an increment of intracellular Ca^{2+} concentration ($[Ca^{2+}]_i$) mediated by stimulation of P2Y or P2X receptors in the airway smooth muscle. In the guinea-pig airway smooth muscle the contraction is mediated by bronchoconstrictor prostaglandins (thromboxane) and involves $P2Y_4$ and $P2Y_6$ receptors on the airway

L.M. Montaño (✉) • E. Flores-Soto
Departamento de Farmacología, Edificio de Investigación, sexto piso, laboratorio 3, Facultad de Medicina, Universidad Nacional Autónoma de México, Ciudad Universitaria, México, DF CP 04510, Mexico
e-mail: lmmr@unam.mx

C. Barajas-López
División de Biología Molecular, Instituto Potosino de Investigación Científica y Tecnológica, San Luis Potosí, SLP, Mexico

Y.-X. Wang (ed.), *Calcium Signaling In Airway Smooth Muscle Cells*,
DOI 10.1007/978-3-319-01312-1_21,

epithelium. However, epithelium removal prevents ATP-induced contraction and, paradoxically, induces relaxation without affecting the Ca^{2+} signaling. This ATP-induced relaxation occurs by smooth muscle prostaglandin production (PGE_2), which involves a G_s protein activation and the subsequent enhancement of voltage-dependent K^+ and Ca^{2+}-dependent K^+ channels. Activation of these K^+ channels occurs through adenylyl cyclase-cAMP pathway. The physiological role of P2X and P2Y receptors in airway smooth muscle is complex and species dependent, and it would be expected to change during inflammation. Thus, further research is required in asthma to determine the purinoceptors' role during this pathology.

Keywords Airway smooth muscle • ATP • Purinergic receptors • P2X receptors • P2Y receptors

1 Introduction

Adenosine 5′-triphosphate (ATP) acts as an extracellular mediator with direct biological effects mediated by purinergic 2 (P2) receptors in different tissues, including the respiratory system. P2 receptors are classified in $P2X_{1-7}$ and $P2Y_{1, 2, 4, 6, 11, 12, 13, 14}$ subunits [8]. P2X receptors are ligand-gated channels [31], and P2Y receptors are coupled to G proteins linked to IP_3 or cAMP second messengers [5, 8, 17].

ATP is hydrolyzed by ectonucleotidases, enzymes bound to the plasma membrane. There are three major families of ATP-hydrolyzing ectonucleotidases: the ectonucleotide triphosphate diphosphohydrolases (NTPDase), the ectonucleotide pyrophosphatase/phosphodiesterases (NPP), and the alkaline phosphatases (AP). Eight different genes have been cloned for the members of NTPDases. In four of them (NTPDase1, 2, 3, 8) the catalytic sites are oriented to the extracellular surface of the plasma membrane and metabolize ATP and ADP at physiological pH. The remaining four nucleotidases (NTPDase4, 5, 6, 7) also hydrolyze nucleotides other than ATP. NTPDase5 and 6 are also secreted as soluble enzymes [36]. The NPP family includes seven members, three (NPP1, 2, 3) are also attached to plasma membrane with their catalytic sites facing the extracellular space and metabolize ATP and other nucleotides [23, 38]. Finally, four isoenzymes recognized as AP include the intestinal AP (IAP), the tissue nonspecific AP (TNAP or NSAP), the placental AP (PLAP), and the germlike AP; all of them are known to metabolize ATP, ADP, and AMP [26]. TNAP and PLAP are localized in human airway epithelium; TNAP is found in the mucosal and serosal surface of epithelial cells from the trachea through the alveoli [34]. NTPDase1 and 3 are found along the whole respiratory epithelial barrier, except for the alveolar region, which only expresses NTPDase1 [12, 33]. NTPDases5 and 6 are secreted as soluble enzymes and have been found in cultured human bronchial epithelial cells [33]. Finally, NPP2 and 3 have been found in lung parenchyma [20, 37]; NPP1, 2, and 3 were described to be present in cultures from bronchial epithelial cells [33]; indeed, NPP2 is secreted by the epithelium as a soluble enzyme [18].

The hydrolysis of ATP seems to play an essential role in allergic bronchospasm in guinea pigs. We found recently that ectonucleotidase inhibition with ARL-67156 greatly augmented the lung resistance after an antigenic challenge. Thus, these results indicate that the ATP amount released during the antigenic challenge could be enough to produce a severe bronchospasm, but in the absence of ARL-67156, the airway obstruction is prevented by the fast hydrolysis of ATP through ectonucleotidases [9]. Therefore, these findings point out the need to investigate whether ectonucleotidase activity is related to asthma severity.

2 ATP Induces Contraction

It is well documented that ATP induces a biphasic response (a contraction followed by a relaxation) in the airway smooth muscle of different species (guinea pig and rat), including human bronchial rings [19, 29].

Most authors have associated the contractile effect of ATP to an increment of intracellular Ca^{2+} concentration ($[Ca^{2+}]_i$) mediated by the stimulation of P2Y or P2X receptors in airway smooth muscle [5, 17, 24, 29]. More recently, the ATP-induced contraction of the guinea-pig trachea was found to be mediated by bronchoconstrictor prostaglandins, such as thromboxane production, and to involve $P2Y_4$ and $P2Y_6$ receptors on the airway epithelium [1, 9, 13, 15]. Thus, the contraction induced by ATP is an indirect effect that does not involve a direct stimulation of purinergic receptors on smooth muscle cells [15].

3 P2X Receptors

P2X receptors are mainly activated by extracellular ATP but not by ADP, AMP, adenosine, or uridine nucleotides. These receptors are ligand-gated ion channels, and when they are activated, their nonselective cation channel opens rapidly (microseconds). These channels are permeable to Na^+, K^+, and Ca^{2+} and are modulated by Ca^{2+}, Mg^{2+}, H^+, or metallic ions such as Zn^{2+} or Cu^{2+} [31]. As a result of their rapid response and their relative low ATP affinity (in the micromolar range), P2X receptors mediate fast ATP signaling over short distances [21]. These purinoceptors are organized as trimers to form functional channels [22] and can be formed by the same (homomeric) or different (heteromeric) subunits, giving rise to a diverse population of channels with specific pharmacological and physiological properties for each combination; this population of channels becomes even more diverse if one considers alternative splicing and species differences [3]. The properties of P2X receptors have been determined generally in studies using recombinant receptors, but a detailed characterization of P2X native receptors has been hindered by a lack of selective agonists and antagonists for receptor subtypes [3, 22, 31]. Nevertheless, other molecular biology techniques are expected to be useful tools in identifying the role of purinoceptors in airway smooth muscle.

The physiological role of P2X receptors in airway smooth muscle has been scantily studied. The first publication on this topic, by Mounkaïla et al., appeared in 2005 [29]. These researchers proposed that P2X receptors in rat and human airway smooth muscle were involved in the contraction induced by ATP based on the fact that α-β-methylene ATP, a P2X agonist, produced a similar contractile response to ATP. They suggested that stimulation of P2X receptors triggered external Na^+ entry and depolarized the plasma membrane, activating L-type voltage-operated Ca^{2+} channels ($Ca_v1.2$) to induce contraction. Later on, it was proposed that P2X receptors were involved in airway hyperresponsiveness through Ca^{2+} sensitization Rho/Rho-kinase pathway [32]. Subsequently, in 2009, Nagaoka et al. [30], using immunohistochemistry, reverse transcription/polymerase chain reaction analysis (RT-PCR), and Western blot analysis, showed the presence of $P2X_4$ receptors in porcine airway smooth muscle. In this work, the authors found that ATP induced a sustained inward current, an increase in intracellular Ca^{2+} concentration $[Ca^{2+}]_i$, and smooth muscle contraction, suggesting that Ca^{2+} influx through $P2X_4$ receptors plays an important role in airway smooth muscle contraction (Fig. 1).

Recently, in guinea-pig tracheal smooth muscle we found, by Western blot, the presence of $P2X_1$ and $P2X_3$ and, to a lesser extent, $P2X_2$ receptors. Activation of these purinoceptors by ATP induces a $[Ca^{2+}]_i$ increment mostly by two mechanisms: the opening of ligand-gated channels and by activation of the reverse mode of the Na^+/Ca^{2+} exchanger (NCX). This exchanger is an integral membrane protein that mediates Ca^{2+} influx and Na^+ extrusion in its reverse mode [11]. One of the requirements to activate the reverse mode of the NCX is the Na^+ entry through nonselective cation channels, leading to a large subplasmalemmal rise in Na^+ concentration near the NCX [6, 10]. $P2X_1$, $P2X_2$, and $P2X_3$ receptors are ATP-gated nonselective cation channels permeable to Ca^{2+} and Na^+, and when activated they favor Na^+ accumulation in the subplasmalemmal space. It is essential to point out that the purpose of the $[Ca^{2+}]_i$ increment induced by both the activation of P2X receptors and the reverse mode of the NCX in guinea-pig airway smooth muscle is to refill the sarcoplasmic reticulum (SR) (and not to produce contraction), facilitating Ca^{2+} availability for bronchoconstrictor agonists that act through the PLC-IP_3 pathway (Fig. 1) [15].

Finally, the physiological role of P2X receptors in airway smooth muscle seems to be species dependent, and further research is required to determine what subunits are present and to explore their functional roles in human airway smooth muscle.

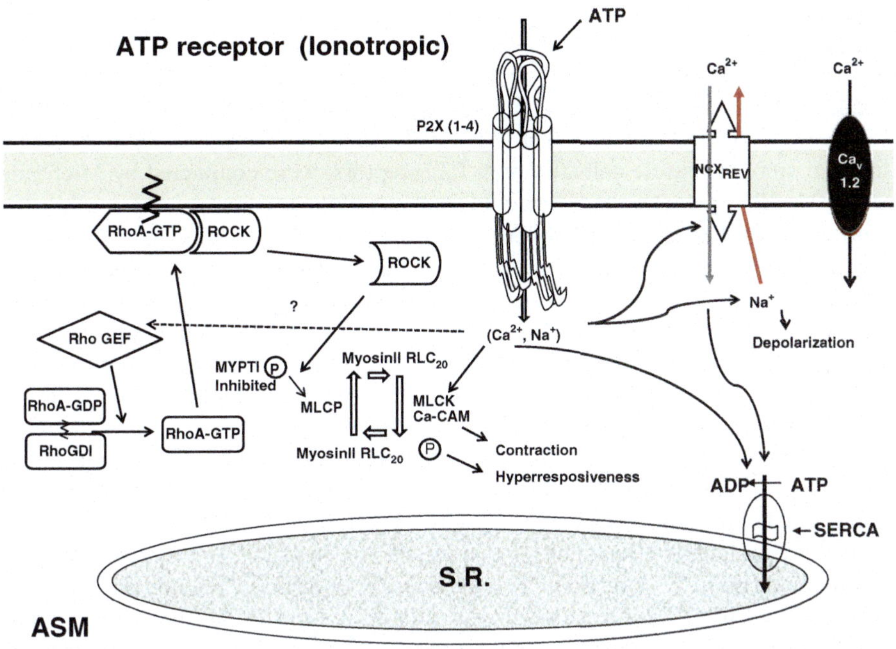

Fig. 1 P2X receptors induce contraction in airway smooth muscle. P2X receptor stimulation with ATP in human and rat airway smooth muscle (ASM) increases $[Na^+]_i$ and induces a cell depolarization. L-type voltage-dependent Ca^{2+} channels ($Ca_v1.2$) are opened, augmenting the $[Ca^{2+}]_i$, inducing ASM contraction. Activation of P2X receptors in guinea-pig airways also participates in ASM hyperresponsiveness through the RhoA–Rho-kinase/ROCK pathway. The $P2X_4$ receptor has been defined as being responsible for $[Ca^{2+}]_i$ increments in pig ASM contraction induced by ATP. Recently, $P2X_1$, $P2X_2$, and $P2X_3$ receptors were found to activate the Na^+/Ca^{2+} exchanger reverse mode (NCX_{REV}) in guinea-pig ASM to promote sarcoplasmic reticulum (SR) Ca^{2+} refilling (for details see text). Abbreviations: *RhoGDI* GDP dissociation inhibitor, *Rho GEF* guanine nucleotide exchange factor, *RhoA-GTP* monomeric G protein, *ROCK* Rho-kinase, *MYPT1* myosin phosphate regulator, *MLCP* myosin light chain phosphate, *RLC_{20}* regulatory light chain of myosin II, *MLCK Ca-CAM* activation of myosin light chain kinase by Ca^{2+} binding to calmodulin, *SERCA* Ca^{2+}ATPase from sarcoplasmic reticulum

4 P2Y Receptors

P2Y receptors are activated in the nanomolar range of purine derivates such as ATP and ADP and by pyrimidine nucleotides like UTP and UDP. $P2Y_{1,6,12}$ are mainly activated by nucleoside diphosphates, whereas $P2Y_{2,4}$ are activated by nucleoside triphosphates. In this context, $P2Y_{2,4,6}$ are activated by both purine and pyrimidine nucleotides, and other purinoceptors such as $P2Y_{1,11,12}$ are activated only by purine nucleotides [8].

Metabotropic P2Y receptors, which have seven transmembranal domains, are coupled to G proteins. $P2Y_{1,2,4,6}$ bind to $G_{q/11}$ and activate the PLC-IP_3 pathway and Ca^{2+} release from the SR. $P2Y_{11}$ can couple to both $G_{q/11}$ and G_s, which stimulate adenylyl cyclase, while $P2Y_{12}$, $P2Y_{13}$, and $P2Y_{14}$ couple to G_i,, which inhibits adenylyl cyclase [8]. Because P2Y receptors trigger second-messenger cascade and can detect low ATP concentrations far from the site of release, they are able to serve a long-lasting modulatory function [21].

The first studies showing that ATP induced a Ca^{2+} transient in cultured rat tracheal smooth muscle cells, through P2 receptors, were conducted by Micheaud et al. [24, 25]; however, the contractile response of these cells was not investigated until 2002 by Bergner and Sanderson [5], who found that ATP induced $[Ca^{2+}]_i$ increase with oscillation and contraction in mice airway smooth muscle and also proposed that $P2Y_2$ and $P2Y_4$ receptors were involved based on a pharmacological approach. It was not until 2005 that Govindaraju et al. [17] reported, in human airway smooth muscle, using RT-PCR and Western blot analysis, the presence of the $P2Y_{1,2,4,6}$ receptors. These authors also reported that the stimulation of human smooth muscle cells with ATP, UTP, ADP, and UDP produced an increase of $[Ca^{2+}]_i$ and contraction, demonstrating that these nucleotides directly stimulate purinoceptors in smooth muscle to induce contraction (Fig. 2). In the context of $[Ca^{2+}]$i increment induced by purinergic receptors, it has been reported in cultured human bronchial epithelial cells [4] that the activation of $P2Y_2$ receptors stimulates the capacitative Ca^{2+} entry pathway as a result of SR Ca^{2+} depletion. Thus, the ATP-induced $[Ca^{2+}]_i$ increase in airway smooth muscle of rat, mice, and human, involving SR Ca^{2+} depletion, could promote capacitative Ca^{2+} entry through nonselective cation channels (store-operated Ca^{2+} channels), allowing external Na^+ and Ca^{2+} entry. This ionic flux through the membrane could induce depolarization and the subsequent activation of $Ca_v1.2$. Nevertheless, ATP-induced contraction in airway smooth muscle seems to involve different mechanisms that depend on the species studied. As previously described, ATP-induced contraction in guinea-pig airway smooth muscle is related to bronchoconstrictor prostaglandins such as thromboxane production, which involves $P2Y_4$ and $P2Y_6$ receptors on the airway epithelium [1, 9, 13, 15]. Various observations indicate that ATP-induced contraction in this species is an indirect effect and does not involve a direct stimulation of purinergic receptors on smooth muscle cells (Fig. 3). Thus, we found in guinea-pig tracheal strips that ATP induced a $[Ca^{2+}]_i$ increment followed by a sustained contraction during simultaneous measurements. However, when tissues were incubated with indomethacin, the contraction induced by ATP was abolished but the Ca^{2+} transient did not change. Additionally, epithelium removal from the strips prevented ATP-induced contraction and, paradoxically, induced relaxation without affecting the Ca^{2+} signaling [15].

5 ATP Induces Relaxation

ATP has been reported to induce relaxation in tracheal smooth muscle from guinea pig, rabbit, mouse, and rat, and human bronchial rings. In the majority of these studies, the ATP-induced relaxation was tested in tissues previously contracted with acetylcholine, methacholine, or histamine. This response has been proposed to be

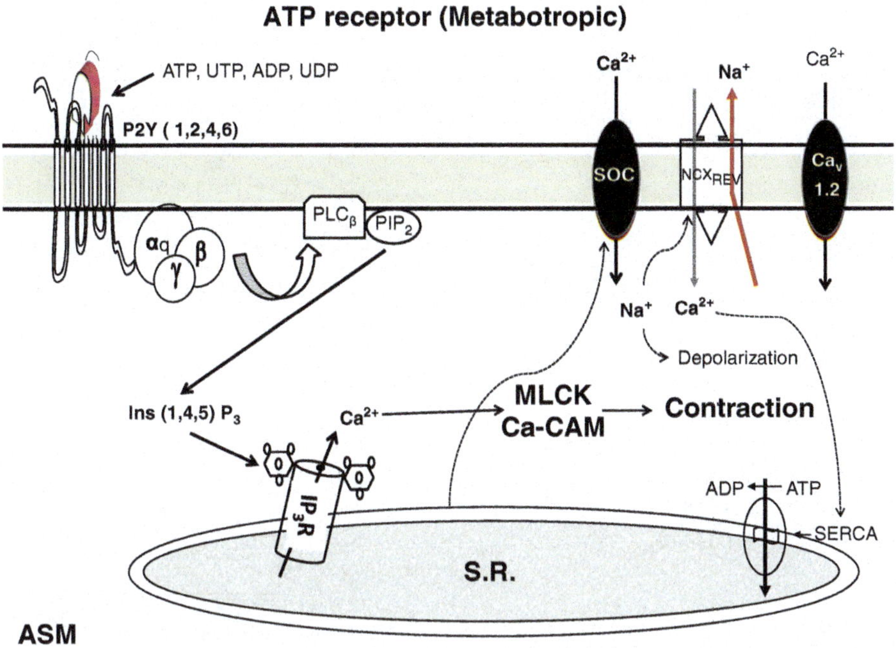

Fig. 2 P2Y receptors induce contraction in airway smooth muscle. P2Y (1,2,4,6) receptor stimulation with ATP, UTP, ADP, or UDP in human and rat ASM activates the PLC_{β}-IP_3 signaling pathway, increasing $[Ca^{2+}]_i$. SR Ca^{2+} depletion could cause the opening of nonspecific cationic channels (capacitative Ca^{2+} entry), allowing external Na^+ and Ca^{2+} entry. This ionic flux through the membrane could induce depolarization, activating the L-type voltage-dependent Ca^{2+} channels ($Ca_v1.2$). Additionally, increment of Na^+ could favor the reverse mode of the Na^+/Ca^{2+} exchanger. Abbreviations: PLC_{β}, phospholipase C_{β}, PIP_2, phosphatidylinositol 4,5- bisphosphate, IP_3 inositol 1,4,5- trisphosphate, *MLCK Ca-CAM* activation of myosin light chain kinase by Ca^{2+} binding to calmodulin, *SERCA* Ca^{2+}ATPase from sarcoplasmic reticulum, *SOC* store-operated channel, NCX_{REV} reverse mode of Na^+/Ca^{2+} exchanger

dependent on purinoceptor desensitization, prostaglandin E_2 production, and PKA activation [2, 14, 16, 19, 28, 29]. However, it was not until 1994 that Aksoy and Kelsen [2] proposed for the first time, in rabbit tracheal smooth muscle, that P2 receptors on the airway epithelium were involved in ATP-induced relaxation, an effect that was mediated in part by the release of one or more cyclooxygenase products [2]. Nevertheless, in 1996 Pipper and Hollingsworth [35] suggested that guinea-pig tracheal smooth muscle relaxation at higher ATP concentrations occurred through enzymatic hydrolysis of this nucleotide to adenosine involving P1 receptors.

Recently, our research group, using Western blot and immunofluorescence, found that during the relaxation induced by ATP in guinea-pig tracheal smooth muscle, $P2Y_2$ and $P2Y_4$ were the main purinergic receptors involved in this response. We also found that this ATP-induced relaxation occurred by epithelium-independent smooth muscle prostaglandin production (PGE_2), which

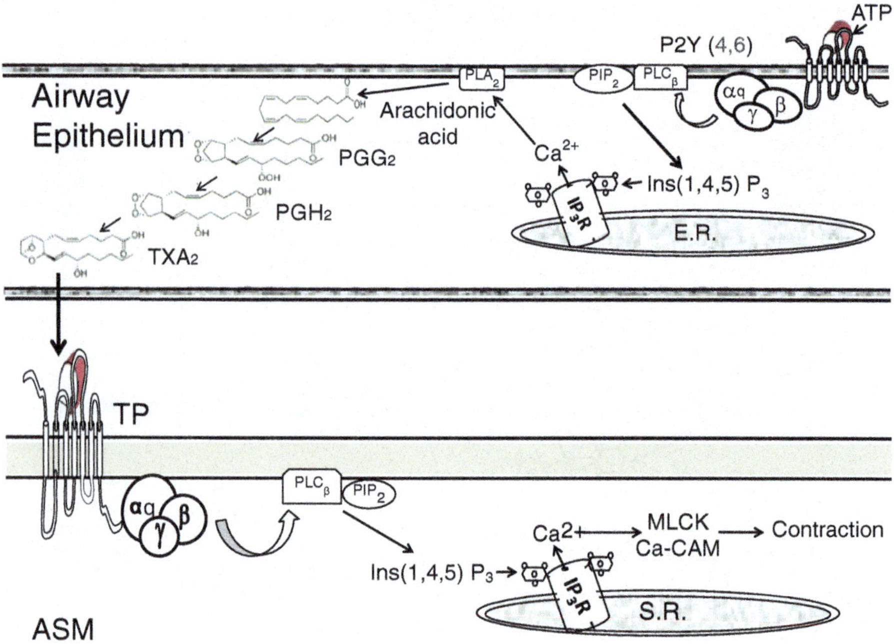

Fig. 3 Contraction pathways activated by ATP in guinea pig airway smooth muscle. In guinea-pig airway smooth muscle, ATP induces an indirect contraction through epithelial production of TXA_2. Intracellular Ca^{2+} increases in the epithelium activate PLA_2, favoring TXA_2 synthesis; hence this prostanoid will induce contraction of smooth muscle via the PLC-IP_3 pathway. Abbreviations: *PLC_β* phospholipase C_β, *PIP_2* phosphatidylinositol 4,5-bisphosphate, *IP_3* inositol 1,4,5-trisphosphate, *MLCK Ca-CAM* activation of myosin light chain kinase by Ca^{2+} binding to calmodulin, *PLA_2* phospholipase A_2, *PGG_2*, prostaglandin G_2, *PGH_2* prostaglandin H_2, *ER* endoplasmic reticulum, *TP* prostanoid receptor, *SR* sarcoplasmic reticulum

involves a G_s protein activation and the subsequent enhancement of voltage-dependent K^+ and Ca^{2+}-dependent K^+ channels. Activation of these K^+ channels occurs through the adenylyl cyclase-cAMP pathway [28].

As described earlier, we found that ATP-induced relaxation in guinea-pig tracheal smooth muscle occurs without affecting Ca^{2+} signaling [15]. A possible explanation of this phenomenon is as follows: ATP stimulates $P2Y_2$ and $P2Y_4$ receptors in airway smooth muscle, and these purinergic receptors are coupled to $G_{q/11}$ and activate the PLC-IP_3 pathway, inducing a $[Ca^{2+}]_i$ increment. Boeynaems and Pearson [7] postulated, in vascular smooth muscle, that the activation of P2Y receptors with ATP induced a $[Ca^{2+}]_i$ increment, which in turn activated the membrane phospholipase A_2 (PLA_2), resulting in an enhanced synthesis and release of prostacyclin (PGI_2). Therefore, the ATP-induced $[Ca^{2+}]_i$ increment due to the stimulation of P2Y receptors in airway smooth muscle could also activate PLA_2 and then favor the production of PGE_2 to induce relaxation (Fig. 4).

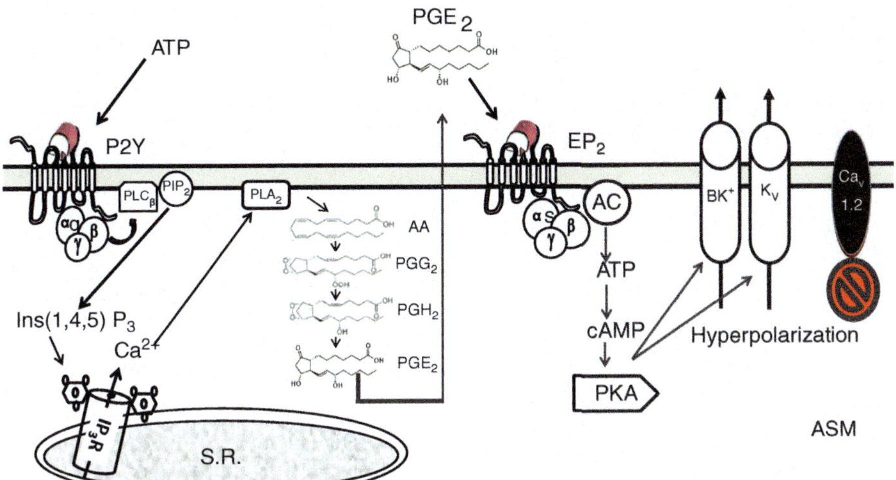

Fig. 4 Relaxation pathways activated by ATP in guinea pig airway smooth muscle. In airway smooth muscle (ASM), the $G\alpha_q$-coupled P2Y receptors are stimulated by ATP, inducing $[Ca^{2+}]_i$ increases through the PLC_β-IP_3 pathway. This $[Ca^{2+}]_i$ increment activates PLA_2 that releases arachidonic acid (AA) from plasma membrane phospholipids. AA is the substrate for the synthesis of prostaglandin E_2 (PGE_2). $G\alpha_s$-coupled PGE_2 receptors (EP_2) on ASM are autocrinely stimulated by PGE_2, generating cyclic AMP (cAMP) through activation of the adenylyl cyclase (AC) enzyme. cAMP then activates protein kinase A (PKA) responsible for phosphorylating the Ca^{2+} dependent (BK^+) and the delayed rectifier K^+ channels (K_v). These PKA-mediated phosphorylations increase the channels' opening probabilities, and when K^+ efflux augments, the ASM is hyperpolarized. During this stage, $Ca_v1.2$ remains closed. Abbreviations: *PLC_β* phospholipase C_β, *PIP_2* phosphatidylinositol 4,5-bisphosphate, *IP_3* inositol 1,4,5-trisphosphate, *PLA_2* phospholipase A_2, *PGG_2* prostaglandin G_2, *PGH_2* prostaglandin H_2, *SR* sarcoplasmic reticulum, *$Ca_v1.2$*,L-type voltage-dependent Ca^{2+} channels

The physiological role of P2X and P2Y receptors in airway smooth muscle is complex and species dependent, and it would be expected to change during inflammation. In this context, in a guinea-pig model of allergic airway inflammation, we recently found that the sensitization procedure to the antigen caused an upregulation of $P2Y_4$ and $P2Y_6$ receptors in the airway epithelium. The increased expression of these receptors would favor a more intense response to ATP or other nucleotides released during antigenic challenge [9].

The role of ATP in asthma has scarcely been investigated. Airway hyperresponsiveness is a cardinal feature of asthma, which is a functional abnormality favoring exaggerated smooth muscle contraction in response to several stimuli. In a recent study in our laboratory, we found that a cholinergic agonist, histamine, and serotonin, all three well known as contractile agonists of airway smooth muscle, induced airway hyperresponsiveness to ATP [27]. These results were surprising because these agonists are not considered as mediators causing airway hyperresponsiveness. Thus, these findings revealed that ATP exerts multiple interactions with airway smooth muscle contraction mediators in asthma, and further research on this topic is of paramount importance.

Acknowledgments This study was partly supported by a grant from DGAPA-UNAM (IN200613-3) to Dr Luis M. Montaño.

References

1. Advenier C, Bidet D, Floch-Saint-Aubin A, Renier A (1982) Contribution of prostaglandins and thromboxanes to the adenosine and ATP-induced contraction of guinea-pig isolated trachea. Br J Pharmacol 77:39–44
2. Aksoy MO, Kelsen SG (1994) Relaxation of rabbit tracheal smooth muscle by adenine nucleotides: mediation by P2-purinoceptors. Am J Respir Cell Mol Biol 10:230–236
3. Anderson CM, Nedergaard M (2006) Emerging challenges of assigning $P2X_7$ receptor function and immunoreactivity in neurons. Trends Neurosci 29:257–262
4. Bahra P, Mesher J, Li S, Poll CT, Danahay H (2004) $P2Y_2$-receptor-mediated activation of a contralateral, lanthanide-sensitive calcium entry pathway in the human airway epithelium. Br J Pharmacol 143:91–98
5. Bergner A, Sanderson MJ (2002) ATP stimulates Ca^{2+} oscillations and contraction in airway smooth muscle cells of mouse lung slices. Am J Physiol Lung Cell Mol Physiol 283: L1271–1279
6. Blaustein MP, Lederer WJ (1999) Sodium/calcium exchange: its physiological implications. Physiol Rev 79:763–854
7. Boeynaems JM, Pearson JD (1990) P2 purinoceptors on vascular endothelial cells: physiological significance and transduction mechanisms. Trends Pharmacol Sci 11:34–37
8. Burnstock G (2007) Physiology and pathophysiology of purinergic neurotransmission. Physiol Rev 87:659–797
9. Chavez J, Vargas MH, Rebollar-Ayala DC, Diaz-Hernandez V, Cruz-Valderrama JE, Flores-Soto E, Flores-Garcia M, Jimenez-Vargas NN, Barajas-Lopez C, Montano LM (2013) Inhibition of extracellular nucleotides hydrolysis intensifies the allergic bronchospasm. A novel protective role of ectonucleotidases. Allergy, DOI: 10.1111/all.12113
10. Dai JM, Kuo KH, Leo JM, van Breemen C, Lee CH (2006) Mechanism of ACh-induced asynchronous calcium waves and tonic contraction in porcine tracheal muscle bundle. Am J Physiol Lung Cell Mol Physiol 290:L459–469
11. Eisner DA, Lederer WJ (1985) Na-Ca exchange: stoichiometry and electrogenicity. Am J Physiol 248:C189–202
12. Fausther M, Pelletier J, Ribeiro CM, Sevigny J, Picher M (2010) Cystic fibrosis remodels the regulation of purinergic signaling by NTPDase1 (CD39) and NTPDase3. Am J Physiol Lung Cell Mol Physiol 298:L804–818
13. Fedan JS, Belt JJ, Yuan LX, Frazer DG (1993) Contractile effects of nucleotides in guinea pig isolated, perfused trachea: involvement of respiratory epithelium, prostanoids and Na^+ and Cl^- channels. J Pharmacol Exp Ther 264:210–216
14. Fedan JS, Belt JJ, Yuan LX, Frazer DG (1993) Relaxant effects of nucleotides in guinea pig isolated, perfused trachea: lack of involvement of prostanoids, Cl^- channels and adenosine. J Pharmacol Exp Ther 264:217–220
15. Flores-Soto E, Carbajal V, Reyes-Garcia J, Garcia-Hernandez LM, Figueroa A, Checa M, Barajas-Lopez C, Montano LM (2011) In airways ATP refills sarcoplasmic reticulum via P2X smooth muscle receptors and induces contraction through P2Y epithelial receptors. Pflugers Arch 461:261–275

16. Fortner CN, Breyer RM, Paul RJ (2001) EP2 receptors mediate airway relaxation to substance P, ATP, and PGE_2. Am J Physiol Lung Cell Mol Physiol 281:L469–474
17. Govindaraju V, Martin JG, Maghni K, Ferraro P, Michoud MC (2005) The effects of extracellular purines and pyrimidines on human airway smooth muscle cells. J Pharmacol Exp Ther 315:941–948
18. Jansen S, Stefan C, Creemers JW, Waelkens E, Van Eynde A, Stalmans W, Bollen M (2005) Proteolytic maturation and activation of autotaxin (NPP2), a secreted metastasis-enhancing lysophospholipase D. J Cell Sci 118:3081–3089
19. Kamikawa Y, Shimo Y (1976) Mediation of prostaglandin E_2 in the biphasic response to ATP of the isolated tracheal muscle of guinea-pigs. J Pharm Pharmacol 28:294–297
20. Kawagoe H, Soma O, Goji J, Nishimura N, Narita M, Inazawa J, Nakamura H, Sano K (1995) Molecular cloning and chromosomal assignment of the human brain-type phosphodiesterase I/nucleotide pyrophosphatase gene (PDNP2). Genomics 30:380–384
21. Khakh BS (2001) Molecular physiology of P2X receptors and ATP signalling at synapses. Nat Rev Neurosci 2:165–174
22. Khakh BS, Burnstock G, Kennedy C, King BF, North RA, Seguela P, Voigt M, Humphrey PP (2001) International union of pharmacology. XXIV. Current status of the nomenclature and properties of P2X receptors and their subunits. Pharmacol Rev 53:107–118
23. Kukulski F, Levesque SA, Sevigny J (2011) Impact of ectoenzymes on P2 and P1 receptor signaling. Adv Pharmacol 61:263–299
24. Michoud MC, Tao FC, Pradhan AA, Martin JG (1999) Mechanisms of the potentiation by adenosine of adenosine triphosphate-induced calcium release in tracheal smooth-muscle cells. Am J Respir Cell Mol Biol 21:30–36
25. Michoud MC, Tolloczko B, Martin JG (1997) Effects of purine nucleotides and nucleoside on cytosolic calcium levels in rat tracheal smooth muscle cells. Am J Respir Cell Mol Biol 16:199–205
26. Millan JL (2006) Alkaline Phosphatases : Structure, substrate specificity and functional relatedness to other members of a large superfamily of enzymes. Purinergic Signal 2:335–341
27. Montano LM, Carbajal V, Vargas MH, Garcia-Hernandez LM, Diaz-Hernandez V, Checa M, Barajas-Lopez C (2013) Histamine, carbachol and serotonin induce hyperresponsiveness to ATP in guinea pig tracheas. Involvement of COX-2 pathway. Pflugers Arch 465:1171–1179
28. Montano LM, Cruz-Valderrama JE, Figueroa A, Flores-Soto E, Garcia-Hernandez LM, Carbajal V, Segura P, Mendez C, Diaz V, Barajas-Lopez C (2011) Characterization of P2Y receptors mediating ATP induced relaxation in guinea pig airway smooth muscle: involvement of prostaglandins and K^+ channels. Pflugers Arch 462:573–585
29. Mounkaila B, Marthan R, Roux E (2005) Biphasic effect of extracellular ATP on human and rat airways is due to multiple P2 purinoceptor activation. Respir Res 6:143
30. Nagaoka M, Nara M, Tamada T, Kume H, Oguma T, Kikuchi T, Zaini J, Moriya T, Ichinose M, Tamura G, Hattori T (2009) Regulation of adenosine 5′-triphosphate (ATP)-gated $P2X_4$ receptors on tracheal smooth muscle cells. Respir Physiol Neurobiol 166:61–67
31. North RA (2002) Molecular physiology of P2X receptors. Physiol Rev 82:1013–1067
32. Oguma T, Ito S, Kondo M, Makino Y, Shimokata K, Honjo H, Kamiya K, Kume H (2007) Roles of P2X receptors and Ca^{2+} sensitization in extracellular adenosine triphosphate-induced hyperresponsiveness in airway smooth muscle. Clin Exp Allergy 37:893–900
33. Picher M, Boucher RC (2001) Metabolism of extracellular nucleotides in human airways by multienzyme system Drug Dev Res 52:66–75
34. Picher M, Burch LH, Hirsh AJ, Spychala J, Boucher RC (2003) Ecto 5′-nucleotidase and nonspecific alkaline phosphatase. Two AMP-hydrolyzing ectoenzymes with distinct roles in human airways. J Biol Chem 278:13468–13479
35. Piper AS, Hollingsworth M (1996) ATP and beta,gamma-methylene ATP produce relaxation of guinea-pig isolated trachealis muscle via actions at P1 purinoceptors. Eur J Pharmacol 307:183–189

36. Robson SC, Sevigny J, Zimmermann H (2006) The E-NTPDase family of ectonucleotidases: Structure function relationships and pathophysiological significance. Purinergic Signal 2:409–430
37. Stefan C, Gijsbers R, Stalmans W, Bollen M (1999) Differential regulation of the expression of nucleotide pyrophosphatases/phosphodiesterases in rat liver. Biochim Biophys Acta 1450:45–52
38. Stefan C, Jansen S, Bollen M (2006) Modulation of purinergic signaling by NPP-type ectophosphodiesterases. Purinergic Signal 2:361–370

Calcium Signaling in Airway Smooth Muscle Remodeling

Tengyao Song, Yun-Min Zheng, and Yong-Xiao Wang

Contents

Abstract Asthma is characterized by airway remodeling. One of the most important manifestations of airway remodeling is an increased mass of airway smooth muscle cells (ASMCs), which is mainly due to hyperplasia (increased cell number). Hyperplasia can result from an increase in cell proliferation, an increase in migration, or a decrease in cell apoptosis. Understanding of the mechanisms underlying ASMC remodeling may give rise to novel avenues for the therapy of asthma and other respiratory diseases. Recent studies from our group and others suggest that Ca^{2+} signaling is important in airway remodeling. Gene and protein expression of Ca^{2+} pump sarco/endoplasmic reticulum Ca^{2+} ATPase 2 (SERCA2) has been shown to be reduced in ASMCs from patients with moderately severe asthma. Conversely, transient receptor potential canonical 3 (TRPC3) channel expression and activity are significantly increased in asthmatic ASMCs, resulting

T. Song • Y.-M. Zheng (✉) • Y.-X. Wang (✉)
Center for Cardiovascular Sciences (MC-8), Albany Medical College, 47 New Scotland Avenue, Albany, NY 12203, USA
e-mail: zhengy@mail.amc.edu; wangy@mail.amc.edu

Y.-X. Wang (ed.), *Calcium Signaling In Airway Smooth Muscle Cells*,
DOI 10.1007/978-3-319-01312-1_22,

in increased extracellular Ca^{2+} influx, activation of the protein tyrosine kinase 2 (PTK2)/nuclear factor kappa-light-chain-enhancer of activated B cells (NF-κB)/calcineurin signaling axis, increased cyclin D1 transcription and proliferation, and, ultimately, airway remodeling. Orai1-encoded Ca^{2+}-release-activated Ca^{2+} (CRAC) channels and stromal interacting molecule 1 (STIM1) also appear to participate in mediating platelet-derived grow factor (PDGF)-dependent store-operated Ca^{2+} entry (SOCE) and the proliferation and migration of ASMCs in asthma. This increased Ca^{2+} signaling may augment the activity of Ca^{2+}/calmodulin-dependent protein kinase II (CaMKII) to phosphorylate cyclic adenosine monophosphate (cAMP) response element binding (CREB) and activate various genes that regulate the cell cycle, thereby leading to asthmatic airway remodeling. Airway remodeling may also be associated with an increased size of ASMCs (hypertrophy). This cellular response is mediated by phosphoinositide 3-kinase (PI3K)/mammalian target of rapamycin (mTOR)/eukaryotic translation initiation factor 4E (eIF-4F), ribosomal S6 kinase (RSK)/S6-related kinase, and glycogen synthase kinase 3β (GSK3β) signaling pathways, all of which are Ca^{2+} dependent. Evidently, continual studies in the field will significantly improve our current knowledge regarding the molecular mechanisms of asthma and other airway disorders.

Keywords TRPC3 channels • Calcium signaling • NF-κB • Airway remodeling • Asthma

Abbreviations

4E-BP	4E-binding protein
Ach	Acetylcholine
ASM	Airway smooth muscle
ASMC	ASM cell
CaM	Calmodulin
CaMK	Ca^{2+}/CaM-dependent kinase
cAMP	Cyclic adenosine monophosphate
CICR	Ca^{2+}-induced Ca^{2+} release
CRAC	Ca^{2+}-release-activated Ca^{2+}
CREB	cAMP response element binding
DAG	Diacylglycerol
EGR1	Early growth response protein 1
eIF2B	Eukaryotic translation initiation factor 2B
eIF-4F	Eukaryotic translation initiation factor 4E
ERK	Extracellular signal-regulated kinase
GPCR	G-protein-coupled receptor
GSK	Glycogen synthase kinase
IP_3	Inositol trisphosphate

IP_3R	IP_3 receptor
$[Ca^{2+}]_i$	Intracellular Ca^{2+} concentration
JNK2	C-Jun amino-terminal kinase 2
mACH	Methacholine
M3AChR	Muscarinic M3 acetylcholine receptor
MKP1	Mitogen-activated protein kinases phosphatase 1
mTOR	Mammalian target of rapamycin
NFAT	Nuclear factor of activated T-cells
NF-κB	Nuclear factor kappa-light-chain-enhancer of activated B cells
NSCC	Nonselective cation channel
PAI1	Plasminogen activator inhibitor-1
PDGF	Platelet-derived growth factor
PI3K	Phosphoinositide 3-kinase
PIP_2	Phosphatidylinositol 4, 5-bisphosphate
RSK	Ribosomal S6 kinase
RyR	Ryanodine receptor
SERCA	Sarco/endoplasmic reticulum Ca^{2+} ATPase
SOCC	Store-operated Ca^{2+} channel
SOCE	Store-operated Ca^{2+} entry
SR	Sarcoplasmic reticulum
STIM	Stromal interacting molecule
TRPC	Transient receptor potential canonical
VDCC	Voltage-dependent Ca^{2+} channel

1 Introduction

Asthma is a common chronic lung disease characterized by airway obstruction, which may occur due to airway remodeling, which refers to a combination of functional and structural changes. A key feature of asthmatic structural changes is an increase in the mass of airway smooth muscle cells (ASMCs) [1]. In asthma, the increased mass of ASMCs can result in enhanced contraction following exposure to certain environmental stimuli and thus further restrict air flow through the airway [1]. Moreover, ASMCs can release inflammatory factors that induce leukocyte infiltration and cause immune injury in the local airway [2, 3]. The increased mass of ASMCs may primarily result from the increased number or proliferation of ASMCs (termed hyperplasia) and size of ASMCs (hypertrophy). Although much recent experimental evidence indicates that the imbalance between survival and apoptosis of ASMCs may play a role in the increased mass of ASMCs [4–7], one study revealed that survival and apoptosis of these cells were not significantly different between subjects with and without asthma [8]. Increased migration of ASMCs is also thought to be involved in asthmatic airway remodeling. Indeed, a number of in vitro experiments have demonstrated that inflammatory factors that are involved in asthma can promote migration of ASMCs, and myofibroblasts derived from ASMCs contribute to the excessive ASMC mass [9–13].

It is well established that Ca^{2+} signaling, i.e., an increase in intracellular Ca^{2+} concentration ($[Ca^{2+}]_i$), is important for multiple cellular responses, including gene expression, cell proliferation, hypertrophy, and contraction in a variety of cells. In ASMCs, stimulation of G-protein-coupled receptors (GPCRs) such as the muscarinic M3 acetylcholine receptor (M3AChR) induces activation of phospholipase C (PLC), hydrolysis of phosphatidylinositol 4, 5-bisphosphate (PIP_2), and inositol trisphosphate (IP_3) and diacylglycerol (DAG) production [14, 15]. IP_3 binds to a specific receptor (IP_3R) located on the sarcoplasmic reticulum (SR) membrane and triggers Ca^{2+} release [16]. Other Ca^{2+} store release factors also contribute to the increased $[Ca^{2+}]_i$ in ASMCs. It has been reported that agonist-induced increase in $[Ca^{2+}]_i$ is not abolished by heparin, an IP_3R antagonist [17], and M3AChR-mediated oscillations in $[Ca^{2+}]_i$ are inhibited by antagonists of ryanodine receptors (RyRs)/ Ca^{2+} release channels, such as ryanodine and ruthenium red [18]. These findings demonstrate that Ca^{2+} release from RyRs cooperates with IP_3Rs to mediate agonist-induced Ca^{2+} release. Furthermore, Ca^{2+} released from IP_3Rs can induce Ca^{2+} release from RyRs, termed Ca^{2+}-induced Ca^{2+} release (CICR) [19]. CICR could increase Ca^{2+} levels in the cytosol and thus further amplify IP_3R function in ASMCs. In addition to Ca^{2+} store release, extracellular Ca^{2+} influx is an important part of ASMC signaling. Nonselective cation channels (NSCCs) play an important role in Ca^{2+} influx. Using permeabilized patch clamp recording in combination with simultaneous measurement of cytosolic Ca^{2+} in equine ASMCs, a constitutively activated single channel is recorded during application of acetylcholine (Ach) or methacholine (mACH). This channel is permeable to both sodium and Ca^{2+}, with a higher permeability to the former ion [20]. It has also been demonstrated that NSCCs are sensitive to Ca^{2+} store depletion, GqPCR activation, membrane stretch, intracellular Ca^{2+}, pH, phospholipids, and other factors [21]. More recently, our group has found that NSCCs are observed in freshly isolated mouse ASMCs using excised inside-out single-channel recording [22]. In addition to NSCCs, other factors contribute to the sustained rise in cytosolic Ca^{2+}. First, the reverse mode of Na^+/Ca^{2+} exchanger provides a source of Ca^{2+} influx from the extracellular space following agonist stimulation [23, 24]. As mentioned earlier, agonists cause the opening of NSCCs and then accumulation of Na^+ within the cell, which forces the activation of the Na^+/Ca^{2+} exchanger into the reverse mode, leading to the influx of Ca^{2+} from the extracellular domain. Activation of voltage-dependent Ca^{2+} channels (VDCCs) can induce Ca^{2+} influx in ASMCs as well [25, 26]. STIM/Orai-mediated store-operated Ca^{2+} entry (SOCE) contributes to the sustained rise in $[Ca^{2+}]_i$. SOCE is the process of replenishing the Ca^{2+} of SR by agonist-induced Ca^{2+} influx, which has been identified as being driven by both Orai family proteins [27–29] and STIM1 [30–32]. Orai1 proteins form a highly Ca^{2+}-sensitive store-operated channel in plasma membrane that is responsible for Ca^{2+}-release-activated Ca^{2+} current (ICRAC). STIM1 is a Ca^{2+}-binding membrane protein in the SR where it appears to function as a Ca^{2+} sensor monitoring and modulating the Ca^{2+} content of SR.

A fall in SR Ca^{2+} leads to the aggregation of STIM1 into punctuate structures within the SR, but close to plasma membrane, where it can interact with Orai1 [33, 34].

The importance of Ca^{2+} signaling in airway remodeling has not been recognized. Thus, in this chapter, we discuss how Ca^{2+} signaling mediates airway remodeling, particularly focusing on proliferation and hypertrophy of ASMCs.

2 Role of Sarco/Endoplasmic Reticulum Ca^{2+} ATPase in Airway Smooth Muscle Remodeling

The sarco/endoplasmic reticulum Ca^{2+} ATPase (SERCA) pump plays a critical role in rapid reuptake of Ca^{2+} back into intracellular store. Interestingly, Mahn et al. reported that SERCA2 expression is decreased in ASMC from patients with moderately severe asthma, which is strongly related to the increased $[Ca^{2+}]_i$ in asthmatic ASMCs, potentially leading to ASMC proliferation in asthma [35]. Moreover, reports from Lewis's group suggest that the generation of Ca^{2+} oscillations by SERCA2 could activate transcription factors such as nuclear factor of activated T-cells (NFAT) and nuclear factor kappa-light-chain-enhancer of activated B cells (NF-κB), providing a potential mechanism for asthmatic ASMC proliferation [36].

3 Contribution of TRPC3 Channels to Enhanced Ca^{2+} Influx in Airway Smooth Muscle Remodeling

Transient receptor potential canonical (TRPC) channels are a subfamily of the transient receptor potential (TRP) channels family, which is a group of ion channels permeable to cations. Among seven members of the TRPC family, TRPC3 channels are not only believed to be a store-operated Ca^{2+} channel (SOCC), but it can also be activated by DAG via the GqPCR signaling pathway described earlier.

It has been reported that TRPC3 channels play a predominant role among other TRPC members in inducing Ca^{2+} influx in ASMCs [22]. In asthmatic ASMCs, increased protein expression and mRNA levels of TRPC3 channels are observed [22]. Our unpublished data indicate that mouse TRPC3 channel gene knockdown leads to significantly reduced ASMC proliferation in a model of remodeling compared with wild-type mice as assessed by immunostaining. How do TRPC3 channels affect ASMC remodeling? On the one hand, increased $[Ca^{2+}]_i$ as a result of the enhanced expression of TRPC3 channels in ASMCs is likely to be a critical factor in ASMC remodeling. Increased $[Ca^{2+}]_i$ activates Ca^{2+}-sensitive transcription factors, which initiate proliferation gene expression in asthma. It has been shown that Ca^{2+} influx through TRPC channels can directly activate Pyk2 [37], which is critical in the activation of extracellular signal-regulated kinases (ERKs) in vascular smooth muscle cells [38]. Hence, following phosphorylation of ERK induced by Pyk2, it translocates to the nucleus, binding with transcription factor SP1 and resulting in cyclin D1 expression. Because SOCCs can directly activate ERK (thapsigargin-induced SOCC phosphorylates ERK) in vascular smooth muscle cells [39, 40] and TRPC3 channels may play a functional role as SOCCs in ASMCs, TRPC3 channels might activate ERK as well. Activated ERK can also enhance CREB phosphorylation by activating ribosomal S6 kinase 2 (RSK2) [41]; phosphorylated CREB plays a role as a transcriptional factor of many SMC proliferation genes. Ca^{2+} may also bind to calmodulin (CaM) to form a complex that activates calcineurin to start this signaling pathway. Activated calcineurin dephosphorylates NFAT, leading to a proliferation mode by starting pro-proliferation gene transcription. Taken together, increased $[Ca^{2+}]_i$ induced by enhanced expression of TRPC3 channels in asthma could initiate ASMC proliferation by activating pro-proliferative transcription factors. On the other hand, TRPC channels have been reported to be involved in the activation of NF-κB through the CaM-CaMKK-β-AMPK-IKKβ/p38-phosphorylated p65 pathway [42], and p65 can increase cyclin D1 expression by binding to its promoter region in a variety of cells [43, 44]. Published data also show that in TRPC3 channel-knockdown Hek293 cells, TNFα significantly induces IκB expression level; in other words, NF-κB activity decreases [45]. Therefore, overexpression of TRPC3 channels might produce less TNFα-induced IκB, which indicates that overexpression of TRPC3 channels enhances activation of NF-κB. Because our unpublished data indicate that NF-κB activity is significantly increased in asthmatic ASMC remodeling, we believe that TRPC3 channel/NF-κB could play an important role in Ca^{2+}-induced ASMC remodeling. Taken together, we provide a diagram (Fig. 1) showing all the aforementioned three potential signaling pathways that are involved in the role of TRPC3 channels in ASMC proliferation and remodeling.

4 Potential Involvement of Orai1 and STIM1 in Airway Smooth Muscle Remodeling

Store-operated Ca^{2+} entry (SOCE) is another important Ca^{2+} influx in a variety of cells and appears to be particularly important in ASMC. Generally speaking, Ca^{2+} store depletion may activate stromal interaction molecule 1 (STIM1), located on the sarco/endoplasmic reticulum. The activated STIM1 is translocated to the cell plasma membrane, where it interacts with Orai1, a Ca^{2+} channel that permits Ca^{2+} influx from the extracellular space [36, 46–48]. STIM1 and Orai1 work together as mediators of SOCE in ASMCs. SOCE has been shown to increase in proliferative ASMCs accompanied by a modest increase in STIM1 mRNA expression and significant increase in Orai1 mRNA expression [46]. Gene knockdown of STIM1 or Orai1 by shRNAs significantly attenuate proliferation of ASMCs [46]. Moreover, it has been reported that STIM1 and Orai1 are involved in PDGF-mediated SOCE

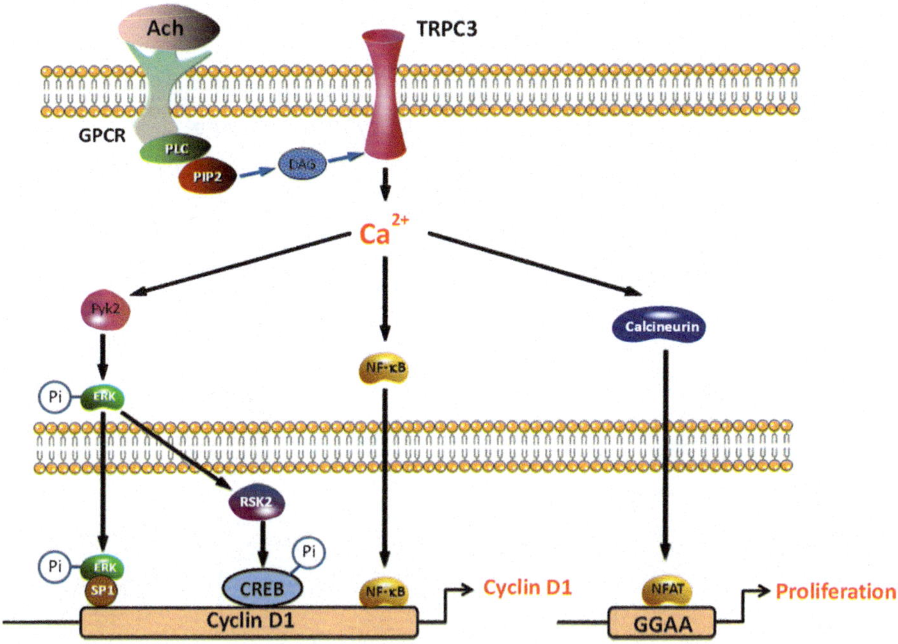

Fig. 1 *TRPC3 channels mediate the role of* Ca^{2+} *signaling in airway smooth muscle cell proliferation.* Ca^{2+} influx through TRPC3 channels may play an important role in airway smooth muscle cell proliferation via three Ca^{2+}-dependent (Pyk2-, NF-κB-, and calcineurin-mediated) signaling pathways

and Ca^{2+}-release-activated Ca^{2+} (CRAC) channels [47]. Gene knockdown of STIM1 or Orai1 inhibits PDGF-activated CRAC currents, potentially leading to inhibition of ASMC proliferation and migration in response to PDGF [47].

5 Calcineurin/NFAT Signaling Axis in Airway Smooth Muscle Remodeling

In addition to acting directly on target proteins, for example, PKC, the effects of Ca^{2+} can be mediated via intracellular Ca^{2+}-binding proteins such as CaM. CaM is a 148-amino-acid, highly conserved Ca^{2+}-binding protein that contains four EF-hand Ca^{2+}-binding motifs [48]. As illustrated in Fig. 2, CaM can bind to calcineurin and Ca^{2+}/CaM-dependent kinases (CaMKs). Calcineurin is a heterodimer composed of a catalytic subunit, calcineurin A, and a Ca^{2+}-binding regulatory subunit, calcineurin B [49–51]. These two subunits are tightly bound, being dissociated only by denaturation, and both subunits are essential for calcineurin function. Calcineurin A and B complex has the major function as Ca^{2+}/CaM-dependent phosphatase, which

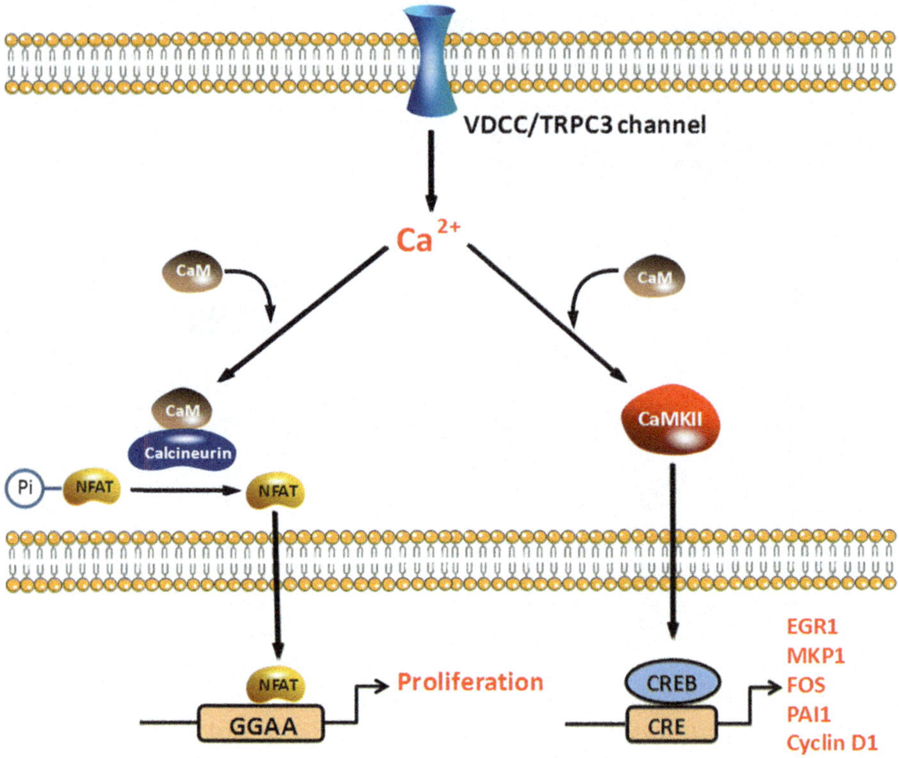

Fig. 2 *The role of calcium signaling in controlling proliferation of airway smooth muscle cells is mediated by* Ca^{2+}*/CaM/calcineurin/NFAT and* Ca^{2+}*/CaM/CaMKII/CREB pathways.* Intracellular Ca^{2+} entered from VDCC and TRPC3 channel can be bound to CaM and then activate the calcineurin/NFAT pathway, leading to proliferation of ASMCs. In addition, Ca^{2+} may also activate CaMKII, which phosphorylates CREB, mediating of ASMC proliferation gene expression, and then proliferation

dephosphorylates NFAT, leading to a translocation of NFAT to nucleus as a transcriptional factor. Many factors regulate NFAT activation and translocation. For example, Ca^{2+} influx through TRP, Orai, or VDCCs can activate calcineurin in many types of cells [52–55]. C-Jun amino-terminal kinase2 (JNK2) can phosphorylate nuclear NFAT and promote its nuclear export in freshly isolated cerebral artery smooth muscle cells [56]. It has been shown that NFAT regulates Ca^{2+}-dependent gene expression in smooth muscle cells. Interestingly, recently published data suggest that NFAT has the same function in airway hyperresponsiveness and pulmonary hypertension, where smooth muscle cells play a predominant role [56] and an NFAT1 knockout mouse shows reduced airway hyperresponsiveness after agonist stimulation [57]. Collectively, NFAT may exert the same function in AMSCs as in vascular smooth muscle cells, triggering remodeling.

6 CaMKII/CREB Signaling Pathway in Airway Smooth Muscle Remodeling

The multifunctional Ca^{2+}/CaM-dependent kinases (CaMKs) are a family of serine/threonine protein kinases that include CaMKI, CaMKII, and CaMKIV [58, 59]. These kinases have an amino-terminal catalytic domain followed by a carboxyl-terminal regulatory domain. The regulatory domain consists of overlapping autoinhibitory and Ca^{2+}/CaM binding domains. Similar to calcineurin, the autoinhibition is relieved upon Ca^{2+}/CaM binding. Ca^{2+}-cyclic AMP response element binding protein (CREB) regulates transcription through binding to Ca^{2+}-cyclic AMP-response elements (CREs) in the promoter of many genes [60, 61]. Phosphorylation of CREB at serine-133 promotes its binding to CREs [62]. CREB activation induced by Ca^{2+} influx from VDCCs has been demonstrated in both cultured SMCs and cerebral arteries [63].

Activation of CaMKII through Ca^{2+}-bound CaM can phosphorylate CREB [61]. Phosphorylated CREB can bind to the promoter region of cyclin D1 and increase cyclin D1 expression [64, 65]. Because cyclin D1 is one of the most important cell cycle mediators that have profound effects in the ASMC cycle, increased expression of cyclin D1 by Ca^{2+}/CaM/CaMKII/CREB pathway could play an important role in ASMC remodeling. Furthermore, published data suggest that CREB has the function of activating pro-proliferative genes such as early growth response protein 1 (EGR1), MAPK phosphatase 1 (MKP1), FOS, and plasminogen activator inhibitor-1 (PAI1) when CREB binds to CREs in the promoter of those genes leading to vascular smooth muscle proliferation [66–68], and it is highly possible that ASMCs may have similar mechanisms in their proliferation. Figure 2 summarizes the aforementioned Ca^{2+}-dependent signaling pathways that are involved in proliferation of ASMCs.

7 Phosphoinositide-3 Kinase Signaling in Airway Smooth Muscle Remodeling

Another potentially important aspect of ASMC remodeling is an increase in its size; hypertrophy has been reported in specimens from fatal [69], severe [70], and allergic [71] asthma. By contrast, ASMCs may not exhibit hypertrophy in mild to moderate asthma [72]. As shown in Fig. 3, ASMC hypertrophy [73] may be mediated by three distinct signaling pathways, i.e., PI3 kinase (PI3K)/mammalian target of rapamycin (mTOR)/4E-binding protein (4E-BP)/transcription factor eukaryotic translation initiation factor 4E (eIF4E) [74], S6 kinase/S6 [75], and glycogen synthase kinase (GSK-3β) [76]. Interestingly, the activation of PI3 kinase, S6 kinase, and GSK-3β are all Ca^{2+}-dependent in smooth muscle cells. Su and his colleagues have reported that activation of PI3-kinase is involved in a Ca^{2+}/MLC phosphorylation-dependent and independent pathway for contraction in vascular smooth muscle cells [77].

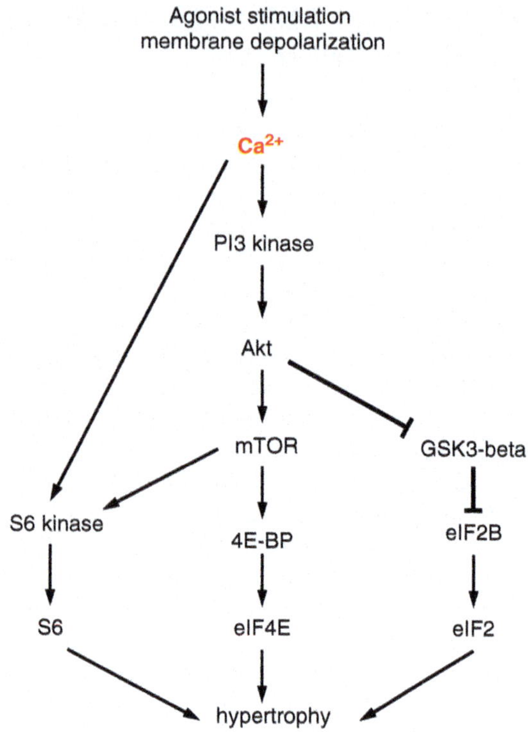

Fig. 3 *Calcium signaling mediates airway smooth muscle cell hypertrophy through three distinct pathways*. This figure is modified from a previous publication [73]

An intracellular Ca^{2+}-activated p70 S6 kinase has been reported in liver epithelial cells [78] and vascular smooth muscle cells [79, 80]. Moreover, a study from Welsh et al. implies that in ASMCs, hypertrophy can take place when GSK-3β is inactivated by Ca^{2+}-activated PI3 kinase/Akt phosphorylation, leading to eukaryotic translation initiation factor 2B (eIF2B) dephosphorylation and then activation, followed by enhancement of translation initiation [81]. Collectively, three major pathways that are involved in ASMC hypertrophy are affected by Ca^{2+}, indicating that a change in Ca^{2+} signaling will change ASMC remodeling in the aspect of cell hypertrophy.

8 Conclusions

Airway remodeling may occur due to ASMC hyperplasia and hypertrophy. This chapter focuses on Ca^{2+}-mediated signaling pathways that affect ASMC proliferation and hypertrophy. TRPC3 channels and SERCA2 are expressed and make important contributions to the control of $[Ca^{2+}]_i$ in ASMCs. These two important

Ca^{2+} signaling molecules appear to impact ASMC proliferation. In terms of ASMC hypertrophy, three signaling pathways known to induce hypertrophy are regulated by Ca^{2+} as well. Therefore, altered $[Ca^{2+}]_i$ in ASMCs is likely to be important for airway remodeling.

References

1. Chung, K.F., and Sterk, P.J. 2000. The airway smooth muscle cell: a major contributor to asthma? *Eur Respir J* 15:438–439.
2. Chung, K.F. 2005. The role of airway smooth muscle in the pathogenesis of airway wall remodeling in chronic obstructive pulmonary disease. *Proc Am Thorac Soc* 2:347–354; discussion 371–342.
3. Damera, G., and Panettieri, R.A., Jr. 2011. Does airway smooth muscle express an inflammatory phenotype in asthma? *Br J Pharmacol* 163:68–80.
4. Bara, I., Ozier, A., Girodet, P.O., Carvalho, G., Cattiaux, J., Begueret, H., Thumerel, M., Ousova, O., Kolbeck, R., Coyle, A.J., *et al.* 2012. Role of YKL-40 in bronchial smooth muscle remodeling in asthma. *Am J Respir Crit Care Med* 185:715–722.
5. Markwick, L.J., Clements, D., Roberts, M.E., Ceresa, C.C., Knox, A.J., and Johnson, S.R. 2012. CCR3 induced-p42/44 MAPK activation protects against staurosporine induced-DNA fragmentation but not apoptosis in airway smooth muscle cells. *Clin Exp Allergy* 42:1040–1050.
6. Plant, P.J., North, M.L., Ward, A., Ward, M., Khanna, N., Correa, J., Scott, J.A., and Batt, J. 2012. Hypertrophic airway smooth muscle mass correlates with increased airway responsiveness in a murine model of asthma. *Am J Respir Cell Mol Biol* 46:532–540.
7. Wei, J.H., Mo, B.W., and Huang, J.W. 2011. [Role of toll-like receptor 4 in the asthmatic rat airway smooth muscle cells proliferation and apoptosis]. *Zhongguo Ying Yong Sheng Li Xue Za Zhi* 27:284–288.
8. Halwani, R., Al-Abri, J., Beland, M., Al-Jahdali, H., Halayko, A.J., Lee, T.H., Al-Muhsen, S., and Hamid, Q. 2011. CC and CXC chemokines induce airway smooth muscle proliferation and survival. *J Immunol* 186:4156–4163.
9. Makinde, T., Murphy, R.F., and Agrawal, D.K. 2007. The regulatory role of TGF-β in airway remodeling in asthma. *Immunol Cell Biol* 85:348–356.
10. Al-Alwan, L.A., Chang, Y., Baglole, C.J., Risse, P.A., Halayko, A.J., Martin, J.G., Eidelman, D.H., and Hamid, Q. 2012. Autocrine-regulated airway smooth muscle cell migration is dependent on IL-17-induced growth-related oncogenes. *J Allergy Clin Immunol* 130:977–985 e976.
11. Chang, Y., Al-Alwan, L., Risse, P.A., Roussel, L., Rousseau, S., Halayko, A.J., Martin, J.G., Hamid, Q., and Eidelman, D.H. 2011. TH17 cytokines induce human airway smooth muscle cell migration. *J Allergy Clin Immunol* 127:1046–1053 e1041-1042.
12. Joubert, P., Lajoie-Kadoch, S., Labonte, I., Gounni, A.S., Maghni, K., Wellemans, V., Chakir, J., Laviolette, M., Hamid, Q., and Lamkhioued, B. 2005. CCR3 expression and function in asthmatic airway smooth muscle cells. *J Immunol* 175:2702–2708.
13. Wright, D.B., Trian, T., Siddiqui, S., Pascoe, C.D., Johnson, J.R., Dekkers, B.G., Dakshinamurti, S., Bagchi, R., Burgess, J.K., Kanabar, V., et al. 2013. Phenotype modulation of airway smooth muscle in asthma. *Pulm Pharmacol Ther* 26:42–49.
14. Billington, C.K., and Penn, R.B. 2003. Signaling and regulation of G protein-coupled receptors in airway smooth muscle. *Respir Res* 4:2.
15. Jude, J.A., Wylam, M.E., Walseth, T.F., and Kannan, M.S. 2008. Calcium signaling in airway smooth muscle. *Proc Am Thorac Soc* 5:15–22.
16. Berridge, M.J. 1993. Inositol trisphosphate and calcium signalling. *Nature* 361:315–325.

17. Deshpande, D.A., White, T.A., Dogan, S., Walseth, T.F., Panettieri, R.A., and Kannan, M.S. 2005. CD38/cyclic ADP-ribose signaling: role in the regulation of calcium homeostasis in airway smooth muscle. *Am J Physiol Lung Cell Mol Physiol* 288:L773-788.
18. McPherson, P.S., and Campbell, K.P. 1993. The ryanodine receptor/Ca^{2+} release channel. *J Biol Chem* 268:13765–13768.
19. Berridge, M.J. 2008. Smooth muscle cell calcium activation mechanisms. *J Physiol* 586:5047–5061.
20. Fleischmann, B.K., Wang, Y.X., and Kotlikoff, M.I. 1997. Muscarinic activation and calcium permeation of nonselective cation currents in airway myocytes. *Am J Physiol* 272:C341-349.
21. Beech, D.J., Muraki, K., and Flemming, R. 2004. Non-selective cationic channels of smooth muscle and the mammalian homologues of Drosophila TRP. *J Physiol* 559:685–706.
22. Xiao, J.H., Zheng, Y.M., Liao, B., and Wang, Y.X. 2010. Functional role of canonical transient receptor potential 1 and canonical transient receptor potential 3 in normal and asthmatic airway smooth muscle cells. *Am J Respir Cell Mol Biol* 43:17–25.
23. Hirota, S., Pertens, E., and Janssen, L.J. 2007. The reverse mode of the Na^+/Ca^{2+} exchanger provides a source of Ca^{2+} for store refilling following agonist-induced Ca^{2+} mobilization. *Am J Physiol Lung Cell Mol Physiol* 292:L438-447.
24. Liu, B., Peel, S.E., Fox, J., and Hall, I.P. 2010. Reverse mode Na^+/Ca^{2+} exchange mediated by STIM1 contributes to Ca^{2+} influx in airway smooth muscle following agonist stimulation. *Respir Res* 11:168.
25. Fleischmann, B.K., Wang, Y.X., Pring, M., and Kotlikoff, M.I. 1996. Voltage-dependent calcium currents and cytosolic calcium in equine airway myocytes. *J Physiol* 492 (Pt 2):347–358.
26. Thorneloe, K.S., and Nelson, M.T. 2005. Ion channels in smooth muscle: regulators of intracellular calcium and contractility. *Can J Physiol Pharmacol* 83:215–242.
27. Mercer, J.C., Dehaven, W.I., Smyth, J.T., Wedel, B., Boyles, R.R., Bird, G.S., and Putney, J. W., Jr. 2006. Large store-operated calcium selective currents due to co-expression of Orai1 or Orai2 with the intracellular calcium sensor, Stim1. *J Biol Chem* 281:24979–24990.
28. Prakriya, M., Feske, S., Gwack, Y., Srikanth, S., Rao, A., and Hogan, P.G. 2006. Orai1 is an essential pore subunit of the CRAC channel. *Nature* 443:230–233.
29. Soboloff, J., Spassova, M.A., Tang, X.D., Hewavitharana, T., Xu, W., and Gill, D.L. 2006. Orai1 and STIM reconstitute store-operated calcium channel function. *J Biol Chem* 281:20661–20665.
30. Luik, R.M., Wu, M.M., Buchanan, J., and Lewis, R.S. 2006. The elementary unit of store-operated Ca^{2+} entry: local activation of CRAC channels by STIM1 at ER-plasma membrane junctions. *J Cell Biol* 174:815–825.
31. Wu, M.M., Buchanan, J., Luik, R.M., and Lewis, R.S. 2006. Ca^{2+} store depletion causes STIM1 to accumulate in ER regions closely associated with the plasma membrane. *J Cell Biol* 174:803–813.
32. Zhang, S.L., Yu, Y., Roos, J., Kozak, J.A., Deerinck, T.J., Ellisman, M.H., Stauderman, K.A., and Cahalan, M.D. 2005. STIM1 is a Ca^{2+} sensor that activates CRAC channels and migrates from the Ca^{2+} store to the plasma membrane. *Nature* 437:902–905.
33. Liou, J., Kim, M.L., Heo, W.D., Jones, J.T., Myers, J.W., Ferrell, J.E., Jr., and Meyer, T. 2005. STIM is a Ca^{2+} sensor essential for Ca^{2+}-store-depletion-triggered Ca^{2+} influx. *Curr Biol* 15:1235–1241.
34. Yuan, J.P., Zeng, W., Huang, G.N., Worley, P.F., and Muallem, S. 2007. STIM1 heteromultimerizes TRPC channels to determine their function as store-operated channels. *Nat Cell Biol* 9:636–645.
35. Mahn, K., Hirst, S.J., Ying, S., Holt, M.R., Lavender, P., Ojo, O.O., Siew, L., Simcock, D.E., McVicker, C.G., Kanabar, V., et al. 2009. Diminished sarco/endoplasmic reticulum Ca^{2+} ATPase (SERCA) expression contributes to airway remodelling in bronchial asthma. *Proc Natl Acad Sci U S A* 106:10775–10780.

36. Dolmetsch, R.E., Xu, K., and Lewis, R.S. 1998. Calcium oscillations increase the efficiency and specificity of gene expression. *Nature* 392:933–936.
37. Yao, H., Peng, F., Fan, Y., Zhu, X., Hu, G., and Buch, S.J. 2009. TRPC channel-mediated neuroprotection by PDGF involves Pyk2/ERK/CREB pathway. *Cell Death Differ* 16:1681–1693.
38. Perez, J., Torres, R.A., Rocic, P., Cismowski, M.J., Weber, D.S., Darley-Usmar, V.M., and Lucchesi, P.A. 2011. PYK2 signaling is required for PDGF-dependent vascular smooth muscle cell proliferation. *Am J Physiol Cell Physiol* 301:C242-251.
39. Chuang, T.Y., Au, L.C., Wang, L.C., Ho, L.T., Yang, D.M., and Juan, C.C. 2012. Potential effect of resistin on the ET-1-increased reactions of blood pressure in rats and Ca^{2+} signaling in vascular smooth muscle cells. *J Cell Physiol* 227:1610–1618.
40. Pulver-Kaste, R.A., Barlow, C.A., Bond, J., Watson, A., Penar, P.L., Tranmer, B., and Lounsbury, K.M. 2006. Ca^{2+} source-dependent transcription of CRE-containing genes in vascular smooth muscle. *Am J Physiol Heart Circ Physiol* 291:H97-105.
41. Strachan, R.T., Allen, J.A., Sheffler, D.J., and Roth, B.L. 2010. p90 Ribosomal S6 kinase 2, a novel GPCR kinase, is required for growth factor-mediated attenuation of GPCR signaling. *Biochemistry* 49:2657–2671.
42. Bair, A.M., Thippegowda, P.B., Freichel, M., Cheng, N., Ye, R.D., Vogel, S.M., Yu, Y., Flockerzi, V., Malik, A.B.., and Tiruppathi, C. 2009. Ca^{2+} entry via TRPC channels is necessary for thrombin-induced NF-kappaB activation in endothelial cells through AMP-activated protein kinase and protein kinase Cdelta. *J Biol Chem* 284:563–574.
43. Guo, Z.Y., Hao, X.H., Tan, F.F., Pei, X., Shang, L.M., Jiang, X.L., and Yang, F. 2011. The elements of human cyclin D1 promoter and regulation involved. *Clin Epigenetics* 2:63–76.
44. Hu, J., and Colburn, N.H. 2005. Histone deacetylase inhibition down-regulates cyclin D1 transcription by inhibiting nuclear factor-kappaB/p65 DNA binding. *Mol Cancer Res* 3:100–109.
45. Smedlund, K., Tano, J.Y., and Vazquez, G. 2010. The constitutive function of native TRPC3 channels modulates vascular cell adhesion molecule-1 expression in coronary endothelial cells through nuclear factor kappaB signaling. *Circ Res* 106:1479–1488.
46. Zou, J.J., Gao, Y.D., Geng, S., and Yang, J. 2011. Role of STIM1/Orai1-mediated store-operated Ca^{2+} entry in airway smooth muscle cell proliferation. *J Appl Physiol* 110:1256–1263.
47. Spinelli, A.M., Gonzalez-Cobos, J.C., Zhang, X., Motiani, R.K., Rowan, S., Zhang, W., Garrett, J., Vincent, P.A., Matrougui, K., Singer, H.A., et al. 2012. Airway smooth muscle STIM1 and Orai1 are upregulated in asthmatic mice and mediate PDGF-activated SOCE, CRAC currents, proliferation, and migration. *Pflugers Arch* 464:481–492.
48. Chin, D., and Means, A.R. 2000. Calmodulin: a prototypical calcium sensor. *Trends Cell Biol* 10:322–328.
49. Aramburu, J., Rao, A., and Klee, C.B. 2000. Calcineurin: from structure to function. *Curr Top Cell Regul* 36:237–295.
50. Guerini, D. 1997. Calcineurin: not just a simple protein phosphatase. *Biochem Biophys Res Commun* 235:271–275.
51. Klee, C.B., Ren, H., and Wang, X. 1998. Regulation of the calmodulin-stimulated protein phosphatase, calcineurin. *J Biol Chem* 273:13367–13370.
52. Crabtree, G.R., and Schreiber, S.L. 2009. SnapShot: Ca2+–calcineurin-NFAT signaling. *Cell* 138:210, 210 e211.
53. Jans, R., Mottram, L., Johnson, D.L., Brown, A.M., Sikkink, S., Ross, K., and Reynolds, N.J. 2013. Lysophosphatidic Acid Promotes Cell Migration through STIM1- and Orai1-Mediated $Ca^{2+}{}_i$ Mobilization and NFAT2 Activation. *J Invest Dermatol* 133:793–802.
54. Nakayama, H., Wilkin, B.J., Bodi, I., and Molkentin, J.D. 2006. Calcineurin-dependent cardiomyopathy is activated by TRPC in the adult mouse heart. *FASEB J* 20:1660–1670.
55. Wellman, G.C., and Nelson, M.T. 2003. Signaling between SR and plasmalemma in smooth muscle: sparks and the activation of Ca^{2+}-sensitive ion channels. *Cell Calcium* 34:211–229.

56. Said, S.I., Hamidi, S.A., and Gonzalez Bosc, L. 2010. Asthma and pulmonary arterial hypertension: do they share a key mechanism of pathogenesis? *Eur Respir J* 35:730–734.
57. Fonseca, B.P., Olsen, P.C., Coelho, L.P., Ferreira, T.P., Souza, H.S., Martins, M.A., and Viola, J.P. 2009. NFAT1 transcription factor regulates pulmonary allergic inflammation and airway responsiveness. *Am J Respir Cell Mol Biol* 40:66–75.
58. Hook, S.S., and Means, A.R. 2001. Ca^{2+}/CaM-dependent kinases: from activation to function. *Annu Rev Pharmacol Toxicol* 41:471–505.
59. Means, A.R. 2000. Regulatory cascades involving calmodulin-dependent protein kinases. *Mol Endocrinol* 14:4–13.
60. Barlow, C.A., Rose, P., Pulver-Kaste, R.A., and Lounsbury, K.M. 2006. Excitation-transcription coupling in smooth muscle. *J Physiol* 570:59–64.
61. Shaywitz, A.J., and Greenberg, M.E. 1999. CREB: a stimulus-induced transcription factor activated by a diverse array of extracellular signals. *Annu Rev Biochem* 68:821–861.
62. Gonzalez, G.A., Yamamoto, K.K., Fischer, W.H., Karr, D., Menzel, P., Biggs, W., 3rd, Vale, W.W., and Montminy, M.R. 1989. A cluster of phosphorylation sites on the cyclic AMP-regulated nuclear factor CREB predicted by its sequence. *Nature* 337:749–752.
63. Cartin, L., Lounsbury, K.M., and Nelson, M.T. 2000. Coupling of Ca^{2+} to CREB activation and gene expression in intact cerebral arteries from mouse : roles of ryanodine receptors and voltage-dependent Ca^{2+} channels. *Circ Res* 86:760–767.
64. Musa, N.L., Ramakrishnan, M., Li, J., Kartha, S., Liu, P., Pestell, R.G., and Hershenson, M.B. 1999. Forskolin inhibits cyclin D1 expression in cultured airway smooth-muscle cells. *Am J Respir Cell Mol Biol* 20:352–358.
65. Page, K., Li, J., Wang, Y., Kartha, S., Pestell, R.G., and Hershenson, M.B. 2000. Regulation of cyclin D1 expression and DNA synthesis by phosphatidylinositol 3-kinase in airway smooth muscle cells. *Am J Respir Cell Mol Biol* 23:436–443.
66. Atsuta, H., Uchiyama, T., Kanai, H., Iso, T., Tanaka, T., Suga, T., Maeno, T., Arai, M., Nagai, R., and Kurabayashi, M. 2009. Effects of a stable prostacyclin analogue beraprost sodium on VEGF and PAI-1 gene expression in vascular smooth muscle cells. *Int J Cardiol* 132:411–418.
67. Che, W., Manetsch, M., Quante, T., Rahman, M.M., Patel, B.S., Ge, Q., and Ammit, A.J. 2012. Sphingosine 1-phosphate induces MKP-1 expression via p38 MAPK- and CREB-mediated pathways in airway smooth muscle cells. *Biochim Biophys Acta* 1823:1658–1665.
68. Pulver, R.A., Rose-Curtis, P., Roe, M.W., Wellman, G.C., and Lounsbury, K.M. 2004. Store-operated Ca^{2+} entry activates the CREB transcription factor in vascular smooth muscle. *Circ Res* 94:1351–1358.
69. Ebina, M., Takahashi, T., Chiba, T., and Motomiya, M. 1993. Cellular hypertrophy and hyperplasia of airway smooth muscles underlying bronchial asthma. A 3-D morphometric study. *Am Rev Respir Dis* 148:720–726.
70. Benayoun, L., Druilhe, A., Dombret, M.C., Aubier, M., and Pretolani, M. 2003. Airway structural alterations selectively associated with severe asthma. *Am J Respir Crit Care Med* 167:1360–1368.
71. Begueret, H., Berger, P., Vernejoux, J.M., Dubuisson, L., Marthan, R., and Tunon-de-Lara, J.M. 2007. Inflammation of bronchial smooth muscle in allergic asthma. *Thorax* 62:8–15.
72. Woodruff, P.G., Dolganov, G.M., Ferrando, R.E., Donnelly, S., Hays, S.R., Solberg, O.D., Carter, R., Wong, H.H., Cadbury, P.S., and Fahy, J.V. 2004. Hyperplasia of smooth muscle in mild to moderate asthma without changes in cell size or gene expression. *Am J Respir Crit Care Med* 169:1001–1006.
73. Bentley, J.K., and Hershenson, M.B. 2008. Airway smooth muscle growth in asthma: proliferation, hypertrophy, and migration. *Proc Am Thorac Soc* 5:89–96.
74. Zhou, L., Goldsmith, A.M., Bentley, J.K., Jia, Y., Rodriguez, M.L., Abe, M.K., Fingar, D.C., and Hershenson, M.B. 2005. 4E-binding protein phosphorylation and eukaryotic initiation factor-4E release are required for airway smooth muscle hypertrophy. *Am J Respir Cell Mol Biol* 33:195–202.

75. Deng, H., Hershenson, M.B., Lei, J., Bitar, K.N., Fingar, D.C., Solway, J., and Bentley, J.K. 2010. p70 Ribosomal S6 kinase is required for airway smooth muscle cell size enlargement but not increased contractile protein expression. *Am J Respir Cell Mol Biol* 42:744–752.
76. Deng, H., Dokshin, G.A., Lei, J., Goldsmith, A.M., Bitar, K.N., Fingar, D.C., Hershenson, M. B., and Bentley, J.K. 2008. Inhibition of glycogen synthase kinase-3β is sufficient for airway smooth muscle hypertrophy. *J Biol Chem* 283:10198–10207.
77. Su, X., Smolock, E.M., Marcel, K.N., and Moreland, R.S. 2004. Phosphatidylinositol 3-kinase modulates vascular smooth muscle contraction by calcium and myosin light chain phosphorylation-independent and -dependent pathways. *Am J Physiol Heart Circ Physiol* 286:H657-666.
78. Graves, L.M., He, Y., Lambert, J., Hunter, D., Li, X., and Earp, H.S. 1997. An intracellular calcium signal activates p70 but not p90 ribosomal S6 kinase in liver epithelial cells. *J Biol Chem* 272:1920–1928.
79. Brinson, A.E., Harding, T., Diliberto, P.A., He, Y., Li, X., Hunter, D., Herman, B., Earp, H.S., and Graves, L.M. 1998. Regulation of a calcium-dependent tyrosine kinase in vascular smooth muscle cells by angiotensin II and platelet-derived growth factor. Dependence on calcium and the actin cytoskeleton. *J Biol Chem* 273:1711–1718.
80. Eguchi, S., Iwasaki, H., Ueno, H., Frank, G.D., Motley, E.D., Eguchi, K., Marumo, F., Hirata, Y., and Inagami, T. 1999. Intracellular signaling of angiotensin II-induced p70 S6 kinase phosphorylation at Ser^{411} in vascular smooth muscle cells. Possible requirement of epidermal growth factor receptor, Ras, extracellular signal-regulated kinase, and Akt. *J Biol Chem* 274:36843–36851.
81. Welsh, G.I., Miller, C.M., Loughlin, A.J., Price, N.T., and Proud, C.G. 1998. Regulation of eukaryotic initiation factor eIF2B: glycogen synthase kinase-3 phosphorylates a conserved serine which undergoes dephosphorylation in response to insulin. *FEBS Lett* 421:125–130.

Regulation of Intracellular Calcium by Bitter Taste Receptors on Airway Smooth Muscle

Deepak A. Deshpande and Stephen B. Liggett

Contents

Abstract Recent studies have demonstrated the expression of bitter taste receptors (BTRs) on airway smooth muscle (ASM) using human, mouse, and guinea-pig airways. BTRs, belonging to the *Tas2R* family, are activated by a wide range of synthetic and natural compounds. However, these receptors are evolved with low, but broad-spectrum affinity for their ligands. Stimulation of BTRs with known bitter tastants results in an elevation of basal intracellular calcium concentration in cultured human ASM cells, similar to stimulation with other G_q-coupled G protein coupled receptors. Studies in human and murine ASM cells have demonstrated that bitter tastant-induced calcium elevation is $G_{\beta\gamma}$-, phospholipase C-, inositol trisphosphate (IP_3)-, and IP_3-receptor dependent. Very interestingly, bitter tastants induce efficacious ASM relaxation and bronchodilation in airways obtained from human, mouse, and guinea-pig lungs. At least two potential mechanisms for bitter-tastant-induced ASM relaxation have been proposed: (1) activation of calcium-activated potassium channels by calcium induced by bitter tastants, resulting in membrane hyperpolarization and ASM relaxation, and (2) inhibition of calcium

D.A. Deshpande, Ph.D. (✉)
Department of Medicine, Pulmonary and Critical Care Medicine, School of Medicine, University of Maryland, HSF-II, Rm-S143, 20 Penn Street, Baltimore, MD 21201, USA
e-mail: ddeshpan@medicine.umaryland.edu

S.B. Liggett, M.D.
Morsani College of Medicine, University of South Florida, Tampa FL, USA

Y.-X. Wang (ed.), *Calcium Signaling In Airway Smooth Muscle Cells*, DOI 10.1007/978-3-319-01312-1_23, entry via voltage-dependent calcium channels upon stimulation of ASM with contractile agonists, thereby inhibiting contraction. Nevertheless, the paradoxical effect of bitter tastants on ASM tone is intriguing and establishes *Tas2* receptors as a novel class of therapeutic targets in the treatment of obstructive airway diseases such as asthma and chronic obstructive pulmonary disease.

Keywords Taste receptors • Bitter • Tas2 • Gustducin • Chloroquine

1 Introduction

Airway smooth muscle (ASM) cells express a variety of receptors, and activation of these cell surface receptors by cognate ligands, released endogenously at nerve terminal or by inflammatory cells, results in functional changes in ASM. G protein coupled receptors (GPCRs) play an important role in the regulation of contraction, relaxation, and proliferation of ASM cells. Most importantly, GPCRs are the primary target of most medications used to treat airway obstructive diseases such as asthma and chronic obstructive pulmonary disease (COPD) [11]. Recent studies have identified expression of a new class of receptors on ASM and determined the functional role of these receptors. These include bitter taste receptors (BTRs) [12, 29], proton-sensing receptors [33], and γ-aminobutyric acid (GABA) [22–24] receptors. This chapter describes the signal transduction, calcium regulation, and functional role of bitter taste receptors, a novel class of receptors on ASM cells.

2 Bitter Taste Receptors

Chemosensory cells primarily responsible for detecting the taste of food or inhaled chemicals are involved in gustatory and olfactory functions in the body. Aggregates of specialized cells form taste buds distributed throughout the oral cavity. These cells are endowed with the responsibility of chemosensation. The activation of taste cells by a variety of chemical agents elicits a series of neuronal reflexes that dictate ingestion or aversion of food and coordinate various effector functions necessary to either digest food or expel. The chemosensory cells express receptor and signal transduction machinery necessary to mediate chemosensation. Recent studies have identified the presence of chemosensory receptors and components of signal transduction in a variety of nongustatory cells and organs [4, 16].

Taste qualities are divided into five different categories: salt, sour, sweet, bitter, and umami [5, 17]. Salt and sour tastes are believed to be detected by change in ion channel activities, resulting in changes in membrane potential. Sweet, bitter, and umami (meaning "delicious" in Japanese and involves the detection of the taste of

Table 1 Subtypes of bitter taste receptors expressed in ASM

Human	Guinea pig	Mice
TAS2R10	TAS2R3	TAS2R107
TAS2R31	TAS2R4	
TAS2R5	TAS2R14	
TAS2R4	TAS2R16	
TAS2R19		
TAS2R3		
TAS2R20		
TAS2R45		

glutamate) tastes involve activation of GPCRs. Two families of taste receptors exist: the *Tas1R* (T1R), for sweet and umami, and *Tas2R*s (T2R), for bitter. T1Rs are classified as T1R1, T1R2, and T1R3. Sweet and umami receptors form heterodimers, T1R2/T1R3 for sweet and T1R1/T1R3 for umami, and heterodimer formation is necessary for the function of T1Rs. A detailed discussion of sweet and umami receptor physiology, biochemistry, and pharmacology is beyond the scope of this chapter.

Bitter taste is mediated by the receptors expressed by the *Tas2R* gene family [1]. Functionally, T2Rs are expressed as monomers and evolved as low-affinity (micromolar to millimolar range) receptors compared with other GPCRs [21, 25]. Nearly 30 subtypes of T2Rs are identified in mammals with a broad range of agonistic activity. Although recent studies have identified the role of a carboxy terminal in agonist selectivity, more studies are needed to establish structure-function relationship in T2Rs [7, 30, 35]. T2Rs are largely expressed on taste buds in the oral cavity. However, expression of T2Rs in the respiratory system has been recently identified [5, 16, 38]. Presumably, activation of respiratory T2Rs evokes reflex responses to avert inhalation of toxic chemicals and support chemoprotection. Several lines of experimental evidence suggest the expression of T2Rs on vomeronasal neuroepithelium [42], solitary chemosensory cells (SCC) [36], ciliary epithelium [34], and tracheal brush cells [38]. Recently we identified the expression of T2Rs on human ASM cells, and subsequently two different groups of investigators demonstrated expression of BTRs on mouse [44] and guinea-pig [29] ASM (Table 1). Using real-time PCR we have shown expression of six different subtypes of T2R in isolated human ASM at levels greater than or equal to that of the β2AR adrenergic receptor (AR). Three subtypes are highly expressed (T2R10, T2R14, T2R31), and three more are moderately expressed (T2R5, T2R4, T2R19) in ASM [12]. Expression of some of these subtypes was further confirmed by immunofluorescence studies. Mouse ASM cells express T2R107 on their membrane [44]. The mechanisms that regulate expression of BTRs are not clearly understood. Future studies are needed to ascertain why different subtypes of T2Rs are expressed at different levels in ASM. Multiple members of the T2R family are present in each taste-responsive cell, thereby making these cells respond to multiple agonists and not restrict the responsiveness

Table 2 Bitter taste receptor subtypes expressed on human ASM and examples of cognate agonists (Source: http://bitterdb.agri.huji.ac.il/bitterdb/dbbitter.php#Home). A number of potential BTR agonists (e.g., 31 agonists for T2R10) have been identified

Receptor	Agonists (number, examples)	
TAS2R10	31	Quinine, denatonium, camphor, coumarin, papaverine, chlorpheniramine
TAS2R14	43	Quinine, aristolochic acid, papaverine, dihydramine
TAS2R31	8	Quinine, aristolochic acid, saccharin, diphenidol
TAS2R5	1	Phenanthroline
TAS2R4	15	Quinine, camphor, yohimbine, colchicine
TAS2R19	0	
TAS2R3	1	Chloroquine
TAS2R20	2	Diphenidol, cromolyn
TAS2R45	0	

to specific agonists [21]. For example, quinine activates T2R10, 14, 31, 4, whereas chloroquine activates only 1 receptor, i.e., T2R10, on ASM cells. A list of potential BTR agonists has been given in Table 2. A recent study found a correlation between the expression of T2Rs and asthma severity using leukocytes obtained from asthmatics and healthy controls [28]. These findings are valuable in exploiting T2Rs as potentially new therapeutic targets in asthma.

3 Intracellular Signaling by T2R

The signaling components that are necessary for functions evoked by T2Rs include G protein such as α-gustducin, phospholipase Cβ2, IP3 receptor, and cation channel TrpM5. Expression of T2Rs is typically associated with the expression of signal transduction components. Therefore, the expression of α-gustducin or TrpM5 is used as a marker of chemoresponsive cells. In human ASM cells, we demonstrated that T2R expression is associated with the expression of α-gustducin. A recent study using murine ASM confirmed the expression of α-gustducin, Gβ3, and Gγ13 in ASM [44]. Mice that lack α-gustducin show significantly attenuated bitter taste sensory response, and lack of α-gustducin is also associated with decreased expression of T2Rs in the taste cells. These studies demonstrate the pivotal role of α-gustducin in chemosensation. However, the functional role of α-gustducin in the signal transduction of bitter taste receptors in ASM cells needs further experimental evidence based on either knockdown of expression or using α-gustducin null mice. It is important to note that several cell types that do not express α-gustducin are found to be positive for other components of taste receptor signal transduction [20, 36]. Therefore, cells that are negative for α-gustducin do not necessarily lack an ability to respond to bitter tastants. Gα14 and Gα-transducins are another closely related family of G proteins that are thought to be involved in taste receptor signaling [14, 20, 37].

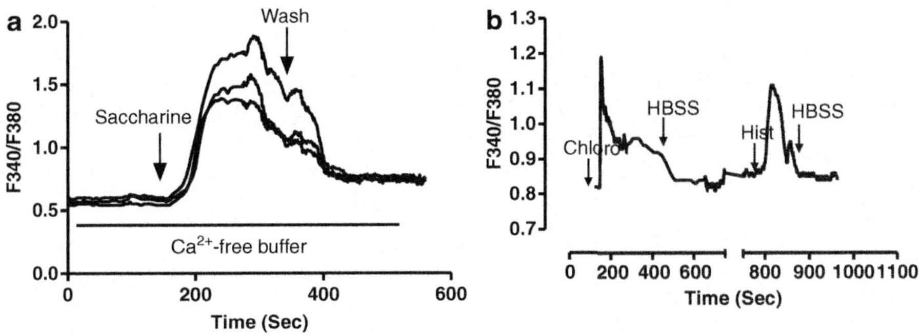

Fig. 1 Global calcium rise in human ASM cells maintained in culture by saccharine when the cells were maintained in calcium-free buffer (**a**). Stimulation of cultured human ASM cells with chloroquine results in the elevation of intracellular calcium concentration. Application of histamine to cells after washing out chloroquine results in an expected rise in the intracellular calcium concentration (**b**). These data suggest that bitter tastants do not damage ASM cell membranes

Taste receptor binding leads to activation of a heterotrimeric G protein, which consists of Gα-gustducin (α-gust) [8, 19] and its βγ partners, β3γ13 [15]. When a bitter tastant binds to its receptor, the heterodimer, β3γ13, is released from tastant-bound receptors separating itself from α-gustducin. The effector molecule that drives signaling and function in taste cells is the β3γ13 complex. β3γ13 stimulates phospholipase Cβ2 (PLCβ2) [8, 31, 32] and produces inositol trisphosphate (IP_3), which in turn activates the IP_3 receptor (IP_3R) to release stored calcium (Ca^{2+}). Expression of Gβ3, Gγ13, and PLCβ2 in ASM was recently demonstrated using murine ASM [44]. Identification of G-protein subunits and effector enzyme, PLCβ2, in ASM reiterates the presence of canonical BTR signaling in ASM.

Studies from our laboratory demonstrated that stimulation of human ASM cells with a panel of known bitter tastants (chloroquine, quinine, saccharin, denatonium, colchicine, aristolochic acid, and yohimbine) results in a dose-dependent increase in intracellular calcium concentration. Furthermore, rechallenging cells with other Gq-coupled GPCR agonists (e.g., histamine) after washing bitter tastants evokes calcium responses suggesting that the bitter tastants do not damage cell membrane or alter calcium homeostasis (Fig. 1b). In freshly isolated murine ASM cells, Zhang et al. demonstrated a modest increase in intracellular calcium upon stimulation with BTR agonist, chloroquine [44]. These investigations also revealed no damage to ASM cells, although bitter tastants were used at a relatively high concentration. Human ASM cells maintained in the calcium-free buffer were also able to elicit bitter-tastant-induced calcium responses, suggesting that calcium responses evoked upon T2R activation results from the release of calcium from intracellular stores such as the sarcoplasmic reticulum (SR) (Fig. 1a). Similarly, in native murine ASM cells, chloroquine elicited calcium responses in ASM cells in the absence of extracellular calcium [44]. Furthermore, a bitter tastant, saccharin, failed to evoke any calcium response in human ASM cells after depletion of SR calcium by

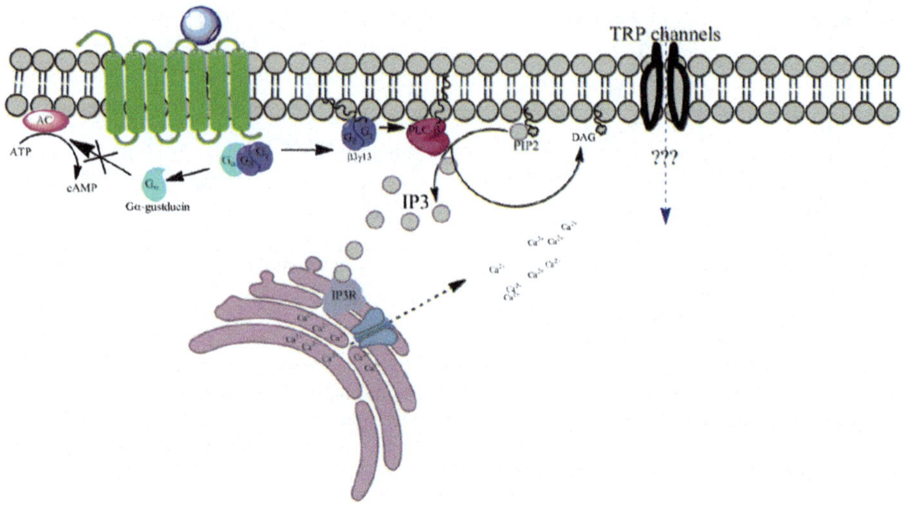

Fig. 2 Calcium regulation by bitter taste receptors. Stimulation of TAS2Rs on cultured human ASM cells results in the activation of trimeric G protein and separation of Gα-gustducin from Gβγ complex. Gβ3γ13 activates phospholipase C and converts phosphoinositide bisphosphate (*PIP2*) into inositol trisphosphate (*IP_3*), which in turn binds to IP_3 receptors on the sarcoplasmic reticulum to release calcium. Interestingly, TAS2R activation does not modulate cyclic AMP levels in ASM cells. Bitter taste receptor activation is also known to activate TRP channels in gustatory cells of the oral cavity. The role of TRP channels in TAS2R signaling in ASM has not been established

thapsigargin pretreatment. These experiments further confirmed our assertion that bitter tastant-induced calcium response involves the release of calcium from intracellular stores. Pretreating human ASM cells with βγ-inhibitor, gallein, PLC inhibitor, U73122, or an IP_3 receptor antagonist, Xestospongin, significantly inhibited bitter tastant-induced calcium responses. These data suggest that T2R-induced calcium signaling in ASM cells involves activation of PLC in a βγ-dependent manner and release of calcium via IP3R channels on the SR. In freshly dissociated murine ASM cells, BTR agonist-induced calcium responses were found to be sensitive to pretreatment of ASM cells with pertussis toxin, gallein, Gβγ blocking peptide, U73122, and 2-aminoethoxydiphenyl borate (2-AP) [44]. Collectively, studies from cultured human ASM and freshly isolated murine ASM cells confirm the presence of canonical taste receptor signaling in ASM (Fig. 2).

In taste cells, IP_3-mediated calcium release by bitter tastants is followed by calcium-dependent activation of a cation channel, transient receptor potential channel M5 (TrpM5) [26, 27, 45]. This leads to membrane depolarization and generation of action potential [16, 40]. The role of TrpM5 in T2R signal transduction in ASM is not established. The expression and functional role of TrpC and TrpV channels in the regulation of ASM calcium homeostasis and membrane potential have been investigated [13]. These channels are believed to play a role in store-operated calcium entry or receptor-operated calcium entry [13]. However, very little is known with regard to the expression or function of TrpMs (TrpM2, TrpM3, TrpM4, and TrpM5). If expressed in ASM, they could modulate calcium influx via both voltage-dependent and voltage-independent calcium-permeable channels. TrpM5 expression was not detected in human airway ciliated epithelial cells in which T2R activation by bitter tastants increased ciliary beat frequency [34]. TrpM5-deficient mice maintain responsiveness to various tastes including bitter taste [10]. These data suggest that the functional effects of T2Rs are not exclusively dependent on the expression of TrpM5 channels.

Using high-resolution confocal microscopy we determined the spatial and temporal distribution characteristics of calcium responses elicited by bitter tastants. In fluo-3-loaded human ASM cells, saccharin exposure resulted in a rapid and localized calcium elevation near the cell membrane. Exposure of ASM cells to histamine, a Gq-coupled contractile GPCR agonist, resulted in a slower but global increase in intracellular calcium. Furthermore, line-scan images confirmed the occurrence of localized calcium events near the membrane upon stimulation of ASM cells with saccharin. These studies suggested the unique spatial distribution of calcium events in ASM by bitter tastants. However, studies of murine ASM cells did not reveal localized calcium events such as sparks or puffs [44]. In all the cells tested, the authors observed a global increase in calcium upon stimulation with bitter tastants like chloroquine. These differences could be due to differences in species or cell preparation (native versus cultured) used in the studies. Taste receptor signaling has been shown to be different among different species [36, 38].

4 Dual Signaling and the Role of the Gα-subunit

α-Gustducin activated upon bitter tastant stimulation is believed to alter cyclic-adenosine monophosphate (cAMP) levels in taste cells. Therefore, taste receptor signal transduction involves the rapid metabolism of dual second messenger systems. Bitter tastants inhibit cAMP levels in vallate and foliate tissue in a α-gustducin-dependent manner [9, 41]. Several studies have established the role of cAMP levels in taste signal transduction, including the potential activation of cyclic nucleotide-gated (CNG) channels and activation of phosphodiesterase [16]. In human ASM cells, bitter tastants did not increase cAMP levels and activate protein kinase A (PKA). Similar findings were observed in guinea-pig ASM (Fig. 2). The role, if any, that α-subunit-mediated signaling plays in ASM is not clearly established. cAMP/PKA-mediated intracellular signaling is the predominant mechanism by which known Gs-coupled GPCR agonists relax ASM. Bitter tastants, in contrast, relax ASM but do not invoke cAMP/PKA activation.

5 Functional Effects of Bitter Taste Receptor Activation: ASM Relaxation

Based on the elevation of calcium by bitter tastants in human ASM cells, it is reasonable to hypothesize that T2R activation results in the contraction of ASM cells, airways, and bronchoconstriction. This phenomenon has been demonstrated in studies involving bronchial ciliated epithelial cells. Exposure of bronchial ciliated epithelial cells to bitter tastants results in elevation of calcium and increased ciliary beat frequency [34]. However, the addition of bitter agonists such as chloroquine, quinine, or denatonium to tracheal or bronchial rings precontracted with methacholine resulted in a robust relaxation. Similar effects were observed when airways were precontracted with serotonin, another Gq-coupled GPCR agonist. We also further demonstrated that the relaxation effect of BTR agonists is completely reversible. Using magnetic twisting cytometry we determined the effects of bitter tastants on ASM cell stiffness. T2R agonists inhibited ASM cell stiffness, suggesting that the relaxation of airways observed in ex vivo experiments was due to the direct effect of bitter tastants on ASM cells. Fourier transform traction microscopy was used to measure spatiotemporal changes in the contractile stress generated by an individual human ASM cell. When ASM cells were exposed to the T2R agonist chloroquine, the traction forces in most of the intracellular areas were reduced [2]. Using two complementary techniques we confirmed the direct effect of T2R agonists on ASM cells. In freshly isolated murine ASM cells, Zhang et al. demonstrated the reversal of methacholine-induced cell shortening by BTR agonists chloroquine and denatonium [44]. The authors also observed a tendency of BTR agonists to decrease the basal tension of isolated murine airways. Pulkkinen and coinvestigators carried out similar studies using airways obtained from guinea-pig lungs [29]. A panel of bitter tastants robustly relaxed guinea-pig airways precontracted with multiple contractile agents. Three independent laboratories have confirmed the airway relaxation effects of T2R agonists using mouse [43], guinea-pig [29], and human [6] airways.

Studies from our laboratory and Pulkkinen et al. demonstrated that T2R-mediated relaxation of airways is not dependent on cAMP generation and activation of PKA in ASM cells [12, 29]. PKA plays a central role in ASM relaxation by other airway relaxing agents such as β2-adrenergic receptor agonists [11]. The presence of indomethacin, a cyclooxygenase inhibitor, had no effect on the relaxation responses of T2R agonists in mice airways, whereas the relaxation effects of denatonium and thiamine, but not chloroquine, were enhanced in guinea-pig airways [29]. Furthermore, the removal of epithelium did not influence airway relaxation by T2R agonists, suggesting no significant contribution of paracrine effect of epithelium-derived airway relaxing agents. Collectively, studies using isolated airway rings and cultured ASM cells demonstrated that the T2R agonists act directly on ASM cells to relax airways.

Pretreating mice and human airways with a large-conductance potassium channel (BK_{Ca}) inhibitor, iberiotoxin, resulted in the partial inhibition of

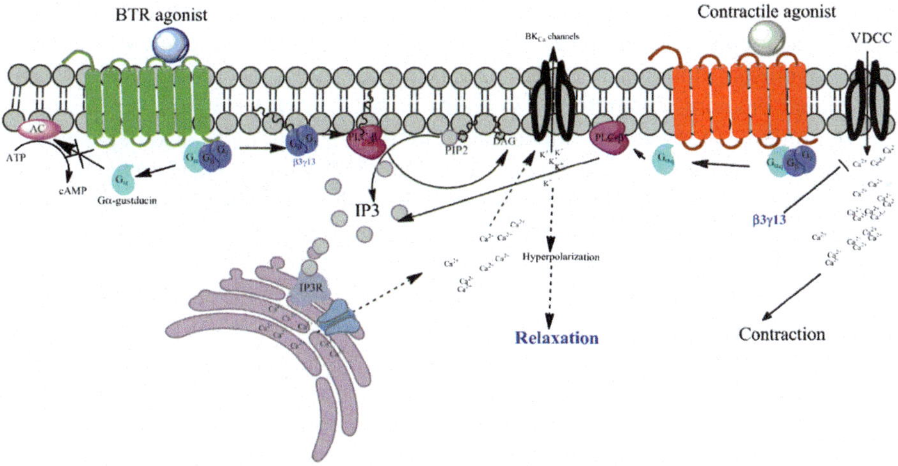

Fig. 3 Two proposed models of ASM relaxation by BTR agonists. Bitter taste receptor activation (*left side*) results in elevation of intracellular calcium, which in turn activates large-conductance calcium-dependent potassium (BK_{Ca}) channels, hyperpolarization of ASM membrane, and relaxation. In a recent study using freshly isolated murine ASM cells, Zhang et al. demonstrated that contractile agents (*right side*) elevate intracellular calcium (followed by contraction) via release of calcium from sarcoplasmic reticulum stores and influx of calcium from extracellular space through voltage-dependent calcium channels (*VDCCs*). BTR activation inhibits calcium entry via VDCC in a βγ-dependent fashion that results in the inhibition of ASM contraction

chloroquine-induced airway relaxation [2, 12, 29] (Fig. 3). In guinea-pig airways, iberiotoxin partially inhibited denatonium, but not chloroquine-induced airway relaxation [29]. In mechanics studies on isolated human ASM cells using magnetic twisting cytometry and Fourier transform traction microscopy, it was demonstrated that BK_{Ca} channel inhibitor partially ablated a chloroquine-induced decrease in ASM cell stiffness or intracellular contractile force [2]. More studies are needed to ascertain how hyperpolarization of ASM induced by BTR agonists results in ASM relaxation, especially when the ASM is stimulated with BTR agonists in the presence of other G_q-coupled GPCR agonists. However, Zhang et al., using murine airways, did not observe the inhibitory effect of iberiotoxin on chloroquine-mediated airway relaxation [43]. Electrophysiological recording in murine ASM cells did not reveal BK_{Ca} currents upon exposure of murine ASM cells to chloroquine. Therefore, T2R-mediated effects on ASM are more complex and involve multiple mechanisms, although BK_{Ca} activation by localized Ca^{2+} events promoted by T2R agonists is one part of a complex mechanism by which these receptors relax ASM. Additional studies will be required to ascertain other partners in this multifactorial airway relaxation response. In fact, a recent study by Zhang et al. demonstrated that stimulation of murine ASM cells with BTR agonists inhibited methacholine-induced increase in intracellular calcium elevation, although BTR agonists by themselves increased basal calcium levels modestly [44]. Using electrophysiology and imaging techniques the investigators demonstrated that BTR agonists inhibited methacholine-induced activation of L-type voltage-dependent calcium channels and calcium influx in ASM (Fig. 3). Interestingly, the inhibition of L-type calcium channels by BTR agonists was also dependent on Gβγ [44]. Although two potential mechanisms (activation of BK_{ca} channels and inhibition of L-type calcium channels) of BTR-mediated ASM relaxation have been proposed, it is plausible that BTRs utilize multiple targets in ASM to induce relaxation.

In vivo lung function studies in mice revealed an efficacious bronchodilatory effect of T2R agonists when administered by an aerosol route [12]. We further carried out experiments using ovalbumin-sensitized and challenged mice to determine whether the bitter taste receptor pathway is compromised by airway inflammation. The T2R agonists denatonium and chloroquine effectively relieved methacholine-induced bronchoconstriction in normal and asthmatic mice [12]. These studies suggest that agonists of BTRs can be exploited as a new class of antiasthma drugs. In fact, several features of BTRs and BTR agonists are favorable in exploiting BTRs as potential antiasthma drug targets: multiple receptors are expressed, a wide range of natural and synthetic pharmacological agents with established pharmacological and toxicological properties are available, they are effective in asthma (at least in animal models), have high efficacy, may play a role in mucociliary clearance, and induce effective airway relaxation under β2-adrenergic receptor agonist tachyphylaxis. Long-term use of bronchodilatory β2-adrenergic receptor agonists results in tachyphylaxis and a gradual loss of bronchoprotective effect. Under conditions of β2-adrenergic tachyphylaxis, T2R-mediated ASM relaxation effects are well maintained [3]. These studies support the idea that bitter tastants could be used to treat asthmatics as an adjunct therapy when β2-adrenergic agonists (mainstay antiasthma therapeutics) fail to provide relief from bronchoconstriction.

Taste sensation is a sophisticated, complex, and well-appreciated biological phenomenon. The primary role of gustation is to accept or reject chemicals for ingestion, and hence it acts as a protective mechanism in preventing the consumption of toxic agents. Nongustatory (gastrointestinal and respiratory) taste receptors are also believed to be involved in evoking protective reflexes in response to toxic stimuli. Bitter taste receptors expressed on different parts of airways, including SCCs, are innervated by cholinergic neurons, and the activation of taste receptors by bitter tastants decreases the rate of respiration [36, 39]. The other potential reflexes include a change in ciliary beat frequency and activation of brush cells [34]. Airway brush cells are predominantly connected to cholinergic innervations. Brush cells also initiate respiratory reflexes when exposed to bitter tastants [18]. Recent studies on bitter taste receptors on ASM, and specifically the efficacious bronchodilatory effects of bitter tastants, underscore the clinical use of bitter tastants as potential therapies in airway diseases.

References

1. Adler E, Hoon MA, Mueller KL, Chandrashekar J, Ryba NJ and Zuker CS. A novel family of mammalian taste receptors. *Cell* 100: 693–702, 2000.
2. An SS, Robinett KS, Deshpande DA, Wang WC and Liggett SB. Reply to: Activation of BK channels may not be required for bitter tastant-induced bronchodilation. *Nat Med* 18: 650–651, 2012.
3. An SS, Wang WC, Koziol-White CJ, Ahn K, Lee DY, Kurten RC, Panettieri RA, Jr. and Liggett SB. TAS2R activation promotes airway smooth muscle relaxation despite beta(2)-adrenergic receptor tachyphylaxis. *Am J Physiol Lung Cell Mol Physiol* 303: L304-L311, 2012.
4. Behrens M and Meyerhof W. Oral and extraoral bitter taste receptors. *Results Probl Cell Differ* 52: 87–99, 2010.
5. Behrens M, Reichling C, Batram C, Brockhoff A and Meyerhof W. Bitter taste receptors and their cells. *Ann N Y Acad Sci* 1170: 111–115, 2009.
6. Belvisi MG, Dale N, Birrell MA and Canning BJ. Bronchodilator activity of bitter tastants in human tissue. *Nat Med* 17: 776–778, 2011.
7. Brockhoff A, Behrens M, Niv MY and Meyerhof W. Structural requirements of bitter taste receptor activation. *Proc Natl Acad Sci U S A* 107: 11110–11115, 2010.
8. Caicedo A, Pereira E, Margolskee RF and Roper SD. Role of the G-protein subunit alpha-gustducin in taste cell responses to bitter stimuli. *J Neurosci* 23: 9947–9952, 2003.
9. Clapp TR, Trubey KR, Vandenbeuch A, Stone LM, Margolskee RF, Chaudhari N and Kinnamon SC. Tonic activity of Galpha-gustducin regulates taste cell responsivity. *FEBS Lett* 582: 3783–3787, 2008.
10. Damak S, Rong M, Yasumatsu K, Kokrashvili Z, Perez CA, Shigemura N, Yoshida R, Mosinger B, Jr., Glendinning JI, Ninomiya Y and Margolskee RF. Trpm5 null mice respond to bitter, sweet, and umami compounds. *Chem Senses* 31: 253–264, 2006.
11. Deshpande DA and Penn RB. Targeting G protein-coupled receptor signaling in asthma. *Cell Signal* 18: 2105–2120, 2006.
12. Deshpande DA, Wang WC, McIlmoyle EL, Robinett KS, Schillinger RM, An SS, Sham JS and Liggett SB. Bitter taste receptors on airway smooth muscle bronchodilate by localized calcium signaling and reverse obstruction. *Nat Med* 16: 1299–1304, 2010.
13. Gosling M, Poll C and Li S. TRP channels in airway smooth muscle as therapeutic targets. *Naunyn Schmiedebergs Arch Pharmacol* 371: 277–284, 2005.
14. He W, Yasumatsu K, Varadarajan V, Yamada A, Lem J, Ninomiya Y, Margolskee RF and Damak S. Umami taste responses are mediated by alpha-transducin and alpha-gustducin. *J Neurosci* 24: 7674–7680, 2004.
15. Huang L, Shanker YG, Dubauskaite J, Zheng JZ, Yan W, Rosenzweig S, Spielman AI, Max M and Margolskee RF. Ggamma13 colocalizes with gustducin in taste receptor cells and mediates IP3 responses to bitter denatonium. *Nat Neurosci* 2: 1055–1062, 1999.
16. Kinnamon SC. Taste receptor signalling - from tongues to lungs. *Acta Physiol (Oxf)* 204: 158–168, 2012.
17. Kinnamon SC and Margolskee RF. Mechanisms of taste transduction. *Curr Opin Neurobiol* 6: 506–513, 1996.
18. Krasteva G, Canning BJ, Hartmann P, Veres TZ, Papadakis T, Muhlfeld C, Schliecker K, Tallini YN, Braun A, Hackstein H, Baal N, Weihe E, Schutz B, Kotlikoff M, Ibanez-Tallon I and Kummer W. Cholinergic chemosensory cells in the trachea regulate breathing. *Proc Natl Acad Sci U S A* 2011.
19. McLaughlin SK, McKinnon PJ and Margolskee RF. Gustducin is a taste-cell-specific G protein closely related to the transducins. *Nature* 357: 563–569, 1992.
20. McLaughlin SK, McKinnon PJ, Robichon A, Spickofsky N and Margolskee RF. Gustducin and transducin: a tale of two G proteins. *Ciba Found Symp* 179: 186–196, 1993.

21. Meyerhof W, Batram C, Kuhn C, Brockhoff A, Chudoba E, Bufe B, Appendino G and Behrens M. The molecular receptive ranges of human TAS2R bitter taste receptors. *Chem Senses* 35: 157–170, 2010.
22. Mizuta K, Mizuta F, Xu D, Masaki E, Panettieri RA, Jr. and Emala CW. Gi-coupled gamma-aminobutyric acid-B receptors cross-regulate phospholipase C and calcium in airway smooth muscle. *Am J Respir Cell Mol Biol* 45: 1232–1238, 2011.
23. Mizuta K, Xu D, Pan Y, Comas G, Sonett JR, Zhang Y, Panettieri RA, Jr., Yang J and Emala CW, Sr. GABAA receptors are expressed and facilitate relaxation in airway smooth muscle. *Am J Physiol Lung Cell Mol Physiol* 294: L1206-L1216, 2008.
24. Osawa Y, Xu D, Sternberg D, Sonett JR, D'Armiento J, Panettieri RA and Emala CW. Functional expression of the GABAB receptor in human airway smooth muscle. *Am J Physiol Lung Cell Mol Physiol* 291: L923-L931, 2006.
25. Palmer K.R. The Pharmacology and Signaling of bitter, sweet, and Umami taste signaling. *Molecular Interventions* 7: 87–98, 2007.
26. Perez CA, Huang L, Rong M, Kozak JA, Preuss AK, Zhang H, Max M and Margolskee RF. A transient receptor potential channel expressed in taste receptor cells. *Nat Neurosci* 5: 1169–1176, 2002.
27. Perez CA, Margolskee RF, Kinnamon SC and Ogura T. Making sense with TRP channels: store-operated calcium entry and the ion channel Trpm5 in taste receptor cells. *Cell Calcium* 33: 541–549, 2003.
28. Pietras CO, James A, Konradsen JR, Nordlund B, Soderhall C, Pulkkinen V, Pedroletti C, Daham K, Kupczyk M, Dahlen B, Kere J, Dahlen SE, Hedlin G and Melen E. Transcriptome analysis reveals upregulation of bitter taste receptors in severe asthmatics. *Eur Respir J* 2012.
29. Pulkkinen V, Manson ML, Safholm J, Adner M and Dahlen SE. The bitter taste receptor (TAS2R) agonists denatonium and chloroquine display distinct patterns of relaxation of the guinea pig trachea. *Am J Physiol Lung Cell Mol Physiol* 303: L956-L966, 2012.
30. Pydi SP, Upadhyaya J, Singh N, Pal BR and Chelikani P. Recent advances in structure and function studies on human bitter taste receptors. *Curr Protein Pept Sci* 13: 501–508, 2012.
31. Rossler P, Boekhoff I, Tareilus E, Beck S, Breer H and Freitag J. G protein beta-gamma complexes in circumvallate taste cells involved in bitter transduction. *Chem Senses* 25: 413–421, 2000.
32. Rossler P, Kroner C, Freitag J, Noe J and Breer H. Identification of a phospholipase C beta subtype in rat taste cells. *Eur J Cell Biol* 77: 253–261, 1998.
33. Saxena H, Deshpande DA, Tiegs BC, Yan H, Battafarano RJ, Burrows WM, Damera G, Panettieri RA, DuBose TD, Jr., An SS and Penn RB. The GPCR OGR1 (GPR68) mediates diverse signalling and contraction of airway smooth muscle in response to small reductions in extracellular pH. *Br J Pharmacol* 166: 981–990, 2012.
34. Shah AS, Ben-Shahar Y, Moninger TO, Kline JN and Welsh MJ. Motile cilia of human airway epithelia are chemosensory. *Science* 325: 1131–1134, 2009.
35. Singh N, Pydi SP, Upadhyaya J and Chelikani P. Structural basis of activation of bitter taste receptor T2R1 and comparison with Class A G-protein-coupled receptors (GPCRs). *J Biol Chem* 286: 36032–36041, 2011.
36. Tizzano M, Cristofoletti M, Sbarbati A and Finger TE. Expression of taste receptors in solitary chemosensory cells of rodent airways. *BMC Pulm Med* 11: 3, 2011.
37. Tizzano M, Dvoryanchikov G, Barrows JK, Kim S, Chaudhari N and Finger TE. Expression of Galpha14 in sweet-transducing taste cells of the posterior tongue. *BMC Neurosci* 9: 110, 2008.
38. Tizzano M and Finger TE. Chemosensors in the nose: guardians of the airways. *Physiology (Bethesda)* 28: 51–60, 2013.
39. Tizzano M, Gulbransen BD, Vandenbeuch A, Clapp TR, Herman JP, Sibhatu HM, Churchill ME, Silver WL, Kinnamon SC and Finger TE. Nasal chemosensory cells use bitter taste signaling to detect irritants and bacterial signals. *Proc Natl Acad Sci U S A* 107: 3210–3215, 2010.

40. Vandenbeuch A and Kinnamon SC. Why do taste cells generate action potentials? *J Biol* 8: 42, 2009.
41. Yan W, Sunavala G, Rosenzweig S, Dasso M, Brand JG and Spielman AI. Bitter taste transduced by PLC-beta(2)-dependent rise in IP(3) and alpha-gustducin-dependent fall in cyclic nucleotides. *Am J Physiol Cell Physiol* 280: C742-C751, 2001.
42. Zancanaro C, Caretta CM, Merigo F, Cavaggioni A and Osculati F. alpha-Gustducin expression in the vomeronasal organ of the mouse. *Eur J Neurosci* 11: 4473–4475, 1999.
43. Zhang CH, Chen C, Lifshitz LM, Fogarty KE, Zhu MS and ZhuGe R. Activation of BK channels may not be required for bitter tastant-induced bronchodilation. *Nat Med* 18: 648–650, 2012.
44. Zhang CH, Lifshitz LM, Uy KF, Ikebe M, Fogarty KE and ZhuGe R. The cellular and molecular basis of bitter tastant-induced bronchodilation. *PLoS Biol* 11: e1001501, 2013.
45. Zhang Z, Zhao Z, Margolskee R and Liman E. The transduction channel TRPM5 is gated by intracellular calcium in taste cells. *J Neurosci* 27: 5777–5786, 2007.

Modulation of Airway Smooth Muscle Contractile Function by TNFα and IL-13 and Airway Hyperresponsiveness in Asthma

Yassine Amrani

Contents

Abstract There is no doubt that airway smooth muscle (ASM) is a key player in asthma pathophysiology, as demonstrated by its heightened sensitivity to both direct and indirect contractile stimuli, which leads to exaggerated airway narrowing and airflow obstruction. The therapeutic benefit in severe asthmatics provided by bronchial thermoplasty, a therapy that attenuates bronchoconstriction via reduction of ASM mass, has provided additional support for the concept that ASM function

Supported by the National Institute for Health Research Leicester Respiratory Biomedical Research Unit. The views expressed are those of the author and not necessarily those of the NHS or the NIHR.

Y. Amrani (✉)
Department of Infection, Immunity and Inflammation,
University of Leicester, University Road, Leicester LE1 7RH, UK

Institute for Lung Health, Department of Respiratory Medicine, Clinical Sciences Building, Glenfield Hospital, Groby Rd, Leicester LE1 7RH, UK
e-mail: ya26@le.ac.uk

Y.-X. Wang (ed.), *Calcium Signaling In Airway Smooth Muscle Cells*,
DOI 10.1007/978-3-319-01312-1_24,

could be abnormally contracting in asthma. The mechanisms in ASM leading to this exaggerated sensitivity to G-protein-coupled receptor (GPCR) agonists, known as bronchial hyper-responsiveness, are still unknown. A number of studies, however, have demonstrated that a direct action of two key pro-asthmatic cytokines, TNFα and IL-13 on ASM, leads to exaggerated ASM contractility via the modulation of GPCR-associated calcium signalling. This chapter reviews the evidence demonstrating a role of both TNFα and IL-13 in driving airway hyperresponsiveness at three different levels: in experimental animal models of asthma, in isolated airway preparations and in isolated ASM cells.

Keywords Cytokines • Calcium metabolism • G-protein-coupled receptor • Rho pathway • Allergen challenge

1 Introduction

The advent of monoclonal antibody therapy has allowed preclinical and clinical trials to strongly support the role of TNFα and IL-13 in driving many features of allergic asthma, including airway inflammation, mucus hypersecretion and airway hyperresponsiveness (AHR) [1–3]. These studies have shown that a direct action of TNFα and IL-13 on airway resident cells, including airway smooth muscle (ASM) cells, could play a role in asthma pathogenesis. In addition, countless reports using cultured ASM cells have demonstrated the expression of functional receptors for IL-13 and TNFα and their ability when activated to stimulate a variety of pro-asthmatic responses, including inflammatory gene expression and ASM hyperresponsiveness [4–6]. In this chapter, we will discuss the reports from the literature describing how cytokines play a role in driving ASM hyperresponsiveness seen in asthma. We will show strong evidence of a direct effect of IL-13 and TNFα on ASM sensitivity to different G-protein-coupled receptor (GPCR) agonists and the underlying potential mechanisms. This chapter will strongly support the current concept that cytokine changes in ASM function, an effector cell thought to solely regulate bronchomotor tone [7], may be regarded as a key player in the development of bronchial hyperresponsiveness, the main feature of allergic asthma. The molecular mechanisms by which both cytokines increase ASM sensitivity and contractility to GPCR agonists have not been clearly established. We will also review the different cytokine-associated pathways in ASM cells implicated in changes in GPCR responsiveness. It is likely that these studies may provide new insight into the design of novel therapeutic approaches for abnormal lung function in asthma.

2 Evidence Suggesting a Role of TNFα in Allergen-Induced Airway Hyperresponsiveness in Experimental Models of Asthma

The report by Renzetti and colleagues was the first pharmacological evidence to support a role of TNFα in allergic asthma [8]. The authors showed the therapeutic value of a soluble TNFα antagonist, called Ro 45-2081, in two different models of allergic asthma, i.e. guinea pigs and brown Norway rats sensitised and challenged with ovalbumin [8]. Ro45-2081 (used 1–3 mg/kg i.p.) is a TNFα receptor 2 fusion protein which suppressed allergen-induced AHR to substance P in sensitised guinea pigs. Ro45-2081 also had a marked effect on allergen-associated infiltration of inflammatory cells, including eosinophils and neutrophils, in the airways. Subsequent studies using monoclonal antibodies as a blocking strategy provided additional evidence for a role of TNFα in the development of key allergic features in asthma. In an animal model that displays both early and late asthmatic responses to ovalbumin, Choi and colleagues reported that an anti-TNFα monoclonal antibody called MP6.XT22.11 markedly inhibited late AHR and lung eosinophilia responses. Because the authors showed that activation of cPLA2 activity in the allergic lungs was dependent on TNFα, it was suggested that cPLA2-dependent metabolites were playing a central role in the allergic responses driven by TNFα [9]. Additional studies using infliximab (a monoclonal antibody) or etanercept (a fusion protein) confirmed the implication of TNFα in the development of allergen-induced AHR in different animal models [10–14]. The mechanisms underlying the beneficial action of anti-TNFα therapy in asthma have not been completely elucidated. The inhibitory action of TNFα blockade on allergen-induced lung eosinophilia and IgE production was shown to be dependent on the stage of the disease, with etanercept treatment being effective only in the animal model with severe asthma while no effect was observed in the mild/moderate model [11]. A reduction of allergen-associated airway mucus hypersecretion has been described as one beneficial effect of TNFα blockade [15]. A similar suppressive action of anti-TNFα therapy on ovalbumin-induced mucus hypersecretion and airway inflammation was recently confirmed in mice treated with TNFα antisense oligonucleotide [16]. Interestingly, the authors showed that TNFα blockade led to an enhancement of regulatory T-cell-mediated immune tolerance, potentially explaining the therapeutic benefit of anti-TNFα therapy in asthma. In guinea-pigs sensitised and challenged ovalbumin, the inhibitory effect of etanercept on allergen-induced AHR was attributed to the downregulation of muscarinic M2 receptor in the parasympathetic neurons within the airways [14]. In this study, antigen-induced AHR to methacholine was not reduced by etanercept suggesting a direct effect of TNFα on the nerves. Another report confirmed earlier findings that therapeutic action of etanercept is linked to the inhibition of allergen-induced TH2 cytokines and infiltration of inflammatory cells

Table 1 Anti-TNFα strategies in preclinical studies

TNFα blockers	Experimental model	Outcome	Author
Ro 45–2081 TNFR1-FcIgG1 fusion protein	Ovalbumin-sensitized and challenged male guinea-pigs and rats	↓ **AHR to substance P** ↓ Neut., Eos. in the BAL	Renzetti et al. [8]
MP6-XT22.11 Anti-TNFα monoclonal antibody	Ovalbumin-sensitized and challenged male BALB/C mice	↓ **Late AHR to methacholine** ↓ Eos. in BAL	Choi et al. [9]
Etanercept, XENP1595, or XENP346	Ovalbumin-sensitized and challenged C57BL/6 wild-type	↓ **AHR to methacholine** ↓ Eos. and Lymphocytes in BAL	Maillet et al. [13]
Etanercept	Ovalbumin-sensitized and challenged guinea-pigs	↓ **AHR to electric vagal stimulation** ↓ Eos. in the BAL, blood, nerves and within ASM	Nie et al. [14]
Etanercept	Short term (ST) and long term (LT) model of ovalbumin-sensitized and challenged mice	↓ Eos., and IL-5 in the BAL (ST) ↓ serum IgE (ST) ↓ IL-5 in the BAL (LT)	Hutchison et al. [11]
Infliximab	Ovalbumin-sensitized and challenged male mice	↓ Eos. in the BAL ↓ Cytokines in BAL and lung tissues	Deveci et al. [10]
Rat anti-TNFα monoclonal antibody	House dust-sensitized and challenged female BALB/C mice	↓ **AHR to methacholine** ↑ Th2 cytokines ↓ Eos., Neut., and lymphocytes in BAL	Kim et al. [12]
Anti-TNFα antibody	Ovalbumin-sensitized and challenged BALB/c mice	↓ **AHR to methacholine** ↓MUC-5AC expression ↓Goblet cell number	Busse et al. [15]

Eos eosinophils, *Neut* neutrophils, *BAL* brochoalveolar lavage, *AHR* airway hyper-responsiveness, *ST* short term model of asthma, *LT* long term model of asthma, ∅ no effect

(eosinophils) within the airways [13]. Together, anti-TNFα strategies are effective in various animal models of allergic asthma induced by different types of allergen including ovalbumin [10, 13, 14] or house dust extracts [12].

The therapeutic action of anti-TNFα therapy described in preclinical studies cannot be attributed to one particular mechanism. The use of anti-TNFα drugs demonstrated that TNFα exerts strong pro-inflammatory actions in allergic responses that include an induction of TH2 responses (possibly via the activation of regulatory T-cells), recruitment of eosinophils in the airways and regulation of mucus hypersecretion and AHR (studies that have used anti-TNFα strategies in experimental asthma are summarised in Table 1).

3 Evidence Suggesting a Direct Action of TNFα on Airway Structural Tissues in Inducing Airway Hyperresponsiveness

There is undeniable evidence from preclinical studies that TNFα is a key player in asthma pathogenesis. The contribution of TNFα in driving AHR has also been investigated ex vivo in isolated airway preparations (mostly tracheal rings) from different animal species. Studies using guinea-pig tracheal rings, for example, showed that the effect of TNFα on methacholine-evoked contractile responses was dependent on treatment duration. It was shown that a short-term treatment (30 min) with TNFα was sufficient to significantly increase the contractile responses to methacholine, in part via the autocrine action of platelet-activating factor (PAF) [17]. A more recent report using tracheal tissues treated with TNFα for 1, 2 and 4 days failed to notice any effects on methacholine, although responses to serotonin (5-HT) and electric field stimulation (EFS) were markedly augmented [18]. Only two studies performed in human bronchial rings, possibly due to the difficulty in obtaining human tissues, reported that TNFα alone or in combination with IL-1β exerted similar effects in human bronchi by enhancing EFS- or acetylcholine-associated contractility [19, 20]. These studies suggest that short-term treatment with TNFα involves an effect on ASM, while long-term treatment may involve an action on ASM and nerve cells. In ovine tracheal airway preparations, TNFα combined with IL-1β augmented the responses to acetylcholine and neurokinin A by involving an effect dependent on changes in calcium metabolism in ASM [21]. Most of the evidence showing an enhancing action of TNFα on ASM contractility comes from studies performed in mice. Our original demonstration that murine isolated tracheal rings with TNFα led to enhanced contractile responses to muscarinic M3 agonist [6, 22, 23] was recently confirmed by other groups [24, 25], while others showed that TNFα also enhanced responses to other GPCR agonists such as bradykinin [26] or serotonin [27]. Despite this growing evidence of ex vivo effect of TNFα on airway structural tissues, the precise mechanism by which TNFα contributes to the development of AHR in vivo has not been yet fully investigated. Increased expression or activity of the ectoenzyme CD38, in part mediated via autocrine action of IFNβ, or up-regulation of caveolin-1 have been described as potential mechanisms mediating TNFα-enhancing effects from studies using isolated airways [23, 28, 29]. We also showed the implication of pertussis-toxin-sensitive pathways in TNFα effects [22], while others demonstrated that increased expression of bradykinin B2 receptor was key in mediating TNFα action on ASM contractility [26]. Additional studies are still needed to pinpoint the mechanisms by which TNFα increases ASM contractility in isolated airways.

4 Direct Effect of TNFα on ASM Cells

Because Ca^{2+} plays a central role in regulating ASM contractile function, it is likely that alterations in Ca^{2+} regulatory mechanisms induced by TNFα will lead to impaired ASM contractility (reviewed in [7]). In the early 1990s, we provided the first evidence of a deleterious effect of TNFα on calcium handling in both animal and human cultured ASM cells [30–33]. A number of recent studies have indeed confirmed this hypothesis and provided potential mechanisms underlying the potentiation of GPCR-agonist-evoked calcium transients induced by TNFα. In recent years, Prakash's group at the Mayo Clinic made a number of original findings about TNFα-enhancing mechanisms by describing the implication of different cellular factors, including the calcium-buffering action of mitochondria [34], the activity of Na(+)/Ca(2+) exchanger (NCX) [35] or the expression levels of caveolin-1 [29] or sarco-/endoplasmic reticulum Ca2+-ATPase (SERCA) [36]. The latter finding is interesting because a reduced SERCA expression in ASM was shown to play a role in airway remodelling in severe asthmatics [37]. White and colleagues showed that calcium influx was altered in TNFα-treated cells as a consequence of the up-regulation of the transient receptor potential C3 (TRPC3) [38]. Other studies have shown the implication of small monomeric G-protein RhoA, which can enhance ASM contractility to any agonists by increasing levels of MLC phosphorylation via the Rho-activated kinase (ROCK)-dependent suppression of MLC phosphatase [39]. Importantly, this Rho-dependent calcium sensitization can be activated by TNFα in human ASM cells [40–43]. An alternative mechanism by which TNFα modulates GPCR-evoked calcium signalling is via the up-regulation of CD38, an ectoenzyme that stimulates calcium signalling via cADPr [44]. How these different pathways contribute to the overall effects of TNFα on ASM contractile responses is still an open question. Nonetheless, these studies have highlighted a number of potential mechanisms that can increase ASM responsiveness to GPCR agonists.

5 Evidence Suggesting a Role of IL-13 in Allergen-Induced Airway Hyperresponsiveness in Experimental Models of Asthma

Since the first demonstration of a role of IL-13 in mediating allergen-induced AHR [45, 46], a number of subsequent studies using different blocking strategies and various animal models have supported the implication of IL-13 in allergic asthma [2]. In mice where sensitization and challenge were performed solely in the upper airways (intranasal instillation), a polyclonal anti-IL-13 antibody had a suppressive effect on allergen-associated inflammation of both nasal and lung compartments and on the development of AHR [47]. Additional studies on ovalbumin-challenged BALB/c mice have validated the therapeutic value of other forms of inhibitory strategies such as anti-IL-13 Fab fragment, which blocked not only AHR but other key asthmatic features such as airway inflammation and airway remodelling [48, 49]. Anti-IL-13 therapy seems to be beneficial in treating mice chronically challenged with the house dust mite [50] or ovalbumin [51] for periods greater than 7 weeks. In these models, IL-13 blockade could be effective when given either *preventively* and *therapeutically* on an established disease, with the exception of AHR, which was not affected by the prophylactic treatment [50]. In addition to preventing AHR, IL-13 blockade also reduced markers of airway remodelling such as subepithelial fibrosis, mucus hyperplasia and progression of inflammation [51]. This was a landmark finding as current anti-asthma therapies appear to have little effect on airway remodelling in asthma [52, 53]. The concept that anti-IL-13 therapy was effective in chronic models of asthma, thought to closely resemble human asthmatic features, has been somewhat challenged by two reports. The study by Goplen and colleagues showed that anti-IL-13 treatment had little therapeutic effect in mice sensitised and challenged with multiple allergens (dust mite, ragweed and Aspergillus species) [54], where the allergic features, including lung structural changes and hyperresponsiveness, persist for longer periods after allergen exposure. The lack of anti-IL-13 actions was attributed to the heightened dendritic-cell-dependent activation of T-cells seen in the lungs of mice exposed to multiple allergens. Another study comparing acute and chronic challenges but with a single allergen also failed to notice a beneficial effect of anti-IL-13 blockade using a fusion protein on AHR in the chronic model [55]. The opposite effect of IL-13 was also reported where IL-13 blockade using a similar approach prevented the late asthmatic responses but not the early asthmatic reactions in previously sensitised animals [56]. The differential effects of anti-IL-13 therapy in these preclinical studies could be attributed to a variety of experimental factors that include types of animal strain, the nature/number of allergens, the treatment/routes of administration and the approach to assessing lung function.

IL-13 seems to be a key player in driving the severity of AHR induced by the respiratory syncytial virus (RSV) [57, 58]. This study suggests anti-IL-13 therapy could be beneficial in preventing asthma exacerbations in patients experiencing respiratory infections by viruses. Also, IL-13 was also involved in mediating rhinovirus (RV)-induced experimental AHR [59]. Together, these different studies provide undeniable evidence supporting a central role of IL-13 in driving both allergen- and virus-induced AHR. It is important to mention that the contribution of anti-IL-13 therapy seems to be highly dependent on the stage or intensity of the disease. The precise mechanism by which IL-13 regulates AHR in asthma has not been yet clearly established, and possible scenarios have been proposed including an effect on lung tissues (studies that have used anti-IL-3 strategies in experimental asthma are summarised in Table 2).

Table 2 Anti-IL-13 strategies in preclinical studies

Anti-IL-13 strategy	Animal model	Targeted allergic responses	Authors
Soluble IL-13Rα2-human Fc fusion protein	OVA-sensitized and challenged BALB/c mice	↓ **AHR to acetylcholine,** ↓ Eosinophil recruitment, ↓ Mucus production	Grunig et al. [46]
Soluble IL-13Rα2-IgGFc fusion protein	OVA-sensitized and challenged male A/J mice	↓ **AHR to acetylcholine**, ↓ mucus overproduction, ∅ Lung Eos. or serum IgE	Wills-Karp et al. [45]
Monoclonal anti-mouse IL-13	OVA-sensitized and challenged male BALB/c mice	↓ **AHR to methacholine**, ↓ Mucous production, ↓ Eos. and Lymphocyte infiltration	Wang and McCusker [47]
Monoclonal anti–IL-13 Fab fragment, (CA154_582)	OVA-sensitized and challenged male BALB/c mice	↓ **AHR to methacholine,** ↓ mucous cell hyperplasia, ↓ Lung mast cells and Eos., ↓ ASM hypertrophia	Hacha et al. [48]
Monoclonal anti-mouse IL-13 antibody	OVA-sensitized and challenged male BALB/c mice	↓ **AHR to methacholine** (multiple challenge), ↓ BAL Eos and Neut. (to a single challenge)	Eum et al. [49]
Monoclonal anti-mouse IL-13 mAb (CA154_582)	House Dust mite-sensitized and challenged male BALB/c mice	↓ **AHR to methacholine** (by prophylactic TRT only), ↓ Peribronchial collagen deposition, ↓ Epithelial goblet cell number.	Tomlinson et al. [50]
Rat anti-TNFα monoclonal antibody (CNTO 134)	OVA-sensitized and challenged female BALB/C mice	↓ **AHR to methacholine**, ↓ subepithelial fibrosis, ↓ mucus hyperplasia, ↓ BAL eosinophilia, ↓ lung MMP-9	Yang et al. [51]
Rat anti-IL-13 monoclonal antibody	Female BALB/C mice sensitized and challenged with Dust Mites, Ragweed and Aspergillus (*chronic model*) or OVA (*acute model*)	↓ **AHR to methacholine** (acute model only), ↓ airway inflammation (acute model only)	Goplen et al. [54]
Soluble murine IL-13rα2-humanIgG fusion protein (sIL-13Rα2.Fc)	OVA-sensitized and challenged female BALB/c mice : 21 days (*acute model*) and 120 days (*chronic model*)	↓ **AHR to methacholine** (acute model only) and ↓ total lung cell count (acute model only), ∅ lung Eos.,	Leigh et al. [55]
Soluble IL-13Rα2-IgG fusion protein	OVA-sensitized and challenged female BALB/cByJ mice	↓ **AHR to methacholine** (in pre-sensitized animals), ∅ serum IgE, ↓ goblet cell hyperplasia, ↓ mucus secretion, ↓ BAL Eos. And Neut.	Taube et al. [56]

Eos eosinophils, *Neut* neutrophils, *BAL* brochoalveolar lavage, *AHR* airway hyper-responsiveness, *ST* short term model of asthma, *LT* long term model of asthma, ∅ no effect

6 Evidence Suggesting a Direct Action of IL-13 on Airway Structural Tissues in Inducing Airway Hyperresponsiveness

In 2003, we tested the hypothesis that IL-13 could contribute to the exaggerated airway responsiveness seen in asthma by acting directly on lung structural cells/tissues. Indeed, we provided the first evidence that murine tracheal rings incubated with IL-13 exhibited increased contractile responses to carbachol [60]. Since then, a number of investigators have provided additional evidence supporting this novel concept of a direct action of IL-13 on airway structural tissues as a potential mechanism underlying the abnormal airway responsiveness seen in allergic asthma. Kannan's group provided a number of original findings regarding the mechanisms involved in driving IL-13-induced AHR by uncovering a unique role of CD38-dependent pathways [61, 62]. CD38 is an approximately 45 kDa transmembrane glycoprotein that catalyses the cyclization of the intermediary metabolite nicotinamide adenine dinucleotide to putative second Ca2+-mobilising messengers cADPr and nicotinic acid adenine dinucleotide phosphate, thus impacting on calcium signalling (reviewed in [5]). Work done earlier by Grunstein's group in rabbit airways showed that in addition to impairing relaxation to beta-2-adrenergic agonist, IL-13 significantly augments ASM contractility to muscarinic agonists via an IL-5-dependent mechanism [63]. This observation suggests that the enhancing effect of IL-13 in isolated airways may be indirect via the action of secreted factors which act in an autocrine manner. Recently, the same group reported that IL-13-induced increased contractility to acetylcholine was sensitive to cortisol, in part because of an increased expression of 11beta-hydroxysteroid dehydrogenase (11beta-HSD)-1; 11beta-HSD-1 is responsible for the conversion of inactive cortisone into cortisol, the bioactive hormone [64]. A recent study added another level of complexity regarding the underlying mechanisms that modulate cytokine-induced AHR. By comparing wild-type and knockout murine tracheal rings, Kudo and colleagues identified the milk fat globule-EGF factor 8 protein (mfge8), a member of the discoidin family, as a novel counter-regulatory factor capable of suppressing the augmenting effect of IL-13 on methacholine-induced contractile responses [24]. This protective effect exerted by mfge8 was, at least in part, due to the inhibition of Rho-dependent calcium sensitization induced by IL-13. The implication of Rho-kinase pathways in mediating IL-13-induced ex vivo AHR was previously reported using the Rho-kinase inhibitor Y-27632 [65]. The fact that the Rho pathway also regulates key features of airway remodelling [66] could explain the therapeutic action of IL-13 blockade on airway remodelling reported in preclinical studies [51]. The decreased expression of mfge8 in airway biopsies of asthmatic patients raises the possibility that loss of mfge8 might play a significant role in the exaggerated airway sensitivity seen in asthma [24]. Another factor recently described to suppress IL-13-induced increased contractile responses to methacholine is a mast cell product called *mouse mast cell protease-4* (mMCP-4) [67]. This is a surprising finding knowing the emerging role of mast cells in the pathogenesis of severe asthma, including the development of AHR [68]. The observation that mMCP-4 inhibitory action was only seen in epithelium-denuded isolated trachea strongly suggests that the bronchoprotector action of mMCP-4 is due to a direct effect of mMCP-4 on ASM.

In addition to isolated tracheal rings, the increased sensitivity of the airways to GPCR agonists induced by IL-13 has also been described using a more elegant experimental approach called the precision-cut lung slice model (PCLS) (elegantly reviewed in [69]). Similar to the effect seen in isolated airway preparations, incubation of murine or human PCLS with IL-13 leads to increased contractility to different GPCR agonists, including methacholine [70–72] or serotonin [73]. In this model, PI3Kγ has been described as an important player in mediating IL-13-induced augmented bronchoconstrictor responses to acetylcholine [73].

7 Direct Effect of IL-13 on ASM Cells

As described for TNFα, cultured ASM cells incubated with IL-13 exhibit enhanced calcium signals to a variety of GPCR agonists, including muscarinic M3 agonists [60, 62, 74], histamine [29, 34, 60, 75, 76], LTD4 [77], bradykinin [29, 60] and oxytocin [78]. The underlying mechanisms of IL-13-inducible non-specific ASM hyperresponsiveness have not been elucidated. Recent studies have uncovered that de novo protein synthesis was indispensable in IL-13 action. In addition to the increased expression of cell surface proteins such as GPCR receptors (OXTR or CysLT1R) [77, 78], IL-13 also increases the protein expression of caveolin-1 [29] and sodium-calcium exchanger NCX1, which regulates Ca^{2+} fluxes [35]. siRNA strategies have indeed confirmed the implication of both proteins in mediating IL-13-induced enhanced GPCR-evoked calcium signals. This is an important finding as caveolin-1 was recently described as essential for ASM to maintain a contractile phenotype [79]. Up-regulation of the ectoenzyme CD38 was also shown to be critical for IL-13 effects on calcium signalling [62]. IL-13 also regulates the intracellular expression of key calcium regulatory proteins such as ryanodine receptors, IP_3 receptor or the calcium ATPase SERCA in murine ASM [74], although one study performed in human ASM cells found that SERCA2 expression was, in contrast, decreased following IL-13 treatment [36]. The exact pathway by which these different Ca^{2+} regulatory proteins contribute to the overall augmented Ca^{2+} responses is a challenging question that remains to be further investigated.

8 Conclusion

There is now increasing evidence to suggest that both TNFα and IL-13 are key players in asthma pathogenesis. These cytokines participate in the pathogenesis of asthma through multiple mechanisms, including the recruitment or activation of

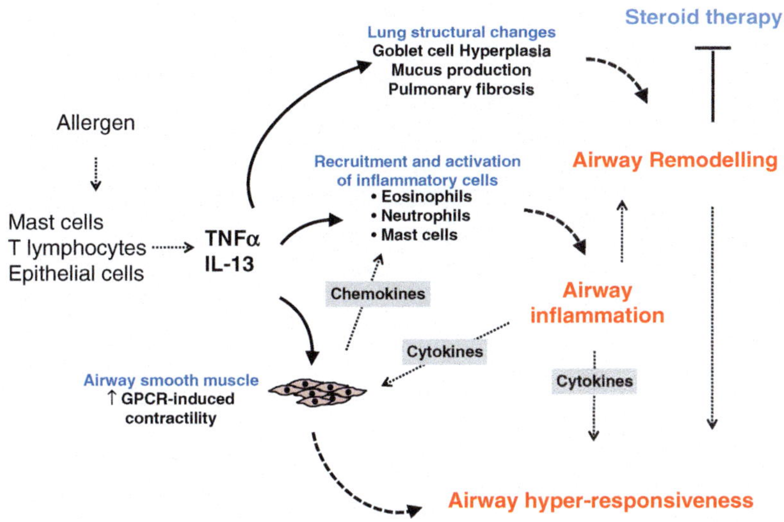

Fig. 1 Role of TNFα and IL-13 in the pathogenesis of severe asthma: Potential cellular sources of TNFα and IL-13 include allergen-induced IgE-dependent activation of mast cells, activation of lung infiltrated of inflammatory cells, or epithelial cells. Following release, both cytokines contributes to asthma pathogenesis through the regulation of airway inflammation possibly as a consequence of the recruitment and/or activation of different inflammatory cells including eosinophils, neutrophils and T lymphocytes. Products released by these activated infiltrated cells could promote airway hyper-responsiveness in part by acting on airway smooth muscle. A direct modulatory action of TNFα and IL-13 on airway smooth muscle could perpetuate the inflammatory process via additional secretion of a variety of chemokines. TNFα and IL-13 could play a role in mediating allergen-induced airway hyper-responsiveness by increasing the responsiveness of airway smooth muscle to a variety of contractile G-Protein Coupled Receptor (GPCR) agonists. By their ability to regulate mucous production, goblet cell hyperplasia and lung fibrosis, anti-IL-13 blockade and some extend anti-TNFα therapy may be useful in treating some features of airway remodelling which are known to be resistant to current therapies and involved in the development of persistent airflow obstruction

inflammatory cells and induction of structural changes in the lungs known as airway remodelling (Fig. 1). Current preclinical and clinical studies support the fact that TNFα and IL-13 also contribute to the development of AHR via mechanisms that have yet to be characterised. We have here provided supporting evidence that a local action of TNFα and IL-13 on airway structural cells represents one plausible mechanism contributing to exaggerated airway responsiveness. Although the overall effect of cytokines is likely to result from an action on different airway tissues/cell types, a direct action on ASM, the main effector regulating bronchomotor tone, is now regarded as one of the most plausible mechanisms driving impaired airway responsiveness. Our lab has provided the first demonstration that both TNFα and IL-13 directly modulate GPCR-induced contractility in isolated airways and calcium signalling in cultured ASM cells. Since then, a number of elegant reports have indeed confirmed this attractive concept that modulation of calcium metabolism is one common mechanism driving cytokine-induced AHR in asthma (Fig. 2).

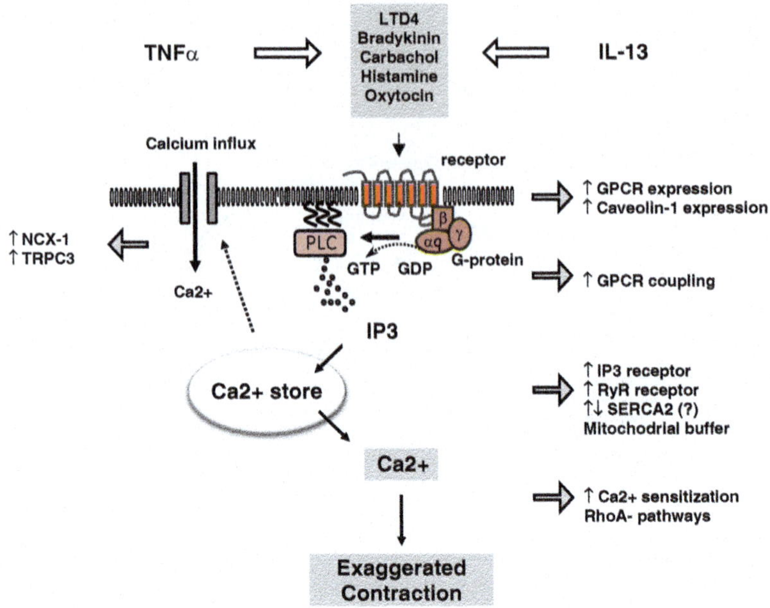

Fig. 2 Pathways activated by TNFα and IL-13 affecting GPCR-associated responsiveness: Different studies have confirmed that airway smooth muscle cells treated with TNFα and IL-13 exhibit an exaggerated calcium responses to a variety of GPCR agonists. Although the underlying mechanisms have not been completely elucidated, a number of recent reports have shown that cytokines act on different targets. Increase in GPCR receptor expression (bradykinin, LTD4 and oxytocin) or coupling with PLCb (bradykinin) has been reported. Changes in Ca^{2+} fluxes across the plasma membrane via the upregulation of activity of $Na^{(+)}/Ca^{(2+)}$ exchanger (NCX) or the transient receptor potential C3 (TRPC3) are induced by TNFα and IL-13. Expression of intracellular Ca^{2+} regulatory proteins such as receptors for IP3 or ryanodine (RYR) or the Ca^{2+}-ATPase SERCA2 have been implicated in the enhanced GPCR-induced ASM contractility. Finally, changes in Ca^{2+} sensitization via the RhoA pathway has been reported as one mechanism driving TNFα and IL-13-induced exaggerated contractility to muscarinic M3 activation.

Additional studies are clearly needed to better characterise the common targets that are altered by TNFα and IL-13; this will likely lead to the design of novel therapies for the treatment of allergic asthma.

References

1. Brightling, C., M. Berry, and Y. Amrani. 2008. Targeting TNF-alpha: a novel therapeutic approach for asthma. *The Journal of allergy and clinical immunology* 121: 5–10; quiz 11–12.
2. Ingram, J. L., and M. Kraft. 2012. IL-13 in asthma and allergic disease: asthma phenotypes and targeted therapies. *The Journal of allergy and clinical immunology* 130: 829–842; quiz 843–824.

3. Wills-Karp, M. 2001. IL-12/IL-13 axis in allergic asthma. *The Journal of allergy and clinical immunology* 107: 9–18.
4. Shore, S. A. 2004. Direct effects of Th2 cytokines on ASM. *Current opinion in pharmacology* 4: 235–240.
5. Amrani, Y., O. Tliba, D. A. Deshpande, T. F. Walseth, M. S. Kannan, and R. A. Panettieri, Jr. 2004. Bronchial hyperresponsiveness: insights into new signaling molecules. *Current opinion in pharmacology* 4: 230–234.
6. Amrani, Y., H. Chen, and R. A. Panettieri, Jr. 2000. Activation of tumor necrosis factor receptor 1 in ASM: a potential pathway that modulates bronchial hyper-responsiveness in asthma? *Respiratory research* 1: 49–53.
7. Amrani, Y. 2006. ASM modulation and airway hyper-responsiveness in asthma: new cellular and molecular paradigms. *Exp Rev Clin Immunol* 2: 353–364.
8. Renzetti, L. M., P. M. Paciorek, S. A. Tannu, N. C. Rinaldi, J. E. Tocker, M. A. Wasserman, and P. R. Gater. 1996. Pharmacological evidence for tumor necrosis factor as a mediator of allergic inflammation in the airways. *The Journal of pharmacology and experimental therapeutics* 278: 847–853.
9. Choi, I. W., K. Sun, Y. S. Kim, H. M. Ko, S. Y. Im, J. H. Kim, H. J. You, Y. C. Lee, J. H. Lee, Y. M. Park, and H. K. Lee. 2005. TNF-alpha induces the late-phase airway hyperresponsiveness and airway inflammation through cytosolic phospholipase A(2) activation. *The Journal of allergy and clinical immunology* 116: 537–543.
10. Deveci, F., M. H. Muz, N. Ilhan, G. Kirkil, T. Turgut, and N. Akpolat. 2008. Evaluation of the anti-inflammatory effect of infliximab in a mouse model of acute asthma. *Respirology* 13: 488–497.
11. Hutchison, S., B. S. Choo-Kang, R. V. Bundick, A. J. Leishman, J. M. Brewer, I. B. McInnes, and P. Garside. 2008. Tumour necrosis factor-alpha blockade suppresses murine allergic airways inflammation. *Clinical and experimental immunology* 151: 114–122.
12. Kim, J., L. McKinley, S. Natarajan, G. L. Bolgos, J. Siddiqui, S. Copeland, and D. G. Remick. 2006. Anti-tumor necrosis factor-alpha antibody treatment reduces pulmonary inflammation and methacholine hyper-responsiveness in a murine asthma model induced by house dust. *Clinical and experimental allergy : journal of the British Society for Allergy and Clinical Immunology* 36: 122–132.
13. Maillet, I., S. Schnyder-Candrian, I. Couillin, V. F. Quesniaux, F. Erard, R. Moser, S. Fleury, A. Kanda, D. Dombrowicz, D. E. Szymkowski, and B. Ryffel. 2011. Allergic lung inflammation is mediated by soluble tumor necrosis factor (TNF) and attenuated by dominant-negative TNF biologics. *American journal of respiratory cell and molecular biology* 45: 731–739.
14. Nie, Z., D. B. Jacoby, and A. D. Fryer. 2009. Etanercept prevents airway hyperresponsiveness by protecting neuronal M2 muscarinic receptors in antigen-challenged guinea pigs. *British journal of pharmacology* 156: 201–210.
15. Busse, P. J., T. F. Zhang, B. Schofield, S. Kilaru, S. Patil, and X. M. Li. 2009. Decrease in airway mucous gene expression caused by treatment with anti-tumor necrosis factor alpha in a murine model of allergic asthma. *Annals of allergy, asthma & immunology : official publication of the American College of Allergy, Asthma, & Immunology* 103: 295–303.
16. Luo, Y., Z. Pang, Q. Zhu, X. Cai, Y. Yin, M. Wang, J. Zhu, J. Chen, K. Zeng, C. Zhang, and J. Zhang. 2012. Locally instilled tumor necrosis factor-alpha antisense oligonucleotide inhibits allergic inflammation via the induction of Tregs. *The journal of gene medicine* 14: 374–383.
17. Pennings, H. J., K. Kramer, A. Bast, W. A. Buurman, and E. F. Wouters. 1998. Tumour necrosis factor-alpha induces hyperreactivity in tracheal smooth muscle of the guinea-pig in vitro. *The European respiratory journal : official journal of the European Society for Clinical Respiratory Physiology* 12: 45–49.
18. Makwana, R., N. Gozzard, D. Spina, and C. Page. 2012. TNF-alpha-induces airway hyperresponsiveness to cholinergic stimulation in guinea pig airways. *British journal of pharmacology* 165: 1978–1991.

19. Anticevich, S. Z., J. M. Hughes, J. L. Black, and C. L. Armour. 1995. Induction of human airway hyperresponsiveness by tumour necrosis factor-alpha. *Eur J Pharmacol* 284: 221–225.
20. Sukkar, M. B., J. M. Hughes, C. L. Armour, and P. R. Johnson. 2001. Tumour necrosis factor-alpha potentiates contraction of human bronchus in vitro. *Respirology* 6: 199–203.
21. Reynolds, A., M. Holmes, and R. Scicchitano. 2000. Cytokines enhance ASM contractility in response to acetylcholine and neurokinin A. *Respirology*. 5: 153–160.
22. Chen, H., O. Tliba, C. R. Van Besien, R. A. Panettieri, Jr., and Y. Amrani. 2003. Selected Contribution: TNF-{alpha} modulates murine tracheal rings responsiveness to G-protein-coupled receptor agonists and KCl. *Journal of applied physiology* 95: 864–872.
23. Jain, D., S. Keslacy, O. Tliba, Y. Cao, S. Kierstein, K. Amin, R. A. Panettieri, Jr., A. Haczku, and Y. Amrani. 2008. Essential role of IFNbeta and CD38 in TNFalpha-induced ASM hyper-responsiveness. *Immunobiology* 213: 499–509.
24. Kudo, M., S. M. Khalifeh Soltani, S. A. Sakuma, W. McKleroy, T. H. Lee, P. G. Woodruff, J. W. Lee, K. Huang, C. Chen, M. Arjomandi, X. Huang, and K. Atabai. 2013. Mfge8 suppresses airway hyperresponsiveness in asthma by regulating smooth muscle contraction. *Proceedings of the National Academy of Sciences of the United States of America* 110: 660–665.
25. Kudo, M., A. C. Melton, C. Chen, M. B. Engler, K. E. Huang, X. Ren, Y. Wang, X. Bernstein, J. T. Li, K. Atabai, X. Huang, and D. Sheppard. 2012. IL-17A produced by alphabeta T cells drives airway hyper-responsiveness in mice and enhances mouse and human ASM contraction. *Nature medicine* 18: 547–554.
26. Zhang, Y., M. Adner, and L. O. Cardell. 2004. Up-regulation of bradykinin receptors in a murine in-vitro model of chronic airway inflammation. *Eur J Pharmacol* 489: 117–126.
27. Adner, M., A. C. Rose, Y. Zhang, K. Sward, M. Benson, R. Uddman, N. P. Shankley, and L. O. Cardell. 2002. An assay to evaluate the long-term effects of inflammatory mediators on murine ASM: evidence that TNFalpha up-regulates 5-HT(2A)-mediated contraction. *British journal of pharmacology* 137: 971–982.
28. Guedes, A. G., J. A. Jude, J. Paulin, H. Kita, F. E. Lund, and M. S. Kannan. 2008. Role of CD38 in TNF-alpha-induced airway hyperresponsiveness. *American journal of physiology. Lung cellular and molecular physiology* 294: L290–299.
29. Sathish, V., A. J. Abcejo, S. K. VanOosten, M. A. Thompson, Y. S. Prakash, and C. M. Pabelick. 2011. Caveolin-1 in cytokine-induced enhancement of intracellular Ca(2+) in human ASM. *American journal of physiology. Lung cellular and molecular physiology* 301: L607–614.
30. Amrani, Y., and C. Bronner. 1993. Tumor necrosis factor alpha potentiates the increase in cytosolic free calcium induced by bradykinin in guinea-pig tracheal smooth muscle cells. *C R Acad Sci III* 316: 1489–1494.
31. Amrani, Y., V. Krymskaya, C. Maki, and R. A. Panettieri, Jr. 1997. Mechanisms underlying TNF-alpha effects on agonist-mediated calcium homeostasis in human ASM cells. *Am. J. Physiol.* 273: L1020–1028.
32. Amrani, Y., A. L. Lazaar, R. Hoffman, K. Amin, S. Ousmer, and R. A. Panettieri, Jr. 2000. Activation of p55 Tumor Necrosis Factor-alpha Receptor-1 Coupled to Tumor Necrosis Factor Receptor-Associated Factor 2 Stimulates Intercellular Adhesion Molecule-1 Expression by Modulating a Thapsigargin-Sensitive Pathway in Human Tracheal Smooth Muscle Cells. *Molecular pharmacology* 58: 237–245.
33. Amrani, Y., N. Martinet, and C. Bronner. 1995. Potentiation by tumour necrosis factor-alpha of calcium signals induced by bradykinin and carbachol in human tracheal smooth muscle cells. *British journal of pharmacology* 114: 4–5.
34. Delmotte, P., B. Yang, M. A. Thompson, C. M. Pabelick, Y. S. Prakash, and G. C. Sieck. 2012. Inflammation alters regional mitochondrial Ca(2)+ in human ASM cells. *American journal of physiology. Cell physiology* 303: C244–256.

35. Sathish, V., P. F. Delmotte, M. A. Thompson, C. M. Pabelick, G. C. Sieck, and Y. S. Prakash. 2011. Sodium-calcium exchange in intracellular calcium handling of human ASM. *PloS one* 6: e23662.
36. Sathish, V., M. A. Thompson, J. P. Bailey, C. M. Pabelick, Y. S. Prakash, and G. C. Sieck. 2009. Effect of proinflammatory cytokines on regulation of sarcoplasmic reticulum Ca2+ reuptake in human ASM. *American journal of physiology. Lung cellular and molecular physiology* 297: L26–34.
37. Mahn, K., S. J. Hirst, S. Ying, M. R. Holt, P. Lavender, O. O. Ojo, L. Siew, D. E. Simcock, C. G. McVicker, V. Kanabar, V. A. Snetkov, B. J. O'Connor, C. Karner, D. J. Cousins, P. Macedo, K. F. Chung, C. J. Corrigan, J. P. Ward, and T. H. Lee. 2009. Diminished sarco/endoplasmic reticulum Ca2+ ATPase (SERCA) expression contributes to airway remodelling in bronchial asthma. *Proceedings of the National Academy of Sciences of the United States of America* 106: 10775–10780.
38. White, T. A., A. Xue, E. N. Chini, M. Thompson, G. C. Sieck, and M. E. Wylam. 2006. Role of transient receptor potential C3 in TNF-alpha-enhanced calcium influx in human airway myocytes. *American journal of respiratory cell and molecular biology* 35: 243–251.
39. Amrani, Y., and R. A. Panettieri, Jr. 2002. Modulation of calcium homeostasis as a mechanism for altering smooth muscle responsiveness in asthma. *Current opinion in allergy and clinical immunology* 2: 39–45.
40. Hunter, I., H. J. Cobban, P. Vandenabeele, D. J. MacEwan, and G. F. Nixon. 2003. Tumor Necrosis Factor-alpha-Induced Activation of RhoA in ASM Cells: Role in the Ca(2+) Sensitization of Myosin Light Chain(20) Phosphorylation. *Molecular pharmacology* 63: 714–721.
41. Hunter, I., and G. F. Nixon. 2006. Spatial compartmentalization of tumor necrosis factor (TNF) receptor 1-dependent signaling pathways in human ASM cells. Lipid rafts are essential for TNF-alpha-mediated activation of RhoA but dispensable for the activation of the NF-kappaB and MAPK pathways. *The Journal of biological chemistry* 281: 34705–34715.
42. Sakai, H., S. Otogoto, Y. Chiba, K. Abe, and M. Misawa. 2004. Involvement of p42/44 MAPK and RhoA protein in augmentation of ACh-induced bronchial smooth muscle contraction by TNF-alpha in rats. *Journal of applied physiology* 97: 2154–2159.
43. Sakai, H., S. Otogoto, Y. Chiba, K. Abe, and M. Misawa. 2004. TNF-alpha augments the expression of RhoA in the rat bronchus. *J Smooth Muscle Res* 40: 25–34.
44. Deshpande, D. A., T. F. Walseth, R. A. Panettieri, and M. S. Kannan. 2003. CD38-cyclic ADP-ribose-mediated Ca2+ signaling contributes to ASM hyperresponsiveness. *The FASEB journal : official publication of the Federation of American Societies for Experimental Biology*: FASEB Journal Express Article 10.1096/fj.1002-0450fje.
45. Wills-Karp, M., J. Luyimbazi, X. Xu, B. Schofield, T. Y. Neben, C. L. Karp, and D. D. Donaldson. 1998. Interleukin-13: central mediator of allergic asthma. *Science* 282: 2258–2261.
46. Grunig, G., M. Warnock, A. E. Wakil, R. Venkayya, F. Brombacher, D. M. Rennick, D. Sheppard, M. Mohrs, D. D. Donaldson, R. M. Locksley, and D. B. Corry. 1998. Requirement for IL-13 independently of IL-4 in experimental asthma. *Science* 282: 2261–2263.
47. Wang, Y., and C. T. McCusker. 2005. Interleukin-13-dependent bronchial hyper-responsiveness following isolated upper-airway allergen challenge in a murine model of allergic rhinitis and asthma. *Clinical and experimental allergy : journal of the British Society for Allergy and Clinical Immunology* 35: 1104–1111.
48. Hacha, J., K. Tomlinson, L. Maertens, G. Paulissen, N. Rocks, J. M. Foidart, A. Noel, R. Palframan, M. Gueders, and D. D. Cataldo. 2012. Nebulized anti-IL-13 monoclonal antibody Fab' fragment reduces allergen-induced asthma. *American journal of respiratory cell and molecular biology* 47: 709–717.
49. Eum, S. Y., K. Maghni, B. Tolloczko, D. H. Eidelman, and J. G. Martin. 2005. IL-13 may mediate allergen-induced hyperresponsiveness independently of IL-5 or eotaxin by effects on ASM. *American journal of physiology. Lung cellular and molecular physiology* 288: L576–584.

50. Tomlinson, K. L., G. C. Davies, D. J. Sutton, and R. T. Palframan. 2010. Neutralisation of interleukin-13 in mice prevents airway pathology caused by chronic exposure to house dust mite. *PloS one* 5.
51. Yang, G., L. Li, A. Volk, E. Emmell, T. Petley, J. Giles-Komar, P. Rafferty, M. Lakshminarayanan, D. E. Griswold, P. J. Bugelski, and A. M. Das. 2005. Therapeutic dosing with anti-interleukin-13 monoclonal antibody inhibits asthma progression in mice. *The Journal of pharmacology and experimental therapeutics* 313: 8–15.
52. Poon, A. H., D. H. Eidelman, J. G. Martin, C. Laprise, and Q. Hamid. 2012. Pathogenesis of severe asthma. *Clinical and experimental allergy : journal of the British Society for Allergy and Clinical Immunology* 42: 625–637.
53. Vrugt, B., R. Djukanovic, A. Bron, and R. Aalbers. 1996. New insights into the pathogenesis of severe corticosteroid-dependent asthma. *The Journal of allergy and clinical immunology* 98: S22–26; discussion S33–40.
54. Goplen, N., M. Z. Karim, Q. Liang, M. M. Gorska, S. Rozario, L. Guo, and R. Alam. 2009. Combined sensitization of mice to extracts of dust mite, ragweed, and Aspergillus species breaks through tolerance and establishes chronic features of asthma. *The Journal of allergy and clinical immunology* 123: 925–932 e911.
55. Leigh, R., R. Ellis, J. Wattie, D. D. Donaldson, and M. D. Inman. 2004. Is interleukin-13 critical in maintaining airway hyperresponsiveness in allergen-challenged mice? *American journal of respiratory and critical care medicine* 170: 851–856.
56. Taube, C., C. Duez, Z. H. Cui, K. Takeda, Y. H. Rha, J. W. Park, A. Balhorn, D. D. Donaldson, A. Dakhama, and E. W. Gelfand. 2002. The role of IL-13 in established allergic airway disease. *J Immunol* 169: 6482–6489.
57. Lukacs, N. W., K. K. Tekkanat, A. Berlin, C. M. Hogaboam, A. Miller, H. Evanoff, P. Lincoln, and H. Maassab. 2001. Respiratory syncytial virus predisposes mice to augmented allergic airway responses via IL-13-mediated mechanisms. *J Immunol* 167: 1060–1065.
58. Tekkanat, K. K., H. F. Maassab, D. S. Cho, J. J. Lai, A. John, A. Berlin, M. H. Kaplan, and N. W. Lukacs. 2001. IL-13-induced airway hyperreactivity during respiratory syncytial virus infection is STAT6 dependent. *J Immunol* 166: 3542–3548.
59. Schneider, D., J. Y. Hong, A. P. Popova, E. R. Bowman, M. J. Linn, A. M. McLean, Y. Zhao, J. Sonstein, J. K. Bentley, J. B. Weinberg, N. W. Lukacs, J. L. Curtis, U. S. Sajjan, and M. B. Hershenson. 2012. Neonatal rhinovirus infection induces mucous metaplasia and airways hyperresponsiveness. *J Immunol* 188: 2894–2904.
60. Tliba, O., Deshpande D, Van Besien C, Chen H, Kannan M, Panettieri RA, Amrani Y. 2003. IL-13 enhances agonist-evoked calcium signals and contractile responses in ASM. *British journal of pharmacology* 140: 1159–1162.
61. Guedes, A. G., J. Paulin, L. Rivero-Nava, H. Kita, F. E. Lund, and M. S. Kannan. 2006. CD38-deficient mice have reduced airway hyperresponsiveness following IL-13 challenge. *American journal of physiology. Lung cellular and molecular physiology* 291: L1286–1293.
62. Deshpande, D. A., S. Dogan, T. F. Walseth, S. M. Miller, Y. Amrani, R. A. Panettieri, and M. S. Kannan. 2004. Modulation of calcium signaling by interleukin-13 in human ASM: role of CD38/cyclic adenosine diphosphate ribose pathway. *American journal of respiratory cell and molecular biology* 31: 36–42.
63. Grunstein, M. M., H. Hakonarson, J. Leiter, M. Chen, R. Whelan, J. S. Grunstein, and S. Chuang. 2002. IL-13-dependent autocrine signaling mediates altered responsiveness of IgE-sensitized ASM. *American journal of physiology. Lung cellular and molecular physiology* 282: L520–528.
64. Hu, A., S. Fatma, J. Cao, J. S. Grunstein, G. Nino, Y. Grumbach, and M. M. Grunstein. 2009. Th2 cytokine-induced upregulation of 11beta-hydroxysteroid dehydrogenase-1 facilitates glucocorticoid suppression of proasthmatic ASM function. *American journal of physiology. Lung cellular and molecular physiology* 296: L790–803.
65. Chiba, Y., S. Nakazawa, M. Todoroki, K. Shinozaki, H. Sakai, and M. Misawa. 2009. Interleukin-13 augments bronchial smooth muscle contractility with an up-regulation of RhoA protein. *American journal of respiratory cell and molecular biology* 40: 159–167.

66. Possa, S. S., H. T. Charafeddine, R. F. Righetti, P. A. da Silva, R. Almeida-Reis, B. M. Saraiva-Romanholo, A. Perini, C. M. Prado, E. A. Leick-Maldonado, M. A. Martins, and F. Tiberio Ide. 2012. Rho-kinase inhibition attenuates airway responsiveness, inflammation, matrix remodeling, and oxidative stress activation induced by chronic inflammation. *American journal of physiology. Lung cellular and molecular physiology* 303: L939–952.
67. Sugimoto, K., M. Kudo, A. Sundaram, X. Ren, K. Huang, X. Bernstein, Y. Wang, W. W. Raymond, D. J. Erle, M. Abrink, G. H. Caughey, X. Huang, and D. Sheppard. 2012. The alphavbeta6 integrin modulates airway hyperresponsiveness in mice by regulating intraepithelial mast cells. *The Journal of clinical investigation* 122: 748–758.
68. Carter, R. J., and P. Bradding. 2011. The role of mast cells in the structural alterations of the airways as a potential mechanism in the pathogenesis of severe asthma. *Current pharmaceutical design* 17: 685–698.
69. Sanderson, M. J. 2011. Exploring lung physiology in health and disease with lung slices. *Pulmonary pharmacology & therapeutics* 24: 452–465.
70. Cooper, P. R., C. T. Poll, P. J. Barnes, and R. G. Sturton. 2010. Involvement of IL-13 in tobacco smoke-induced changes in the structure and function of rat intrapulmonary airways. *American journal of respiratory cell and molecular biology* 43: 220–226.
71. Cooper, P. R., R. Lamb, N. D. Day, P. J. Branigan, R. Kajekar, L. San Mateo, P. J. Hornby, and R. A. Panettieri, Jr. 2009. TLR3 activation stimulates cytokine secretion without altering agonist-induced human small airway contraction or relaxation. *American journal of physiology. Lung cellular and molecular physiology* 297: L530–537.
72. Bergner, A., J. Kellner, A. K. Silva, F. Gamarra, and R. M. Huber. 2006. Ca2+-signaling in ASM cells is altered in T-bet knock-out mice. *Respiratory research* 7: 33.
73. Jiang, H., Y. Xie, P. W. Abel, M. L. Toews, R. G. Townley, T. B. Casale, and Y. Tu. 2012. Targeting phosphoinositide 3-kinase gamma in ASM cells to suppress interleukin-13-induced mouse airway hyperresponsiveness. *The Journal of pharmacology and experimental therapeutics* 342: 305–311.
74. Kellner, J., J. Tantzscher, H. Oelmez, M. Edelmann, R. Fischer, R. M. Huber, and A. Bergner. 2008. Mechanisms altering ASM cell Ca+ homeostasis in two asthma models. *Respiration; international review of thoracic diseases* 76: 205–215.
75. Risse, P. A., T. Jo, F. Suarez, N. Hirota, B. Tolloczko, P. Ferraro, P. Grutter, and J. G. Martin. 2011. Interleukin-13 inhibits proliferation and enhances contractility of human ASM cells without change in contractile phenotype. *American journal of physiology. Lung cellular and molecular physiology* 300: L958–966.
76. Moynihan, B., B. Tolloczko, M. C. Michoud, M. Tamaoka, P. Ferraro, and J. G. Martin. 2008. MAP kinases mediate interleukin-13 effects on calcium signaling in human ASM cells. *American journal of physiology. Lung cellular and molecular physiology* 295: L171–177.
77. Matsumoto, H., Y. Hirata, K. Otsuka, T. Iwata, A. Inazumi, A. Niimi, I. Ito, E. Ogawa, S. Muro, H. Sakai, K. Chin, Y. Oku, and M. Mishima. 2012. Interleukin-13 enhanced Ca2+ oscillations in ASM cells. *Cytokine* 57: 19–24.
78. Amrani, Y., F. Syed, C. Huang, K. Li, V. Liu, D. Jain, S. Keslacy, M. W. Sims, H. Baidouri, P. R. Cooper, H. Zhao, S. Siddiqui, C. E. Brightling, D. Griswold, L. Li, and R. A. Panettieri, Jr. 2010. Expression and activation of the oxytocin receptor in ASM cells: Regulation by TNFalpha and IL-13. *Respiratory research* 11: 104.
79. Gosens, R., G. L. Stelmack, S. T. Bos, G. Dueck, M. M. Mutawe, D. Schaafsma, H. Unruh, W. T. Gerthoffer, J. Zaagsma, H. Meurs, and A. J. Halayko. 2011. Caveolin-1 is required for contractile phenotype expression by ASM cells. *Journal of cellular and molecular medicine* 15: 2430–2442.

Airway Smooth Muscle Malfunction in COPD

Yunchao Su

Contents

Abstract Airway smooth muscle has been conventionally regarded as a contractile partner in bronchoconstriction. It also interacts dynamically with its environment, especially under inflammatory conditions, modulates the pathological processes in the development of chronic obstructive pulmonary disease (COPD). Airway smooth muscle cells are able to proliferate, to secrete cytokines, growth factors, prostanoid, and extracellular matrix proteins, and to adapt to these functions by

---

Y. Su, M.D., Ph.D. (✉)
Department of Pharmacology and Toxicology, Department of Medicine, Vascular Biology Center, Center for Biotechnology and Genomic Medicine, Medical College of Georgia, Georgia Regents University, 1120 15th Street, Augusta, GA 30912, USA
e-mail: ysu@gru.edu

Y.-X. Wang (ed.), *Calcium Signaling In Airway Smooth Muscle Cells*,
DOI 10.1007/978-3-319-01312-1_25,

changing its phenotype from contractile to synthetic. In COPD, smooth muscle in the small airways exhibit hyperresponsiveness and higher contractility in calcium-dependent and -independent mechanisms. Airway smooth muscle cell hyperplasia and hypertrophy as well as increased deposition of extracellular matrix (ECM) proteins contribute to airway remodeling and thickening. Many inflammatory mediators and growth factors, including interleukins, tumor necrosis factor-α (TNF-α), leukotrienes, prostaglandins, acetylcholine, platelet-derived growth factor (PDGF), transform growth factor β (TGFβ), connective tissue growth factor (CTGF), endothelin, and reactive oxygen species (ROS), contribute to the alterations of airway smooth muscle in COPD. Investigating the mechanism and regulatory pathways of airway smooth muscle malfunction may provide novel options for the prevention and treatment for airflow obstruction in COPD.

Keywords Airway • Smooth muscle • Inflammation • Bronchoconstriction • Remodeling

1 Introduction

Chronic obstructive pulmonary disease (COPD) is a progressive and debilitating inflammatory disease that affects between 10 and 24 million adults and is the third leading cause of death in the United States [1]. The most important pathological features in the pathogenesis of COPD are airflow limitation, which is caused by airway obstruction due to small airway remodeling and increased airway smooth muscle (ASM) contractility, and structural changes of the lung, including emphysema. The alterations of smooth muscle of the small airways in COPD have been ignored for a long time. ASM has been conventionally regarded as a contractile partner in bronchoconstriction. Overwhelming evidences indicate that ASM interacts dynamically with its environment, especially under inflammatory conditions, and orchestrates the pathological processes in the development of COPD. ASM cells are able to proliferate, to migrate, to secrete chemokines, cytokines, growth factors and extracellular matrix proteins, and, most importantly, to adapt to these functions by changing its phenotype from contractile to proliferative/synthetic. This chapter will review the recent advances in the alterations of ASM cells and the inflammatory factors that affect ASM functions in the pathogenesis of COPD.

2 Alterations of Airway Smooth Muscle Cells in COPD

2.1 Increased Contractility of Airway Smooth Muscle in COPD

Altered ASM contractility contributes to the pathogenesis of COPD [2, 3]. Peripheral airways of COPD have an increased ability to generate force [2]. Airway hyperresponsiveness is present in a considerable number of COPD patients, though it is regarded as a hallmark of asthma [4, 5]. Calcium-dependent and calcium-independent mechanisms contribute to the increased contraction of ASM in COPD.

Calcium-dependent mechanism: calcium mediates smooth muscle contraction. The response of ASM to diverse stimuli including inflammatory mediators is controlled by changes in the concentration of free cytosolic calcium ($[Ca^{2+}]_i$). Elevation of $[Ca^{2+}]_i$ can result from increased Ca^{2+} influx from extracellular fluid and Ca^{2+} release from the sarcoplasmic reticulum (SR). In smooth muscle cells, excitation-contraction (E-C) coupling involves depolarization of the plasma membrane to open voltage-gated calcium channels. However, the voltage-dependent calcium channels do not appear to strongly influence ASM contraction, though maintaining agonist-induced contraction requires the presence of extracellular calcium [6, 7]. Agonist-induced calcium release is largely from the SR [6, 7]. Ca^{2+} efflux from the SR is mediated by the SR-Ca^{2+} release channel, known as ryanodine receptor (RyR) [8]. Cyclic-ADP-ribose (cADPR) is an endogenous ligand and a potent activator of RyR [9]. A 12.6-kDa FK506-binding protein (FKBP12.6) is associated with type 2 ryanodine receptors (RyR2) protein [10]. FKBP12.6 stabilizes the RyR2 channel in the closed state and reduces its activity. Binding of cADPR results in the activation of RyR2 and Ca^{2+} efflux from the SR by dissociating FKBP12.6 from the RyR2 complex [11]. RyR, cADPR, and FKBR could be therapeutic target to reverse bronchoconstriction in COPD.

Calcium-independent mechanism: calcium-independent mechanisms are characterized by augmented contraction at a fixed Ca^{2+} concentration, which is referred to as Ca^{2+} sensitization. The RhoA/Rho-kinase cascade, a key regulatory pathway of Ca^{2+} sensitivity in ASM, can be activated by G-protein-coupled receptors such as muscarinic M2, muscarinic M3 , and endothelin ETA and ETB receptors in COPD [12–14]. RhoA and Rho-kinase augment agonist-induced contraction by inactivating myosin light chain phosphatase (MLCP) or directly affecting myosin light chain phosphorylation and actin cytoskeletal dynamics [15]. MLCP is inhibited by the direct phosphorylation of its regulatory myosin-binding binding subunit by Rho-kinase. MLCP is also inhibited by binding to the phosphoprotein CPI-17, which is targeted for phosphorylation by both Rho-kinase and PKC [16].

2.2 *Airway Smooth Muscle Remodeling in COPD*

One of the key pathological features of COPD is the thickening of airway walls. This thickening is largely due to the result of ASM cell hyperplasia and hypertrophy as well as increased deposition of extracellular matrix (ECM) proteins such as collagens, elastin, laminin, and proteoglycans around the smooth muscle. The increases in ASM mass observed in COPD could in part be mediated by inflammatory mediators and peptide growth factors, which are released from inflammatory cells, such as eosinophils and macrophages as well as epithelium, extravasated plasma, and the ASM itself. In response to inflammatory mediators and growth factors, several signal pathways are activated in ASM cells. The most investigated pathways include mitogen-activated protein kinase (MAPK), nuclear factor-kappa B (NF-kB), Janus kinases and signal transducers and activators of transcription (JAK/STATs), SMADs, and β-catenin pathways [17–22]. Activation of these pathways leads to ASM cell proliferation, migration, and ECM protein production and secretion of cytokine and chemokines. In turn, these secretory mediators may affect ASM proliferation and contractility in a vicious cycle. Excessive deposition of collagen in the airway contributes to thickening and increased stiffness of the bronchial wall, predisposing to airway obstruction in COPD [23, 24].

2.3 Contractile Versus Synthetic Phenotype of Airway Smooth Muscle in COPD

Mature ASM cells retain their capacity for phenotype plasticity to shift between contractile and synthetic phenotype. ASM phenotype modulation is regulated by inflammatory mediators, growth factors, and acetylcholine [25, 26]. In vitro, ASM cells and intact tissue strips switch to a less contractile, more proliferative phenotype when exposed to high concentrations of fetal bovine serum or growth factors [27]. Such a phenotype shift is accompanied by a dramatic decrease in the expression of specific contractile proteins, such as smooth muscle myosin heavy chain, calponin, and α-actin [27]. In contrast, induction of a contractile (mature) phenotype in cultured ASM cells is characterized by an accumulation of contractile proteins and contraction regulatory proteins [27, 28]. RhoA/Rho-kinase signaling plays an essential role in driving the transcription of genes encoding for contractile proteins in ASM [29]. Serum response factor (SRF) has emerged as a pivotal transcriptional activator of these smooth-muscle-specific promoters [30]. Localization and activation of SRF in the nucleus and subsequent induction of smooth-muscle-specific genes are importantly regulated by RhoA/Rho-kinase signaling in ASM [29], presumably through Rho/Rho-kinase-induced actin polymerization, a process that is required for the activation of SRF [31].

In COPD, small airways exhibit hyperresponsiveness and higher contractility while ASM mass is increased [5, 32–34]. These findings seem paradoxical in regards to whether ASM cells in COPD exhibit contractile or synthetic phenotype. It might be inappropriate to refer higher contractility to contractile phenotype and hyperplasia to synthetic phenotype. Contractile versus synthetic phenotype is a manifestation of ASM cells under culture conditions that can be manipulated in vitro. In the airways of COPD, the dynamics of phenotypic switching of smooth muscle depends on its dynamic interaction with the inflammatory environment. The in vivo behavior of ASM in COPD deserves continued study and discussion.

3 Mediators That Are Synthesized/Secreted by ASM or Affect ASM Function in COPD

3.1 Interleukins

The interleukins that are involved in airway inflammation include interleukin-1β (IL-1β), IL-6, IL-8, and IL-10. These interleukins are produced by a number of cell populations, including activated bronchial epithelial cells, macrophages, neutrophils, and ASM cells [35, 36].

IL-1β is one of the major cytokines involved in the initiation and persistence of inflammation in COPD and its exacerbations. The macrophage/monocyte is a primary source of IL-1, but many other cells, including fibroblasts, T-cells, neutrophils, and bronchial and alveolar epithelial cells, also produce IL-1β. IL-1β is synthesized as 31-kD precursor, pro-IL-1β. The active, mature IL-1β is produced upon cleavage of pro-IL-1β by a specific IL-1β–converting enzyme (ICE or caspase-1) or by proteases such as matrix metalloprotease-9 (MMP-9 or gelatinase B). IL-1β gene polymorphisms are associated with susceptibility to COPD [37]. IL-1β enhances the production of neutrophil-attractant CXC chemokines KC (CXCL1) and MIP-2 (CXCL2) as well as of matrix metalloproteases MMP-9 and MMP-12 in the lungs [38]. Over-expression of IL-1β is sufficient to induce lung inflammation, enlargement of distal airspaces, mucus metaplasia, and ASM thickening and fibrosis in adult mice [38].

IL-6 is a potent proinflammatory cytokine produced by a diverse set of cell populations including ASM cells [32]. In animal models, its expression can be induced by LPS, rhinovirus 16 (RV16), and respiratory syncytial virus [39, 40]. IL-6 exerts proinflammatory effects by activating both leukocytes and structural cells including pulmonary epithelial cells. The levels of IL-6 are increased in the induced sputum, bronchoalveolar lavage, and blood of patients with COPD [41]. Its levels have been reported to be even higher during exacerbation [41]. IL-6 levels in plasma are consistently associated with impaired lung function [42]. In a murine model, IL-6 overexpression resulted in emphysema like airspace enlargement, airway inflammation, airway hyperresponsiveness, and bronchial wall thickening [43].

IL-8, which is also called chemokine (C-X-C motif) ligand 8 (CXCL-8), is a potent chemoattractant for neutrophils and monocytes and, depending on the anatomical location, may contribute to the mobilization of these leukocyte populations into the lung [44]. In some studies, the levels of IL-8 are increased in the induced sputum of patients with COPD, and the levels appear to correlate with the proportion of neutrophils and are also increased in the sputum during exacerbations [45, 46].

3.2 *Tumor Necrosis Factor-α (TNF-α)*

Tumor Necrosis Factor-α (TNF-α) is a potent proinflammatory mediator in COPD [47, 48]. TNF-α is produced by many different cell types. The main sources in vivo are stimulated monocytes, fibroblasts, and endothelial cells. Macrophages, T-cells, B-lymphocytes, granulocytes, smooth muscle cells, and eosinophils also produce TNF-α after stimulation. The synthesis of TNF-α can be stimulated by IL-1β, bacterial endotoxins, and platelet-derived growth factor (PDGF). The activation of TNF-α receptors induces the production and release of a number of inflammatory mediators such as reactive oxygen species (ROS), CXCL10, vascular cell adhesion molecule one (VCAM-1), and intercellular adhesion molecule one (ICAM-1), which in turn induce additional inflammatory effects [49, 50]. In addition, TNF-α contributes to airway hyperresponsiveness by altering ASM Ca^{2+} responses to agonist stimulation such as acetylcholine and bradykinin [51–53]. TNF-α induces the expression of transient receptor potential (TRP) channel family proteins that function as channels through which receptor-operated and store-operated Ca^{2+} entry (SOCE) occurs [54]. TNF-α induces Ca^{2+} sensitization of MLC20 phosphorylation in ASM cells by an inhibition of the smooth muscle myosin light chain phosphatase activity via Rho-kinase activation [52].

3.3 *Leukotrienes*

Leukotrienes are synthesized in cells from arachidonic acid (AA) by the enzyme 5-lipoxygenase. 5-lipoxygenase first converts AA into 5-hydroperoxyeicosatetraenoic acid (5-HPETE), which spontaneously reduces to 5-hydroxyeicosatetraenoic acid (5-HETE). The enzyme 5-lipoxygenase acts again on 5-HETE to convert it into leukotriene (LTA_4), an unstable epoxide. In neutrophils and monocytes, LTA_4 is then converted by LTA_4 hydrolase into the dihydroxy acid leukotriene LTB_4. In mast cells and eosinophils, LTA_4 is conjugated with the tripeptide glutathione to form the cysteinyl leukotrienes, LTC_4, under the catalysis of LTC_4 synthase. Outside of cells, LTC_4 can be converted by ubiquitous enzymes to form LTD_4 and LTE_4. The levels of leukotrienes are increased in the lungs, sputum, and exhaled breath condensate in stable COPD and its exacerbations [32, 55, 56]. Leukotrienes are extremely potent mediators of airway hyperresponsiveness and inflammation, producing smooth muscle contraction, especially bronchoconstriction, increased vascular permeability, and migration of leukocytes to areas of inflammation.

3.4 Prostaglandins

Prostaglandins are products of the metabolism of AA by cyclooxygenase (COX). There are two major COX isoenzymes: COX-1 is expressed constitutively (constantly) in most tissues, whereas COX-2 is induced by inflammation. Both COX-1 and COX-2 use AA to generate the same product, prostaglandin H_2. A number of enzymes further modify this product to generate bioactive lipids (prostanoids) such as prostacyclin (PGI_2), thromboxane A_2 (TXA_2), prostaglandin D_2 (PGD_2), prostaglandin E_2 (PGE_2), and prostaglandin $F_{2\alpha}$ ($PGF_{2\alpha}$). PGD_2, PGE_2, $PGF_{2\alpha}$, PGI_2, and TXA_2 bind to prostanoid DP, EP, IP, and TP receptors, respectively, which belong to the family of rhodopsin like seven transmembrane receptors. PGD_2, PGE_2, $PGF_{2\alpha}$, and TXA_2 are mainly produced by monocytes, macrophages, neutrophils, lung parenchyma, and platelets as well as ASM. The levels of prostaglandins are increased in the lungs, sputum, and exhaled breath condensate in stable COPD and its exacerbations [57, 58]. PGD_2, $PGF_{2\alpha}$, and TXA_2 induce strong bronchoconstriction and contribute to airway hyperresponsiveness in COPD [59–61]. In addition, TXA_2 is a potent stimulator of ASM cell proliferation, which leads to bronchial smooth muscle hyperplasia and airway remodeling [62].

3.5 Acetylcholine

Acetylcholine is the primary parasympathetic neurotransmitter in the airways. The major source of acetylcholine in the airways is the vagal nerve. The release of acetylcholine from the vagal nerve is regulated by a variety of prejunctional receptors, including autoinhibitory muscarinic M_2 receptors [3, 63] . Muscarinic M_2 autoreceptor dysfunction is mediated by eosinophils. This mechanism may be more prominent in asthma and COPD acute exacerbations than in stable COPD [64]. Inflammatory mediators such as prostaglandins, histamine, bradykinin, serotonin, adenosine, and endothelin can augment acetylcholine release from cholinergic nerve endings by prejunctional facilitation [65]. Recent studies indicate that acetylcholine is also released from nonneuronal origins. Bronchial epithelial cells, T- and B-lymphocytes, mast cells, monocytes, granulocytes, alveolar macrophages, and ASM cells contain acetylcholine or express its synthesizing enzyme, choline acetyltransferase (ChAT) [66, 67].

Acetylcholine acts through both Gi-coupled muscarinic M_2 and Gq-coupled muscarinic M_3 receptors on ASM. Muscarinic receptor signaling increases ASM contractility, ASM proliferation and differentiation, and production of ECM proteins. Parasympathetic activity is increased in airway inflammation, which is the basis for the use of anticholinergic therapy in COPD. The increased muscarinic activity is caused by exaggerated release of neuronal acetylcholine due to neuronal mechanisms associated with inflammation. Increased expression and enhanced function of signaling molecules essential for muscarinic receptor mediate ASM contraction and proliferation. In COPD, muscarinic receptor-linked signaling pathways that regulate the Ca^{2+} sensitivity of ASM cells are enhanced. The expression and function of RhoA, CPI-17, and CD38 are increased by proinflammatory cytokines in vitro and in animal models of COPD [68]. Muscarinic receptor signals augment the mitogenic responses of ASM to EGF and PDGF, contributing to the increase in ASM mass observed in COPD [69].

3.6 PDGF

PDGFs are dimers of disulfide-linked polypeptide chains (PDGF-AA, PDGF-AB, or PDGF-BB). PDGFs bind via two receptor tyrosine (Tyr) kinase receptors (PDGFR-α and PDGFR-β). PDGFRs contain common domain structures, including five extracellular immunoglobulin loops and a split intracellular Tyr kinase domain. Ligand binding promotes receptor dimerization, which allows for receptor autophosphorylation on Tyr residues in the intracellular domain and initiation of signaling. Autophosphorylation of PDGFR provides docking sites for downstream signaling molecules and propagates signals through protein-protein interactions through specific domains, such as Src homology domain (SH_2) and phosphotyrosine binding domains (PTBs). PDGF binding to PDGFR-α and PDGFR-β engages several signaling pathways, such as Ras-MAPK, PI3K, and PLC-γ [70, 71]. Elevated expressions of PDGFs and PDGFR have been found in COPD [72, 73]. PDGF and acetylcholine act synergistically and induce ASM proliferation in COPD [69].

3.7 TGFβ

TGFβ is produced by a range of cells including platelets, macrophages, epithelial cells, and ASM. TGFβ1 is synthesized as an inactive latent precursor called the small latent TGFβ1 complex, which is composed of an N-terminal latency-associated peptide (LAP) and a C-terminal mature TGFβ1 [74]. After secretion from the cells, the small latent TGFβ1 complex binds to a latent TGFβ1 binding protein (LTBP). Latent TGFβ1 must be activated into a mature form for receptor binding and subsequent activation of signal transduction pathways. Latent TGFβ1 can be activated either by removing LTBP extracellularly through proteolytic cleavage by various proteases such as plasmin, thrombin, plasma transglutaminase, or endoglycosylases [75], or by physical interactions of the LAP with other proteins, such as thrombospondin-1 [76], or by an integrin-induced conformational change in the small latent TGFβ1 complex [77, 78]. Recently, we found that calpain activates TGFβ1 by cleaving latent TGFβ1 [79]. We have reported that PDGF induces activation of intracellular TGFβ1 in the Golgi and triggers an intracrine TGFβ/Smad signal pathway [79].

TGFβ1 first binds to TGFβ1 receptor type II (TGFβRII) and induces the assembly of type I and type II receptors into complexes, within which TGFβRII phosphorylates TGFβ1 receptor type I (TGFβRI). TGFβRI expressed in smooth muscle cells is activin-receptor-like kinase 5 (ALK5) [80]. Phosphorylation of ALK5 activates its kinase and subsequently causes Smad2/3 phosphorylation [81]. Smads have intrinsic transcription-inducing activity. In the basal state, Smad2/3 are found in the cytoplasm, but after phosphorylation/activation, they form a complex with Smad4 and are shuttled to the nuclei, where they bind to a number of transcription factor sites on the *COL1A1* and *COL1A2* gene promoter [82]. Besides inducing *COL1A1* and *COL1A2* gene transcription, TGFβ1 also increases collagen synthesis through posttranscriptional and translational mechanisms. TGFβ1 increases *COL1A1* mRNA by stabilizing the mRNA through p38 MAPK signaling [83, 84]. TGFβ1 also induces the genes that regulate the transport and metabolism of L-arginine and L-proline [85, 86], which provide crucial substrates for collagen synthesis. TGFβ1 stimulates smooth muscle proliferation at low concentrations, whereas at higher concentrations, it induces smooth muscle differentiation and the expression of α-actin and calponin [87–89].

The TGFβ1/Smad pathway is upregulated in COPD [90–92]. Inhibition of TGFβ1 signaling attenuates airway remodeling [93]. TGFβ-induces collagen synthesis and proliferation in the ASM cells [90]. TGFβ stimulates the expression of connective tissue growth factor (CTGF) and IL-6 in ASM [90, 94]. Coordinating with CTGF and inflammatory mediators, the TGFβ1/Smad pathway plays an important role in the pathogenesis of airway remodeling in COPD [90].

3.8 CTGF

CTGF belongs to the family of Cyr61/CTGF/Nov (CCN) proteins, structurally related secreted matricellular proteins with functions in adhesion, migration, proliferation, and extracellular matrix synthesis [95]. The expression of CTGF is potently induced by TGFβ [90]. CTGF mediates TGFβ-induced expression of collagen types I and IV, fibronectin, elastin, and the chondroitin sulfate proteoglycans biglycan and decorin in ASM [90]. CTGF is synthesized in the alveolar epithelial cells, small airway epithelial cells, stromal cells, and inflammatory cells in lung tissues [92]. Its expression is increased in the lungs of COPD patient, and smokers [91, 92, 96]. CTGF stimulates connective tissue deposition in the airway inflammatory response in COPD [22, 91].

3.9 Endothelin

Endothelin (ET) was first identified as a vasoactive peptide. Investigators have unveiled its wider biological activities. ET interacts with various cytokines, cellular elements, and oxidative processes in COPD. Three isoforms of endothelin have been described in humans and other mammals, ET-1, ET-2, and ET-3. The most common ET in humans is ET-1 because it is the only ET to be produced in endothelial cells and is also produced in smooth muscle cells. ET-1 acts locally as a paracrine factor. ET-1 levels in sputum and plasma are elevated in exacerbations of COPD [41]. Endothelins exert their action via receptors, endothelin type A (ETA) and type B (ETB). The ETB receptors are further divided into two subtypes, ETB1 and ETB2. Endothelin receptors are linked to G proteins, which leads to increased levels of inositol 1,4,5-trisphosphate (IP_3) and 1,2 diacylglycerol (DAG), as well as increased levels of cAMP. These processes result in increased cytosolic free calcium concentration, stimulation of phospholipase A2 (PLA2), activation of protein kinase C (PKC), and ROS production [97–99]. Activation of ETA and ETB2 receptors induces SMC contraction, proliferation, and migration [99–104]. ET also participates in inflammatory processes by inducing IL-6 synthesis in ASM cells [105].

3.10 ROS

In COPD, ROS are released from activated inflammatory cells or structural cells such as epithelial, endothelial, and smooth muscle cells [106, 107]. ROS may also be present in tobacco smoke [108]. The ROS formed in COPD include superoxide, H_2O_2, hydroxyl, and peroxynitrite. Superoxide anions are converted to H_2O_2 by superoxide dismutase (SOD). H_2O_2 is then dismuted to water by catalase. Superoxide and H_2O_2 may interact in the presence of free iron to form the highly reactive hydroxyl radical. The increased ROS formation is from mitochondria, NADPH oxidase, or xanthine oxidase, three major sources of ROS in cells [94, 109].

ROS play an important role and underlie the pathological processes in COPD. Oxidative stress directly damages cellular components such as lipids, proteins, and DNA, which may result in the death of bronchial epithelial cells and capillary endothelial cells [110, 111]. Oxidative stress can cause the activation of metalloproteases and inactivation of antiproteases (such as α1-antitrypsin or secretory leukoprotease inhibitor), leading to degradation of the ECM surrounding the peripheral airspace [112]. ROS activate redox-sensitive transcription factor such as nuclear factor-kappaB, which switches on multiple inflammatory genes, resulting in amplification of the inflammation in the terminal airway [113]. More importantly, ROS participate in ASM cell proliferation and deposition of ECM, such as collagen [114]. The signal transduction of ROS is related to the activation of extracellular signal-regulated kinases, c-Src, and MAPK [109].

4 Summary

The secretory mediators from inflammatory cells, airway epithelial cells, and ASM cells per se affect ASM proliferation and contractility in a vicious cycle in COPD pathogenesis. Targeting the signaling pathways associated with these mediators may hold promise as a therapeutic intervention in airway hyperresponsiveness and remodeling in COPD.

References

1. Sherry J, Murohy BS, Xu J, Kenneth D, Kochanek MA. Deaths: Preliminary Data for 2010 . National Vital Statistics Reports 2012 Jan 11;60(4):1-52.
2. Opazo Saez AM, Seow CY, Pare PD. Peripheral airway smooth muscle mechanics in obstructive airways disease. Am.J.Respir.Crit Care Med. 2000 Mar;161(3 Pt 1):910-7.
3. Gosens R, Zaagsma J, Meurs H, Halayko AJ. Muscarinic receptor signaling in the pathophysiology of asthma and COPD. Respir.Res. 2006;7:73.
4. Kanner RE, Connett JE, Altose MD, Buist AS, Lee WW, Tashkin DP, Wise RA. Gender difference in airway hyperresponsiveness in smokers with mild COPD. The Lung Health Study. Am.J.Respir.Crit Care Med. 1994 Oct;150(4):956-61.
5. Scichilone N, Battaglia S, La SA, Bellia V. Clinical implications of airway hyperresponsiveness in COPD. Int.J.Chron.Obstruct.Pulmon.Dis. 2006;1(1):49-60.
6. Sanderson MJ, Delmotte P, Bai Y, Perez-Zogbhi JF. Regulation of airway smooth muscle cell contractility by Ca2+ signaling and sensitivity. Proc.Am.Thorac.Soc. 2008 Jan 1;5(1):23-31.
7. Janssen LJ. Ionic mechanisms and Ca(2+) regulation in airway smooth muscle contraction: do the data contradict dogma? Am.J.Physiol Lung Cell Mol.Physiol 2002 Jun;282(6):L1161-L1178.
8. Tazzeo T, Zhang Y, Keshavjee S, Janssen LJ. Ryanodine receptors decant internal Ca2+ store in human and bovine airway smooth muscle. Eur.Respir.J. 2008 Aug;32(2):275-84.
9. Prakash YS, Kannan MS, Walseth TF, Sieck GC. Role of cyclic ADP-ribose in the regulation of [Ca2+]i in porcine tracheal smooth muscle. Am.J.Physiol 1998 Jun;274(6 Pt 1):C1653-C1660.
10. Wang YX, Zheng YM, Mei QB, Wang QS, Collier ML, Fleischer S, Xin HB, Kotlikoff MI. FKBP12.6 and cADPR regulation of Ca2+ release in smooth muscle cells. Am.J.Physiol Cell Physiol 2004 Mar;286(3):C538-C546.
11. Fliegert R, Gasser A, Guse AH. Regulation of calcium signalling by adenine-based second messengers. Biochem.Soc.Trans. 2007 Feb;35(Pt 1):109-14.
12. Croxton TL, Lande B, Hirshman CA. Role of G proteins in agonist-induced Ca2+ sensitization of tracheal smooth muscle. Am.J.Physiol 1998 Oct;275(4 Pt 1):L748-L755.
13. Hirshman CA, Lande B, Croxton TL. Role of M2 muscarinic receptors in airway smooth muscle contraction. Life Sci. 1999;64(6-7):443-8.
14. Lutz S, Freichel-Blomquist A, Yang Y, Rumenapp U, Jakobs KH, Schmidt M, Wieland T. The guanine nucleotide exchange factor p63RhoGEF, a specific link between Gq/11-coupled receptor signaling and RhoA. J.Biol.Chem. 2005 Mar 25;280(12):11134-9.
15. Somlyo AP, Somlyo AV. Ca2+ sensitivity of smooth muscle and nonmuscle myosin II: modulated by G proteins, kinases, and myosin phosphatase. Physiol Rev. 2003 Oct;83 (4):1325-58.
16. Sakai H, Chiba Y, Misawa M. Role of Rho kinase in endothelin-1-induced phosphorylation of CPI-17 in rat bronchial smooth muscle. Pulm.Pharmacol.Ther. 2007;20(6):734-9.

17. Baarsma HA, Menzen MH, Halayko AJ, Meurs H, Kerstjens HA, Gosens R. beta-Catenin signaling is required for TGF-beta1-induced extracellular matrix production by airway smooth muscle cells. Am.J.Physiol Lung Cell Mol.Physiol 2011 Dec;301(6):L956-L965.
18. Gosens R, Baarsma HA, Heijink IH, Oenema TA, Halayko AJ, Meurs H, Schmidt M. De novo synthesis of {beta}-catenin via H-Ras and MEK regulates airway smooth muscle growth. FASEB J. 2010 Mar;24(3):757-68.
19. Pera T, Gosens R, Lesterhuis AH, Sami R, Toorn M, Zaagsma J, Meurs H. Cigarette smoke and lipopolysaccharide induce a proliferative airway smooth muscle phenotype. Respir.Res. 2010;11:48.
20. Gosens R, Roscioni SS, Dekkers BG, Pera T, Schmidt M, Schaafsma D, Zaagsma J, Meurs H. Pharmacology of airway smooth muscle proliferation. Eur.J.Pharmacol. 2008 May 13;585 (2-3):385-97.
21. Oenema TA, Kolahian S, Nanninga JE, Rieks D, Hiemstra PS, Zuyderduyn S, Halayko AJ, Meurs H, Gosens R. Pro-inflammatory mechanisms of muscarinic receptor stimulation in airway smooth muscle. Respir.Res. 2010;11:130.
22. Ogawa E, Elliott WM, Hughes F, Eichholtz TJ, Hogg JC, Hayashi S. Latent adenoviral infection induces production of growth factors relevant to airway remodeling in COPD. Am. J.Physiol Lung Cell Mol.Physiol 2004 Jan;286(1):L189-L197.
23. Shapiro SD. Vascular atrophy and VEGFR-2 signaling: old theories of pulmonary emphysema meet new data. J.Clin.Invest 2000 Dec;106(11):1309-10.
24. Wright JL. Emphysema: concepts under change–a pathologist's perspective. Mod.Pathol. 1995 Oct;8(8):873-80.
25. Halayko AJ, Stelmack GL, Yamasaki A, McNeill K, Unruh H, Rector E. Distribution of phenotypically disparate myocyte subpopulations in airway smooth muscle. Can.J.Physiol Pharmacol. 2005 Jan;83(1):104-16.
26. Schaafsma D, McNeill KD, Stelmack GL, Gosens R, Baarsma HA, Dekkers BG, Frohwerk E, Penninks JM, Sharma P, Ens KM, et al. Insulin increases the expression of contractile phenotypic markers in airway smooth muscle. Am.J.Physiol Cell Physiol 2007 Jul;293(1): C429-C439.
27. Halayko AJ, Salari H, Ma X, Stephens NL. Markers of airway smooth muscle cell phenotype. Am.J.Physiol 1996 Jun;270(6 Pt 1):L1040-L1051.
28. Ma X, Wang Y, Stephens NL. Serum deprivation induces a unique hypercontractile phenotype of cultured smooth muscle cells. Am.J.Physiol 1998 May;274(5 Pt 1):C1206-C1214.
29. Liu HW, Halayko AJ, Fernandes DJ, Harmon GS, McCauley JA, Kocieniewski P, McConville J, Fu Y, Forsythe SM, Kogut P, et al. The RhoA/Rho kinase pathway regulates nuclear localization of serum response factor. Am.J.Respir.Cell Mol.Biol. 2003 Jul;29(1):39-47.
30. Camoretti-Mercado B, Liu HW, Halayko AJ, Forsythe SM, Kyle JW, Li B, Fu Y, McConville J, Kogut P, Vieira JE, et al. Physiological control of smooth muscle-specific gene expression through regulated nuclear translocation of serum response factor. J.Biol. Chem. 2000 Sep 29;275(39):30387-93.
31. Mack CP, Somlyo AV, Hautmann M, Somlyo AP, Owens GK. Smooth muscle differentiation marker gene expression is regulated by RhoA-mediated actin polymerization. J.Biol.Chem. 2001 Jan 5;276(1):341-7.
32. Koziol-White CJ, Panettieri RA, Jr. Airway smooth muscle and immunomodulation in acute exacerbations of airway disease. Immunol.Rev. 2011 Jul;242(1):178-85.
33. Kuwano K, Bosken CH, Pare PD, Bai TR, Wiggs BR, Hogg JC. Small airways dimensions in asthma and in chronic obstructive pulmonary disease. Am.Rev.Respir.Dis. 1993 Nov;148 (5):1220-5.
34. Churg A, Tai H, Coulthard T, Wang R, Wright JL. Cigarette smoke drives small airway remodeling by induction of growth factors in the airway wall. Am.J.Respir.Crit Care Med. 2006 Dec 15;174(12):1327-34.

35. Baarsma HA, Meurs H, Halayko AJ, Menzen MH, Schmidt M, Kerstjens HA, Gosens R. Glycogen synthase kinase-3 regulates cigarette smoke extract- and IL-1beta-induced cytokine secretion by airway smooth muscle. Am.J.Physiol Lung Cell Mol.Physiol 2011 Jun;300(6):L910-L919.
36. Kim V, Rogers TJ, Criner GJ. New concepts in the pathobiology of chronic obstructive pulmonary disease. Proc.Am.Thorac.Soc. 2008 May 1;5(4):478-85.
37. Asada M, Yamaya M, Ebihara S, Yasuda H, Tomita N, Kubo H, Sasaki H. Interleukin-1beta gene polymorphisms associated with COPD. Chest 2005 Aug;128(2):1072-3.
38. Lappalainen U, Whitsett JA, Wert SE, Tichelaar JW, Bry K. Interleukin-1beta causes pulmonary inflammation, emphysema, and airway remodeling in the adult murine lung. Am.J.Respir.Cell Mol.Biol. 2005 Apr;32(4):311-8.
39. Clarke D, Damera G, Sukkar MB, Tliba O. Transcriptional regulation of cytokine function in airway smooth muscle cells. Pulm.Pharmacol.Ther. 2009 Oct;22(5):436-45.
40. Elias JA, Wu Y, Zheng T, Panettieri R. Cytokine- and virus-stimulated airway smooth muscle cells produce IL-11 and other IL-6-type cytokines. Am.J.Physiol 1997 Sep;273(3 Pt 1):L648-L655.
41. Roland M, Bhowmik A, Sapsford RJ, Seemungal TA, Jeffries DJ, Warner TD, Wedzicha JA. Sputum and plasma endothelin-1 levels in exacerbations of chronic obstructive pulmonary disease. Thorax 2001 Jan;56(1):30-5.
42. Walter RE, Wilk JB, Larson MG, Vasan RS, Keaney JF, Jr., Lipinska I, O'Connor GT, Benjamin EJ. Systemic inflammation and COPD: the Framingham Heart Study. Chest 2008 Jan;133(1):19-25.
43. Kuhn C, III, Homer RJ, Zhu Z, Ward N, Flavell RA, Geba GP, Elias JA. Airway hyperresponsiveness and airway obstruction in transgenic mice. Morphologic correlates in mice overexpressing interleukin (IL)-11 and IL-6 in the lung. Am.J.Respir.Cell Mol.Biol. 2000 Mar;22(3):289-95.
44. Mortaz E, Henricks PA, Kraneveld AD, Givi ME, Garssen J, Folkerts G. Cigarette smoke induces the release of CXCL-8 from human bronchial epithelial cells via TLRs and induction of the inflammasome. Biochim.Biophys.Acta 2011 Sep;1812(9):1104-10.
45. Yamamoto C, Yoneda T, Yoshikawa M, Fu A, Tokuyama T, Tsukaguchi K, Narita N. Airway inflammation in COPD assessed by sputum levels of interleukin-8. Chest 1997 Aug;112 (2):505-10.
46. Gompertz S, O'Brien C, Bayley DL, Hill SL, Stockley RA. Changes in bronchial inflammation during acute exacerbations of chronic bronchitis. Eur.Respir.J. 2001 Jun;17(6):1112-9.
47. Churg A, Wang RD, Tai H, Wang X, Xie C, Wright JL. Tumor necrosis factor-alpha drives 70% of cigarette smoke-induced emphysema in the mouse. Am.J.Respir.Crit Care Med. 2004 Sep 1;170(5):492-8.
48. Kersul AL, Iglesias A, Rios A, Noguera A, Forteza A, Serra E, Agusti A, Cosio BG. Molecular mechanisms of inflammation during exacerbations of chronic obstructive pulmonary disease. Arch.Bronconeumol. 2011 Apr;47(4):176-83.
49. Hardaker EL, Bacon AM, Carlson K, Roshak AK, Foley JJ, Schmidt DB, Buckley PT, Comegys M, Panettieri RA, Jr., Sarau HM, et al. Regulation of TNF-alpha- and IFN-gamma-induced CXCL10 expression: participation of the airway smooth muscle in the pulmonary inflammatory response in chronic obstructive pulmonary disease. FASEB J. 2004 Jan;18(1):191-3.
50. Mukhopadhyay S, Hoidal JR, Mukherjee TK. Role of TNFalpha in pulmonary pathophysiology. Respir.Res. 2006;7:125.
51. Amrani Y. TNF-alpha and calcium signaling in airway smooth muscle cells: a never-ending story with promising therapeutic relevance. Am.J.Respir.Cell Mol.Biol. 2007 Mar;36(3):387-8.
52. Hunter I, Cobban HJ, Vandenabeele P, MacEwan DJ, Nixon GF. Tumor necrosis factor-alpha-induced activation of RhoA in airway smooth muscle cells: role in the Ca2+ sensitization of myosin light chain20 phosphorylation. Mol.Pharmacol. 2003 Mar;63(3):714-21.

53. Sukkar MB, Hughes JM, Armour CL, Johnson PR. Tumour necrosis factor-alpha potentiates contraction of human bronchus in vitro. Respirology. 2001 Sep;6(3):199-203.
54. White TA, Xue A, Chini EN, Thompson M, Sieck GC, Wylam ME. Role of transient receptor potential C3 in TNF-alpha-enhanced calcium influx in human airway myocytes. Am.J.Respir. Cell Mol.Biol. 2006 Aug;35(2):243-51.
55. Profita M, Giorgi RD, Sala A, Bonanno A, Riccobono L, Mirabella F, Gjomarkaj M, Bonsignore G, Bousquet J, Vignola AM. Muscarinic receptors, leukotriene B4 production and neutrophilic inflammation in COPD patients. Allergy 2005 Nov;60(11):1361-9.
56. Antczak A, Ciebiada M, Pietras T, Piotrowski WJ, Kurmanowska Z, Gorski P. Exhaled eicosanoids and biomarkers of oxidative stress in exacerbation of chronic obstructive pulmonary disease. Arch.Med.Sci. 2012 May 9;8(2):277-85.
57. Montuschi P, Kharitonov SA, Ciabattoni G, Barnes PJ. Exhaled leukotrienes and prostaglandins in COPD. Thorax 2003 Jul;58(7):585-8.
58. Davi G, Basili S, Vieri M, Cipollone F, Santarone S, Alessandri C, Gazzaniga P, Cordova C, Violi F. Enhanced thromboxane biosynthesis in patients with chronic obstructive pulmonary disease. The Chronic Obstructive Bronchitis and Haemostasis Study Group. Am.J.Respir.Crit Care Med. 1997 Dec;156(6):1794-9.
59. Hardy CC, Robinson C, Tattersfield AE, Holgate ST. The bronchoconstrictor effect of inhaled prostaglandin D2 in normal and asthmatic men. N.Engl.J.Med. 1984 Jul 26;311 (4):209-13.
60. Coleman RA, Sheldrick RL. Prostanoid-induced contraction of human bronchial smooth muscle is mediated by TP-receptors. Br.J.Pharmacol. 1989 Mar;96(3):688-92.
61. Devillier P, Bessard G. Thromboxane A2 and related prostaglandins in airways. Fundam. Clin.Pharmacol. 1997;11(1):2-18.
62. Rolin S, Masereel B, Dogne JM. Prostanoids as pharmacological targets in COPD and asthma. Eur.J.Pharmacol. 2006 Mar 8;533(1-3):89-100.
63. Gosens R, Zaagsma J, Grootte BM, Nelemans A, Meurs H. Acetylcholine: a novel regulator of airway smooth muscle remodelling? Eur.J.Pharmacol. 2004 Oct 1;500(1-3):193-201.
64. On LS, Boonyongsunchai P, Webb S, Davies L, Calverley PM, Costello RW. Function of pulmonary neuronal M(2) muscarinic receptors in stable chronic obstructive pulmonary disease. Am.J.Respir.Crit Care Med. 2001 May;163(6):1320-5.
65. Undem BJ, Kollarik M. The role of vagal afferent nerves in chronic obstructive pulmonary disease. Proc.Am.Thorac.Soc. 2005;2(4):355-60.
66. Kawashima K, Fujii T. The lymphocytic cholinergic system and its biological function. Life Sci. 2003 Mar 28;72(18-19):2101-9.
67. Wessler I, Kilbinger H, Bittinger F, Unger R, Kirkpatrick CJ. The non-neuronal cholinergic system in humans: expression, function and pathophysiology. Life Sci. 2003 Mar 28;72 (18-19):2055-61.
68. Chiba Y, Takada Y, Miyamoto S, MitsuiSaito M, Karaki H, Misawa M. Augmented acetylcholine-induced, Rho-mediated Ca2+ sensitization of bronchial smooth muscle contraction in antigen-induced airway hyperresponsive rats. Br.J.Pharmacol. 1999 Jun;127 (3):597-600.
69. Krymskaya VP, Orsini MJ, Eszterhas AJ, Brodbeck KC, Benovic JL, Panettieri RA, Jr., Penn RB. Mechanisms of proliferation synergy by receptor tyrosine kinase and G protein-coupled receptor activation in human airway smooth muscle. Am.J.Respir.Cell Mol.Biol. 2000 Oct;23 (4):546-54.
70. Krymskaya VP, Goncharova EA, Ammit AJ, Lim PN, Goncharov DA, Eszterhas A, Panettieri RA, Jr. Src is necessary and sufficient for human airway smooth muscle cell proliferation and migration. FASEB J. 2005 Mar;19(3):428-30.
71. Thomas SM, Brugge JS. Cellular functions regulated by Src family kinases. Annu.Rev.Cell Dev.Biol. 1997;13:513-609.
72. Aubert JD, Hayashi S, Hards J, Bai TR, Pare PD, Hogg JC. Platelet-derived growth factor and its receptor in lungs from patients with asthma and chronic airflow obstruction. Am.J.Physiol 1994 Jun;266(6 Pt 1):L655-L663.

73. Xing AP, Hu XY, Shi YW, DU YC. Implication of PDGF signaling in cigarette smoke-induced pulmonary arterial hypertension in rat. Inhal.Toxicol. 2012 Jul;24(8):468-75.
74. Zhu HJ, Burgess AW. Regulation of transforming growth factor-beta signaling. Mol.Cell Biol.Res.Commun. 2001 Nov;4(6):321-30.
75. Verrecchia F, Mauviel A. Transforming growth factor-beta and fibrosis. World J. Gastroenterol. 2007 Jun 14;13(22):3056-62.
76. Annes JP, Munger JS, Rifkin DB. Making sense of latent TGFbeta activation. J.Cell Sci. 2003 Jan 15;116(Pt 2):217-24.
77. Munger JS, Huang X, Kawakatsu H, Griffiths MJ, Dalton SL, Wu J, Pittet JF, Kaminski N, Garat C, Matthay MA, et al. The integrin alpha v beta 6 binds and activates latent TGF beta 1: a mechanism for regulating pulmonary inflammation and fibrosis. Cell 1999 Feb 5;96(3):319-28.
78. Annes JP, Chen Y, Munger JS, Rifkin DB. Integrin alphaVbeta6-mediated activation of latent TGF-beta requires the latent TGF-beta binding protein-1. J.Cell Biol. 2004 Jun 7;165 (5):723-34.
79. Ma W, Han W, Greer PA, Tuder RM, Toque HA, Wang KKW, Caldwell RW, Su Y. Calpain mediates pulmonary vascular remodeling in rodent models of pulmonary hypertension and its inhibition attenuates pathologic features of disease. J.Clin.Invest 2011 Nov 1;121 (11):4548-66.
80. Bertolino P, Deckers M, Lebrin F, ten DP. Transforming growth factor-beta signal transduction in angiogenesis and vascular disorders. Chest 2005 Dec;128(6 Suppl):585S-90S.
81. Lebrin F, Deckers M, Bertolino P, ten DP. TGF-beta receptor function in the endothelium. Cardiovasc.Res. 2005 Feb 15;65(3):599-608.
82. Qing J, Zhang Y, Derynck R. Structural and functional characterization of the transforming growth factor-beta -induced Smad3/c-Jun transcriptional cooperativity. J.Biol.Chem. 2000 Dec 8;275(49):38802-12.
83. Tsukada S, Westwick JK, Ikejima K, Sato N, Rippe RA. SMAD and p38 MAPK signaling pathways independently regulate alpha1(I) collagen gene expression in unstimulated and transforming growth factor-beta-stimulated hepatic stellate cells. J.Biol.Chem. 2005 Mar 18;280(11):10055-64.
84. Penttinen RP, Kobayashi S, Bornstein P. Transforming growth factor beta increases mRNA for matrix proteins both in the presence and in the absence of changes in mRNA stability. Proc.Natl.Acad.Sci.U.S.A 1988 Feb;85(4):1105-8.
85. Durante W, Liao L, Reyna SV, Peyton KJ, Schafer AI. Transforming growth factor-beta(1) stimulates L-arginine transport and metabolism in vascular smooth muscle cells: role in polyamine and collagen synthesis. Circulation 2001 Feb 27;103(8):1121-7.
86. Ensenat D, Hassan S, Reyna SV, Schafer AI, Durante W. Transforming growth factor-beta 1 stimulates vascular smooth muscle cell L-proline transport by inducing system A amino acid transporter 2 (SAT2) gene expression. Biochem.J. 2001 Dec 1;360(Pt 2):507-12.
87. Sturrock A, Cahill B, Norman K, Huecksteadt TP, Hill K, Sanders K, Karwande SV, Stringham JC, Bull DA, Gleich M, et al. Transforming growth factor-beta1 induces Nox4 NAD(P)H oxidase and reactive oxygen species-dependent proliferation in human pulmonary artery smooth muscle cells. Am.J.Physiol Lung Cell Mol.Physiol 2006 Apr;290(4):L661-L673.
88. Bobik A. Transforming growth factor-betas and vascular disorders. Arterioscler.Thromb. Vasc.Biol. 2006 Aug;26(8):1712-20.
89. Wang Y, it-Oufella H, Herbin O, Bonnin P, Ramkhelawon B, Taleb S, Huang J, Offenstadt G, Combadiere C, Renia L, et al. TGF-beta activity protects against inflammatory aortic aneurysm progression and complications in angiotensin II-infused mice. J.Clin.Invest 2010 Feb 1;120(2):422-32.
90. Black JL, Burgess JK, Johnson PR. Airway smooth muscle–its relationship to the extracellular matrix. Respir.Physiol Neurobiol. 2003 Sep 16;137(2-3):339-46.
91. Chung KF. Cytokines as targets in chronic obstructive pulmonary disease. Curr.Drug Targets. 2006 Jun;7(6):675-81.

92. Ning W, Li CJ, Kaminski N, Feghali-Bostwick CA, Alber SM, Di YP, Otterbein SL, Song R, Hayashi S, Zhou Z, et al. Comprehensive gene expression profiles reveal pathways related to the pathogenesis of chronic obstructive pulmonary disease. Proc.Natl.Acad.Sci.U.S.A 2004 Oct 12;101(41):14895-900.
93. Le AV, Cho JY, Miller M, McElwain S, Golgotiu K, Broide DH. Inhibition of allergen-induced airway remodeling in Smad 3-deficient mice. J.Immunol. 2007 Jun 1;178 (11):7310-6.
94. Michaeloudes C, Sukkar MB, Khorasani NM, Bhavsar PK, Chung KF. TGF-beta regulates Nox4, MnSOD and catalase expression, and IL-6 release in airway smooth muscle cells. Am. J.Physiol Lung Cell Mol.Physiol 2011 Feb;300(2):L295-L304.
95. Chaqour B, Goppelt-Struebe M. Mechanical regulation of the Cyr61/CCN1 and CTGF/CCN2 proteins. FEBS J. 2006 Aug;273(16):3639-49.
96. Llinas L, Peinado VI, Ramon GJ, Rabinovich R, Pizarro S, Rodriguez-Roisin R, Barbera JA, Bastos R. Similar gene expression profiles in smokers and patients with moderate COPD. Pulm.Pharmacol.Ther. 2011 Feb;24(1):32-41.
97. Zhang Y, Edvinsson L, Xu CB. Up-regulation of endothelin receptors induced by cigarette smoke–involvement of MAPK in vascular and airway hyper-reactivity. Scientific-WorldJournal. 2010;10:2157-66.
98. Wedgwood S, Dettman RW, Black SM. ET-1 stimulates pulmonary arterial smooth muscle cell proliferation via induction of reactive oxygen species. Am.J.Physiol Lung Cell Mol. Physiol 2001 Nov;281(5):L1058-L1067.
99. Maxwell MJ, Goldie RG, Henry PJ. Ca2+ signalling by endothelin receptors in rat and human cultured airway smooth muscle cells. Br.J.Pharmacol. 1998 Dec;125(8):1768-78.
100. Xu CB, Lei Y, Chen Q, Pehrson C, Larsson L, Edvinsson L. Cigarette smoke extracts promote vascular smooth muscle cell proliferation and enhances contractile responses in the vasculature and airway. Basic Clin.Pharmacol.Toxicol. 2010 Dec;107(6):940-8.
101. Akhtar S, Podgoreanu M, Harrison BA, Brull SJ. Effect of lidocaine on endothelin-1-mediated airway smooth muscle contraction in the rat trachea. Minerva Anestesiol. 2008 Nov;74 (11):643-50.
102. McWhinnie R, Pechkovsky DV, Zhou D, Lane D, Halayko AJ, Knight DA, Bai TR. Endothelin-1 induces hypertrophy and inhibits apoptosis in human airway smooth muscle cells. Am.J.Physiol Lung Cell Mol.Physiol 2007 Jan;292(1):L278-L286.
103. Fehr JJ, Hirshman CA, Emala CW. Cellular signaling by the potent bronchoconstrictor endothelin-1 in airway smooth muscle. Crit Care Med. 2000 Jun;28(6):1884-8.
104. Carratu P, Scuri M, Styblo JL, Wanner A, Glassberg MK. ET-1 induces mitogenesis in ovine airway smooth muscle cells via ETA and ETB receptors. Am.J.Physiol 1997 May;272(5 Pt 1): L1021-L1024.
105. Iwata S, Ito S, Iwaki M, Kondo M, Sashio T, Takeda N, Sokabe M, Hasegawa Y, Kume H. Regulation of endothelin-1-induced interleukin-6 production by Ca2+ influx in human airway smooth muscle cells. Eur.J.Pharmacol. 2009 Mar 1;605(1-3):15-22.
106. MacNee W. Oxidants and COPD. Curr.Drug Targets.Inflamm.Allergy 2005 Dec;4(6):627-41.
107. van der TM, Rezayat D, Kauffman HF, Bakker SJ, Gans RO, Koeter GH, Choi AM, van Oosterhout AJ, Slebos DJ. Lipid-soluble components in cigarette smoke induce mitochondrial production of reactive oxygen species in lung epithelial cells. Am.J.Physiol Lung Cell Mol.Physiol 2009 Jul;297(1):L109-L114.
108. Epperlein MM, Nourooz-Zadeh J, Noronha-Dutra AA, Woolf N. Nitric oxide in cigarette smoke as a mediator of oxidative damage. Int.J.Exp.Pathol. 1996 Oct;77(5):197-200.
109. Taille C, El-Benna J, Lanone S, Boczkowski J, Motterlini R. Mitochondrial respiratory chain and NAD(P)H oxidase are targets for the antiproliferative effect of carbon monoxide in human airway smooth muscle. J.Biol.Chem. 2005 Jul 8;280(27):25350-60.
110. Deslee G, ir-Kirk TL, Betsuyaku T, Woods JC, Moore CH, Gierada DS, Conradi SH, Atkinson JJ, Toennies HM, Battaile JT, et al. Cigarette Smoke Induces Nucleic Acid Oxidation in Lung Fibroblasts. Am.J.Respir.Cell Mol.Biol. 2009 Dec 11.

111. Kovacic P, Somanathan R. Pulmonary toxicity and environmental contamination: radicals, electron transfer, and protection by antioxidants. Rev.Environ.Contam Toxicol. 2009;201:41-69.
112. Shapiro SD, Senior RM. Matrix metalloproteinases. Matrix degradation and more. Am.J. Respir.Cell Mol.Biol. 1999 Jun;20(6):1100-2.
113. Adcock IM, Cosio B, Tsaprouni L, Barnes PJ, Ito K. Redox regulation of histone deacetylases and glucocorticoid-mediated inhibition of the inflammatory response. Antioxid.Redox.Signal. 2005 Jan;7(1-2):144-52.
114. Ravasi S, Citro S, Viviani B, Capra V, Rovati GE. CysLT1 receptor-induced human airway smooth muscle cells proliferation requires ROS generation, EGF receptor transactivation and ERK1/2 phosphorylation. Respir.Res. 2006;7:42.

Index

Y.-X. Wang (ed.), *Calcium Signaling In Airway Smooth Muscle Cells*,
DOI 10.1007/978-3-319-01312-1, © Springer International Publishing Switzerland 2014

D

T

The manufacturer's authorised representative in the EU is Springer Nature Customer Service Centre GmbH, Europaplatz 3, 69115 Heidelberg, Germany. If you have any concerns regarding our products, please contact ProductSafety@springernature.com

Printed and bound by CPI Group (UK) Ltd, Croydon, CR0 4YY
22/09/2026
02220980-0005